T0181830

Communications in Computer and Information Science 1969

Rationale

The CCIS series is devoted to the publication of proceedings of computer science conferences. Its aim is to efficiently disseminate original research results in informatics in printed and electronic form. While the focus is on publication of peer-reviewed full papers presenting mature work, inclusion of reviewed short papers reporting on work in progress is welcome, too. Besides globally relevant meetings with internationally representative program committees guaranteeing a strict peer-reviewing and paper selection process, conferences run by societies or of high regional or national relevance are also considered for publication.

Topics

The topical scope of CCIS spans the entire spectrum of informatics ranging from foundational topics in the theory of computing to information and communications science and technology and a broad variety of interdisciplinary application fields.

Information for Volume Editors and Authors

Publication in CCIS is free of charge. No royalties are paid, however, we offer registered conference participants temporary free access to the online version of the conference proceedings on SpringerLink (http://link.springer.com) by means of an http referrer from the conference website and/or a number of complimentary printed copies, as specified in the official acceptance email of the event.

CCIS proceedings can be published in time for distribution at conferences or as post-proceedings, and delivered in the form of printed books and/or electronically as USBs and/or e-content licenses for accessing proceedings at SpringerLink. Furthermore, CCIS proceedings are included in the CCIS electronic book series hosted in the SpringerLink digital library at http://link.springer.com/bookseries/7899. Conferences publishing in CCIS are allowed to use Online Conference Service (OCS) for managing the whole proceedings lifecycle (from submission and reviewing to preparing for publication) free of charge.

Publication process

The language of publication is exclusively English. Authors publishing in CCIS have to sign the Springer CCIS copyright transfer form, however, they are free to use their material published in CCIS for substantially changed, more elaborate subsequent publications elsewhere. For the preparation of the camera-ready papers/files, authors have to strictly adhere to the Springer CCIS Authors' Instructions and are strongly encouraged to use the CCIS LaTeX style files or templates.

Abstracting/Indexing

CCIS is abstracted/indexed in DBLP, Google Scholar, EI-Compendex, Mathematical Reviews, SCImago, Scopus. CCIS volumes are also submitted for the inclusion in ISI Proceedings.

How to start

To start the evaluation of your proposal for inclusion in the CCIS series, please send an e-mail to ccis@springer.com.

Biao Luo · Long Cheng · Zheng-Guang Wu ·
Hongyi Li · Chaojie Li
Editors

Neural Information Processing

30th International Conference, ICONIP 2023
Changsha, China, November 20–23, 2023
Proceedings, Part XV

 Springer

Editors
Biao Luo 🆔
School of Automation
Central South University
Changsha, China

Long Cheng 🆔
Institute of Automation
Chinese Academy of Sciences
Beijing, China

Zheng-Guang Wu 🆔
Institute of Cyber-Systems and Control
Zhejiang University
Hangzhou, China

Hongyi Li 🆔
School of Automation
Guangdong University of Technology
Guangzhou, China

Chaojie Li 🆔
School of Electrical Engineering
and Telecommunications
UNSW Sydney
Sydney, NSW, Australia

ISSN 1865-0929   ISSN 1865-0937 (electronic)
Communications in Computer and Information Science
ISBN 978-981-99-8183-0   ISBN 978-981-99-8184-7 (eBook)
https://doi.org/10.1007/978-981-99-8184-7

This Springer imprint is published by the registered company Springer Nature Singapore Pte Ltd.
The registered company address is: 152 Beach Road, #21-01/04 Gateway East, Singapore 189721, Singapore

Paper in this product is recyclable.

Preface

Welcome to the 30th International Conference on Neural Information Processing (ICONIP2023) of the Asia-Pacific Neural Network Society (APNNS), held in Changsha, China, November 20–23, 2023.

The mission of the Asia-Pacific Neural Network Society is to promote active interactions among researchers, scientists, and industry professionals who are working in neural networks and related fields in the Asia-Pacific region. APNNS has Governing Board Members from 13 countries/regions – Australia, China, Hong Kong, India, Japan, Malaysia, New Zealand, Singapore, South Korea, Qatar, Taiwan, Thailand, and Turkey. The society's flagship annual conference is the International Conference of Neural Information Processing (ICONIP). The ICONIP conference aims to provide a leading international forum for researchers, scientists, and industry professionals who are working in neuroscience, neural networks, deep learning, and related fields to share their new ideas, progress, and achievements.

ICONIP2023 received 1274 papers, of which 394 papers were accepted for publication in Communications in Computer and Information Science (CCIS), representing an acceptance rate of 30.93% and reflecting the increasingly high quality of research in neural networks and related areas. The conference focused on four main areas, i.e., "Theory and Algorithms", "Cognitive Neurosciences", "Human-Centered Computing", and "Applications". All the submissions were rigorously reviewed by the conference Program Committee (PC), comprising 258 PC members, and they ensured that every paper had at least two high-quality single-blind reviews. In fact, 5270 reviews were provided by 2145 reviewers. On average, each paper received 4.14 reviews.

We would like to take this opportunity to thank all the authors for submitting their papers to our conference, and our great appreciation goes to the Program Committee members and the reviewers who devoted their time and effort to our rigorous peer-review process; their insightful reviews and timely feedback ensured the high quality of the papers accepted for publication. We hope you enjoyed the research program at the conference.

October 2023

Biao Luo
Long Cheng
Zheng-Guang Wu
Hongyi Li
Chaojie Li

Organization

Honorary Chair

Weihua Gui   Central South University, China

Advisory Chairs

Jonathan Chan   King Mongkut's University of Technology Thonburi, Thailand
Zeng-Guang Hou   Chinese Academy of Sciences, China
Nikola Kasabov   Auckland University of Technology, New Zealand
Derong Liu   Southern University of Science and Technology, China
Seiichi Ozawa   Kobe University, Japan
Kevin Wong   Murdoch University, Australia

General Chairs

Tingwen Huang   Texas A&M University at Qatar, Qatar
Chunhua Yang   Central South University, China

Program Chairs

Biao Luo   Central South University, China
Long Cheng   Chinese Academy of Sciences, China
Zheng-Guang Wu   Zhejiang University, China
Hongyi Li   Guangdong University of Technology, China
Chaojie Li   University of New South Wales, Australia

Technical Chairs

Xing He   Southwest University, China
Keke Huang   Central South University, China
Huaqing Li   Southwest University, China
Qi Zhou   Guangdong University of Technology, China

Local Arrangement Chairs

Wenfeng Hu	Central South University, China
Bei Sun	Central South University, China

Finance Chairs

Fanbiao Li	Central South University, China
Hayaru Shouno	University of Electro-Communications, Japan
Xiaojun Zhou	Central South University, China

Special Session Chairs

Hongjing Liang	University of Electronic Science and Technology, China
Paul S. Pang	Federation University, Australia
Qiankun Song	Chongqing Jiaotong University, China
Lin Xiao	Hunan Normal University, China

Tutorial Chairs

Min Liu	Hunan University, China
M. Tanveer	Indian Institute of Technology Indore, India
Guanghui Wen	Southeast University, China

Publicity Chairs

Sabri Arik	Istanbul University-Cerrahpaşa, Turkey
Sung-Bae Cho	Yonsei University, South Korea
Maryam Doborjeh	Auckland University of Technology, New Zealand
El-Sayed M. El-Alfy	King Fahd University of Petroleum and Minerals, Saudi Arabia
Ashish Ghosh	Indian Statistical Institute, India
Chuandong Li	Southwest University, China
Weng Kin Lai	Tunku Abdul Rahman University of Management & Technology, Malaysia
Chu Kiong Loo	University of Malaya, Malaysia
Qinmin Yang	Zhejiang University, China
Zhigang Zeng	Huazhong University of Science and Technology, China

Publication Chairs

Zhiwen Chen	Central South University, China
Andrew Chi-Sing Leung	City University of Hong Kong, China
Xin Wang	Southwest University, China
Xiaofeng Yuan	Central South University, China

Secretaries

Yun Feng	Hunan University, China
Bingchuan Wang	Central South University, China

Webmasters

Tianmeng Hu	Central South University, China
Xianzhe Liu	Xiangtan University, China

Program Committee

Rohit Agarwal	UiT The Arctic University of Norway, Norway
Hasin Ahmed	Gauhati University, India
Harith Al-Sahaf	Victoria University of Wellington, New Zealand
Brad Alexander	University of Adelaide, Australia
Mashaan Alshammari	Independent Researcher, Saudi Arabia
Sabri Arik	Istanbul University, Turkcy
Ravneet Singh Arora	Block Inc., USA
Zeyar Aung	Khalifa University of Science and Technology, UAE
Monowar Bhuyan	Umeå University, Sweden
Jingguo Bi	Beijing University of Posts and Telecommunications, China
Xu Bin	Northwestern Polytechnical University, China
Marcin Blachnik	Silesian University of Technology, Poland
Paul Black	Federation University, Australia
Anoop C. S.	Govt. Engineering College, India
Ning Cai	Beijing University of Posts and Telecommunications, China
Siripinyo Chantamunee	Walailak University, Thailand
Hangjun Che	City University of Hong Kong, China

Wei-Wei Che	Qingdao University, China
Huabin Chen	Nanchang University, China
Jinpeng Chen	Beijing University of Posts & Telecommunications, China
Ke-Jia Chen	Nanjing University of Posts and Telecommunications, China
Lv Chen	Shandong Normal University, China
Qiuyuan Chen	Tencent Technology, China
Wei-Neng Chen	South China University of Technology, China
Yufei Chen	Tongji University, China
Long Cheng	Institute of Automation, China
Yongli Cheng	Fuzhou University, China
Sung-Bae Cho	Yonsei University, South Korea
Ruikai Cui	Australian National University, Australia
Jianhua Dai	Hunan Normal University, China
Tao Dai	Tsinghua University, China
Yuxin Ding	Harbin Institute of Technology, China
Bo Dong	Xi'an Jiaotong University, China
Shanling Dong	Zhejiang University, China
Sidong Feng	Monash University, Australia
Yuming Feng	Chongqing Three Gorges University, China
Yun Feng	Hunan University, China
Junjie Fu	Southeast University, China
Yanggeng Fu	Fuzhou University, China
Ninnart Fuengfusin	Kyushu Institute of Technology, Japan
Thippa Reddy Gadekallu	VIT University, India
Ruobin Gao	Nanyang Technological University, Singapore
Tom Gedeon	Curtin University, Australia
Kam Meng Goh	Tunku Abdul Rahman University of Management and Technology, Malaysia
Zbigniew Gomolka	University of Rzeszow, Poland
Shengrong Gong	Changshu Institute of Technology, China
Xiaodong Gu	Fudan University, China
Zhihao Gu	Shanghai Jiao Tong University, China
Changlu Guo	Budapest University of Technology and Economics, Hungary
Weixin Han	Northwestern Polytechnical University, China
Xing He	Southwest University, China
Akira Hirose	University of Tokyo, Japan
Yin Hongwei	Huzhou Normal University, China
Md Zakir Hossain	Curtin University, Australia
Zengguang Hou	Chinese Academy of Sciences, China

Lu Hu	Jiangsu University, China
Zeke Zexi Hu	University of Sydney, Australia
He Huang	Soochow University, China
Junjian Huang	Chongqing University of Education, China
Kaizhu Huang	Duke Kunshan University, China
David Iclanzan	Sapientia University, Romania
Radu Tudor Ionescu	University of Bucharest, Romania
Asim Iqbal	Cornell University, USA
Syed Islam	Edith Cowan University, Australia
Kazunori Iwata	Hiroshima City University, Japan
Junkai Ji	Shenzhen University, China
Yi Ji	Soochow University, China
Canghong Jin	Zhejiang University, China
Xiaoyang Kang	Fudan University, China
Mutsumi Kimura	Ryukoku University, Japan
Masahiro Kohjima	NTT, Japan
Damian Kordos	Rzeszow University of Technology, Poland
Marek Kraft	Poznań University of Technology, Poland
Lov Kumar	NIT Kurukshetra, India
Weng Kin Lai	Tunku Abdul Rahman University of Management & Technology, Malaysia
Xinyi Le	Shanghai Jiao Tong University, China
Bin Li	University of Science and Technology of China, China
Hongfei Li	Xinjiang University, China
Houcheng Li	Chinese Academy of Sciences, China
Huaqing Li	Southwest University, China
Jianfeng Li	Southwest University, China
Jun Li	Nanjing Normal University, China
Kan Li	Beijing Institute of Technology, China
Peifeng Li	Soochow University, China
Wenye Li	Chinese University of Hong Kong, China
Xiangyu Li	Beijing Jiaotong University, China
Yantao Li	Chongqing University, China
Yaoman Li	Chinese University of Hong Kong, China
Yinlin Li	Chinese Academy of Sciences, China
Yuan Li	Academy of Military Science, China
Yun Li	Nanjing University of Posts and Telecommunications, China
Zhidong Li	University of Technology Sydney, Australia
Zhixin Li	Guangxi Normal University, China
Zhongyi Li	Beihang University, China

Ziqiang Li	University of Tokyo, Japan
Xianghong Lin	Northwest Normal University, China
Yang Lin	University of Sydney, Australia
Huawen Liu	Zhejiang Normal University, China
Jian-Wei Liu	China University of Petroleum, China
Jun Liu	Chengdu University of Information Technology, China
Junxiu Liu	Guangxi Normal University, China
Tommy Liu	Australian National University, Australia
Wen Liu	Chinese University of Hong Kong, China
Yan Liu	Taikang Insurance Group, China
Yang Liu	Guangdong University of Technology, China
Yaozhong Liu	Australian National University, Australia
Yong Liu	Heilongjiang University, China
Yubao Liu	Sun Yat-sen University, China
Yunlong Liu	Xiamen University, China
Zhe Liu	Jiangsu University, China
Zhen Liu	Chinese Academy of Sciences, China
Zhi-Yong Liu	Chinese Academy of Sciences, China
Ma Lizhuang	Shanghai Jiao Tong University, China
Chu-Kiong Loo	University of Malaya, Malaysia
Vasco Lopes	Universidade da Beira Interior, Portugal
Hongtao Lu	Shanghai Jiao Tong University, China
Wenpeng Lu	Qilu University of Technology, China
Biao Luo	Central South University, China
Ye Luo	Tongji University, China
Jiancheng Lv	Sichuan University, China
Yuezu Lv	Beijing Institute of Technology, China
Huifang Ma	Northwest Normal University, China
Jinwen Ma	Peking University, China
Jyoti Maggu	Thapar Institute of Engineering and Technology Patiala, India
Adnan Mahmood	Macquarie University, Australia
Mufti Mahmud	University of Padova, Italy
Krishanu Maity	Indian Institute of Technology Patna, India
Srimanta Mandal	DA-IICT, India
Wang Manning	Fudan University, China
Piotr Milczarski	Lodz University of Technology, Poland
Malek Mouhoub	University of Regina, Canada
Nankun Mu	Chongqing University, China
Wenlong Ni	Jiangxi Normal University, China
Anupiya Nugaliyadde	Murdoch University, Australia

Shanchuan Wan	University of Tokyo, Japan
Tao Wan	Beihang University, China
Ying Wan	Southeast University, China
Bangjun Wang	Soochow University, China
Hao Wang	Shanghai University, China
Huamin Wang	Southwest University, China
Hui Wang	Nanchang Institute of Technology, China
Huiwei Wang	Southwest University, China
Jianzong Wang	Ping An Technology, China
Lei Wang	National University of Defense Technology, China
Lin Wang	University of Jinan, China
Shi Lin Wang	Shanghai Jiao Tong University, China
Wei Wang	Shenzhen MSU-BIT University, China
Weiqun Wang	Chinese Academy of Sciences, China
Xiaoyu Wang	Tokyo Institute of Technology, Japan
Xin Wang	Southwest University, China
Xin Wang	Southwest University, China
Yan Wang	Chinese Academy of Sciences, China
Yan Wang	Sichuan University, China
Yonghua Wang	Guangdong University of Technology, China
Yongyu Wang	JD Logistics, China
Zhenhua Wang	Northwest A&F University, China
Zi-Peng Wang	Beijing University of Technology, China
Hongxi Wei	Inner Mongolia University, China
Guanghui Wen	Southeast University, China
Guoguang Wen	Beijing Jiaotong University, China
Ka-Chun Wong	City University of Hong Kong, China
Anna Wróblewska	Warsaw University of Technology, Poland
Fengge Wu	Institute of Software, Chinese Academy of Sciences, China
Ji Wu	Tsinghua University, China
Wei Wu	Inner Mongolia University, China
Yue Wu	Shanghai Jiao Tong University, China
Likun Xia	Capital Normal University, China
Lin Xiao	Hunan Normal University, China
Qiang Xiao	Huazhong University of Science and Technology, China
Hao Xiong	Macquarie University, Australia
Dongpo Xu	Northeast Normal University, China
Hua Xu	Tsinghua University, China
Jianhua Xu	Nanjing Normal University, China

Xinyue Xu   Hong Kong University of Science and Technology, China

Yong Xu   Beijing Institute of Technology, China

Ngo Xuan Bach   Posts and Telecommunications Institute of Technology, Vietnam

Hao Xue   University of New South Wales, Australia

Yang Xujun   Chongqing Jiaotong University, China

Haitian Yang   Chinese Academy of Sciences, China

Jie Yang   Shanghai Jiao Tong University, China

Minghao Yang   Chinese Academy of Sciences, China

Peipei Yang   Chinese Academy of Science, China

Zhiyuan Yang   City University of Hong Kong, China

Wangshu Yao   Soochow University, China

Ming Yin   Guangdong University of Technology, China

Qiang Yu   Tianjin University, China

Wenxin Yu   Southwest University of Science and Technology, China

Yun-Hao Yuan   Yangzhou University, China

Xiaodong Yue   Shanghai University, China

Paweł Zawistowski   Warsaw University of Technology, Poland

Hui Zeng   Southwest University of Science and Technology, China

Wang Zengyunwang   Hunan First Normal University, China

Daren Zha   Institute of Information Engineering, China

Zhi-Hui Zhan   South China University of Technology, China

Baojie Zhang   Chongqing Three Gorges University, China

Canlong Zhang   Guangxi Normal University, China

Guixuan Zhang   Chinese Academy of Science, China

Jianming Zhang   Changsha University of Science and Technology, China

Li Zhang   Soochow University, China

Wei Zhang   Southwest University, China

Wenbing Zhang   Yangzhou University, China

Xiang Zhang   National University of Defense Technology, China

Xiaofang Zhang   Soochow University, China

Xiaowang Zhang   Tianjin University, China

Xinglong Zhang   National University of Defense Technology, China

Dongdong Zhao   Wuhan University of Technology, China

Xiang Zhao   National University of Defense Technology, China

Xu Zhao   Shanghai Jiao Tong University, China

Liping Zheng	Hefei University of Technology, China
Yan Zheng	Kyushu University, Japan
Baojiang Zhong	Soochow University, China
Guoqiang Zhong	Ocean University of China, China
Jialing Zhou	Nanjing University of Science and Technology, China
Wenan Zhou	PCN&CAD Center, China
Xiao-Hu Zhou	Institute of Automation, China
Xinyu Zhou	Jiangxi Normal University, China
Quanxin Zhu	Nanjing Normal University, China
Yuanheng Zhu	Chinese Academy of Sciences, China
Xiaotian Zhuang	JD Logistics, China
Dongsheng Zou	Chongqing University, China

Contents – Part XV

Applications

Towards Deeper and Better Multi-view Feature Fusion for 3D Semantic Segmentation

Chaolong Yang[1], Yuyao Yan[2], Weiguang Zhao[1], Jianan Ye[2], Xi Yang[2], Amir Hussain[3], Bin Dong[4], and Kaizhu Huang[1](✉)

[1] Duke Kunshan University, Suzhou 215000, China
{chaolong.yang,weiguang.zhao,kaizhu.huang}@dukekunshan.edu.cn
[2] Xi'an Jiaotong-Liverpool University, Suzhou 215000, China
{yuyao.yan,xi.yang01}@xjtlu.edu.cn, Jianan.Ye20@student.xjtlu.edu.cn
[3] Edinburgh Napier University, Edinburgh EH11 4BN, UK
A.Hussain@napier.ac.uk
[4] Ricoh Software Research Center (Beijing) Co., Ltd., Beijing 100000, China
Bin.Dong@srcb.ricoh.com

Abstract. 3D point clouds are rich in geometric structure information, while 2D images contain important and continuous texture information. Combining 2D information to achieve better 3D semantic segmentation has become a mainstream in 3D scene understanding. Albeit the success, it still remains elusive how to fuse and process the cross-dimensional features from these two distinct spaces. Existing state-of-the-art usually exploit bidirectional projection methods to align the cross-dimensional features and realize both 2D & 3D semantic segmentation tasks. However, to enable bidirectional mapping, this framework often requires a symmetrical 2D-3D network structure, thus limiting the network's flexibility. Meanwhile, such dual-task settings may distract the network easily and lead to overfitting in the 3D segmentation task. As limited by the network's inflexibility, fused features can only pass through a decoder network, which affects model performance due to insufficient depth. To alleviate these drawbacks, in this paper, we argue that despite its simplicity, projecting unidirectionally multi-view 2D deep semantic features into the 3D space aligned with 3D deep semantic features could lead to better feature fusion. On the one hand, the unidirectional projection enforces our model focused more on the core task, i.e., 3D segmentation; on the other hand, unlocking the bidirectional to unidirectional projection enables a deeper cross-domain semantic alignment and enjoys the flexibility to fuse better and complicated features from very different spaces. In joint 2D-3D approaches, our proposed method achieves superior performance on the ScanNetv2 benchmark for 3D semantic segmentation.

Keywords: Point cloud · Semantic segmentation · Multi-view fusion

B. Luo et al. (Eds.): ICONIP 2023, CCIS 1969, pp. 3–15, 2024.
https://doi.org/10.1007/978-981-99-8184-7_1

1   Introduction

Semantic understanding of scenes is essential in numerous fields, including robot navigation, automatic driving systems, and medical diagnosis. While early researchers focused on 2D images to achieve scene understanding, fixed-view 2D images lack spatial structure information and suffer from object occlusion, limiting their use in spatially location-sensitive downstream tasks. In contrast, 3D point clouds offer a complete spatial structure without object occlusion. However, traditional 2D neural network-based methodologies cannot be used directly to deal with 3D data. To address this issue, point-based [21, 22, 29] and voxel-based [4, 7, 33] neural networks have been explored for 3D point cloud recognition and understanding. Nonetheless, 3D point clouds have low resolution and lack rich texture information. Thus, a promising solution to jointly understand complex scenes is to combine 2D images with detailed texture information and 3D point clouds with rich knowledge of geometric structures.

2D-3D fusion schemes for 3D semantic segmentation tasks can be categorized as bidirectional and unidirectional projection. A comparison of network frameworks between these two schemes is depicted in Fig. 1.

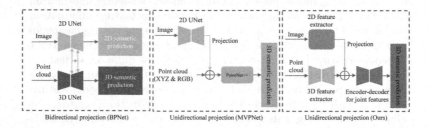

Fig. 1. Comparison of bidirectional & unidirectional projection

The pioneering BPNet [11] utilizes bidirectional projection to allow 2D and 3D features to flow between networks. Nevertheless, in order to mutually fuse information from 2D to 3D and 3D to 2D, it usually has to exploit a symmetrical decoder network. This makes its framework less flexible and could not take advantage of network depth, thus limiting its performance. Additionally, in complicated scenes, the 2D semantic component may distract from the core 3D task. To illustrate, we implement the idea of bidirectional projection on our proposed unidirectional projection framework. Namely, 3D features are also projected into 2D space combined with 2D features and then input to a complete 2D encoder-decoder network. On a large-scale complex indoor scene ScanNetv2 [5], we compare in Fig. 2 the 3D semantic loss on the validation set for unidirectional and bidirectional projection ideas during model training. Clearly, on the complicated scene of ScanNetv2, the bidirectional projection scheme causes distraction in the 3D task, where the loss goes up as the training continues. In comparison, the uni-projection implementation would lead to more stable performance with the focus mainly on the 3D task. Motivated by these findings, we argue that projecting unidirectionally multi-view 2D deep semantic features into the 3D space aligned with 3D deep semantic features can result in better feature fusion and more potential for downstream tasks like scene understanding.

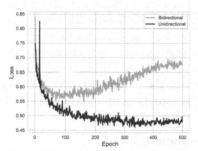

Fig. 2. Validation loss for unidirectional & bidirectional projection on validation set of the benchmark ScanNetv2 data

Previous unidirectional projection methods [3, 13] have been proposed in the literature to fuse 2D deep semantics and 3D shallow information (XYZ & RGB) for 3D semantic segmentation tasks (see the middle graph in Fig. 1). Direct connection of deep and shallow semantics from different data domains could however result in the misalignment of the semantic space. To this end, we design a novel unidirectional projection framework, called the Deep Multi-view Fusion Network (DMF-Net), to more effectively fuse 2D & 3D features (see the right graph in Fig. 1). On the implementation front, we evaluate our model on the 3D semantic segmentation datasets: ScanNetv2 [5] and NYUv2 [25]. DMF-Net not only achieves top performance for joint 2D-3D methods on the ScanNetv2 benchmark, but also achieves state-of-the-art performance on the NYUv2 dataset. Our contributions can be summarized as follows.

- We argue that the unidirectional projection mechanism is not only more focused on 3D semantic understanding tasks than bidirectional projection but also facilitates deeper feature fusion. To this end, we design a method for uni-directional cross-domain semantic feature fusion to extract 2D & 3D deep features for alignment simultaneously.
- We propose a novel framework named Deep Multi-view Fusion Network (DMF-Net) for 3D scene semantic understanding. For the joint 2D-3D approaches, DMF-Net obtains top mIOU performance on the 3D Semantic Label Benchmark of ScanNetv2 [5], while it reaches the state-of-the-art on NYUv2 [25] datasets.
- We demonstrate the flexibility of DMF-Net, where all backbone modules in the network framework can be replaced. Specifically, 3D semantic segmentation performance will be stronger with a powerful backbone. Compared with common U-Net34 [23], the advanced Swin-UNet [2] shows a relative 4.4% improvement on mIOU in our framework.

2   Related Work

2.1   2D Semantic Segmentation

Image semantic segmentation has been significantly improved by the development of deep-learning [12] models. In the field of 2D semantic segmentation, Fully Convolution Network (FCN) [18] is a landmark work although it has some limitations. Several Encoder-Decoder-based [23, 26, 30] structures combined multi-level information to fine segmentation. Besides, attention-based [2] models capable of extracting long-range contextual information were introduced into the image segmentation task. However, the lack of 3D spatial geometry information in 2D images hinders semantic comprehension of scenes.

2.2   3D Semantic Segmentation

To cope with the structuring problem of point cloud data, one popular research is to apply projection-based techniques [1, 16, 27]. However, multi-viewpoint projections are quite sensitive to the viewpoints chosen. Voxelized point clouds can be processed by 3D convolution in the same way as pixels in 2D neural networks. But, high-resolution voxels result in high memory and computational costs, while lower resolutions cause loss of detail. Consequently, 3D sparse convolutional networks [4, 7] are designed to overcome these computational inefficiencies. Direct processing of point clouds [21, 22] to achieve semantic segmentation has become a popular research topic in recent years. However, sparse 3D point clouds lack continuous texture information, resulting in limited recognition performance of 3D scenes.

2.3   3D Semantic Segmentation Based on Joint 2D-3D Data

There has been some research in recent years on 2D and 3D data fusion, which can be broken down into unidirectional projection networks [3, 6, 13, 15] and bidirectional projection networks [11]. The bidirectional projection network, typified by BPNet [11], focuses on both 2D and 3D semantic segmentation tasks. Due to the mutual flow of its 2D and 3D information, its network framework has to rely on a symmetrical decoding network. In addition to the inflexibility of its framework, the 2D task introduced by the bi-projection idea will distract the network from the 3D segmentation task. This is our motivation for choosing a framework based on the uni-projection idea.

In terms of view selection, 3DMV [6] and MVPNet [13] adopted a scheme with a fixed number of views, which means the views may not cover the entire 3D scene. In contrast, VMFusion [15] solves narrow viewing angle and occlusion issues by creating virtual viewpoints, but this approach has high computational costs that increase with the number of views. With an excellent balance, our work employs a dynamic view scheme that selects views based on the greedy algorithm of MVPNet until the view covers more than 90% of the 3D scene while keeping the scene uncut.

3 Methodology

3.1 Overview

An overview of our DMF-Net pipeline is illustrated in Fig. 3. Each scene data consist of a sequence of video frames and one point cloud scene. The input point cloud is called the original point cloud, while the point cloud formed by back-projecting all 2D feature maps into 3D space is called a back-projected point cloud with 2D features. DMF-Net consists of three U-shaped sub-networks, where the 2D feature extractor is 2D U-Net [23], and the 3D feature extractor is 3D MinkowskiUNet [4] (M.UNet). Moreover, the encoder-decoder for joint features is also the 3D M.UNet. Although we set a specific network backbone in our implementation, different network backbones can also be utilised in our DMF-Net. The view selection and back-projection modules will be elaborated in Sect. 3.2 and Sect. 3.3, respectively. As for the feature integration module, similar to MVPNet [13], it finds k nearest neighbour 2D features for each point of the original point cloud. Subsequently, it directly concatenates with the deep 3D semantic features of the original point cloud and input to 3D M.UNet for further learning to predict the semantic results of the entire scene.

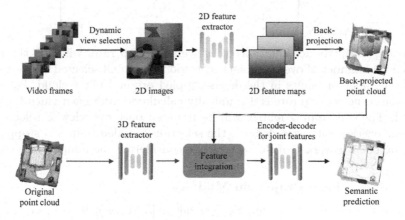

Fig. 3. Overview of the proposed DMF-Net

3.2 Dynamic View Selection

Previous work fixed the number of views, which would however cause insufficient overlaps between all the back-projected RGB-D frames and the scene point cloud. To make all the back-projected images cover as much of the scene point cloud as possible, many methods, e.g. MVPNet [13], generally choose to cut the point cloud scene. This will affect the recognition accuracy of cutting-edge objects. To this end, we propose a method for dynamic view selection, which sets a threshold for overlaps to ensure that the scene coverage is greater than 90% so that the number of views selected for each scene is different. The entire dynamic view selection algorithm is divided into two stages. The first stage is to construct the overlapping matrix. In the second stage, the view with the highest degree of overlap is dynamically selected according to the overlapping matrix cycle.

First, we define an overlapping matrix between the point cloud and video frames, as shown in Equation (1). This overlapping matrix indicates the relationship between each point and video frame. The first column lists the indices of the points, while the first row provides the indices of the frames. The entries in the matrix are either 0 or 1, representing non-overlapping and overlapping between points and frames respectively. Specifically, if a point of the original point cloud can find the back-projection point of the video frame within a certain range (1cm), it is determined that the point is an overlapping point. Considering that each point cloud contains hundreds of thousands of points, we randomly sample the original point cloud to reduce the computation.

$$
\begin{array}{c}
\begin{array}{cccc} Frame\ 1 & Frame\ 2 & \cdots & Frame\ V \end{array} \\
\begin{array}{c} Point\ 1 \\ Point\ 2 \\ Point\ 3 \\ \vdots \\ Point\ N \end{array}
\left(\begin{array}{cccc}
1 & 0 & \cdots & 1 \\
0 & 1 & \cdots & 1 \\
1 & 0 & \cdots & 0 \\
\vdots & \vdots & \cdots & \vdots \\
1 & 0 & \cdots & 1
\end{array}\right)
\end{array}
\tag{1}
$$

Second, we introduce the concept of scene overlap rate, which is the ratio between the number of overlap points corresponding to all selected video frames and the number of points in the down-sampled original point cloud. For each scene, the scene overlap rate is dynamically calculated after each video frame is selected. The overlapping matrix will be updated once one view is selected. If the scene overlap rate exceeds 90%, the selection of video frames is stopped to ensure excellent coverage of the scene while considering the number of views.

3.3   Unidirectional Projection Module

The video frames in the benchmark ScanNetv2 [5] dataset used in our experiments are captured by a fixed camera and reconstructed into a 3D scene point cloud. Therefore, we establish a mapping relationship between multi-view images and 3D point clouds based on depth maps, camera intrinsics, and poses. The world coordinate system is located where the point cloud scene is located, while the multi-view pictures belong to the pixel coordinate system. $(x_w, y_w, z_w)^T$ denotes a point in the world coordinate system and $(u, v)^T$ denotes a pixel point in the pixel coordinate system. Thus, the formula for converting the pixel coordinate system to the world coordinate system is shown as follows [32].

$$
\begin{bmatrix} x_w \\ y_w \\ z_w \\ 1 \end{bmatrix} = Z_c K^{-1} \begin{bmatrix} R & t \\ 0^T & 1 \end{bmatrix}^{-1} \begin{bmatrix} u \\ v \\ 1 \end{bmatrix},
\tag{2}
$$

where Z_c is the depth value of the image, K is the camera internal parameter matrix, R is the orthogonal rotation matrix, and t is the translation vector.

To verify our unidirectional projection module, we back-project all the dynamically selected multi-view images into 3D space and put them together with the original point cloud. The visualization results are shown in Fig. 4. Each color in the back-projected point cloud represents the projected point set for each view. It is clear to see that all views dynamically selected basically cover the indoor objects in the whole point cloud scene.

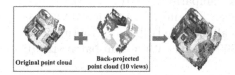 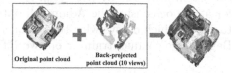

Fig. 4. Back-projection result   **Fig. 5.** Feature integration method

3.4 Feature Integration

The multi-view images are mapped to the space of the original point cloud using the unidirectional projection module in Sect. 3.3 to obtain the back-projected point cloud, each point of which contains 64-dimensional 2D deep semantic feature, denoted as f_j. Here, we define each point of the original point cloud as $p_i(x, y, z)$, as shown in Fig. 5. Specifically, each p_i utilizes the K-Nearest Neighbors (KNN) algorithm to find k back-projected points $p_j (j \in K)$ in the 3D Euclidean distance space. f_j is summed to obtain f_{2d} representing the 2D features of the p_i, while p_i has obtained the 64-dimensional 3D deep semantic feature f_{3d} through the 3D feature extractor. Finally, f_{2d} and f_{3d} are directly concatenated to become a 128-dimensional fusion feature F_i, which is calculated as follows.

$$F_i = \text{Concat} \left[f_{2d}, f_{3d} \right], \ f_{2d} = \sum_{j \in N_k(i)} f_j \tag{3}$$

4 Experiments

4.1 Datasets and Implementation Details

ScanNetv2 [5] is an indoor dataset including 706 different scenes, officially divided into 1201 training and 312 validation scans. Besides, the test set of 100 scans with hidden ground truth is used for benchmark. NYUv2 [25] contains 1449 densely labeled pairs of aligned RGB and depth images. We follow the official split of the dataset, using 795 for training and 654 for testing. Since this dataset has no 3D data, we need to use depth and camera intrinsics to generate 3D point clouds with 2D labels.

The training process can be divided into two stages. In the first stage, 2D images of ScanNetv2 were utilized to train a 2D feature extractor with 2D semantic labels. Noted that the original image resolution was downsampled to 320×240 for model acceleration and memory savings. In the second stage, a 3D network was trained with the frozen 2D feature extractor. The loss function used in the experiment is cross-entropy loss. As for the hyperparameter k in the feature integration module, we followed the previous practice, e.g. MVPNet and set it to 3. In the ablation study, the network structure of the proposed 3D feature extractor was set to M.UNet18A and the voxel size was set to 5 cm, which is consistent with the BPNet setup for a fair comparison. DMF-Net was trained for 500 epochs using the Adam [14] optimizer. The initial learning rate was set to 0.001, which decays with the cosine anneal schedule [19] at the 150th epochs. Besides, we conduct training on two RTX8000 cards with a mini-batch size 16.

4.2 Comparison with SoTAs on ScanNetv2 Benchmark

Quantitative Results. We compare our method with mainstream methods on the test set of ScanNetv2 to evaluate the 3D semantic segmentation performance of DMF-Net. The majority of these methods can be divided into point-based methods [10, 22, 28], convolution-based methods [4, 17, 29, 31]), and 2D-3D fusion-based methods [6, 11, 13]. The results are reported in Table 1.

Table 1. Comparison with typical approaches on ScanNetv2 benchmark, including point-based, convolution-based and 2D-3D fusion-based (marked with *) methods

Methods	mIOU	bath	bed	bkshf	cab	chair	cntr	curt	desk	door	floor	other	pic	fridge	shower	sink	sofa	table	toilet	wall	window
P.Net++ [22]	33.9	58.4	47.8	45.8	25.6	36.0	25.0	24.7	27.8	26.1	67.7	18.3	11.7	21.2	14.5	36.4	34.6	23.2	54.8	52.3	25.2
3DMV* [6]	48.4	48.4	53.8	64.3	42.4	60.6	31.0	57.4	43.3	37.8	79.6	30.1	21.4	53.7	20.8	47.2	50.7	41.3	69.3	60.2	53.9
FAConv [31]	63.0	60.4	74.1	76.6	59.0	74.7	50.1	73.4	50.3	52.7	91.9	45.4	32.3	55.0	42.0	67.8	68.8	54.4	89.6	79.5	62.7
MCCNN [10]	63.3	86.6	73.1	77.1	57.6	80.9	41.0	68.4	49.7	49.1	94.9	46.6	10.5	58.1	64.6	62.0	68.0	54.2	81.7	79.5	61.8
FPConv [17]	63.9	78.5	76.0	71.3	60.3	79.8	39.2	53.4	60.3	52.4	94.8	45.7	25.0	53.8	72.3	59.8	69.6	61.4	87.2	79.9	56.7
MVPNet* [13]	64.1	83.1	71.5	67.1	59.0	78.1	39.4	67.9	64.2	55.3	93.7	46.2	25.6	64.9	40.6	62.6	69.1	66.6	87.7	79.2	60.8
DCM-Net [24]	65.8	77.8	70.2	80.6	61.9	81.3	46.8	69.3	49.4	52.4	94.1	44.9	29.8	51.0	82.1	67.5	72.7	56.8	82.6	80.3	63.7
KP-FCNN [28]	68.4	84.7	75.8	78.4	64.7	81.4	47.3	77.2	60.5	59.4	93.5	45.0	18.1	58.7	80.5	69.0	78.5	61.4	88.2	81.9	63.2
M.UNet [4]	73.6	85.9	81.8	**83.2**	70.9	**84.0**	52.1	85.3	66.0	64.3	95.1	54.4	28.6	73.1	**89.3**	67.5	77.2	68.3	87.4	85.2	72.7
BPNet* [11]	74.9	**90.9**	81.8	81.1	**75.2**	83.9	48.5	84.2	67.3	**64.4**	95.7	52.8	30.5	77.3	85.9	**78.8**	**81.8**	69.3	91.6	**85.6**	72.3
Ours*	**75.2**	90.6	79.3	80.2	68.9	82.5	**55.6**	**86.7**	**68.1**	60.2	**96.0**	**55.5**	**36.5**	**77.9**	85.9	74.7	79.5	**71.7**	**91.7**	**85.6**	**76.4**

DMF-Net achieves a significant mIOU performance improvement compared with point-based methods which are limited by their receptive field range and inefficient local information extraction. For convolution-based methods, such as stronger sparse convolution, M.UNet can expand the range of receptive fields. Our method outperforms M.UNet by a relative 2.2% on mIOU because 2D texture information was utilized. DMF-Net shows a relative improvement of 17.3% on mIOU compared to MVPNet, a baseline unidirectional projection scheme. Such improvement can be attributed to the fact that the feature alignment problem of MVPNet is alleviated. Especially, our unidirectional projection scheme

DMF-Net is significantly better than the bidirectional projection method BPNet, one state-of-the-art in 2D-3D information fusion. The inflexibility of the BPNet framework limits its performance, while the high flexibility of our network framework enables further improvements.

Qualitative Results We compare the pure 3D sparse convolution M.UNet, the joint 2D-3D approach BPNet, and our method DMF-Net to conduct inference on the validation set of ScanNetv2. The visualization results are shown in Fig. 6.

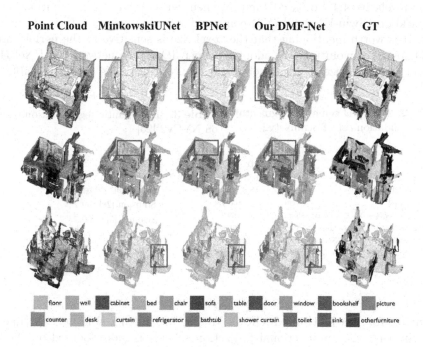

Fig. 6. Qualitative results of 3D semantic segmentation

As indicated by the red boxes, the 3D-only method M.UNet does not discriminate well between smooth planes or objects with insignificant shape differences, such as windows, doors, pictures, and refrigerators. This may due to the low resolution of the 3D point cloud and the lack of texture information for smooth planes. Despite the joint 2D-3D approach used by BPNet, the segmented objects are usually incomplete, owing that the bidirectional projection network distracts the core task, i.e. the 3D semantic segmentation.

4.3   Ablation Study and Analysis

Ablation for 2D-3D Fusion Effectiveness. We first fuse 2D deep semantic features with 3D shallow semantic features (i.e. each point contains XYZ and

RGB), followed by 3D sparse convolution. As shown in Table 2, the 3D semantic segmentation performance mIOU is improved from 66.4 to 70.8, indicating that 2D semantic features can benefit the 3D semantic segmentation task. The direct fusion of 2D deep semantic features with 3D shallow geometric features will cause misalignment in the semantic depth space affecting the network's performance. For this reason, our DMF-Net adds a 3D feature extractor based on the above framework so that 2D & 3D features are fused and aligned in semantic depth. As shown in Table 2, the feature-aligned model (V2) has a relative improvement of 1.3% on mIOU performance compared to the unaligned model (V1). In addition, we get a relative 4.4% on mIOU improvement with the stronger attention-based 2D backbone Swin-UNet [2] (V3) compared with the common U-Net34 model (V2). It is worth mentioning that the voxel size is sensitive to the performance of 3D sparse convolution. We adopt a deeper 3D sparse network, M.UNet34C, and set the voxel size to 2cm to obtain better results, as V4 shown in Table 2.

Table 2. 2D & 3D semantic segmentation on the validation set of ScanNetv2

Methods	Voxel Size	mIOU 2D	mIOU 3D
U-Net34 [23]	–	60.7	–
Swin-UNet [2]	–	68.8	–
M.UNet18A [4]	5 cm	–	66.4
Ours V1 (U-Net34 + XYZ & RGB)	5 cm	–	69.9
Ours V2 (U-Net34 + M.UNet18A)	5 cm	–	70.8
Ours V3 (Swin-UNet + M.UNet18A)	5 cm	–	73.9
Ours V4 (Swin-UNet + M.UNet34C)	2 cm	–	**75.6**

Table 3. 3D Semantic segmentation results on NYUv2 [25]

Methods	Mean Acc
SceneNet [8]	52.5
D3SM [9]	54.3
S.Fusion [20]	59.2
Scannet [5]	60.7
3DMV [6]	71.2
BPNet [11]	73.5
DMF-Net	**78.4**

Ablation for Projection Methods We conduct further ablative experiments to verify that the unidirectional projection scheme is more focused on the 3D semantic segmentation task than the bidirectional projection. Using the same framework in Fig. 3, we project 3D features into the 2D deep semantic feature space. Essentially, we apply a projection method similar to Sect. 3.3, which is an opposite process. Meanwhile, the 2D-3D feature fusion is the same as in Sect. 3.4. After 3D features are fused with multi-view features, semantic labels are output through a U-Net34 Network. Hence 2D cross-entropy loss is introduced on the total loss of the model. To avoid focusing too much on the optimization of 2D tasks, we multiply the 2D loss by a weight of 0.1, the same with BPNet. At this time, the new 2D model parameters for training no longer freeze, and the learning rate of all 2D models is 10 times lower than that of the 3D model.

Our experiments show that the bidirectional projection model overfits when it is trained to the 200th epoch, as seen from the 3D validation loss in Fig. 2. Meanwhile, the 3D mIOU of bidirectional projection only reaches 70.6, which is lower than the performance of the simple unidirectional projection (70.8). As the 2D task is also introduced, the increased learning parameters and the difficulty in adjusting the hyperparameters made it difficult for the model to focus more on 3D tasks. In this sense, unidirectional projection can focus more on 3D semantic segmentation tasks than bidirectional projection, leading to better flexibility.

4.4 DMF-Net on NYUv2

To verify the generalization ability, we conduct experiments on another popular RGB-D dataset, NYUv2 [25]. We report a dense pixel classification mean accuracy for DMF-Net, obtaining a significant performance improvement compared to other typical methods, especially joint 2D-3D methods, e.g. 3DMV [6] and BPNet [11]. As seen in Table 3, our DMF-Net gains a relative 6.6% performance improvement compared to the state-of-the-art BPNet [11]. This result demonstrates the strong generalization capability of the DMF-Net.

5   Conclusions and Future Work

In our work, we propose a Deep Multi-view Fusion Network (DMF-Net) based on a unidirectional projection method to perform 3D semantic segmentation utilizing 2D continuous texture information and 3D geometry information. Compared with the previous 2D-3D fusion methods, DMF-Net enjoys a deeper and more flexible network. Thus DMF-Net enables improved segmentation accuracy for objects with little variation in shape, effectively compensating for the limitations of pure 3D methods. In addition, DMF-Net achieves the superior performance of the joint 2D-3D method in the ScanNetv2 benchmark. Moreover, we obtain significant performance gains over previous approaches on the NYUv2 dataset. Currently, the number of dynamically selected multi-view images in DMF-Net is relatively large in order to cover the full 3D scene. In the future, we will explore efficient view selection algorithms so that even a few image inputs could achieve the full coverage of the 3D scene.

Acknowledgement. The work was partially supported by the following: National Natural Science Foundation of China under no. 62376113; Jiangsu Science and Technology Programme (Natural Science Foundation of Jiangsu Province) under no. BE2020006-4, UK Engineering and Physical Sciences Research Council (EPSRC) Grants Ref. EP/M026981/1, EP/T021063/1, EP/T024917/.

References

1. Boulch, A., Le Saux, B., Audebert, N.: Unstructured point cloud semantic labeling using deep segmentation networks. In: 3DOR (2017)
2. Cao, H., et al.: Swin-Unet: Unet-like pure transformer for medical image segmentation. arXiv preprint arXiv:2105.05537 (2021)
3. Chiang, H.Y., Lin, Y.L., Liu, Y.C., Hsu, W.H.: A unified point-based framework for 3D segmentation. In: 3DV, pp. 155–163 (2019)

4. Choy, C., Gwak, J., Savarese, S.: 4D spatio-temporal ConvNets: minkowski convolutional neural networks. In: IEEE/CVF Conference on Computer Vision and Pattern Recognition (CVPR), pp. 3075–3084 (2019)
5. Dai, A., Chang, A.X., Savva, M., Halber, M., Funkhouser, T., Nießner, M.: ScanNet: richly-annotated 3D reconstructions of indoor scenes. In: IEEE/CVF Conference on Computer Vision and Pattern Recognition (CVPR), pp. 5828–5839 (2017)
6. Dai, A., Nießner, M.: 3DMV: joint 3D-multi-view prediction for 3D semantic scene segmentation. In: Ferrari, V., Hebert, M., Sminchisescu, C., Weiss, Y. (eds.) ECCV 2018. LNCS, vol. 11214, pp. 458–474. Springer, Cham (2018). https://doi.org/10.1007/978-3-030-01249-6_28
7. Graham, B., Engelcke, M., Van Der Maaten, L.: 3D semantic segmentation with submanifold sparse convolutional networks. In: IEEE/CVF Conference on Computer Vision and Pattern Recognition (CVPR), pp. 9224–9232 (2018)
8. Handa, A., Patraucean, V., Badrinarayanan, V., Stent, S., Cipolla, R.: SceneNet: Understanding real world indoor scenes with synthetic data. arXiv preprint arXiv:1511.07041 (2015)
9. Hermans, A., Floros, G., Leibe, B.: Dense 3D semantic mapping of indoor scenes from RGB-D images. In: 2014 IEEE International Conference on Robotics and Automation (ICRA), pp. 2631–2638 (2014)
10. Hermosilla, P., Ritschel, T., Vázquez, P.P., Vinacua, À., Ropinski, T.: Monte Carlo convolution for learning on non-uniformly sampled point clouds. ACM Trans. Graph. (TOG) **37**(6), 1–12 (2018)
11. Hu, W., Zhao, H., Jiang, L., Jia, J., Wong, T.T.: Bidirectional projection network for cross dimension scene understanding. In: IEEE/CVF Conference on Computer Vision and Pattern Recognition (CVPR), pp. 14373–14382 (2021)
12. Huang, K., Hussain, A., Wang, Q., Zhang, R.: Deep learning: fundamentals, theory and applications, vol. 2. Springer (2019). https://doi.org/10.1007/978-3-030-06073-2
13. Jaritz, M., Gu, J., Su, H.: Multi-view pointNet for 3D scene understanding. In: 2019 IEEE/CVF International Conference on Computer Vision Workshop (ICCVW) (2019)
14. Kingma, D.P., Ba, J.: Adam: a method for stochastic optimization. In: International Conference on Learning Representations (ICLR) (2015)
15. Kundu, A., et al.: Virtual multi-view fusion for 3D semantic segmentation. In: ECCV, pp. 518–535 (2020)
16. Lawin, F.J., Danelljan, M., Tosteberg, P., Bhat, G., Khan, F.S., Felsberg, M.: Deep projective 3D semantic segmentation. In: Felsberg, M., Heyden, A., Krüger, N. (eds.) CAIP 2017. LNCS, vol. 10424, pp. 95–107. Springer, Cham (2017). https://doi.org/10.1007/978-3-319-64689-3_8
17. Lin, Y., et al.: FPConv: learning local flattening for point convolution. In: IEEE/CVF Conference on Computer Vision and Pattern Recognition (CVPR), pp. 4293–4302 (2020)
18. Long, J., Shelhamer, E., Darrell, T.: Fully convolutional networks for semantic segmentation. In: IEEE/CVF Conference on Computer Vision and Pattern Recognition (CVPR), pp. 3431–3440 (2015)
19. Loshchilov, I., Hutter, F.: SGDR: stochastic gradient descent with warm restarts. In: International Conference on Learning Representations (ICLR) (2017)
20. McCormac, J., Handa, A., Davison, A., Leutenegger, S.: SemanticFusion: dense 3D semantic mapping with convolutional neural networks. In: 2017 IEEE International Conference on Robotics and Automation (ICRA), pp. 4628–4635 (2017)

21. Qi, C.R., Su, H., Mo, K., Guibas, L.J.: PointNet: deep learning on point sets for 3D classification and segmentation. In: IEEE/CVF Conference on Computer Vision and Pattern Recognition (CVPR), pp. 652–660 (2017)
22. Qi, C.R., Yi, L., Su, H., Guibas, L.J.: PointNet++ deep hierarchical feature learning on point sets in a metric space. In: NeurIPS, pp. 5105–5114 (2017)
23. Ronneberger, O., Fischer, P., Brox, T.: U-Net: convolutional networks for biomedical image segmentation. In: Navab, N., Hornegger, J., Wells, W.M., Frangi, A.F. (eds.) MICCAI 2015. LNCS, vol. 9351, pp. 234–241. Springer, Cham (2015). https://doi.org/10.1007/978-3-319-24574-4_28
24. Schult, J., Engelmann, F., Kontogianni, T., Leibe, B.: DualConvMesh-Net: joint geodesic and Euclidean convolutions on 3D meshes. In: IEEE/CVF Conference on Computer Vision and Pattern Recognition (CVPR), pp. 8612–8622 (2020)
25. Silberman, N., Hoiem, D., Kohli, P., Fergus, R.: Indoor segmentation and support inference from RGBD images. In: Fitzgibbon, A., Lazebnik, S., Perona, P., Sato, Y., Schmid, C. (eds.) ECCV 2012. LNCS, vol. 7576, pp. 746–760. Springer, Heidelberg (2012). https://doi.org/10.1007/978-3-642-33715-4_54
26. Sun, K., Xiao, B., Liu, D., Wang, J.: Deep high-resolution representation learning for human pose estimation. In: IEEE/CVF Conference on Computer Vision and Pattern Recognition (CVPR), pp. 5693–5703 (2019)
27. Tatarchenko, M., Park, J., Koltun, V., Zhou, Q.Y.: Tangent convolutions for dense prediction in 3D. In: IEEE/CVF Conference on Computer Vision and Pattern Recognition (CVPR), pp. 3887–3896 (2018)
28. Thomas, H., Qi, C.R., Deschaud, J.E., Marcotegui, B., Goulette, F., Guibas, L.J.: KPConv: flexible and deformable convolution for point clouds. In: International Conference on Computer Vision (ICCV), pp. 6411–6420 (2019)
29. Wu, W., Qi, Z., Fuxin, L.: PointConv: deep convolutional networks on 3D point clouds. In: IEEE/CVF Conference on Computer Vision and Pattern Recognition (CVPR), pp. 9621–9630 (2019)
30. Xiao, X., Lian, S., Luo, Z., Li, S.: Weighted res-UNet for high-quality retina vessel segmentation. In: International Conference on Information Technology in Medicine and Education (ITME), pp. 327–331 (2018)
31. Zhang, J., Zhu, C., Zheng, L., Xu, K.: Fusion-aware point convolution for online semantic 3D scene segmentation. In: conference on Computer Vision and Pattern Recognition (CVPR), pp. 4534–4543 (2020)
32. Zhang, Z.: A flexible new technique for camera calibration. IEEE Trans. Pattern Anal. Mach. Intell. (TPAMI) 22(11), 1330–1334 (2000)
33. Zhao, W., Yan, Y., Yang, C., Ye, J., Yang, X., Huang, K.: Divide and conquer: 3D point cloud instance segmentation with point-wise binarization. In: International Conference on Computer Vision (ICCV) (2023)

RF-Based Drone Detection with Deep Neural Network: Review and Case Study

Norah A. Almubairik[1,3] and El-Sayed M. El-Alfy[2,3(✉)]

[1] Networks and Communications Department, Imam Abdulrahman Bin Faisal University, Dammam, Saudi Arabia
[2] SDAIA-KFUPM Joint Research Center for Artificial Intelligence, IRC for Intelligent Secure Systems (IRC-ISS), KFUPM, Dhahran, Saudi Arabia
[3] College of Computing and Mathematics, King Fahd University of Petroleum and Minerals (KFUPM), Dhahran, Saudi Arabia
alfy@kfupm.edu.sa

Abstract. Drones have been widely used in many application scenarios, such as logistics and on-demand instant delivery, surveillance, traffic monitoring, firefighting, photography, and recreation. On the other hand, there is a growing level of misemployment and malicious utilization of drones being reported on a local and global scale. Thus, it is essential to employ security measures to reduce these risks. Drone detection is a crucial initial step in several tasks such as identifying, locating, tracking, and intercepting malicious drones. This paper reviews related work for drone detection and classification based on deep neural networks. Moreover, it presents a case study to compare the impact of utilizing magnitude and phase spectra as input to the classifier. The results indicate that prediction performance is better when the magnitude spectrum is used. However, the phase spectrum can be more resilient to errors due to signal attenuation and changes in the surrounding conditions.

Keywords: Border security · Drone detection · Radio-frequency signals · FFT spectrum · Deep learning

1 Introduction

Unmanned Aerial Vehicles (UAV) or pilotless or uncrewed aircraft, commonly known as drones, were once considered a restricted technology that only official authorities, such as the military and government, could use. Currently, UAVs are also being used at a growing scale in commercial and personal services, e.g. to distribute goods and services in different industries. There are various potential use cases for drones, including, but not limited to, logistics, monitoring traffic, monitoring and fertilizing crop fields, building safety inspection, surveillance and border control, photography, and recreational services [12]. Despite the variety of beneficial applications of drones, their misuse threatens national and international security and public safety. There is a growing amount of reported incidents on local and global scales. For instance, Saudi Arabia has cut oil and gas production due to drone attacks on two major oil facilities run by the state-owned

B. Luo et al. (Eds.): ICONIP 2023, CCIS 1969, pp. 16–27, 2024.
https://doi.org/10.1007/978-981-99-8184-7_2

company, Aramco, in 2019 [BBC News[1]]. In the UK, a serious incident occurred in London between the 19th and 21st of December 2018, when Gatwick Airport was forced to shut down due to a drone strike. Around 1000 flights were either diverted or canceled, affecting an estimated number of 140,000 passengers [The telegraph[2]].

Drone detection is crucial for border security and public safety. It is an important step for identifying, locating, tracking, altering, and intercepting unauthorized drones. Different approaches have been proposed for drone detection based on various types of sensors used: (1) Radar-based, (2) Optical or video-based, (3) Radio Frequency (RF) based, and (4) Acoustic-based sensors. Radar-based detection systems have a fast-tracking mechanism and 360-degree coverage; however, they fail to detect small UAV objects [18,19]. In the same manner, video-based detection systems are unable to detect drones in long-range scenarios and foggy conditions [6]. Moreover, acoustic-based techniques are affected by noisy environments and have a short range of detection. Detecting drones using radio frequency, on the other hand, is not affected by the size of the UAV, its distance, foggy conditions, or noisy environments. In addition, it is considered a relatively reliable and low-cost solution. RF fingerprinting techniques depend on the particular characteristics of the radio frequency waveform emitted from the drone and/or its controllers. Experiments have shown that the majority of commercial UAVs have distinct RF signatures as a result of the electronics design, modulation techniques, and body vibration. Consequently, RF fingerprints obtained from the UAV or its remote controller signals can be used to identify and classify UAVs and their activities [8].

Over the past few years, deep learning (DL), a sub-field of machine learning, has gained popularity and has been a driving force behind several recent innovations. In comparison to other paradigms, deep learning techniques are widely recognized as being one of the most efficient and effective end-to-end modeling techniques that embody feature analysis and extraction from raw data, relaxing human experts from the tedious process of feature engineering. Over time, deep learning has been able to solve increasingly complex applications in natural language processing and computer vision with high accuracy [5].

The aim of this paper is to first present a review of work related to deep learning for RF-based drone detection and classification. Additionally, it provides a case study by extending the work done in [2] to compare the performance of a deep neural network model with the magnitude and phase spectra of RF signals. The RF signals are transformed using the Fast Fourier Transform (FFT) then the magnitude and phase spectra are computed and normalized. After that, two sets of experiments are conducted using neural networks of multiple layers, and the results are analyzed using various types of features of segmented signals in order to: (i) detect drone presence, and (ii) detect the drone and recognize its type.

The remainder of this paper is arranged as follows. Section 2 reviews work related to drone detection systems using Radio Frequency and deep learning

---

[1] https://cutt.ly/0hAlsjZ.
[2] https://cutt.ly/2bxdUaO.

techniques. Section 3 describes a case study including the experimental setup and the methodology followed as well as a description of various conducted experiments and analysis of the results. Finally, Sect. 4 concludes the paper and highlights recommendations some potential issues for future work.

2   Background

A drone is a form of aircraft that does not have a human pilot on board. Its main parts include drone body, remote control device, and energy device. It can be remotely or autonomously controlled. It has become a widely-used technology due to the significant reduction in costs and sizes. It has several potential applications, such as express shipping and package delivery, aerial photography for journalism and film, weather forecasting, crop spraying, entertainment, etc. However, the misuse of drone technology can have major impacts on public safety and national security. They can threaten flight safety, engage in criminal acts, and invade personal privacy as they are supplemented with high-quality cameras [10,17]. This includes, but is not limited to, offensive reconnaissance and monitoring of individuals.

Drones can be classified into the following main categories:

- Fixed-Wing Systems: A term used specifically in the aviation industry to describe aircraft that use fixed rigid wings to produce lift in conjunction with forwarding airspeed. Yaacoub et al. describe fixed-wing systems as follows: "They are based on the Vertical Take-Off and Landing (VTOL) principle" [23]. Traditional airplanes, surface-attached kites, and various kinds of gliders, such as hang gliders or para-gliders, are examples of this type of aircraft [7,21]
- Multi-Rotor Systems: Airplanes that produce lift using rotary wings. A traditional helicopter is a common example of a rotorcraft, which can have one or many rotors. Multiple small rotors, which are required for their stability, are often fitted with drones using rotary systems [7,21]. DJI Phantom and Parrot Bebop are considered commercial multi-rotor drones.
- Hybrid-Wing Drones: These types of drones have been developed with fixed or rotary wings to reach the intended location faster and hover over the air using their rotor wings [23]. They can cover a maximum of 150km in a single plane and are easy to control.
- Ornithopter Drones: This category includes unmanned aircraft that fly by imitating insect or bird wing motions (i.e. flapping their wings). The majority of these ornithopters are scaled in relation to the birds or insects they represent [7,21]. The flapping wing mechanism converts the motor's rotary motion into the ornithopter's reciprocating motion, allowing it to provide the necessary lift and thrust to travel steadily [11]. Delfly explorer and the micro-mechanical flying insect are two examples of ornithopter drones.

The existence of various types of drones with a variety of characteristics makes their detection and the design of anti-drone systems a daunting task.

Al-Sa'd et al. [2] developed a drone detection mechanism. They captured a large number of drone's RF signals and created a dataset called DroneRF. Then, the RF signals have been transformed using FFT magnitude for different frequency segments and fed into three separate Deep Neural Networks (DNNs) to detect drones and recognize their types and operational states. The overall accuracy of the designed system decreased when the number of classes was increased. The classification accuracy reached 99.7%, 84.5%, and 46.8% for binary, four-class, and ten-class classification problems, respectively.

Al-Emad and Al-Senaid [1] have also utilized the same dataset DroneRF and proposed a drone detection system using a Convolutional Neural Network (CNN) instead of DNN. The results confirmed that drone detection mode identification using CNN outperformed the drone detection solutions performed using DNN. Similarly, Allahham et al. [4] applied CNN for the DroneRF dataset to develop a drone detection, identification, and classification approach. Their analysis shows that Drone and Background activity spectra are significantly distinguishable. However, the RF spectra of different types of drones as well as different operation modes are either identical or overlapped. For that reason, they channelized the spectrum into multiple channels and considered each one as a separate input into the classifier. The results showed perfect drone detection but the accuracy is reduced to 94.6% and 87.4% for identifying drone types and states, respectively.

Nguyen et al. [15] developed a MATTHAN algorithm that detects drone's body movements, namely body shifting and body vibration. The algorithm analyzes the radio frequencies emitted from the communication between the drone and its remote controller. The algorithm gathers evidence from multiple sources (e.g. moving object detection, body shifting patterns, body vibration). Then, the algorithm combines these sources of evidence to form a binary classifier. The MATTHAN algorithm is evaluated across seven different drones and three different environments. The results showed that the more time the drone stays in the coverage area, the more accurate the results are. Furthermore, the detection accuracy increased when relying on several sources of evidence. In addition, the findings revealed that when the distances between the drone and the anti-drone system increased, the detection percentage decreased. To illustrate, when the drone was 10 m away from the detection system, the accuracy reached 96.5% but when the distance increased, the detection accuracy reduced to 89.4%

Nguyen et al. [14] examined a drone detection system that is developed to autonomously detect and characterize drones via RF signals. They combined two techniques: Passive (i.e. Radio Frequency) and Active (i.e. Radar), to identify the presence of potential invading drones. The proposed system depends mainly on three characteristics: the drone's rotating propellers, the drone's communication, and the drone's body vibration. As for the rotating propellers, the detection system uses a WiFi receiver (e.g. Alfa WiFi Network Adapter) and analyzes the significance of the signal reflected from its propellers. Regarding the drone's communication, it is noticed that the communication link between the wireless mobile devices and their connected access points is reaching 10 times per second whereas drones and their controllers reach 30 cycles per second. This is because of the importance of automatically changing the drone status. The third feature is the drone's body vibration in which the receiver monitors any modifications in the reflected signal intensity generated by the vibration of the drone body. The distance between the drone and the receiver can be measured using either phase variations or Received Signal Strength (RSS).

Xiao and Zhange [22] worked on drone detection and classification based on RF signal buried in ambient noise (e.g., WiFi signal). Their classification technique focused on the RF signature extracted from the down-link communication between the drone and its controller rather than the up-link communication. The RF signature consists of cyclostationarity features (i.e. a signal having statistical characteristics that differ cyclically over time), kurtosis (i.e. monitoring tailedness of a Gaussian scattered signal and the impact of central-limit such as conditions), and spectrum factors. The created RF signature is fed into two machine-learning classifiers: Support Vector Machine (SVM) and K-Nearest Neighbor (KNN). Different Signal-to-Noise Ratios (SNRs) and feature selection methods were considered while testing the drone signals. The testing results showed that the KNN classifier outperformed SVM.

AirID dataset was developed by Mohanti at Genesys Lab at the Northeastern University. It includes raw Interleaved Quadrature (IQ) samples taken from over-the-air transmissions of four USRP B200mini radios, each of which was mounted as a transmitter on a DJI M100 UAV and transmitted with a different IQ imbalance. Every recording in this dataset is made up of two files: a metadata file and a dataset file. The metadata file holds information about the dataset while the dataset file is a binary file containing digital samples. The metadata and data are in compliance with the Signal Metadata Format (SigMF) [13].

DroneRC is another RF-based dataset generated and utilized in some other works for drone detection and classification. For instance, Ezuma et al. [9] utilized the DroneRC dataset to develop a micro-UAV detection and classification system using RF fingerprints of signals emitted from the UAV remote controllers (RCs) and captured through an antenna and a high-frequency oscilloscope. The utilized dataset contained 100 RF signals collected from each of 14 different types of micro-UAV controllers. The duration of each signal was 0.25 ms and was represented by a vector of 5 million samples. The dataset was fed into wavelet analysis to reduce possible bias. Then, several machine learning classifiers have been applied, including SVM, KNN, and neural network (NN). Using KNN classification, all of the micro-UAVs were correctly identified, with an overall precision of 96.3%.

Ozturk et al. [16] used the DroneRC dataset to look at the issue of classifying UAVs based on RF fingerprints at low SNRs. They used CNNs trained on RF time-series images and spectrograms of 15 separate off-the-shelf drone controller RF signals. They reached a classification accuracy ranging from 92% to 100% for an SNR range of −10 dB to 30 dB, which outperformed current methods substantially. Similarly, Ezuma et al. [8] investigated drone detection and classification of UAVs in the presence of Bluetooth and WiFi interference signals. At 25 dB SNR, the KNN classifier achieved a classification accuracy of

98.13%. The efficiency of classification was also studied for a group of 17 UAV controllers at various SNR stages. In [20], the authors focused on drone detection in the presence of interference. They utilized RSS feature-based detectors to detect the presence of a drone signal buried in the RF interference and thermal noise. Using RSS feature requires a high SNR to ensure that the RF signal is insusceptible to interference with other background signals. The findings showed that the detection probability changes in a non-monotonic pattern.

3   Case Study

This study implements a drone detection and classification system for border security intelligence using Radio Frequency Signals and Deep Learning. The aim is to compare the impact of using the magnitude and phase spectrum of the FFT on the performance. This section describes how the research was conducted in details.

3.1   System Overview and Experimental Setup

Figure 1 illustrates the overall drone detection and recognition methodology. The system is composed of five modules: the drone under analysis, drone remote controller, RF sensing module, RF analyzer, and drone detection and classification module. The first three modules were conducted by Al-Sa'd et al. [2,3] and the DroneRF dataset was created. The last two modules are the focus of our research.

Different drones can emit different RF signals, which can then be used by intelligent systems to detect and identify them. During data collection, three types of drones were used, namely, DJI Phantom 3, Parrot Bebop, and Parrot AR Drone. The drone remote controller can also be a cellphone that sends and receives RF commands to and from the drones under investigation in order to change the drone's flight mode. Controlling the drones with a mobile phone necessitates mobile applications customized to each drone, e.g. "FreeFlight Pro," "AR.FreeFlight," and "DJI Go". Other applications can be used as well; however, for this dataset collection, the drone's official application was utilized.

The drone's communications with the flight control module were intercepted by an RF receiver connected to a laptop via a cable, which runs a program that retrieves, processes, and stores the RF data in a database, named "DroneRF". The aim of this module is to capture all unlicensed RF bands used by drones without making any assumptions about their flight mode. The dataset contains recordings of RF activities of the three types of drones (AR, Bepop and Phantom) as well as background signals (no drones). There are four operating modes: on and connected, hovering, flying but no video recording, and flying while recording video. The total number of segments in the dataset is 227 segments distributed as follows: 84 segments for Bepop (4 modes), 81 segments for AR (4 modes), 21 segments for Phantom (one mode) and 41 segments for Background. Each segment duration is 250 ms and captured using two simultaneous receivers with a sampling rate 40M sample/s per channel: one for the lower half

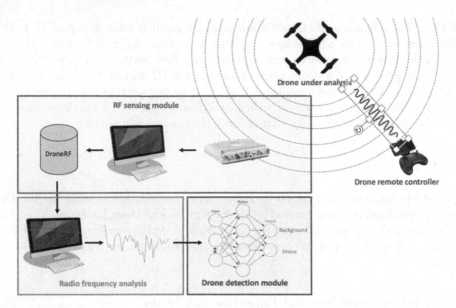

Fig. 1. Illustration of the main modules of the RF-based drone detection and classification system

of the frequency band (10M samples) and the other for the upper half of the frequency band (10M samples). For binary classification, there are two classes: No drone (82 segments) and Drone (372 segments). For 4-class classification, the drone class is divided into three other types: Phantom (42 segments), AR (162 segments), and Bepop (168 segments).

Signal analysis is used to discover hidden information in the recorded RF signals that can be used to improve detection. In this study, we adopted the Fast Fourier transform (FFT) to analyze the frequency-domain spectrum of the recorded RF signals. It is an invertible function, i.e. and an approximate form of the signal can be reconstructed using the inverse FFT (IFFT). FFT provides a fast method for computing the Discrete Fourier Transform (DFT), which is widely used in several other signal-processing applications such as remote sensing, communication, speech, and financial time series. It decomposes a signal into a series of sinusoidal (harmonic) components or vibrations to show how the signal energy is distributed over a particular range of frequencies (signal bandwidth).

Mathematically, a signal uniformly-sampled in time or a sequence of N values $\{x_n : x_0, x_1, x_2, \ldots, x_{N-1}\}$ is transformed into another sequence of N complex numbers in frequency domain $\{X_k : X_0, X_1, X_2, \ldots, X_{N-1}\}$ as follows:

$$X_k = \frac{1}{N} \sum_{n=0}^{N-1} x_n e^{\frac{-j2\pi kn}{N}} \tag{1}$$

where $j = \sqrt{-1}$. Alternatively, using Euler formula $e^{j\theta} = cos\theta + jsin\theta$, the FFT transform can be rewritten as,

$$X_k = \frac{1}{N}\sum_{n=0}^{N-1} x_n \cdot \left[\cos\frac{2\pi}{N}kn - j \cdot \sin\frac{2\pi}{N}kn\right] \tag{2}$$

FFT of a real-time signal is the sum of complex numbers z; each can be represented by a real part $real(z)$ and an imaginary part $img(z)$ or a magnitude part $|z|$ and a phase part $\angle z$. These quantities are mathematically related as follows:

$$|z| = \sqrt{real(z)^2 + img(z)^2} \tag{3}$$

$$\angle z = \tan^{-1}\frac{img(z)}{real(z)} \tag{4}$$

Figure 2 presents a sample of RF signals (segment #1 of the background and drone activities) as well as their magnitude and phase spectra. The FFT of each observed segment is calculated two times since the DroneRF dataset captures the entire 2.4 GHz bandwidth using two receivers (i.e. the first receiver captures the lower half frequency and the second receiver captures the upper half frequency). After extracting both the magnitude and phase spectra, they are used as the inputs to a deep neural network in three sets of experiments. For comparison purpose, each network is composed of three dense layers with Adam optimizer and ReLU activation for inner layers and sigmoid activation for output layer. The results are reported for stratified 10-fold cross validation, batch size = 10, and number of epochs = 200.

3.2   Results and Discussions

For binary classification experiments, the performance results of three DNN models are presented in the confusion matrices shown in Fig. 3 as well as accuracy, recall, precision, and F1 score. In the first two models, the magnitude spectrum and phase spectrum are used separately as input to the DNN model whereas in the third model, they are combined and used as input. The results show that the magnitude alone has the highest performance.

The binary classification problem is further divided into sub-problems: *Background vs AR*, *Background vs Bebop*, and *Background vs Phantom*. This decomposition may help to understand why the signal phase feature performed lower than the signal magnitude feature. The experimental results also show that the performance is consistent among all drones when the magnitude spectrum feature is utilized alone. On the other hand, when phase feature is considered, the detection system effectively classifies Phantom activities from background activities with an accuracy of 99.34%. The predictive performances for Bebop and AR drones are reduced to 79.94% and 66.12% (approx. to 2 decimal places), respectively.

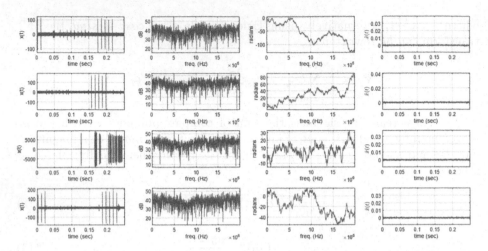

Fig. 2. Sample RF signals (segment#10) and its magnitude and phase spectra (1st and 2nd rows for Background signal low and high band channels, 3rd and 4th rows are for Bepop low and high band channels, 1st column is the time-domain of the original signals, 2nd column is the magnitude spectrum in dB, 3rd column is the unwrapped phase spectrum, 4th column is the reconstructed signal in time-domain from the magnitude only (i.e. zero phase))

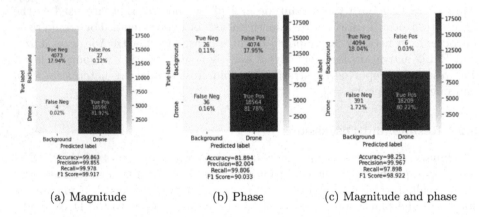

(a) Magnitude   (b) Phase   (c) Magnitude and phase

Fig. 3. Overall performance comparison of binary classification for drone detection (Drone presence vs Background)

The same techniques were also applied to the four-class classification problem and similar results were obtained which confirms that the magnitude spectrum based features is more accurate than the phase spectrum based features, as shown in Fig. 4. As a final conclusion, the magnitude spectrum is sufficient to detect and classify drones. However, the magnitude spectrum is susceptible to noise and other environmental conditions that can degrade the signal quality and hence the models' performance. Yet, more future work is recommended to study the impact of various conditions on the system performance.

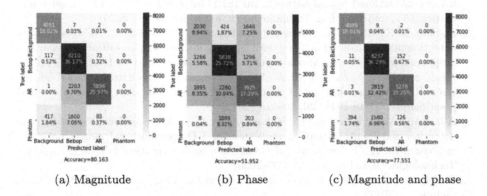

      (a) Magnitude   (b) Phase   (c) Magnitude and phase

Fig. 4. Performance evaluation of four-class classification problem (Background - Bebop - AR - Phantom)

4 Conclusion

UAVs are becoming more common, posing a threat to public safety and personal privacy. In order to reduce these threats, it is critical to efficiently identify invading UAVs. In this research, we used FFT to analyze Radio Frequency emitted by civilian drones, implemented Deep Learning-based models for drone detection and classification, compared signal magnitudes and phases of drone and background activities, and evaluated the effectiveness of the detection system using several evaluation metrics. The experimental results show that using the magnitude of the segmented signal has a different predictive performance than using the phase feature. By using the signal phase based features to solve binary classification problems, the classification accuracy is 81.894%, which is about 16% lower than when using the signal magnitude based features. For future research, it is suggested to add new types of drones such as fixed wings and hybrid wings to the dataset. This work can be also extended by investigating phase and magnitude features operation mode classification with deteriorated RF signals.

Acknowledgments. The authors would like to thank King Fahd University of Petroleum and Minerals for providing support during this work. The second author would also like to acknowledge the fellowship support at SDAIA-KFUPM Joint Research Center for Artificial Intelligence under grant no. JRC-AI-RFP-04.

References

1. Al-Emadi, S., Al-Senaid, F.: Drone detection approach based on radio-frequency using convolutional neural network. In: IEEE International Conference on Informatics, IoT, and Enabling Technologies (ICIoT), pp. 29–34 (2020)
2. Al-Sa'd, M.F., Al-Ali, A., Mohamed, A., Khattab, T., Erbad, A.: RF-based drone detection and identification using deep learning approaches: an initiative towards a large open source drone database. Futur. Gener. Comput. Syst. **100**, 86–97 (2019)
3. Allahham, M.S., Al-Sa'd, M.F., Al-Ali, A., Mohamed, A., Khattab, T., Erbad, A.: DroneRF dataset: a dataset of drones for RF-based detection, classification and identification. Data Brief **26**, 104313 (2019)
4. Allahham, M.S., Khattab, T., Mohamed, A.: Deep learning for RF-based drone detection and identification: a multi-channel 1-D convolutional neural networks approach. In: IEEE International Conference on Informatics, IoT, and Enabling Technologies (ICIoT), pp. 112–117 (2020)
5. Bengio, Y., Goodfellow, I., Courville, A.: Deep Learning, vol. 1. MIT Press, Massachusetts (2017)
6. Bisio, I., Garibotto, C., Lavagetto, F., Sciarrone, A., Zappatore, S.: Unauthorized amateur UAV detection based on WiFi statistical fingerprint analysis. IEEE Commun. Mag. **56**(4), 106–111 (2018)
7. Custers, B. (ed.): The Future of Drone Use. ITLS, vol. 27. T.M.C. Asser Press, The Hague (2016). https://doi.org/10.1007/978-94-6265-132-6
8. Ezuma, M., Erden, F., Anjinappa, C.K., Ozdemir, O., Guvenc, I.: Detection and classification of UAVs using RF fingerprints in the presence of Wi-Fi and Bluetooth interference. IEEE Open J. Commun. Soc. **1**, 60–76 (2019)
9. Ezuma, M., Erden, F., Anjinappa, C.K., Ozdemir, O., Guvenc, I.: Micro-UAV detection and classification from RF fingerprints using machine learning techniques. In: IEEE Aerospace Conference, pp. 1–13 (2019)
10. Liu, Z., Li, Z., Liu, B., Fu, X., Raptis, I., Ren, K.: Rise of mini-drones: applications and issues. In: Workshop on Privacy-Aware Mobile Computing, pp. 7–12 (2015)
11. Mahendran, S., Asokan, R., Kumar, A., Ria, V., Jayadeep, S.: Development of the flapping wing for ornithopters: a numerical modelling. Int. J. Ambient Energy **43**(1), 795–802 (2022). https://doi.org/10.1080/01430750.2019.1662841
12. Merkert, R., Bushell, J.: Managing the drone revolution: a systematic literature review into the current use of airborne drones and future strategic directions for their effective control. J. Air Transp. Manag. **89**, 101929 (2020)
13. Mohanti, S., Soltani, N., Sankhe, K., Jaisinghani, D., Di Felice, M., Chowdhury, K.: AirID: injecting a custom RF fingerprint for enhanced UAV identification using deep learning. In: IEEE GLOBECOM 2020-IEEE Global Communications Conference, pp. 370–378 (2020)
14. Nguyen, P., Ravindranatha, M., Nguyen, A., Han, R., Vu, T.: Investigating cost-effective RF-based detection of drones. In: 2nd Workshop on Micro Aerial Vehicle Networks, Systems, and Applications for Civilian Use, pp. 17–22 (2016)
15. Nguyen, P., Truong, H., Ravindranathan, M., Nguyen, A., Han, R., Vu, T.: Matthan: drone presence detection by identifying physical signatures in the drone's RF communication. In: 15th Annual International Conference on Mobile Systems, Applications, and Services, pp. 211–224 (2017)
16. Ozturk, E., Erden, F., Guvenc, I.: RF-based low-SNR classification of UAVs using convolutional neural networks. arXiv preprint arXiv:2009.05519 (2020)

17. Samland, F., Fruth, J., Hildebrandt, M., Hoppe, T., Dittmann, J.: AR.Drone: security threat analysis and exemplary attack to track persons. In: Intelligent Robots and Computer Vision XXIX: Algorithms and Techniques, vol. 8301, p. 83010G. International Society for Optics and Photonics (2012)
18. Schmidt, M.S., Shear, M.D.: A drone, too small for radar to detect, rattles the white house. New York Times **26** (2015)
19. Shi, X., Yang, C., Xie, W., Liang, C., Shi, Z., Chen, J.: Anti-drone system with multiple surveillance technologies: architecture, implementation, and challenges. IEEE Commun. Mag. **56**(4), 68–74 (2018)
20. Sinha, P., Yapici, Y., Güvenç, İ., Turgut, E., Gursoy, M.C.: RSS-based detection of drones in the presence of RF interferers. In: 2020 IEEE 17th Annual Consumer Communications & Networking Conference (CCNC), pp. 1–6. IEEE (2020)
21. Vergouw, B., Nagel, H., Bondt, G., Custers, B.: Drone technology: types, payloads, applications, frequency spectrum issues and future developments. In: Custers, B. (ed.) The Future of Drone Use. ITLS, vol. 27, pp. 21–45. T.M.C. Asser Press, The Hague (2016). https://doi.org/10.1007/978-94-6265-132-6_2
22. Xiao, Y., Zhang, X.: Micro-UAV detection and identification based on radio frequency signature. In: IEEE 6th International Conference on Systems and Informatics (ICSAI), pp. 1056–1062 (2019)
23. Yaacoub, J.P., Salman, O.: Security analysis of drones systems: attacks, limitations, and recommendations. Internet Things, 100218 (2020)

Effective Skill Learning on Vascular Robotic Systems: Combining Offline and Online Reinforcement Learning

Hao Li[1,2], Xiao-Hu Zhou[1,2(✉)], Xiao-Liang Xie[1,2], Shi-Qi Liu[1,2],
Mei-Jiang Gui[1,2], Tian-Yu Xiang[1,2], De-Xing Huang[1,2],
and Zeng-Guang Hou[1,2(✉)]

[1] State Key Laboratory of Multimodal Artificial Intelligence Systems, Institute of Automation, Chinese Academy of Sciences, Beijing 100190, China
[2] The School of Artificial Intelligence, University of Chinese Academy of Sciences, Beijing 100049, China
{xiaohu.zhou,zengguang.hou}@ia.ac.cn

Abstract. Vascular robotic systems, which have gained popularity in clinic, provide a platform for potentially semi-automated surgery. Reinforcement learning (RL) is a appealing skill-learning method to facilitate automatic instrument delivery. However, the notorious sample inefficiency of RL has limited its application in this domain. To address this issue, this paper proposes a novel RL framework, Distributed Reinforcement learning with Adaptive Conservatism (DRAC), that learns manipulation skills with a modest amount of interactions. DRAC pretrains skills from rule-based interactions before online fine-tuning to utilize prior knowledge and improve sample efficiency. Moreover, DRAC uses adaptive conservatism to explore safely during online fine-tuning and a distributed structure to shorten training time. Experiments in a pre-clinical environment demonstrate that DRAC can deliver guidewire to the target with less dangerous exploration and better performance than prior methods (success rate of 96.00% and mean backward steps of 9.54) within 20k interactions. These results indicate that the proposed algorithm is promising to learn skills for vascular robotic systems.

Keywords: Vascular robotic system · Reinforcement learning · Neural network

This work was supported in part by the National Natural Science Foundation of China under Grant 62003343, Grant 62222316, Grant U1913601, Grant 62073325, Grant U20A20224, and Grant U1913210; in part by the Beijing Natural Science Foundation under Grant M22008; in part by the Youth Innovation Promotion Association of Chinese Academy of Sciences (CAS) under Grant 2020140; in part by the CIE-Tencent Robotics X Rhino-Bird Focused Research Program.

1   Introduction

Percutaneous coronary intervention (PCI) is a widely used treatment for coronary artery diseases, which affects millions of people worldwide and remains a leading cause of mortality [1]. Compared with traditional PCI, robot-assisted intervention has shown less X-ray exposure and higher control precision in experimental and clinical studies, and is gaining popularity [2–4]. In robot-assisted intervention, physicians completes the intervention using vascular robotic systems with master-slave structure. The master console interfaces with physicians and sends commands to the slave console, while the slave console executes commands and delivers instruments to target locations in vessels for subsequent treatment such as stenting and drugs. To advance in tortuous vessels, catheters, guidewires and other instruments are designed as bendable thin wires. The relationship between commands and instruments motion is non-linear and complex, making delivery challenging. Thus, proficient skills requires a long period of training, and physicians may have a heavy cognitive load during intervention. To alleviate the above problems, vascular robotic systems are expected to have greater autonomy to help physicians with instrument delivery.

Reinforcement learning (RL) is a learning-based control method that maximizes reward functions from interactions [5], and has attracted great interest for skill learning on vascular robotic systems [6–10]. With the position and orientation of instruments obtained by an electromagnetic tracking sensor, RL can achieve performance comparable to experts [6]. However, it is difficult to integrate instruments and sensors in clinical, while physicians usually only use x-ray images for navigation. Using images as input is a more clinical approach, but it greatly reduces the sample efficiency of RL and increases the interaction requirement [11]. Input reconstruction with auto-encoder structure is used to assist the training of convolution layers and decrease the interaction requirement [8]. But image reconstruction also requires diverse data, and it still takes over ten hours to train on a vascular model.

Similar to the popular pretraining-fine-tuning paradigm in supervised learning, RL learns skills in a much more shorter time by pretraining skills with offline interactions [12–14]. Offline interactions can been obtained by human demonstrations, sub-optimal policy or even just random exploration, while pretraining methods can be imitation learning or offline RL. As for RL on vascular robotic systems, existing research has demonstrated that human demonstrations can accurate training [6,9]. However, using human demonstrations runs counter to the goal of reducing the burden on physicians since physicians have to provide demonstrations before RL. To minimize the burden on physicians, a simple manually-designed rule is better suited for collecting offline interactions. Due to the difficulty of the task, the delivery rule is not optimal in itself. As offline RL can achieve impressive performance with sub-optimal offline interactions [15], offline RL has great potential to pretrain delivery skills with manually-designed rule.

The main contributions of this paper are as follows:

1) A novel RL framework, Distributed Reinforcement learning with Adaptive Conservatism (DRAC), is proposed for fast skill learning in robot-assisted intervention. To the best of our knowledge, this is the first framework that combines offline and online RL in this field.
2) In online RL fine-tuning, conservatism in Q-function estimation is dynamically adjusted to achieve safe and efficient exploration.
3) Online interaction collection and parameter update are decoupled in a distributed paradigm, further shortening fine-tuning time.
4) Experiments show that the proposed framework can automate guidewire deliveries with a success rate of 96.00% and mean backward steps of 9.54 after 20k interactions and has less dangerous interactions in online fine-tuning, showing advantage in both performance and safety.

The paper is structured as follows: Section 2 introduces the problem definition. Section 3 elaborates the proposed method. Section 4 demonstrates and analyzes the performance. Section 5 summarizes this paper.

2   Problem Definition

A pre-clinical environment is used to validate our framework as shown in Fig. 1(a). In the pre-clinical environment, the slave console of a vascular robotica system [4] (hereinafter referred to as robot) need to delivery the guidewire from the outset to the target in a vascular model. The instrument delivery task is defined as a Partially Observable Markov Decision Process (POMDP) $\langle S, O, A, P, R, p_0, \gamma \rangle$. S, O, A denote state space, action space, and observation space respectively. $P : S \times A \times S \mapsto R_+$ is the transition probability, and $R : S \times A \mapsto \mathbb{R}$ is the reward function. p_0 is the initial state distribution. γ is the discount factor. The instrument delivery starts with initial state $s_0 \sim p_0$, where the distal tip of the guidewire is 10–15 pixels from the outset and the guidewire is rotated randomly. At time step t with state $s_t \in S$, the robot receives an observation $o_t \in O$ and chooses an action $a_t \in A$. After the robot executes a_t, the state

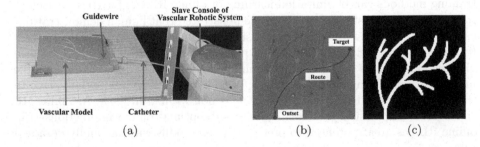

(a)   (b)   (c)

Fig. 1. (a) The preclinical environment. (b) The outset, target and route in delivery. (c) The binary image of the vessel.

changes to s_{t+1} according to the transition probability and the robot receives a reward $r_t = R(s_t, a_t)$. Learning instrument delivery is formalized as maximize the cumulative reward $\sum_t \mathbb{E}_{o_t, a_t}(\gamma^t r_t)$. Details of the problem definition are as follows.

State and Observation: In POMDP, the state refers to a representation that contains all information and is not available. In percutaneous coronary intervention, physicians use images from X-ray fluoroscopy for navigation. Referring to clinical situations, images obtained from a fixed camera are taken as observations, which are the vertical view with size 140×140. The observation is prepossessed into binary images of the vessel and the guidewire as shown in Fig. 1(c).

Action: The robot manipulates instruments through two degrees of freedom, translation and rotation. In the translation freedom, the robot chooses to translate forward or backward 2 mm. In the rotation freedom, the robot chooses from stationary, clockwise and counterclockwise rotation at two speeds. The rotation speeds are 10 and $20°$ per action, respectively. Taken translation and rotation together, there are ten actions in the action space $\mathcal{A}$.

Reward: The reward is designed for safety and efficiency in RL and is shown in Fig. 2(a). Safety is the primary consideration during training. If motor torques exceed a safe threshold, the manipulation is considered unsafe and the delivery is terminated prematurely with reward -100. As long as the safety constraint are met, The reward encourages the robot to deliver the guidewire along the route, i.e. the shortest path from the outset to the target. If the guidewire goes off or back on route then the reward is -20 and 20 respectively. If the guidewire keeps on the route, the reward equal to the decrease in distance to the target Δd. If any of the above conditions are not met, that is, the guidewire keeps in the wrong bifurcation, then the reward is 0. The route and the distances of all points to the target are calculated before training. First, the vessel binary image is skeletonized, and all points on the skeleton are considered to connect to its eight-neighbors with a distance of 1. Then, the Dijkstra algorithm is used to obtain the route and the distance of each point to the target. For other vessel points, whether on the correct path and the distance to the target are decided by the closest point on the skeleton.

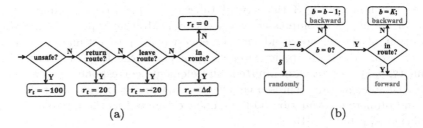

(a)   (b)

Fig. 2. (a) The reward in POMDP. (b) The delivery rule to choose translate direction.

3   Method

Our framework starts by collecting rule-based interactions, and then learns skills from rule-based interactions by imitation and offline RL. After offline RL, the skill is fine-tuned by online RL.

3.1   Rule-Based Interactions

A simple delivery rule is used to collect offline data to speed up RL. The delivery rule is designed according to the following prior knowledge to ensure effectiveness for RL:

1) When the guidewire is far from bifurcations, the robot directly translates the guidewire forward without considering backward.
2) When the guidewire is near bifurcations, the robot needs to adjust the guidewire direction for the correct branch.
3) Corresponding to the necessary exploration in RL, the delivery rule requires randomness to collect diverse data.

Based on the above prior knowledge, the delivery rule randomly selects sub-actions on the rotation freedom degree, while choosing subactions on the translation freedom degree as shown in Fig. 2(b). The rule has a probability δ of randomly choosing forward or backward. Otherwise, subactions are chosen based on variable b and whether the guidewire is on the right path. b represents the number of backward subactions that should be taken before reattempting to cross branches after entering the wrong branches. If the guidewire is on the right path and b is 0, the delivery rule selects the forward subaction, or the backward subaction is chosen. b is initialized to 0 and set to hyperparameter K when the guidewire enters the wrong branches and the reward is -20. In the following, $\mathbb{E}_{\mathbb{D}}\{\cdot\}$ means the expectation that the variable obeys the distribution in rule-based interactions $\mathbb{D}$.

3.2   Network Structure

The neural network follows the structure in our previous work about offline RL [10]. As shown in Fig. 3, the network takes a_{t-1}, o_{t-1} and o_t as inputs. Image inputs o_{t-1} and o_t are compressed by the encoder. Multilayer perceptions (MLP) use output of the encoder and one-hot encoded a_{t-1} to estimate Q-function $Q(o_t, o_{t-1}, a_{t-1})$ and the policy $\pi(o_t, o_{t-1}, a_{t-1})$. In the following, for writing simplicity, o_{t-1} and a_{t-1} are omitted, and elements corresponding to action a_t in $Q(o_t)$ and $\pi(o_t)$ are represented by $Q(o_t, a_t)$ and $\pi(o_t, a_t)$. To reduce the error in Q-function estimation, Q-function is estimated as the minimum of two estimations Q_1 and Q_2 [16]:

$$Q(o_t, a_t) = \min\left\{Q_1\left(o_t, a_t\right), Q_2\left(o_t, a_t\right)\right\} \tag{1}$$

To deal with over-fitting caused by high-dimensional image inputs, the encoder uses Adaptive Local Signal Mixing layer(A-LIX) [17], to smooth the gradient adaptively. A-LIX performs random shifts with adaptive range on feature maps. At training, A-LIX forwards as follows. For input feature maps, A-LIX first samples two uniform continuous random variables $\delta_h, \delta_w \sim U[-S, S]$ representing shifts in the height and weight coordinates respectively, where S denotes the shift range. The output is computed through bilinear interpolation. A-LIX does not work at testing and outputs the input directly.

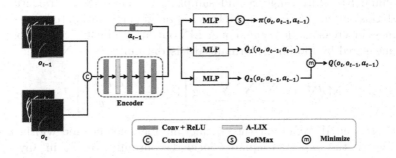

Fig. 3. The network structure.

3.3   Offline Pretraining

In offline pretraining, DRAC first imitate offline interactions and then uses offline RL to update the policy. The policy imitates the delivery rule by minimizing cross-entropy:

$$J_{\text{im}}(\pi) = \mathbb{E}_{\mathbb{D}} \{\log [\pi (o_t, a_t)]\}. \tag{2}$$

Q-function is optimized by loss function $J_{\text{offline}}(Q)$, which is combination of Bellman error $J_{\text{B}}(Q)$ and the advantage of off-line data $J_{\text{A}}(Q)$ as in [18]:

$$J_{\text{B}}(Q) = \mathbb{E}_{\mathbb{D}} \left\{ \sum_{i=1,2} [r_t + \gamma V^\pi (o_{t+1}) - Q_i (o_t, a_t)]^2 \right\},$$

$$J_{\text{A}}(Q) = \mathbb{E}_{\mathbb{D}} \{Q (o_t, a_t) - V^\pi (o_t)\}, \tag{3}$$

$$J_{\text{offline}}(Q) = J_{\text{B}}(Q) + \beta J_{\text{A}}(Q).$$

Hyperparemeter β balances J_{B} and J_{A}. Value function V^π is computed as follows:

$$V^\pi(o_t) = w \pi (o_t)^{\text{T}} \bar{Q} (o_t) + (1 - w) \pi (o_t)^{\text{T}} Q (o_t). \tag{4}$$

$\bar{Q}$ denotes the target Q-function whose parameters are the exponentially moving average of Q-function parameters, and w is a hyperparemeter to balance $\bar{Q}$ and Q.

After imitation with fixed steps, the policy π is trained by maximizing Q function and keeping reasonable entropy with policy loss function $J(\pi)$:

$$J(\pi) = \mathbb{E}_{\mathbb{D}} \{\alpha \mathcal{H}[\pi(o_t)] - V^\pi (o_t)\}, \tag{5}$$

where $\mathcal{H}(\cdot)$ stands for entropy. α is a non-negative parameter updated by minimizing the following loss function:

$$J(\alpha) = \mathbb{E}_{\mathbb{D}}\left\{\alpha\mathcal{H}[\pi(o_t)] - \alpha\bar{\mathcal{H}}\right\}, \tag{6}$$

where hyperparameter $\bar{\mathcal{H}}$ is the target entropy. Since the gradient of $J(\pi)$ is unstable, the gradient of $J(\pi)$ is prevented to update the encoder [19].

In offline RL, shift range S and sampling probabilities of transitions are updated. S is automatically adjusted by a Lagrangian relaxation to keep gradient smoothness in a reasonable range. For A-LIX output gradient $\nabla\hat{z}$, its smoothness can be measured by Modified Normalized Discontinuity Score $\widetilde{ND}$ [17]:

$$\widetilde{ND}(\nabla\hat{z}) = \sum_{c=1}^{C}\sum_{h=1}^{H}\sum_{w=1}^{W}\log\left[1 + \frac{D_{\nabla\hat{z}}(c,h,w)}{\nabla\hat{z}(c,h,w)^2}\right], \tag{7}$$

where C, H, and W represent the channel number, height, and width, respectively. $D_{\nabla\hat{z}}$ is the expected squared local discontinuity of $\nabla\hat{z}$ in any spatial direction v:

$$D_{\nabla\hat{z}} = \mathbb{E}_v\left[\left(\frac{\partial\nabla\hat{z}}{\partial v}\right)^2\right]. \tag{8}$$

Smaller $\widetilde{ND}(\nabla\hat{z})$ means that $\nabla\hat{z}$ is smoother, and S is updated by minimizing the following loss function:

$$J(S) = -S \times \mathbb{E}_{\mathbb{D}}\left[\widetilde{ND}(\nabla\hat{z}) - \overline{ND}\right], \tag{9}$$

where hyperparameter $\overline{ND}$ is the target of $\widetilde{ND}(\nabla\hat{z})$. To pay more attention to difficult manipulations, the sampling probability of transition i are proportional to weight f_i:

$$p(i) = \frac{f_i}{\sum_{j=1}^{N}f_j}, \tag{10}$$

where N denotes the number of transitions, and f_i is updated as follows after transition i is sampled:

$$\begin{aligned}
f_i &= (1-\tau)f_i + \tau\min\left(|\delta_{\text{TD}}^i| + \epsilon_1, \epsilon_2\right), \\
\delta_{\text{TD}}^i &= r_i + \gamma V^\pi(o_{i+1}) - Q(o_i, a_i).
\end{aligned} \tag{11}$$

δ_{TD}^i is TD error of transition i, and hyperparameter τ controls update speed. ϵ_1 and ϵ_2 are positive constants to avoid too small or too large weights. Transition weights are initlized as ϵ_2.

3.4   Online Fine-Tuning

After offline pretraining with rule-based interactions, online RL is used to fine-tune parameters. Compared with offline RL described above, there are two differences in online RL, namely the interactions used and Q-function update.

In addition to offline rule-based interactions, online RL also uses the interactions collected with the learned policy. To shorten the training time, online RL decouples interaction collection and parameter update. Online RL is deployed in a distributed structure containing an actor and a learner as shown in Fig. 4. The actor chooses action using the encoder and the policy and interacts with the environment. After each interaction, the actor packs an interaction in form of $(o_{t-1}, o_t, a_{t-1}, a_t, r_t, o_{t+1})$ and sends the interaction to the learner. The learner contains all received interactions in a replay buffer, and samples interactions from the replay buffer to update parameters. The replay buffer is initialized with rule-based interactions, and replaces the oldest interactions with new ones when capacity is reached. After every 50 interactions, the actor copies parameters from the learner. The actor and the learner run asynchronously and use Ray for communication [20].

An intuitive idea is to directly use J_B for Q-function update in online fine-tuning. However, due to the incompleteness of offline interactions, Q-function obtained by J_B often overestimates some actions and leads to dangerous exploration, which may cause harm to the robot and instruments. To enforce safety, DRAC estimates Q-function with adaptive conservatism in online fine-tuning. As in offline RL, DRAC uses $J_A(Q)$ for conservatism. With the increase of online interactions, the robot performs better and can explore with less conservatism. Thus, DRAC updates Q-function in online fine-tuning as following:

$$J_{\text{online}}(Q) = J_B + k^n \beta J_A, \tag{12}$$

where k denotes the ratio of offline interactions in the replay buffer and controls conservatism, and n is a hyperparameter. Other parts are updated in the same way as offline RL.

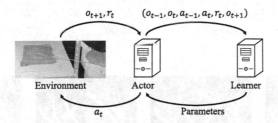

Fig. 4. The distributed structure in online fine-tuning.

4   Results

In experiments, hyperparameter n are setted as 1, and we uses 10k rule-based interactions and 10k online interactions. Other hyperparameters of DRAC are set as in our previous work that uses fully offline RL [10]. All experiments are trained on an NVIDIA 3090 GPU with 3 random seeds and tested in 50 deliveries unless otherwise stated.

DRAC is compared with the following baselines, which use the same network structure if necessary:

1) **Offline RL**: Using imitation and offline RL as in Sect. 3.3 to train parameters without online RL fine-tuning. For comparison, offline RL are trained with 20k rule-based interactions, which is equal to the sum of offline and online interactions for our method.
2) **Online RL**: Directly using online RL as described in Sect. 3.4 without pre-training or rule-based interactions. Online RL utilizes 20k interactions during training.
3) **DRAC w/o adaptive conservatism (DRAC w/o A-C)**: Same as DRAC, except that the method updates Q-function using J_B without J_A in online RL fine-tuning.
4) **Rule-based**: Using the rule designed in Sect. 3.1 and being tested in 400 deliveries.

All methods are measured in success rate, backward steps, and episode reward. Backward steps refer to the mean number of backward translation in successful deliveries. Episode reward is the mean of reward sum in deliveries.

Table 1. Performance of DRAC and baselines

Method	Success Rate	Backward Steps	Episode Reward
DRAC (ours)	**96.00% ± 1.63%**	**9.54 ± 3.62**	**126.97 ± 1.32**
Online RL	49.33% ± 11.12%	34.70 ± 5.90	104.47 ± 4.02
Offline RL	69.33% ± 3.40%	41.33 ± 4.00	113.86 ± 2.39
DRAC w/o A-C	84.67% ± 2.49%	23.40 ± 5.35	122.72 ± 2.95
Rule-based	19.00%	45.00	37.49

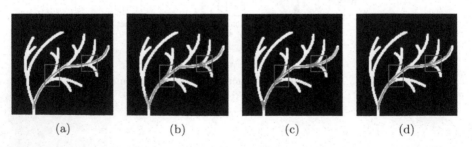

(a)   (b)   (c)   (d)

Fig. 5. Guidewire motions of different methods in ten deliveries. Each color represents a delivery. Red rectangles highlight triple bifurcations that are more difficult than other parts. (a) DRAC. (b) Online RL. (c) Offline RL. (d) DRAC w/o D-C. (Color figure online)

The results are shown in the Table 1. DRAC has the highest success rate (96.00%), the fewest backward steps (9.54) and the highest reward (126.97), outperforming all baselines. This result indicates the efficiency of our methods. Besides, DRAC has the smallest variance in all metrics, which shows the stability of the proposed method. Rule-based delivery has a poor performance with a success rate of about 20%, which is significantly worse than other methods. With the poorly-performing rule, offline RL and DRAC both perform better than online RL. This shows that the designed delivery rule can significantly speed up RL despite its poorer quality. Offline RL with 20k interactions performs worse than DRAC and DRAC w/o A-C that use both offline and online interactions, showing that online fine-tuning is beneficial to improve performance.

As for time efficiency, DRAC complete training within 8.79 h, where offline sample collection, offline pre-training, and online fine-tuning take 2.52, 2.92, and 3.35 h, respectively. In addition, duo to the shallow neural network, DRAC is able to choose an action within 1 ms without GPU acceleration.

To visualize the delivery process, Fig. 5 shows motions of the distal tip in ten deliveries. As shown in Fig. 5(a), DRAC goes off the route less than other methods, especially near triple bifurcations (red rectangles in Fig. 5) that require finer manipulations. This result indicates that DRAC has higher accuracy in delivery and requires less backward steps to correct the orientation. Moreover, motions with DRAC are more concentrated, which shows that our method can maintain better stability under the random initialization described in Sect. 2.

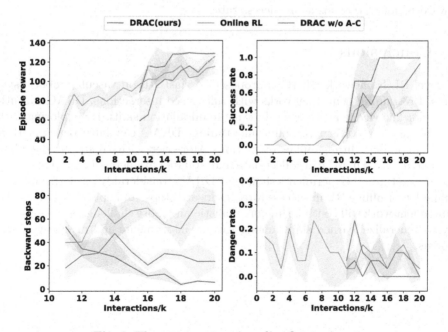

Fig. 6. The metric curves in online fine-tuning.

The performance during online fine-tuning is visualized in Fig. 6. All metrics are calculated from the last ten deliveries of every thousand interactions. Due to the utilization of 10k offline interactions, curves of DRAC and DRAC w/o A-C are started from 11k interactions. The episode reward of DRAC rises rapidly and remains stable, demonstrating a great advantage over the other two methods. Online RL has a higher episode reward than DRAC w/o A-C in the early stage of training but is surpassed by DRAC w/o A-C later. Moreover, after 10k interactions, the success rate of online RL does not increase significantly. However, there is a clear upward trend in the success rate of DRAC during online interactions, while DRAC w/o A-C has a slower upward trend. As for backward steps, both online RL and DRAC w/o A-C have no obvious decline but fluctuations, while backward steps of DRAC decrease steadily. Furthermore, DRAC has a consistently low danger rate. However, the other two methods have non-negligible danger rates, especially in the initial stage of online interactions. In summary, the proposed method DRAC performs better in the learning speed and stability of various metrics, and is safer in online interactions.

In online fine-tuning, although the success rate of online RL is significantly lower than that of DRAC and DRAC w/o A-C, the gap in episode reward is not so obvious. Since all methods after 20k interactions are with a danger rate of less than 0.1, episode reward is largely dependent on the guidewire position at the end of delivery. Most failed deliveries are stuck in the last triple bifurcation and have completed five-sixths of the route. Therefore, the difference between failed delivery and successful delivery is only about 20, and the difference in episode reward is not as obvious as in success rate.

5   Conclusions

A novel RL framework DRAC is proposed for fast RL on vascular robotic systems. Pretraining neural networks with rule-based interactions, DRAC can utilize prior knowledge and efficiently learn manipulation skills. By exploring with dynamical conservatism in online fine-tuning, DRAC can have better performance and safety during online interactions. Moreover, with a distributed deployment structure, DRAC can collect data and update parameters in parallel during online fine-tuning. Experiments show that DRAC considerably outperforms pure online RL or offline RL in success rate, backward steps, and episode reward. The subsequent work will continue in more realistic and more challenging settings and try to generalize learned skills among various instruments and diverse vascular models.

References

1. Wang, H., et al.: Global, regional, and national life expectancy, all-cause mortality, and cause-specific mortality for 249 causes of death, 1980–2015: a systematic analysis for the global burden of disease study 2015. Lancet (London, England) **388**, 1459–1544 (2016)
2. Granada, J.F., et al.: First-in-human evaluation of a novel robotic-assisted coronary angioplasty system. J. Am. Coll. Cardiol. Intv. **4**(4), 460–465 (2011)
3. Guo, S., et al.: A novel robot-assisted endovascular catheterization system with haptic force feedback. IEEE Trans. Rob. **35**(3), 685–696 (2019)
4. Zhao, H.-L., et al.: Design and performance evaluation of a novel vascular robotic system for complex percutaneous coronary interventions. In: Proceedings of 43rd Annual International Conference of the IEEE Engineering in Medicine and Biology Society, pp. 4679–4682 (2021)
5. Sutton, R.S., Barto, A.G.: Reinforcement Learning: An Introduction. MIT Press (2018)
6. Chi, W., et al.: Collaborative robot-assisted endovascular catheterization with generative adversarial imitation learning. In: Proceedings of 2020 IEEE International Conference on Robotics and Automation, pp. 2414–2420 (2020)
7. Karstensen, L., et al.: Autonomous guidewire navigation in a two dimensional vascular phantom. Current Dir. Biomed. Eng. **6**, 20200007 (2020)
8. Li, H., et al.: Discrete soft actor-critic with auto-encoder on vascular robotic system. Robotica **41**, 1115–1126 (2022)
9. Kweon, J., et al.: Deep reinforcement learning for guidewire navigation in coronary artery phantom. IEEE Access **9**, 166409–166422 (2021)
10. Li, H., Zhou, X.-H., Xie, X.-L., Liu, S.-Q., Feng, Z.-Q., Hou, Z.-G.: CASOG: conservative actor-critic with SmOoth gradient for skill learning in robot-assisted intervention. Arxiv (2020)
11. Yarats, D., Fergus, R., Lazaric, A., Pinto, L.: Mastering visual continuous control: improved data-augmented reinforcement learning. ArXiv, abs/2107.09645 (2021)
12. Nair, A., Dalal, M., Gupta, A., Levine, S.: Accelerating online reinforcement learning with offline datasets. ArXiv, abs/2006.09359 (2020)
13. Lu, Y.: AW-Opt: learning robotic skills with imitation and reinforcement at scale. In: Conference on Robot Learning (2021)
14. Kalashnikov, D.: QT-Opt: scalable deep reinforcement learning for vision-based robotic manipulation. ArXiv, abs/1806.10293 (2018)
15. Fu, J., Kumar, A., Nachum, O., Tucker, G., Levine, S.: D4RL: datasets for deep data-driven reinforcement learning. ArXiv, abs/2004.07219 (2020)
16. Fujimoto, S., van Hoof, H., Meger, D.: Addressing function approximation error in actor-critic methods. In: Proceedings of the 35th International Conference on Machine Learning, pp. 1582–1591 (2018)
17. Cetin, E., Ball, P.J., Roberts, S.J., Çeliktutan, O.: Stabilizing off-policy deep reinforcement learning from pixels. In: Proceedings of the 39th International Conference on Machine Learning, pp. 2784–2810 (2022)
18. Cheng, C.-A., Xie, T., Jiang, N., Agarwal, A.: Adversarially trained actor critic for offline reinforcement learning. In: Proceedings of the 39th International Conference on Machine Learning, pp. 3852–3878 (2022)

19. Yarats, D., et al.: Improving sample efficiency in model-free reinforcement learning from images. In: Proceedings of 35th AAAI Conference on Artificial Intelligence, pp. 10674–10681 (2021)
20. Moritz, P.: Ray: a distributed framework for emerging AI applications. Arxiv, abs/1712.05889 (2017)

Exploring Efficient-Tuned Learning Audio Representation Method from BriVL

Sen Fang[1], Yangjian Wu[2], Bowen Gao[1], Jingwen Cai[1], and Teik Toe Teoh[3(✉)]

[1] Victoria University, Footscray, Australia
{sen.fang,bowen.gao,jingwen.cai}@live.vu.edu.au
[2] Hainan University, Haikou, China
yangjian.wu@hainanu.edu.cn
[3] Nanyang Technological University, Singapore, Singapore
ttteoh@ntu.edu.sg

Abstract. Recently, there has been an increase in the popularity of multimodal approaches in audio-related tasks, which involve using not only the audible modality but also textual or visual modalities in combination with sound. In this paper, we propose a robust audio representation learning method WavBriVL based on Bridging-Vision-and-Language (BriVL). It projects audio, image and text into a shared embedded space, so that multi-modal applications can be realized. We tested it on some downstream tasks and presented the images rearranged by our method and evaluated them qualitatively and quantitatively. The main purpose of this article is to: (1) Explore new correlation representations between audio and images; (2) Explore a new way to generate images using audio. The experimental results show that this method can effectively do a match on the audio image.

Keywords: Audio-Visual · Multimodal Learning · Generative Adversarial Network · Speech Representation Learning

1 Introduction

Data volumes are crucial for training large-scale language models, and the use of vast corpora has been the standard practice [15]. However, the limited availability of high-quality labeled data in both unimodal and multimodal directions has hindered the development of the field [6]. To address this constraint, researchers have turned to zero-shot and few-shot learning approaches that rely on contrastive learning methods using textual descriptions. While the use of additional modalities together with text is common, the combination of more than two modalities in the audio domain is still uncommon. Given that manual data annotation in supervised learning is prohibitively expensive, self-supervised learning holds significant value for training large models. To overcome resource constraints and expand the research field, we present a novel multimodal self-supervised model based on the cutting-edge work of **Bridging-Vision-and-Language** [10].

B. Luo et al. (Eds.): ICONIP 2023, CCIS 1969, pp. 41–53, 2024.
https://doi.org/10.1007/978-981-99-8184-7_4

On the basis of this work, we propose an audio-visual correspondence model that extracts training from the BriVL model. BriVL is a text image correspondence model similar to OpenAI CLIP [21] and Google ALIGN [14], published in Nature Communications. Like CLIP, BriVL can rearrange images based on how well they match text images to find the best match. The principle of Our straightforward procedure is to freeze the BriVL visual encoder and part weight, run video on the visual stream of the model, and train a new model to predict BriVL embedding independently from the audio stream. The method we talked about above is very simple, which can secondary train, output images, and expand on more tasks. In addition, WavBriVL embeddings originate from BriVL, which means they align with text. This makes audio guided image repair [30], audio subtitles and cross mode text/audio to audio/image retrieval possible. We found WavBriVL is easy to extend to large video datasets. We systematically evaluated WavBriVL in a variety of audio tasks, including classification and retrieval, and compared it with other audio representation learning methods, as well as the SOTA results of each task. We showed how to apply WavBriVL to solve multiple multi-modal and zero start tasks.

2   Related Works

Our work is motivated by recent advancements in the field of multimodal learning, particularly in the first half of 2022. BriVL has shown to outperform CLIP [21] in various benchmarks and Microsoft's new WavLM [3] has surpassed their previous Wav2Vec [1] in multiple aspects. We hypothesize that combining these two models could achieve better results than Wav2CLIP [28]. Research on giving AI multi-modal perception and reasoning has been ongoing for many years, however, image generation for audio input is still an emerging field. With the emergence of different generation models, such as Goodfellow introduced GAN in 2014, there has been a lot of excellent work in the field of image generation [5,16]. From single mode to multi-modal, from text guidance about 2015 years later to audio guidance 2020 years later, there are many impressive works [20,32]. Naturally, there exist numerous antecedent endeavors and unacknowledged contributions. After this, there have been significant advancements in the field of image generation. Some of the most popular techniques include Stable Diffusion models [22] and ControlNet [29], a diverse range of applications has utilized these techniques, showcasing their potential in generating high-quality images.

In the work of modal extension for CLIP, AudioCLIP and related works like Wav2CLIP[1] are similar but different from our work. Compared to Wav2CLIP, BriVL has less dependence on specific semantic relationships, resulting in more creativity. Compare with them, we adopted a simple and efficient dual-tower architecture that does not rely on time-consuming target detectors. Finally, BriVL incorporates a cross-modal comparative learning algorithm based on MoCo [12], which differs from the approach in CLIP. Overall, while our work builds upon existing multi-modal learning research, and temporarily verify our

---

[1] https://github.com/descriptinc/lyrebird-wav2clip.

idea on image generation area. We introduce novel techniques and a unique approach that contributes to the field of audio-guided image generation. Furthermore, we conducted a comprehensive series of multimodal task tests that successfully demonstrated the efficacy of our proposed method.

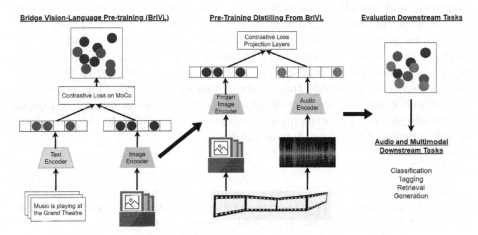

Fig. 1. Bridge Vision-Language Pre-training (BriVL), and our two-stage approaches including pre-training and evaluation.

3   Methodology and Experiments

Our method uses the WavBriVL model to guide the generation of VQGAN [8] output images through adversarial neural networks. This process utilizes meaningful embedding in the embedding space, by calculating the matching score between audio and image to rearrange the image, and this rearrangement idea is like CLIP. Our code is improved from the official model code and similarity calculation tools[2]. We found that this method can generate images that are appropriate for a given audio input, as confirmed by feedback from related experiments. Furthermore, our approach has the advantage of requiring less data compared to other fully supervised models for achieving competitive performance in downstream tasks. Specifically, WavBriVL's pre-training is more effective than competitive methods because it does not need to completely re-learn the visual model; instead, it only needs to train the audio model. This makes the model a promising model for various applications.

Bridging-Vision-and-Language is a model that has been trained on 650 million weak semantic datasets containing both text and images. To achieve this, the researchers created a cross-modal comparison learning algorithm based on

---

[2] https://github.com/BAAI-WuDao/BriVL.

MoCo [12], a monomodal comparison learning method. The Memory Bank mechanism was employed to maintain negative sample queues across different training batches, enabling the generation of a large number of negative samples for use in the comparison learning method. Additionally, BriVL has demonstrated state-of-the-art performance across various multi-modal tasks, including image annotation and zero-shot classification. As shown in Fig. 1, we replace the text encoder with the audio encoder by freezing the visual model of BriVL, running the image through it, and training the new model to predict that only the matching image-embedded content is obtained from the audio. After the audio encoder is trained, we freeze it and used it for experiments related to WavBriVL image generation to evaluate the results of these experiments. The assessment tasks include sound volume assessment and qualitative image quality and quantitative assessment. The rearranged images are all provided by selecting from the 100th epoch of the same 20 text inputs.

Table 1. Downstream tasks, including 1. classification: multi-class (MC), zero-shot (ZS), 2. retrieval: audio (AR) and cross-modal retrieval (CMR), and 3. audio captioning (AC) task, with various of clips, classes, and common metrics.

Dataset	Task	Clip (Split)	Class	Metric
ESC-50 [19]	MC/ZS	2k (5 folds)	50	ACC
UrbanSound8K [23]	MC/ZS	8k (10 folds)	10	ACC
VGGSound [2]	MC/ZS	185k	309	mAP
DESED [26]	AR	2.5k (valid)	10	F1
VGGSound [2]	CMR	15k (test)	309	MRR
Clotho [7]	AC	5k (evaluation)		COCO

3.1   Dataset for WavBriVL Performance Test

Our experiments encompassed a broad range of datasets, incorporating varying numbers of clips and categories. Additionally, we performed diverse tasks such as classification, retrieval, and generation to ensure comprehensive performance evaluation. For evaluation, we use relevant metrics detailed in Table 1 for each task. BriVL needs more than 100 A100 graphics cards to train for 10 days, so we don't consider retraining it. Our training and performance testing are based on the pre-trained model.

3.2   Dataset for VQGAN

In our study, we utilized one video dataset, namely VGG-Sound [2] to conduct the main experiment. The VGG-Sound dataset comprises audio sound clips, comprising 310 video classes and 200,000 audio samples that present challenging acoustic environments and practical noise characteristics. These audio samples are associated with non-man-made videos, and our model randomly selected one image from each sample video, with an audio sampling rate of 16 kHz. We used 20000 video clips for training. Then select another part to evaluate. It is noteworthy that the original BriVL model used Chinese datasets, while VGG-Sound's datasets are in English. Since we only attempted to perform related evaluation tasks based on image generation, the language of the dataset has no significant impact at the moment, which is sufficient for some analysis, and we only consider fine-tuning it. Moreover, we incidentally cleared the original text-related weights to avoid potential impacts.

3.3   Feature Extraction Processing

Our image and audio encoders utilize EfficientNet-B7 [25] as the CNN in the former and WavLM [3] as the basic transformer in the latter. The self-attention block combines four Transformer encoder layers and an MLP block, incorporating two fully connected layers and one ReLU activation layer. Our sorting code conducts forward operations on both the audio and image data before generating an image-audio matrix for sorting and output. Additionally, the code provides similarity scores for the processed vector matrix and the indices sorted by score to sort the images based on relevance.

Image Encoder. We adapted BriVL's approach of using random grayscale in the input image with color jitter for data enhancement. We standardized all videos at 1080P resolution and separated images (or 720P if not available), cropping them down to 480×480 pixels. To capture patch features and extract them via average pooling, we use a Transformer. To better capture patch feature relationships, we employed a self-attention (SA) block containing multiple Transformer encoder layers, each comprising a multi-head attention (MHA) layer and a feedforward network (FFN) layer [10].

$$\mathbf{S}' = \text{LayerNorm}(\mathbf{S} + \text{MHA}(\mathbf{S})) \tag{1}$$

$$\mathbf{S} = \text{LayerNorm}(\mathbf{S}' + \text{FFN}(\mathbf{S}')) \tag{2}$$

Then, they use the average pooling layer to fuse the extracted patch features:

$$\mathbf{r}^{(i)} = \frac{1}{N_p} \sum_{j=1}^{N_p} \mathbf{S}_j \in \mathbb{R}^c \tag{3}$$

where $\mathbf{S}_j$ is the j-th column of $\mathbf{S}$. A two-layer MLP block with a ReLU activation layer is adopted to project $\mathbf{r}^{(i)}$ to the joint cross-modal embedding space, resulting in the final d-dimensional image embedding $\mathbf{z}^{(i)} \in \mathbb{R}^d$.

Audio Encoder. In the case of audio input, our method initially converts the original 1D audio waveform into a 2D spectrum that serves as the input for WavLM. To output an embedding, we pool the entire 512-dimensional audio sequence and calculate the WavLM embedding using a weighted average of outputs from all transformer layers. The WavLM model[3], which was inspired by

---

[3] https://github.com/microsoft/unilm/tree/master/wavlm

HuBERT, comprises a CNN encoder and a Transformer with L blocks. During training, some frames of the CNN encoder output $\mathbf{x}$ are randomly masked and fed to the Transformer as input. The Transformer is trained to predict the discrete target sequence $\mathbf{z}$, in which each $z_t \in [C]$ represents a C-class categorical variable. The distribution over classes is parameterized using Equation (1), where $\mathbf{W}^P$ is a projection matrix, $\mathbf{h}_t^L$ denotes the output hidden state for step t, $\mathbf{e}_c$ represents the embedding for class c, $\text{sim}(a, b)$ calculates the cosine similarity between a and b, and $\tau = 0.1$ scales the logit [3]:

$$p(c|\mathbf{h}_t) = \frac{\exp(\text{sim}(\mathbf{W}^P\mathbf{h}_t^L, \mathbf{e}_c)/\tau)}{\sum_{c'=1}^{C} \exp(\text{sim}(\mathbf{W}^P\mathbf{h}_t^L, \mathbf{e}_{c'})/\tau)} \tag{4}$$

The distribution over classes in extraction method is parameterized using Equation (1), where $\mathbf{W}^P$ represents a projection matrix that is optimized during training. The output hidden state for step t, denoted by $\mathbf{h}_t^L$, is computed by the Transformer with L blocks. The embedding for class c is represented by $\mathbf{e}_c$, and the cosine similarity between a and b is calculated using $\text{sim}(a, b)$. The scaling factor $\tau = 0.1$ is used to adjust the logit score. During fine-tuning, the WavLM embedding as the weighted average of all transformer layer outputs, with the weights learned during this stage. Fine-tuning involves updating the parameters of WavLM to optimize the model for a specific downstream task.

4   Task 1: WavBriVL Performance Test

In this chapter, we begin by discussing the training, development, and evaluation process of the WavBriVL model. We use publicly available datasets of varying sizes and tasks, including classification, retrieval, and audio captioning tasks. We compare WavBriVL with some widely used as strong benchmarks in this field, and evaluate its performance in these tasks. Additionally, we investigate the effect of sound volume on the generated images. We hypothesize that the volume of sounds can influence the generated images. Hence, we explore the influence of sound volume on image features extracted from the sound using the sound correlation model. We also perform quantitative image analysis to evaluate the performance of WavBriVL compared to previous work, such as S2I and Pedersoli et al. We test model with five categories from VEGAS [31] and compare its performance with other methods in terms of generating visually plausible images. In the experiment, the rearranged images are all provided by selecting from the 100th epoch of the same 20 text inputs.

4.1   Training, Development, and Evaluation

We selected publicly available audio classification data of different sizes, which are generally used for evaluation [4], and also included some audio tasks/data, as shown in Table 1, including classification, retrieval and audio captioning. ESC-50 [19] is a simple data set with only 2 thousand samples, while UrbanSound8K

[23] is a large environmental data set with 10 categories. VGGSound [2] is a huge set of audio and video materials as we said before. DESED is used again as an audio extraction (AR) job because DESED can perform sound extraction at the fragment level. Finally, Clotho [7] is a unique set of audio subtitles. In addition, in Subsect. 4.3, we also carried out other image analysis (Table 2).

Table 2. In the subsequent classification and acquisition work, there will be supervised training, other audio representation modes, OpenL3, and the latest SOTA [11,17]. ZS is based on WavBriVL as a zero sample size model, some of which are derived from the original literature.

Model	Classification				Retrieval	
	ESC-50	UrbanSound8K	VGGSound	DESED (AR)	VGGSound (CMR)	
	ACC	ACC	mAP	F1	A→I (MRR)	I→A (MRR)
Supervise	0.5200	0.6179	0.4331			
OpenL3	0.733	0.7588	0.3487	0.1170	0.0169	0.0162
Wav2CLIP	0.8595	0.8101	0.4663	0.3955	0.0566	0.0678
WavBriVL	**0.9117**	**0.8832**	**0.4741**	0.3720	**0.0611**	**0.0608**
SOTA	0.959	0.8949	0.544			
WavBriVL (ZS)	0.412	0.4024	0.1001			

For multi-class (MC) classification problems, an MLP-based classifier is employed, with a corresponding number of classes as output. In DESED, we use the way of simulating WavBriVL and sed_eval[4] to realize audio retrieval (AR). At the same time, we also explore the performance of ours when dealing with multimodal tasks, and how to transfer zero samples to other modalities.

4.2 Comparisons with Previous Work

First, we monitor the benchmark by training from scratch on each downlink (with random initialization of the encoder weights). Next, we compare WavBriVL with other publicly available OpenL3 [4] pre-trained on different pretext tasks in OpenL3. OpenL3 multimodal Self-supervised training with AudioSet. It serves as a strong benchmark for different audio tasks, such as audio classification and retrieval. We extract features from OpenL3 (512 dim) and WavBriVL (512 dim) and apply the same training scheme to all downstream classification and retrieval tasks. In the chart, we can see that in the retrieval of classification, we are slightly better than our previous work, with an average increase of about 0.04, and only some deficiencies in AR. But it's only about 0.02. We approach or slightly outperform our previous work in retrieval tasks.

In sumary, our model has good effects in both data sets of audio retrieval classification, for the source of our strengths: In the Classification tasks, on the four datasets, three of us achieved good results close to or exceeding SOTA.

---

[4] https://github.com/TUT-ARG/sed_eval.

one of reason may be related to our data, and the other may be the effect of BriVL. As for the lack of excellent performance in AR tasks, it may be due to the excessive divergence of the BriVL dataset. If we retrain the basic model on a large scale, we may achieve better results. In the Retrieva tasks, such mrr tasks from A to I, from I to A we have also achieved excellent results, which mainly comes from the excellent training effect of the previous two towers model and the pre-training model, the structure of the brief is useful for general with tasks.

4.3 Quantitative Image Analysis

We have carried out a comparative analysis of our proposed model against relevant prior works (S2I[5] [9,18,24]). It should be noted that although the S2I was not initially designed for sound-to-image conversion, it leverages a VQVAE-based model for generating sound-to-depth or segmentation. For a fair comparison, we ensured that our model and the one proposed by Pedersoli et al. were trained using the same training set up as S2I, including five categories in VEGAS. As reported in Table 3, our proposed model outperforms most of the other models while generating visually enthralling and recognizable images, thanks to the amalgamation of visually enriched audio embeddings and a potent image generator. Our method outperforms the others which using GAN both qualitatively and quantitatively in the VEGAS dataset.

Table 3. Compared to the previous three jobs [9,18,24].

	Method	VEGAS (5 classes)		
		R@1	FID (↓)	IS (↑)
(A)	Baseline	23.10	118.68	1.19
(B)	S2I	39.19	114.84	1.45
(C)	Ours	**53.55**	**96.12**	**3.52**
(D)	SOTA	77.58	34.68	4.01

Fig. 2. The effect of different volume levels on three inputs: water, fireworks, and rain.

5 Task 2: Speech Generation Picture

In this chapter, our objective was to enhance the detectability of the shared embedded space of WavBriVL through visual analysis. To accomplish this, we employed VQGAN [8] to generate images with audio guidance. To generate these

---

[5] https://github.com/leofanzeres/s2i.

images, we matched them against the input audio and assessed whether BriVL "approved" of the output. In the case of a mismatch, feedback was provided to VQGAN[6] to refine subsequent image generation attempts. Both remain were kept frozen during the performance testing process, as is standard practice.

5.1 Sound Volume

As shown in Fig. 2, in order to ascertain the robustness of our approach in acquiring an understanding of the correlation between auditory stimuli and visual representations, we conducted a study investigating the impact of varying sound volumes on the quality of the generated images. To accomplish this, we manipulated the sound volume levels during the test phase and extracted features from the corresponding sound files. Subsequently, these modified sound features were employed as input into our model and improved the similarity calculation tool, which had been trained on a standardized volume scale. The final three sets of images can prove our hypothesis that the magnitude of different volume levels is usually positively correlated with the effects and meanings displayed in the images.

Fig. 3. Images generated from five audio in VGG-Sound [2]. Top: Wav2CLIP, Bottom: WavBriVL - The x-axis is the audio based on which the rearrangement is based, and the rearranged images are generated using the same relevant text.

5.2 Comparison with Previous Work

In previous work, Wav2CLIP also tried to generate images by audio. Figure 3 shows two group of pictures generated by Wav2CLIP and WavBriVL using English audio. We can see that the style of the text generation diagram is similar, which is because they all use the same GAN generation model. However, in

---

[6] https://github.com/CompVis/taming-transformers.

detail, they have their own characteristics: for example, in their understanding of "Blast", the explosion in Chinese videos is often accompanied by an overwhelming number of Chinese New Year videos. Therefore, it can be seen that when the image with high similarity of "audio-image matrix" is finally selected by our model, the preference is to represent the festive orange, red firecracker skin, and other elements. This is not the difference caused by translation, because the input is the same file, which can show that BriVL has more self-characteristics, and the other pictures have similar characteristics; In underwater sound input, we can see that their generated images are similar. After all, natural underwater sound is more common but more coherent. This reflects the argument/advantage of "BriVL generated images are more integrated than CLIP just stacking elements" mentioned in BriVL's original paper.

The audio generation process exhibits two distinct characteristics in the two models. The first characteristic is convergence, as evident from the similarity of images produced by both models. This can be attributed to the training set being composed of English images, which dampens the stylistic differences between the two models. The second characteristic is divergence At the same time, in the process of audio guidance and selection, the final result also shows the characteristics of divergence. As depicted in Fig. 3, where the images are more imaginative compared to those in Fig. 4. This divergence can be explained by two factors: firstly, BriVL's weak semantic text image dataset has a higher imaginative capacity, and secondly, the audio itself has a strong ability to diverge, thereby enhancing the model's associative capabilities.

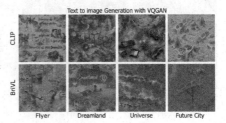

Fig. 4. Examples of CLIP (top) and BriVL (bottom) to image generation from text, BriVL's labels in x-axis are translated.

Table 4. Human scores on correlation between sounds and images, both Audio-CLIP and Wav2BriVL use GAN generated images.

Options	Positive	Negative	Neither
AudioCLIP	78%	14%	8%
WavBriVL	81%	15%	4%

5.3   Correlation Between Sounds and Images

In Fig. 3, We show that we can generate better images and interpretable correlations. However, it only demonstrate their authenticity, but we wanted to assess the relationship between sounds and images. To accomplish this, we conducted a test like previous work [13,27] where participants were asked to select the image that was most closely linked to a given sound from a set of two images. Each of these images was conditioned on a different sound class, so if our model could generate images that were relevant to a particular sound class, then participants would select the corresponding image that was generated from the inputted sound rather than an image generated from a different class of sampled sounds.

Table 4 displays the findings of the study, with the options representing the participants' choices. A positive option indicates that participants selected the image generated from the sound they heard, while a negative option suggests that they chose the image generated from a different class. The "neither" option indicates that participants believed that neither image could represent the sound they heard. The results (repeat three times to take the average integer) reveal that the majority of participants believed that the images generated by our model were correlated with the input sounds, demonstrating our model's capability to produce images that are related to the given sounds.

6 Summary and Conclusions

In this manuscript, we present a method that employs BriVL to extract audio representations for the purpose of audio-guided GAN image generation. Our proposed method has been shown to produce suitable audio representations, as demonstrated by its reliability in sound volume assessment, and quantitative and qualitative image quality assessment. In the future, we plan to study a model suitable for more modalities, develop the model in greater depth, explore its potential in more tasks, and compare it with other advanced models. In addition, we will consider trying other Generative model and further using audio representation (such as directly generating images instead of indirectly generating them now) as the work of the next version.

References

1. Baevski, A., Zhou, Y., Mohamed, A., Auli, M.: wav2vec 2.0: a framework for self-supervised learning of speech representations. Adv. Neural Inf. Process. Syst. **33**, 12449–12460 (2020)
2. Chen, H., Xie, W., Vedaldi, A., Zisserman, A.: Vggsound: a large-scale audio-visual dataset. In: ICASSP, pp. 721–725. IEEE (2020)
3. Chen, S., et al.: WavLM: large-scale self-supervised pre-training for full stack speech processing. IEEE J. Sel. Top. Sig. Process. **16**(6), 1505–1518 (2022)
4. Cramer, J., Wu, H.H., Salamon, J., Bello, J.P.: Look, listen, and learn more: Design choices for deep audio embeddings. In: ICASSP, pp. 3852–3856. IEEE (2019)
5. Cudeiro, D., Bolkart, T., Laidlaw, C., Ranjan, A., Black, M.J.: Capture, learning, and synthesis of 3d speaking styles. In: Proceedings of the IEEE/CVF Conference on Computer Vision and Pattern Recognition, pp. 10101–10111 (2019)
6. Devlin, J., Chang, M.W., Lee, K., Toutanova, K.: Bert: pre-training of deep bidirectional transformers for language understanding. arXiv preprint arXiv:1810.04805 (2018)
7. Drossos, K., Lipping, S., Virtanen, T.: Clotho: an audio captioning dataset. In: ICASSP, May 2020. https://arxiv.org/abs/1910.09387

8. Esser, P., Rombach, R., Ommer, B.: Taming transformers for high-resolution image synthesis. In: Proceedings of the IEEE/CVF Conference on Computer Vision and Pattern Recognition, pp. 12873–12883 (2021)
9. Fanzeres, L.A., Nadeu, C.: Sound-to-imagination: unsupervised crossmodal translation using deep dense network architecture. arXiv preprint arXiv:2106.01266 (2021)
10. Fei, N., et al.: Towards artificial general intelligence via a multimodal foundation model. Nat. Commun. **13**(1), 1–13 (2022)
11. Guzhov, A., Raue, F., Hees, J., Dengel, A.: Audioclip: extending clip to image, text and audio. arXiv preprint arXiv:2106.13043 (2021)
12. He, K., Fan, H., Wu, Y., Xie, S., Girshick, R.: Momentum contrast for unsupervised visual representation learning. In: 2020 IEEE/CVF Conference on Computer Vision and Pattern Recognition (CVPR), pp. 9726–9735 (2020). https://doi.org/10.1109/CVPR42600.2020.00975
13. Ilharco, G., Zhang, Y., Baldridge, J.: Large-scale representation learning from visually grounded untranscribed speech. In: Proceedings of the 23rd Conference on Computational Natural Language Learning (CoNLL), pp. 55–65. Association for Computational Linguistics, Hong Kong, China, November 2019. https://doi.org/10.18653/v1/K19-1006, https://aclanthology.org/K19-1006
14. Jia, C., et al.: Scaling up visual and vision-language representation learning with noisy text supervision. In: International Conference on Machine Learning, pp. 4904–4916. PMLR (2021)
15. Kaplan, J., et al.: Scaling laws for neural language models. arXiv preprint arXiv:2001.08361 (2020)
16. Karras, T., Aila, T., Laine, S., Herva, A., Lehtinen, J.: Audio-driven facial animation by joint end-to-end learning of pose and emotion. ACM Trans. Graph. (TOG) **36**(4), 1–12 (2017)
17. Kazakos, E., Nagrani, A., Zisserman, A., Damen, D.: Slow-fast auditory streams for audio recognition. In: ICASSP, pp. 855–859 (2021). https://doi.org/10.1109/ICASSP39728.2021.9413376
18. Pedersoli, F., Wiebe, D., Banitalebi, A., Zhang, Y., Yi, K.M.: Estimating visual information from audio through manifold learning. arXiv preprint arXiv:2208.02337 (2022)
19. Piczak, K.J.: ESC: dataset for environmental sound classification. In: ACM Multimedia, p. 1015. ACM Press (2015). https://doi.org/10.1145/2733373.2806390, http://dl.acm.org/citation.cfm?doid=2733373.2806390
20. Qiu, Y., Kataoka, H.: Image generation associated with music data. In: Proceedings of the IEEE Conference on Computer Vision and Pattern Recognition Workshops, pp. 2510–2513 (2018)
21. Radford, A., et al.: Learning transferable visual models from natural language supervision. In: ICML (2021)
22. Rombach, R., Blattmann, A., Lorenz, D., Esser, P., Ommer, B.: High-resolution image synthesis with latent diffusion models. In: Proceedings of the IEEE/CVF Conference on Computer Vision and Pattern Recognition (CVPR), pp. 10684–10695, June 2022
23. Salamon, J., Jacoby, C., Bello, J.P.: A dataset and taxonomy for urban sound research. In: ACM Multimedia, pp. 1041–1044. Orlando, FL, USA, Nov 2014
24. Sung-Bin, K., Senocak, A., Ha, H., Owens, A., Oh, T.H.: Sound to visual scene generation by audio-to-visual latent alignment (2023)

25. Tan, M., Le, Q.: EfficientNet: rethinking model scaling for convolutional neural networks. In: International Conference on Machine Learning, pp. 6105–6114. PMLR (2019)
26. Turpault, N., Serizel, R., Parag Shah, A., Salamon, J.: Sound event detection in domestic environments with weakly labeled data and soundscape synthesis. In: DCASE. New York City, United States, October 2019. https://hal.inria.fr/hal-02160855
27. Wan, C.H., Chuang, S.P., Lee, H.Y.: Towards audio to scene image synthesis using generative adversarial network. In: ICASSP 2019–2019 IEEE International Conference on Acoustics, Speech and Signal Processing (ICASSP), pp. 496–500 (2019). https://doi.org/10.1109/ICASSP.2019.8682383
28. Wu, H.H., Seetharaman, P., Kumar, K., Bello, J.P.: Wav2clip: learning robust audio representations from clip. In: ICASSP 2022–2022 IEEE International Conference on Acoustics, Speech and Signal Processing (ICASSP), pp. 4563–4567. IEEE (2022)
29. Zhang, L., Agrawala, M.: Adding conditional control to text-to-image diffusion models (2023)
30. Zhao, P., Chen, Y., Zhao, L., Wu, G., Zhou, X.: Generating images from audio under semantic consistency. Neurocomputing **490**, 93–103 (2022)
31. Zhou, Y., Wang, Z., Fang, C., Bui, T., Berg, T.L.: Visual to sound: generating natural sound for videos in the wild. In: CVPR (2018)
32. Zhu, H., Luo, M.D., Wang, R., Zheng, A.H., He, R.: Deep audio-visual learning: A survey. Int. J. Autom. Comput. **18**(3), 351–376 (2021)

Can You Really Reason: A Novel Framework for Assessing Natural Language Reasoning Datasets and Models

Shanshan Huang[(✉)]

Shanghai Jiao Tong University, Shanghai, China
huangss_33@sjtu.edu.cn

Abstract. Recent studies have illuminated a pressing issue in the domain of natural language understanding (NLU) and reasoning: many of these datasets are imbued with subtle statistical cues. These cues, often unnoticed, provide sophisticated models an unintended edge, allowing them to exploit these patterns, leading to a potentially misleading overestimation of their genuine capabilities. While the existence of these cues has been noted, a precise and systematic identification has remained elusive in existing literature. Addressing this gap, our paper presents a novel lightweight framework. This framework is meticulously designed to not only detect these hidden biases in multiple-choice NLU datasets but also rigorously evaluate the robustness of models that are developed based on these datasets. By unveiling these biases and assessing model integrity, we aim to pave the way for more genuine and transparent advancements in NLU research.

Keywords: Dataset bias · Model bias · Model robustness

1 Introduction

The advancements in neural network models have yielded significant enhancements in a plethora of tasks, including natural language inference [1,20], argumentation [11], commonsense reasoning [9,14,23], reading comprehension [6], question answering [19], and dialogue analysis [7]. However, recent studies [4,12,15] have unveiled that superficial statistical patterns, including sentiment, word repetition, and shallow n-gram tokens in benchmark datasets, can forecast the correct answer. These patterns or features, termed as spurious **cues** when appearing in both training and test datasets with similar distributions. When these cues are neutralized, leading to a "stress test" [8,10,13], models exhibit reduced performance, suggesting an overestimation of their capabilities when evaluated on these datasets.

Several natural language reasoning tasks, exemplified by those in the Stanford Natural Language Inference (SNLI) dataset, can be cast as multiple-choice questions. A typical question can be structured as follows:

B. Luo et al. (Eds.): ICONIP 2023, CCIS 1969, pp. 54–66, 2024.
https://doi.org/10.1007/978-981-99-8184-7_5

Example 1. An instance from SNLI. **Premise**: A swimmer playing in the surf watches a low flying airplane headed inland.
Hypothesis: Someone is swimming in the sea.
Label: a) Entail. b) Contradict. c) Neutral.

Humans approach these questions by examining the logical relations between the premise and the hypothesis. Yet, previous work [10,16] has unveiled that several NLP models can correctly answer these questions by only considering the hypothesis. This observation often traces back to the presence of artifacts in the manually crafted hypotheses within many datasets. Although identifying problematic questions with a "hypothesis-only" test is theoretically sound, this approach often i) relies on specific models like BERT [3], which require costly retraining, and ii) fails to explain why a question is problematic.

This paper puts forth a lightweight framework aimed at identifying simple yet impactful cues in multiple-choice natural language reasoning datasets, enabling the detection of problematic questions. While not all multiple-choice questions in these datasets include a premise, a hypothesis, and a label, we detail a method to standardize them in Sect. 2. We leverage words as fundamental features in crafting spurious cues, since they serve as the foundational units in modeling natural language across most contemporary machine learning methods. Even complex linguistic features, such as sentiment, style, and opinions, are anchored on word features. Subsequent experimental sections will demonstrate that word-based cues can detect statistical bias in datasets as effectively as the more resource-demanding hypothesis-only method.

2   Approach

We evaluate the information leak in the datasets using only statistical features. First, we formulate a number of natural language reasoning (NLR) tasks in a general form. Then, based on the frequency of words associated with each label, we design a number of metrics to measure the correlation between words and labels. Such correlation scores are called "cue scores" because they are indicative of potential cue patterns. Afterward, we aggregate the scores using a number of simple statistical models to make predictions.

2.1   Task Formulation

Given a question instance x of an NLR task dataset X, we formulate it as

$$x = (p, h, l) \in X, \tag{1}$$

where p is the context against which to do the reasoning, and p corresponds to the "premise" in example 1; h is the hypothesis given the context p. $l \in \mathcal{L}$ is the label that depicts the type of relation between p and h. The size of the relation set $\mathcal{L}$ varies between tasks. We argue that most of the discriminative NLR tasks can be formulated into this general form. For example, an NLI question consists of a *premise*, a *hypothesis*, and a *label* on the relation between the premise and hypothesis. $|\mathcal{L}| = 3$ for three different relations: *entailment, contradiction,* and *neutral.* We will discuss how to transform into this form in Sect. 2.4.

2.2   Cue Metric

For a dataset X, we collect a set of all words $\mathcal{N}$ that exist in X. The cue metric for a word measures the disparity of the word's appearance under a specific label. Let w be a word in $\mathcal{N}$, we compute a scalar statistic metric called *cue score*, $f_{\mathcal{F}}^{(w,l)}$, in one of the following eight ways. We categorized the metrics into two genres: the first four use only statistics, and the last four use a notion of angles in the Euclidean space. Let $\mathcal{L}' = \mathcal{L} - L \setminus l$, and we define

$$\#(w, \mathcal{L}') = \sum_{l' \in \mathcal{L}'} \#(w, l'). \tag{2}$$

Frequency (Freq)
The simplest measurement is the co-occurrence of words and labels, where $\#()$ denotes naive counting. This metric aims to capture the raw frequency of words appearing in a particular label.

$$f_{Freq}^{(w,l)} = \#(w, l) \tag{3}$$

Relative Frequency (RF)
Relative Frequency extends the Frequency metric by accounting for the total frequency of the word across all labels. It's defined as follows:

$$f_{RF}^{(w,l)} = \frac{\#(w, l)}{\#(w)} \tag{4}$$

Conditional Probability (CP)
The Conditional Probability of label l given word w is another way to capture the association between a word and a label. This metric is essentially the Relative Frequency as defined above.

$$f_{CP}^{(w,l)} = p(l|w) = \frac{\#(w, l)}{\#(w)} \tag{5}$$

Point-wise Mutual Information (PMI)
PMI is a popular metric used in information theory and statistics. It measures the strength of association between a word and a label. PMI is higher when the word and label co-occur more often than would be expected if they were independent. We define the PMI of word w and label l as follows, where $p(w)$ and $p(l)$ are the probabilities of w and l respectively, and $p(w, l)$ is the joint probability of w and l.

$$f_{PMI}^{(w,l)} = \log \frac{p(w, l)}{p(w)p(l)} \tag{6}$$

Local Mutual Information (LMI)
The LMI is a variant of PMI that weighs the PMI by the joint probability of the word and label. This has the effect of giving more importance to word-label pairs that occur frequently. The LMI of word w with respect to label l is defined as follows.

$$f_{LMI}^{(w,l)} = p(w,l) \log \frac{p(w,l)}{p(w)p(l)}. \tag{7}$$

Ratio Difference (RD)
The Ratio Difference metric measures the absolute difference between the word-label ratio and the overall label ratio. This metric helps identify words that are disproportionately associated with a specific label.

$$f_{RD}^{(w,l)} = \left| \frac{\#(w,l)}{\#(w,\mathcal{L}')} - \frac{\#(l)}{\#(\mathcal{L}')} \right| \tag{8}$$

Angle Difference (AD)
Angle Difference is similar to *Ratio Difference* but accounts for the non-linear relationship between the ratios by taking the arc-tangent function. This metric can be more robust to outliers.

$$f_{AD}^{(w,l)} = \left| \arctan \frac{\#(w,l)}{\#(w,\mathcal{L}')} - \arctan \frac{\#(l)}{\#(\mathcal{L}')} \right| \tag{9}$$

Cosine (Cos)
The Cosine metric considers $v_w = [\#(w,l), \#(w,\mathcal{L}')]$ and $v_l = [(\#(l), \#(\mathcal{L}')]$ as two vectors on a 2D plane. Intuitively, if v_w and v_l are co-linear, w leaks no spurious information. Otherwise, w is suspected to be a spurious cue as it tends to appear more with a specific label l. This metric quantifies the similarity of the word-label relationship in a geometric manner.

$$f_{Cos}^{(w,l)} = \cos(v_w, v_l) \tag{10}$$

Weighted Power (WP)
The Weighted Power metric combines the Cosine metric with a frequency-based weighting, emphasizing the importance of words with higher frequencies. This metric can help prioritize cues that are more likely to impact the model.

$$f_{WP}^{(w,l)} = (1 - f_{Cos}^l)\#(w)^{f_{Cos}^l} \tag{11}$$

In general, we can denote the *cue score* of a word w w.r.t. label l as $f^{(w,l)}$, by dropping the method subscript $\mathcal{F}$.

These metrics provide different perspectives on the association between words and labels, which can help identify potential spurious correlations.

2.3   Aggregation Methods

We can use simple methods $\mathcal{G}$ to aggregate the cue scores of words within a question instance x to make a prediction. These methods are designed to be easily implemented and computationally efficient, given the low-dimensional cue features.

Average and Max

The most straightforward way to predict a label is to select the label with the highest average or maximum *cue score* in an instance.

$$\mathcal{G}average = \arg\max l\frac{\sum_w f^{w,l}}{|x|}, l \in \mathcal{L}, w \in \mathcal{N} \tag{12}$$

$$\mathcal{G}max = \arg\max l\max_w(f^{w,l}), l \in \mathcal{L}, w \in \mathcal{N} \tag{13}$$

Linear Models

To better utilize the *cue score* in making predictions, we employ two simple linear models: SGDClassifier and logistic regression. The input for the models is a concatenated vector of *cue scores* for each label in instance x:

$$input(x) = [f^{w_1,l_1}, , ..., f^{w_d,l_1}, f^{w_1,l_2}, ..., f^{w_d,l_2}, \\ ..., f^{w_1,l_t}, ..., f^{w_d,l_t}]. \tag{14}$$

Here, d denotes the length of x. In practice, input vectors are padded to the same length. The training loss for the linear model is:

$$\hat{\phi}n = \arg\min \phi_n loss(\mathcal{G}_{linear}(input(x); \phi_n)) \tag{15}$$

The loss is calculated between the gold label l_g and the predicted label $\mathcal{G}linear(input(x); \phi_n)$. ϕ_n represents the optimal parameters in $\mathcal{G}linear$ that minimize the loss for label l_g.

2.4   Transformation of MCQs with Dynamic Choices

Until now, we have focused on multiple-choice questions (MCQs) that are classification problems with a fixed set of choices. However, some language reasoning tasks involve MCQs with non-fixed choices, such as the ROCStory dataset. In these cases, we can separate the original story into two unified instances, $u_1 = (context, ending1, false)$ and $u_2 = (context, ending2, true)$. We predict the label probability for each instance, $\mathcal{G}(input(u_1); \phi)$ and $\mathcal{G}(input(u_2); \phi)$, and choose the ending with the higher probability as the prediction.

3   Experiment

We proceed to demonstrate the effectiveness of our framework in this section. We apply our method to detect cues and measure the amount of information leakage in 12 datasets from 6 different tasks, as shown in   Table 1. Our experimental findings are segmented into five sub-sections: Datasets, Quantifying Information Leakage, Bias Evaluation Methods, Comparison with Hypothesis-only Models, and Identifying Problematic Datasets.

Table 1. Dataset examples and normalized version.

Task Name	Datasets	Example			
		Original	"Premise"	"Hypothesis"	label
Natural Language Inference	SNLI, MNLI, QNLI	(SNLI) **Premise**: A woman and a child holding on to the railing while on trolley. **Hypothesis**: The people are not holding on anything. **Label**: contradiction	A woman and a child holding on to the railing while on trolley .	The people are not holding on anything.	contradiction
Argumentation	ARCT, ARCT_adv	(ARCT) **Reason**: Milk isn't a gateway drug even though most people drink it as children. **Claim**: Marijuana is not a gateway drug.	Milk isn't a gateway drug even though most people drink it as children. Marijuana is not a gateway drug.	Milk is similar to marijuana.	true
Reading Comprehension	RACE, RECLOR	**Warrant 1**: Milk is similar to marijuana. **Warrant 2**: Milk is not marijuana.	Milk isn't a gateway drug even though most people drink it as children. Marijuana is not a gateway drug.	Milk is not marijuana.	false
Commonsense Reasoning	ROCStory, COPA, SWAG	(COPA) The woman hummed to herself. What was the cause for this?	The woman hummed to herself. What was the cause for this?	She was in a good mood.	true
Question Answering	CQA	**Alternative1**: She was in a good mood. **Alternative2**: She was nervous.	The woman hummed to herself. What was the cause for this?	She was nervous.	false
Dialogue Analysis	Ubuntu				

3.1 Datasets

In this section, we present the results of our experiments conducted on 12 diverse datasets as outlined in Table 1. The datasets can be broadly classified into two categories based on the tasks they present: NLI classification tasks and multiple-choice problems. The NLI classification tasks constitute the first type. They are, in essence, a specialized variant of multiple-choice datasets. The second type includes datasets like ARCT, ARCT_adv [16], RACE [6], and RECLOR [22]. In these, one of the alternatives is the "hypothesis", and the "premise" contains more than a single context role. As an example, in ARCT, **Reason** and **Claim** act as the "premise", requiring the correct warrant to be chosen. Other datasets like Ubuntu [7], COPA [14], ROCStory, SWAG [23], and CQA [19] belong to the second type as well but have only a single context role in the "premise".

Table 2 outlines how hypotheses are gathered in these datasets. Most datasets utilize human-written hypotheses, barring CQA and SWAG.

3.2 Quantifying Information Leakage

In our effort to effectively measure the severity of information leakage or bias in these datasets, we formulated a measurement expressed as $\mathcal{D} = Acc - Majority$. Here, $Majority$ is the accuracy achieved through majority voting and Acc represents the accuracy of a model that bases its prediction solely on spurious cues.

A high absolute value of $\mathcal{D}$ indicates the existence of more cues in a dataset. However, a smaller $\mathcal{D}$ doesn't necessarily mean less bias in the training data, but rather less "leakage" between the training and test data. If $\mathcal{D}$ is positive, it

Table 2. The methods of hypothesis collection for the datasets. AE = Adversarial Experiment, LM = language model, CD = crowdsourcing, Human represents human performance on the datasets.

Datasets	Data Size	Data source	AE	Human(%)
ROCStory	3.9k	CD	No	100.0
COPA	1k	CD	No	100.0
SWAG	113k	LM	Yes	88.0
SNLI	570K	CD	No	80.0
QNLI	11k	CD	No	80.0
MNLI	413k	CD	No	80.0
RACE	100k	CD	No	94.5
RECLOR	6k	CD	No	63.0
CQA	12k	CD	No	88.9
ARCT	2k	CD	No	79.8
ARCT_adv	4k	CD	Yes	-
Ubuntu	100k	Random Selection	No	-

implies the model is utilizing the cues for its prediction. This evaluation method can be universally applied to any multiple-choice dataset.

3.3   Cue Evaluation Methods

The primary technique we use in our analysis is the hypothesis-only method, which we use as a gold standard for examining the existence of spurious cues. This method assumes that the model can only access the hypothesis and has to make its prediction without considering the premise.

To simplify this process and to find a measure that is as close to the hypothesis only method, we employed four simpler methods to make decisions based solely on spurious statistical cues. These methods include the average value classifier (Ave), the maximum value classifier (Max), SGD classifier (SGDC), and logistic regression (LR). These are outlined in detail in Sect. 2.

The main difference between our methods and the hypothesis-only method lies in the type of cues used. While our method uses word-level cues that are interpretable, the hypothesis-only method uses more complex cues, which are not easily interpretable.

3.4   Comparison with Hypothesis-Only Models

Our research aimed to assess and validate our proposed bias detection methods, chiefly by comparing their performance with hypothesis-only models. The goal was to demonstrate the effectiveness of our method in identifying spurious

Table 3. The Pearson Score of $\mathcal{D}$ on 12 datasets, between our methods and hypothesis-only models, fastText and BERT. P is the average Pearson score of BERT and fast-Text(FT).

	Ave			Max			SGDC			LR		
	FT	BERT	P	FT	BERT	P	FT	BERT	P	FT	BERT	P
PMI	90.87	96.23	93.55	95.37	79.82	87.59	97.81	91.01	94.41	97.14	96.05	96.6
LMI	65.13	49.18	57.16	34.52	30.71	32.62	69.88	79.06	74.47	77.46	81.21	79.33
AD	84.62	72.49	78.56	90.75	73.02	81.89	93.73	76.24	84.98	97.56	86.91	92.24
WP	86.87	73.09	79.98	92.47	79.87	86.17	94.0	22.59	56.53	61.28	75.55	65.86
RD	96.59	93.82	95.21	98.23	91.04	94.63	94.30	93.98	94.14	94.21	95.59	94.90
Cos	94.84	82.94	88.89	92.73	75.40	84.07	98.08	87.86	92.97	87.38	78.44	82.91
Freq	68.00	50.02	59.01	34.45	30.67	32.56	64.08	67.11	65.60	74.58	88.64	81.61
CP	93.09	96.61	94.85	95.29	79.80	87.54	97.19	96.16	96.67	97.17	97.34	**97.26**

statistical cues in multiple-choice datasets, underpinning the contribution we introduced.

In the context of this experimental comparison, we utilized the Pearson Correlation Coefficient (PCC) to measure the similarity between our method and the established hypothesis-only models, specifically fastText and BERT. The analysis encompassed a range of twelve datasets, making use of eight distinct cue score metrics and four aggregation algorithms.

The outcomes of this analysis, as depicted in Table 3, highlight that the CP cue score coupled with the logistic regression model achieved high correlations across all twelve datasets when compared to the gold standard hypothesis-only models. The PCC scores obtained were 97.17% with fastText and 97.34% with BERT. These remarkable results led us to conclude that the combination of CP and logistic regression forms a robust method for evaluating all datasets in subsequent experiments. The detailed data behind this study is comprehensively presented in the Appendix.

Given these findings, we are confident in asserting that our CP based approach is a powerful tool in identifying problematic word features within datasets, through the calculation of a "cueness score" described in Sect. 2. Furthermore, the coupling of CP and logistic regression offers a compelling measure to determine the extent to which multiple-choice datasets are affected by information leakage, a significant contribution to this field of research.

Further, we visualized our findings by plotting $\mathcal{D}$ for our CP+LR method and two hypothesis-only models (fastText and BERT) on 12 datasets in Fig. 1. The close tracking lines in the plot clearly indicate the strong correlation between our method and the hypothesis-only models.

Overall, our method effectively identifies and quantifies biases in the datasets, and the strong correlation with hypothesis-only models demonstrates the validity and effectiveness of our approach.

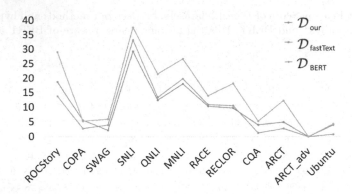

Fig. 1. Deviation scores for three prediction models on all 12 datasets.

3.5 Identifying Problematic Datasets

To better discern problematic datasets, we developed a criterion based on our experiment findings. According to this criterion, if a model's $\mathcal{D}$ exceeds 10% on any cue feature, the dataset is deemed problematic. This straightforward criterion allows for a quick identification of datasets with severe statistical cue issues.

Table 4. Highest accuracy of our 4 simple classification models on 12 datasets and the deviations from majority selection.

Datasets	Majority	Word Cues	
	(%)	Acc.(%)	$\mathcal{D}$(%)
ROCStory	50.0	68.68	**18.68**
COPA	50.0	55.60	5.60
SWAG	25.0	27.23	2.23
SNLI	33.33	62.57	**29.24**
QNLI	50.0	62.49	**12.49**
MNLI	33.33	51.37	**18.04**
RACE	25.0	35.42	**10.42**
RECLOR	25.0	34.80	9.80
CQA	20.0	23.42	3.42
ARCT	50.0	54.95	4.95
ARCT_adv	50.0	50	0.0
Ubuntu	1.0	4.96	3.96

As per this criterion, we identified ROCStories, SNLI, MNLI, QNLI, RACE, and RECLOR as datasets with considerable statistical cue problems. These findings are detailed in Table 4, which highlights the selection results using word cue features on several datasets. For some of these datasets, our methods significantly outperform the random selection probability, showcasing the extent of the statistical cues present. For instance, in the case of the ROCStories dataset, the highest accuracy achieved with our methods exceeds the random selection probability by 20.92%, and even higher for the SNLI dataset by 33.59%. This indicates that the datasets contain substantial spurious statistical cues that the models can exploit.

In the case of manually intervened datasets without adversarial filtering, such as ARCT, we found that they contained more spurious statistical cues. For instance, human adjustments to the ARCT dataset(ARCT_adv) have a notable impact on accuracy (from 54.95% to 50%).

Finally, in Table 4, we report the highest accuracy of our four simple classification models on the 12 datasets, along with the deviations from majority selection. Our findings reveal that deviation $\mathcal{D}$ can effectively identify problematic datasets. We can thus use $\mathcal{D}$ to assess the extent to which a dataset contains word cues.

In conclusion, our analysis and criteria for problematic datasets can help researchers identify datasets with substantial statistical cue issues. This critical insight can improve the development of more robust models that do not rely on superficial cues.

4   Related Work

Our work is related to and, to some extent, comprises elements in three research directions: spurious features analysis, bias calculation.

Spurious Features Analysis has been increasingly studied recently. Much work [17,18,23] has observed that some NLP models can surprisingly get good results on natural language understanding questions in MCQ form without even looking at the stems of the questions. Such tests are called "hypothesis only" tests in some works. Further, some research [15] discovered that these models suffer from insensitivity to certain small but semantically significant alterations in the hypotheses, leading to speculations that the hypothesis-only performance is due to simple statistical correlations between words in the hypothesis and the labels. Spurious features can be classified into lexicalized and unlexicalized [1]: lexicalized features mainly contain indicators of n-gram tokens and cross-ngram tokens, while unlexicalized features involve word overlap, sentence length, and BLUE score between the premise and the hypothesis. [10] refined the lexicalized classification to Negation, Numerical Reasoning, Spelling Error. [8] refined the word overlap features to Lexical overlap, Subsequence, and Constituent which also considers the syntactical structure overlap. [15] provided unseen tokens an extra lexicalized feature.

Bias Calculation is concerned with methods to quantify the severity of the cues. Some work [2,5,21] attempted to encode the cue feature implicitly by hypothesis-only training or by extracting features associated with a certain label from the embeddings. Other methods compute the bias by statistical metrics. For example, [22] used the probability of seeing a word conditioned on a specific label to rank the words by their biasness. LMI [16] was also used to evaluate cues and re-weight in some models. However, these works did not give the reason to use these metrics, one way or the other. Separately, [13] gave a test data augmentation method, without assessing the degree of bias in those datasets.

5   Conclusion and Future Work

We have addressed the critical issue of statistical biases present in natural language understanding and reasoning datasets. We have proposed a lightweight framework that automatically identifies potential biases in multiple-choice NLU-related datasets and assesses the robustness of models designed for these datasets. Our experimental results have demonstrated the effectiveness of this framework in detecting dataset biases and evaluating model performance.

As future work, we plan to further investigate the nature of biases in NLU datasets and explore more sophisticated techniques to detect and mitigate these biases. Additionally, we aim to extend our framework to other types of NLU tasks beyond multiple-choice settings. By continuing to refine our understanding of dataset biases and their impact on model performance, we hope to contribute to the development of more robust, accurate, and reliable NLU models that can better generalize to real-world applications.

References

1. Bowman, S., Angeli, G., Potts, C., Manning, C.D.: A large annotated corpus for learning natural language inference. In: Proceedings of the 2015 Conference on Empirical Methods in Natural Language Processing, pp. 632–642 (2015)
2. Clark, C., Yatskar, M., Zettlemoyer, L.: Don't take the easy way out: ensemble based methods for avoiding known dataset biases. In: Proceedings of the 2019 Conference on Empirical Methods in Natural Language Processing and the 9th International Joint Conference on Natural Language Processing (EMNLP-IJCNLP), pp. 4060–4073 (2019)
3. Devlin, J., Chang, M.W., Lee, K., Toutanova, K.: BERT: pre-training of deep bidirectional transformers for language understanding. arXiv preprint arXiv:1810.04805 (2018)
4. Gururangan, S., Swayamdipta, S., Levy, O., Schwartz, R., Bowman, S., Smith, N.A.: Annotation artifacts in natural language inference data. In: Proceedings of the 2018 Conference of the North American Chapter of the Association for Computational Linguistics: Human Language Technologies, vol. 2 (Short Papers), pp. 107–112 (2018)
5. He, H., Zha, S., Wang, H.: Unlearn dataset bias in natural language inference by fitting the residual. EMNLP-IJCNLP **2019**, 132 (2019)
6. Lai, G., Xie, Q., Liu, H., Yang, Y., Hovy, E.: RACE: large-scale reading comprehension dataset from examinations. In: Proceedings of the 2017 Conference on Empirical Methods in Natural Language Processing, pp. 785–794 (2017)

7. Lowe, R., Pow, N., Serban, I.V., Pineau, J.: The ubuntu dialogue corpus: a large dataset for research in unstructured multi-turn dialogue systems. In: Proceedings of the 16th Annual Meeting of the Special Interest Group on Discourse and Dialogue, pp. 285–294 (2015)
8. McCoy, T., Pavlick, E., Linzen, T.: Right for the wrong reasons: diagnosing syntactic heuristics in natural language inference. In: Proceedings of the 57th Annual Meeting of the Association for Computational Linguistics, pp. 3428–3448 (2019)
9. Mostafazadeh, N., et al.: A corpus and cloze evaluation for deeper understanding of commonsense stories. In: Proceedings of the 2016 Conference of the North American Chapter of the Association for Computational Linguistics: Human Language Technologies, pp. 839–849 (2016)
10. Naik, A., Ravichander, A., Sadeh, N., Rose, C., Neubig, G.: Stress test evaluation for natural language inference. In: Proceedings of the 27th International Conference on Computational Linguistics, pp. 2340–2353 (2018)
11. Niven, T., Kao, H.Y.: Probing neural network comprehension of natural language arguments. In: Proceedings of the 57th Annual Meeting of the Association for Computational Linguistics, pp. 4658–4664 (2019)
12. Poliak, A., Naradowsky, J., Haldar, A., Rudinger, R., Van Durme, B.: Hypothesis only baselines in natural language inference. In: Proceedings of the Seventh Joint Conference on Lexical and Computational Semantics, pp. 180–191 (2018)
13. Ribeiro, M.T., Wu, T., Guestrin, C., Singh, S.: Beyond accuracy: Behavioral testing of NLP models with checklist. In: Proceedings of the 58th Annual Meeting of the Association for Computational Linguistics, ACL 2020, Online, July 5–10, 2020, pp. 4902–4912 (2020)
14. Roemmele, M., Bejan, C.A., Gordon, A.S.: Choice of plausible alternatives: an evaluation of commonsense causal reasoning. In: 2011 AAAI Spring Symposium Series (2011)
15. Sanchez, I., Mitchell, J., Riedel, S.: Behavior analysis of NLI models: Uncovering the influence of three factors on robustness. In: Proceedings of the 2018 Conference of the North American Chapter of the Association for Computational Linguistics: Human Language Technologies, vol. 1 (Long Papers), pp. 1975–1985 (2018)
16. Schuster, T., Shah, D.J., Yeo, Y.J.S., Filizzola, D., Santus, E., Barzilay, R.: Towards debiasing fact verification models. arXiv preprint arXiv:1908.05267 (2019)
17. Sharma, R., Allen, J., Bakhshandeh, O., Mostafazadeh, N.: Tackling the story ending biases in the story cloze test. In: Proceedings of the 56th Annual Meeting of the Association for Computational Linguistics (Volume 2: Short Papers), pp. 752–757 (2018)
18. Srinivasan, S., Arora, R., Riedl, M.: A simple and effective approach to the story cloze test. In: Proceedings of the 2018 Conference of the North American Chapter of the Association for Computational Linguistics: Human Language Technologies, vol. 2 (Short Papers), pp. 92–96 (2018)
19. Talmor, A., Herzig, J., Lourie, N., Berant, J.: CommonsenseQA: a question answering challenge targeting commonsense knowledge. In: Proceedings of the 2019 Conference of the North American Chapter of the Association for Computational Linguistics: Human Language Technologies, vol. 1 (Long and Short Papers), pp. 4149–4158 (2019)
20. Wang, A., Singh, A., Michael, J., Hill, F., Levy, O., Bowman, S.R.: GLUE: a multi-task benchmark and analysis platform for natural language understanding. EMNLP 2018, 353 (2018)
21. Yaghoobzadeh, Y., Tachet, R., Hazen, T., Sordoni, A.: Robust natural language inference models with example forgetting. arXiv preprint arXiv:1911.03861 (2019)

22. Yu, W., Jiang, Z., Dong, Y., Feng, J.: Reclor: A reading comprehension dataset requiring logical reasoning. arXiv preprint arXiv:2002.04326 (2020)
23. Zellers, R., Bisk, Y., Schwartz, R., Choi, Y.: Swag: A large-scale adversarial dataset for grounded commonsense inference. In: Proceedings of the 2018 Conference on Empirical Methods in Natural Language Processing, pp. 93–104 (2018)

End-to-End Urban Autonomous Navigation with Decision Hindsight

Qi Deng[1,2], Guangqing Liu[1], Ruyang Li[1,2(✉)], Qifu Hu[1], Yaqian Zhao[1],
and Rengang Li[1]

[1] Inspur (Beijing) Electronic Information Industry Co., Ltd., Beijing 100085, China
{dengqi01,liugq,liruyang,huqifu}@ieisystem.com,
zhaoyaqian.hsslab@gmail.com, lirengang.hsslab@gmail.com
[2] Shandong Massive Information Technology Research Institute, Jinan 250101, China

Abstract. Urban autonomous navigation has broad application
prospects. Reinforcement Learning (RL) based navigation models can
be continuously optimized through self-exploration, eliminating the need
for human heuristics. However, training effective navigation models faces
challenges due to the dynamic nature of urban traffic conditions and
the exploration-exploitation dilemma in RL. Moreover, the limited vehicle perception and traffic uncertainty introduce potential safety haz-
ards, hampering the real-world application of RL-based navigation models. In this paper, we proposed a novel end-to-end urban navigation
framework with decision hindsight. By formulating the problem of Partially Observable Markov Decision Process (POMDP), we employ a
causal Transformer-based autoregressive modeling approach to process
the historical navigation information as supplementary observations.
We then combine these historical observations with current perceptions
to construct a history-feedforward state representation that enhances
global awareness, improving data availability and decision predictability. Furthermore, by integrating the historical-feedforward state encoding upstream, we develop an end-to-end learning framework based on
RL to obtain a navigation model with decision hindsight, enabling more
reliable navigation. To validate the effectiveness of our proposed method,
we conduct experiments on challenging urban navigation tasks using the
CARLA simulator. The results demonstrate that our method achieves
higher learning efficiency and improved driving performance, taking priority over prior methods on urban navigation benchmarks.

Keywords: Urban Autonomous Navigation · Reinforcement
Learning · Partially Observable Markov Decision Process (POMDP) ·
Decision Hindsight

This work was supported by Shandong Provincial Natural Science Foundation, China
under Grant ZR2022LZH002.

1   Introduction

Urban navigation is a crucial task in autonomous driving, it faces numerous uncertainties such as variable environments, dynamic interactions, and complex traffic conditions, which can lead to potential dangers [9,12]. Traditional modular pipelines, although offering good interpretability, suffer from over-reliance on human heuristics and strong entanglement between modules, resulting in inefficiency and laboriousness in complex urban areas [23]. Recently, Reinforcement Learning (RL) has shown great promise in the field of autonomous driving [2,17,30]. Through self-exploration and reinforcement, the RL-based autonomous navigation model can be continuously optimized without much hand-engineered involvement. Nevertheless, urban scenarios pose challenges that require extensive interactions for sufficient exploration. The dynamic traffic conditions, coupled with the exploration-exploitation dilemma inherent in RL, further exacerbate the data inefficiency [6]. Moreover, the limited perception of on-board sensors and the unpredictable intentions of surrounding objects hinder the vehicle's ability to respond promptly and reasonably, potentially compromising safety and impeding real-world applications.

To enhance learning efficiency and decision reliability, it is crucial to design an effective state representation that balances dimensionality reduction with rich information content. One common approach is extract latent features from raw perceptual data using encoders [1], but these often require pre-training with large amounts of data to ensure accuracy, making them sensitive to environmental variations. To address differences in environment perception, the Bird's-Eye View (BEV) can serve as an intermediate state representation constructed on top of the raw perceptual data [4,17,28], retaining essential information about the driving decisions. Nonetheless, due to the limited sensing range, distant environmental conditions may remain undetected. Some studies have explored the use of high-definition maps or roadside perception as supplements to driving environment information, heavily relying on Vehicle-to-Everything (V2X) technology which is still in an immature stage.

As most RL-based urban navigation researches strictly follow the Markov property, it is inaccurate to describe the driving state solely by the current perception from vehicle's limited perspective. Recently, some studies have incorporated historical behavior into current decision-making based on the Partially Observable Markov Decision Process (POMDP) to address the uncertainty in dynamic environments, and achieved more efficient works [26,27]. Motivated by this, we also formulate the urban navigation problem with the idea of POMDP, and harness the information from history trajectories to provide a stronger description of the driving context. The extraction of historical behavior information can be achieved through sequence modeling. Transformer, as a novel network architecture based on the attention mechanism, has gained popularity in processing sequential data due to its ability to capture higher-order relationships and long-range dependencies. Several studies have extended Transformer to RL frameworks for tasks such as representation learning, capturing multi-step temporal dependencies, environment modeling, and sequential decision-making.

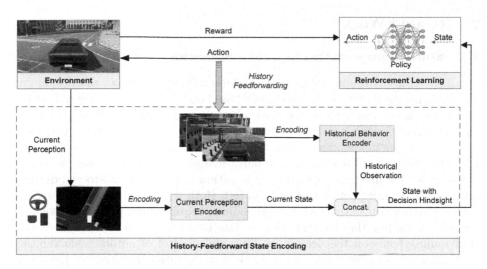

Fig. 1. End-to-end urban navigation framework with decision hindsight.

However, Transformer-based architectures often entail high-performance computing and memory costs, resulting in resource-intensive training and inference.

In this paper, we propose a novel end-to-end urban navigation framework with decision hindsight, as depicted in Fig. 1. Based on the POMDP, we integrate the historical information into current decision by constructing a history-feedforward state representation, so as to improve the data utilization and decision foreseeability. Considering the impressive modeling capabilities of Transformer but its associated training complexity, we utilize it specifically for processing historical behavioral sequences of predefined lengths, in which the driving perception of each timestep has been encoded into the latent feature via a upstream Convolutional Neural Network (CNN). Given the small input scale and straightforward mapping, we seamlessly integrate Transformer into the state encoding module that is then connected to the RL policy, resulting in our comprehensive end-to-end learning framework with decision hindsight. The main contributions of this paper are summarized as follows:

- We formulate the urban navigation task based on the POMDP, and incorporate the historical information to enhance the context of driving perception for better data utilization.
- We design a history-feedforward state encoding module that integrates a causal Transformer to model the historical trajectories, and construct the end-to-end navigation framework with decision hindsight, enabling more reliable action inference.
- Our method improves learning efficiency and convergence, and achieves better driving performance than several state-of-the-art methods in challenging urban navigation scenarios.

2   Related Work

2.1   State Representation in Urban Scenarios

To address the environmental diversity and uncertainty in urban scenarios, many studies generate the semantic BEV mask based on the raw perception of LiDAR and Camera to clearly show the vehicle's surrounding environment [3,15]. In [3], the semantic BEV mask was enforced to connect with certain intermediate properties in the modularized framework to better explain the learned policy. In [15], the predictions from a real-time multi-modal fusion network were incorporated with BEV images as hybrid representation to learn the end-to-end driving policy ensuring safety, efficiency, and comfort. Both studies aim to address vision-centric driving applications with sensor inputs, requiring the exquisite design of perception module that convert the raw data into BEV masks. In those studies that mainly focus on the decision-making and control of autonomous vehicle, BEV image is directly used as the state representation to describe the environment synthetically [17,28,30]. To further reduce the computational complexity, CNN models were applied to learn latent feature from the BEV image in [17,30]. In [4,14], the latent state representation was encoded by Variational Auto-Encoder (VAE) instead.

Although the BEV-form representation offers a highly descriptive view of driving scenarios, its practicality is hindered by the limited sensing range of on-board devices. Without mature V2X technology or well-equipped roadside facilities, vehicles struggle to acquire comprehensive global environmental information. Incomplete perceptions may lead to unreliable driving policies with vague state-action mapping. To address this challenge, RL based on POMDP modeling proves to be effective, where the agent's confidence regarding its own state can be enhanced by integrating historical information, allowing for improved handling of uncertain conditions. In [27], the Long Short-Term Memory (LSTM) based architecture was proposed to capture the historical impact of interactive environment on long-term reward during highway on-ramp merging. In [26], the fast recurrent deterministic policy gradient (Fast-RDPG) framework was developed for making decisions based on historical trajectories in the UAV navigation. In [11], the incorporation of trajectory-based context variable brought significantly improvement, with performance comparable to state-of-the-art meta-RL methods. Inspired by this, we formulate the urban navigation task based on POMDP, and feedforward historical information to the current state, providing hindsight for action inference.

2.2   Transformer in RL

The historical information should be refined to achieve better utilization, which involves the processing of sequential data. Recently, Transformer has shown superior performance over CNN and Recurrent Neural Network (RNN), as it can model long dependencies and has excellent scalability. In the context of RL,

Transformer can be applied for representation learning with self-attention mechanism to better support the policy learning. In [29], as a preliminary attempt, the multi-head dot-product attention was introduced for representing and reasoning about states via relational inductive biases, which was subsequently used to process challenging multi-entity observation [25]. Meanwhile, Transformer can also be used to address the issue of partial observability, by working as a memory architecture to capture multi-step temporal dependencies [20]. In [18], a Transformer-based shortcut recurrence with memory vectors was designed for complex reasoning over long-term dependencies. In [19], a meta-RL agent that mimics the memory reinstatement mechanism was proposed, where the episodic memory is built recursively through the transformer layers.

In addition to being embedded into components of traditional RL, Transformer can directly serve as the sequential decision-making model to generate action sequence that can yield high returns, which is motivated by the idea of offline RL. Decision Transformer (DT) [5] is the first framework to project RL as a sequence modeling problem. Without explicit Temporal Difference (TD) learning or dynamic programming, DT produces the desired trajectory by conditioning an autoregressive model on the desired return (reward), past states, and actions. Another similar method named Trajectory Transformer (TT) [13] also use a Transformer architecture to model distributions over trajectories but use beam search for planning during execution. Thereafter, a growing number of works has focused on the Transformer-based sequential decision-making, with different choices of conditioning or structure [16]. Given the difficulty of defining the return-to-go conditioning in long-term urban navigation, we still follow the TD-learning while use Transformer as a behavior encoder for trajectory modeling, aiming to incorporate the historical information most relevant to action inference via self-attention mechanism.

3 Methodology

3.1 Problem Formulation

Considering the delayed impact of historical behavior on the current decision, we model the autonomous driving task based on the POMDP, which is described by the tuple $\mathcal{S}, \mathcal{A}, \mathcal{T}, \mathcal{R}, \mathcal{O}, \mathcal{Z}, \gamma$, where $\mathcal{S}$ is the state space, $\mathcal{A}$ is the action space, $\mathcal{T}$ is the state transition function, $\mathcal{R}$ is the reward function, $\mathcal{O}$ is the observation space, $\mathcal{Z}$ is the observation function, $\gamma \in [0, 1]$ is the discount factor. We use $s_t \in \mathcal{S}$, $a_t \in \mathcal{A}$, and $r_t = \mathcal{R}(s_t, a_t)$ to denote the state, action, and reward at timestep t, respectively. Based on the historical trajectory $\mathbf{h}_t = (s_{t-1}, a_{t-1}, r_{t-1}, \cdots, s_0, a_0, r_0)$, with the predefined observation function $\mathcal{Z} : \mathcal{S} \times \mathcal{A} \times \mathcal{R} \to \mathcal{O}$, we can get the historical observation $o_t = \mathcal{Z}(\mathbf{h}_t) \in \mathcal{O}$. Then, o_t is incorporated with s_t to get the joint state $\hat{s}_t = [s_t, o_t]$ as the basis for action inference.

In this work, we focus on the end-to-end urban navigation with continuous control signals, where considering the POMDP with continuous state and action, RL aims to learn an optimal stochastic policy for projecting states into actions.

Given the policy $a \sim \pi_\theta(\cdot|\hat{s})$ with parameters θ, at each timestep t, the vehicle chooses the desired action a_t based on $\hat{s}_t$ according to π_θ, then receives a reward r_t and transits to the next state s_{t+1} according to the state transition $\mathcal{T}(\cdot|\hat{s}_t, a_t)$. The goal of RL is to maximize the expected return $R_t = \mathbb{E}\left[\sum_{i=t}^{\infty} \gamma^{i-t} r_i\right]$. To obtain the optimal policy π^*, the value function of states and the action-value function of state-action pairs should be estimated. The value function $V^\pi(s)$ is defined as the expected return starting from s and successively following policy,

$$V^\pi(\hat{s}_t) = \mathbb{E}\left[R_t|\hat{s}_t\right], \tag{1}$$

and the action-value function $V^\pi(s)$ is defined as the expected return when executing action a at $\hat{s}$,

$$Q^\pi(\hat{s}_t, a_t) = \mathbb{E}\left[R_t|\hat{s}_t, a_t\right]. \tag{2}$$

In order to address the continuous action space, we use the RL method based on policy gradient to learn π_θ with the objective as follows,

$$\mathcal{L}(\theta) = \mathbb{E}[A^\pi(\hat{s}, a) \log \pi_\theta(a|\hat{s})], \tag{3}$$

where $A^\pi(\hat{s}, a) := Q^\pi(\hat{s}, a) - V^\pi(\hat{s})$ is the advantage function. By repeatedly estimating the gradient of (3), the optimal π^* could be gradually approached.

3.2   History-Feedforward State Encoding

To improve information utilization and decision reliability, we combine current perceptions and historical behaviors to create driving states as policy inputs. The current perceptions contain environmental perceptions and vehicle measurements. The real-time perceptions of environmental dynamics are described by the BEV image to reduce the complexity, which consists of C grayscale images with dimension $W \times H$. The vehicle measurements, including steering, throttle, brake, gear, lateral and horizontal speed, are represented as a 6-dimensional vector. On the other hand, the historical behavioral sequence refer to the historical trajectory of the most recent K timesteps instead of the entire episode to reduce storage and computational overhead.

Accordingly, we design two encoders to process the current perceptions and historical trajectories respectively, forming a state encoding module with decision hindsight. The current perception encoder uses 6 convolutional layers and 2 fully-connected (FC) layers to encode the BEV image $\mathbf{i} \in [0, 1]^{W \times H \times C}$ and the measurement vector $\mathbf{m} \in \mathbb{R}^6$ separately. And after stitching the outputs from both parts, the current perception features are obtained by passing the additional 2 FC layers. The historical behavior encoder is built on a causal Transformer, aiming to leverage its good autoregressive modeling capabilities to learn meaningful patterns from historical trajectories.

At each time step t, as shown in Fig. 2, we consider the historical trajectory consisting of the states, actions and rewards of the last K time steps,

$$\mathbf{h}_t = (s_{t-K}, a_{t-K}, r_{t-K}, \cdots, s_{t-1}, a_{t-1}, r_{t-1}) \tag{4}$$

where s is the state feature processed by the perception encoder with output dimension of D_s, $a \in [-1, 1]^2$ is the low-level action for steering and acceleration with dimension $D_a = 2$, r is the scalar reward with dimension $D_r = 1$. Following [5], we project the raw inputs of each modality into the embedded tokens through different embedding layer,

$$e_i^s = f_{stat}(s_i), \ e_i^a = f_{act}(a_i), \ e_i^r = f_{rew}(r_i), \tag{5}$$

where e_i^s, e_i^a, e_i^r are the tokens of state, action, reward of timestep i, each with the same dimension D_e. In this work, $f_{stat}(\cdot)$, $f_{act}(\cdot)$, $f_{rew}(\cdot)$ are realized by 3 FC layers with input size of D_s, D_u, D_r separately, and obviously, with the same output size D_e. Meanwhile, the token for each timestep is also learned as the position embedding,

$$e_i^p = f_{pos}(N_e, i), \tag{6}$$

where e_i^p also keeps the dimension D_e, $f_{pos}(\cdot)$ is a simple lookup table, N_e is defined as the maximum length of rollout episode. Then, we add e_i^p to the tokens of state, action and reward, respectively,

$$\hat{e}_i^s = e_i^s + e_i^p, \ \hat{e}_i^a = e_i^a + e_i^p, \ \hat{e}_i^r = e_i^r + e_i^p. \tag{7}$$

After passing through a layer-normalization layer, the token sequence $\mathbf{h}_t^{in} = (\hat{e}_{t-K}^s, \hat{e}_{t-K}^a, \hat{e}_{t-K}^r, \cdots, \hat{e}_{t-1}^s, \hat{e}_{t-1}^a, \hat{e}_{t-1}^r)$ is fed into a Generative Pre-trained Transformer (GPT), which is a modified Transformer architecture with causal self-attention mask that enables autoregressive trajectory modeling,

$$\mathbf{h}_t^{out} = \text{GPT}(\mathbf{h}_t^{in}), \tag{8}$$

where $\mathbf{h}_t^{out} = (h_{t-K}^s, h_{t-K}^a, h_{t-K}^r, \cdots, h_{t-1}^s, h_{t-1}^a, h_{t-1}^r)$ is the hidden state sequence, h_i^s, h_i^a and h_i^r are the encoded features corresponding to s_i, a_i and r_i in $\mathbf{h}_t^{in}$, respectively. Unlike [5] further projecting $\mathbf{h}_{out}$ to the future action tokens, since GPT works as the observation function here, we directly select the last action feature in $\mathbf{h}_t^{out}$ as the historical observation $o_t = h_{t-1}^a$, which is combined with the current perception state s_t to create the policy input $\hat{s}_t$.

3.3 End-to-End Learning with Decision Hindsight

We integrate the history-feedforward state encoding module as upstream support with the RL-based decision module, aiming to build an end-to-end urban navigation framework with decision hindsight, as shown in Fig. 3. Specifically, we follow the model-free manner and use the Proximal Policy Optimization (PPO) with exploration-enhanced objective in [30] for efficient learning.

The architecture of PPO contains a value network V_ϕ parameterized by ϕ and a policy network π_θ parameterized by θ, both of them consist of 2 FC hidden layers and take the same input $\hat{s}$. The V_ϕ output a scalar state-value

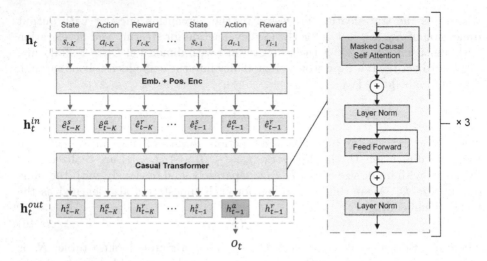

Fig. 2. Structure of historical behavior encoder based on the causal Transformer, where a GPT architecture with 3 layers and 1 attention head is employed here.

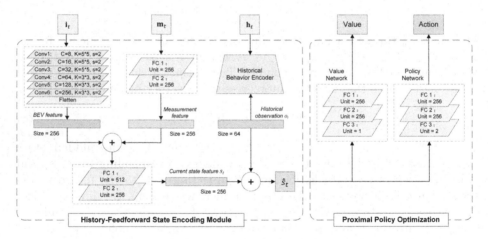

Fig. 3. Architecture of end-to-end urban navigation framework with decision hindsight. For each convolutional (Conv) layer, we specify the output channel (C), kernel size (K) and stride (s). For each fully-connected (FC) layer, we provide the number of units, which is equal to size of output feature. Notably, the FC 3 layer in value network has no activation function, while the FC 3 layer in policy network utilizes Softplus for activation. Furthermore, all other Conv and FC layers are activated by ReLU.

for generalized advantage estimation, which is learned to regress the expected returns via the TD approach,

$$\underset{\phi}{\text{minimize}} \sum_{n=0}^{N} \|V_\phi(\hat{s}_n) - \hat{V}_n\|^2, \tag{9}$$

where n is the index of all timesteps in a batch of trajectories, N is the batch size, $\hat{V}_n = \sum_{i=0}^{\infty} \gamma^i r_{n+i}$ is the discounted sum of rewards. The π_θ project the current input $\hat{s}$ to a distribution of actions, which is updated via the exploration-enhanced objective

$$\underset{\theta}{\text{maximize}}\; \mathbb{E}[\mathcal{L}_{ppo} + \lambda_{ent}\mathcal{L}_{ent} + \lambda_{exp}\mathcal{L}_{exp}], \tag{10}$$

The first item $\mathcal{L}_{ppo}$ is the clipped PPO loss with generalized advantage estimation [22]. The second $\mathcal{L}_{ent}$ is the maximum entropy loss,

$$\mathcal{L}_{ent} = \text{KL}\left(\pi_\theta(\cdot|\hat{s}) \,\|\, \mathcal{U}(-1,1)\right), \tag{11}$$

where $\text{KL}(\cdot)$ refers to the KL-divergence, $\mathcal{U}(-1,1)$ is a uniform distribution. The last $\mathcal{L}_{exp}$ is the exploration loss,

$$\mathcal{L}_{exp} = \mathbb{I}_{\{T-N_z+1,\cdots,T\}}(k)\,\text{KL}\left(\pi_\theta(\cdot|\hat{s}) \,\|\, p_z\right), \tag{12}$$

where z is the terminal event (including collision, running traffic light/sign, route deviation and being blocked), T is the length of an episode, $\mathbb{I}(\cdot)$ is the indicator function within the last N_z steps of an episode, p_z is a predefined exploration prior. Both $\mathcal{L}_{ent}$ and $\mathcal{L}_{exp}$ are added for encourages exploration, with the coefficient λ_{ent} and λ_{exp} for scale balancing. As we attempt to optimize our model in an end-to-end manner, (9) and (10) are combined as one loss function

$$\mathcal{L} = -\mathcal{L}_{policy} + \lambda_{value}\mathcal{L}_{value}, \tag{13}$$

where $\mathcal{L}_{policy}$ and $\mathcal{L}_{value}$ refer to the objective in (9) and (10), respectively, λ_{value} is also a coefficient for scale balancing. By minimizing (13), the networks of V_ϕ and π_θ can be jointly updated, and the state encoding module including perception encoder and history encoder is also updated simultaneously via gradient backpropagation.

We train our end-to-end framework online via environment interactions, and design a trajectory-level replay buffer $\mathcal{R}_{traj}$ with fixed size for rollout collection. To ensure smooth propagation of the computational graph, we store the raw perception data (BEV image $\mathbf{i}$ and measurement vector $\mathbf{m}$) rather than the encoded state features. Moreover, unlike [5] feeding the model with returns-to-go, we still provide the past rewards, and additionally add the timestep as well as the terminal event to our trajectory representation, expressed as

$$\tau_t = (\mathbf{i}_t, \mathbf{m}_t, a_t, r_t, \mathbf{h}_t = \{\mathbf{i}_i, \mathbf{m}_i, a_i, r_i, i\}_{t-K \le i \le t-1}, z_t), \tag{14}$$

where the terminal event z_t should be null if the termination is not triggered at timestep t.

The pseudo-code of our end-to-end learning framework with decision hindsight is given in Algorithm 1. We follow the reward definition and the implementation approach in [30], adding extra penalties for large steering changes and high speed, and setting episode for environment interactions to endless (once the target is reached, a new random target will be chosen). In each episode, the vehicle

---

Algorithm 1 End-to-end learning framework with decision hindsight

---

1: **Initialization**: end-to-end navigation model with policy π_θ, replay buffer $\mathcal{R}_{traj}$.
2: **for** each iteration **do**
3:     **while** replay buffer $\mathcal{R}_{traj}$ is not full **do**
4:         Get BEV image i_t, measurement vector m_t and historical trajectory h_t.
5:         Encode the history-feedforward state $\hat{s}_t$ from i_t, m_t and h_t.
6:         Select action based on the policy: $a_t \sim \pi_\theta(a_t|\hat{s}_t)$.
7:         Execute a_t, recieve reward r_t and terminal event z_t.
8:         Store data in replay buffer: $\mathcal{R}_{traj} \leftarrow \mathcal{R}_{traj} \cup \{(i_t, m_t, a_t, r_t, h_t, z_t)\}$.
9:         **if** z_t is not null **then**
10:             Reset environment.
11:         **end if**
12:     **end while**
13:     **for** every gradient step **do**
14:         Sample minibatch data from replay buffer $\mathcal{R}_{traj}$.
15:         Update the end-to-end navigation model using (13).
16:     **end for**
17:     Reset replay buffer $\mathcal{R}_{traj}$.
18: **end for**

---

begins with randomly assigned starting point and destination for navigation, and the desired waypoints are calculated by A* algorithm. At each timestep, the current perception of vehicle and the historical trajectory are composed as one state feature via the history-feedforward state-encoding module, which is then fed into the RL policy for action reasoning. Once the action is executed, the state transitions to the next timestep and the environment provides a reward. Concurrently, the perception, reward, action and the trajectory of the last K timesteps are combined to form a sample τ, which is then stored into the replay buffer $\mathcal{R}_{traj}$ for training, and the navigation process continues. If any of the terminal conditions are met, the vehicle is immediately reset to a new location, and a new episode for rollout begins. Once $\mathcal{R}_{traj}$ is filled, the episode is paused and the training phase commences. With the importance sampling of PPO, we randomly sample small batches of data from $\mathcal{R}_{traj}$, and update the model using (13). When the update epoch reaches the upper limit, the buffer is reset, and the episode for rollout continues.

4 Experiments

4.1 Simulation Setup

We constructed the urban navigation environments based on the open-source CARLA simulator [10]. Both training and testing experiments consider six towns (Town01–Town06) with different road topologies and navigation routes.

Training Experiments. In the end-to-end training experiments, rollouts are collected parallel from six CARLA servers at 10 FPS, each server corresponds to one of the six towns, reducing the time cost of environment interaction. To simulate the uncertainty and randomness in real urban scenarios, the traffic flow is set to dynamically change during training, with randomly spawned vehicles and pedestrians. Furthermore, the starting point and destination for each episode are randomly generated, adding to the diversity of navigation task.

Testing Experiments. The testing experiments are conducted based on the CARLA benchmarks, where the starting point and destination of each route are predefined, aiming to verify the advantages of our proposed method through comparison. On the CARLA 42 Routes [7] (Route-42) benchmark, we compare our method with state-of-the-art methods in terms of driving score (DS), road completion (RC), and infraction score (IS). Meanwhile, we conduct additional internal evaluations on NoCrash [8] and offline Leaderboard [30] (Offline-LB) benchmarks, taking into account higher traffic density or more diverse routes, with success rate (SR) and DS as metrics.

Implementation Details. Following [30], the input size of BEV image is set to 192 × 192 px with 15 channels, where cyclists and pedestrians are rendered larger than their actual sizes. The main hyperparameters for PPO training are set according to the specifications of [27]. The length of historical trajectory for behavior encoding is set to $K = 10$, which is a tradeoff between performance and computational efficiency after repeated attempts. Each training is implemented on a Tesla A100 GPU with batch size of 256, and our end-to-end framework is updated using the Adam optimizer with initial learning rate of 1e-5, which is then scheduled based on the empirical KL-divergence between the policy before and after the update.

4.2 Result Analysis

Comparison to the State of the Art. Table 1 compares our end-to-end RL framework with Transformer-based decision hindsight (TDH-RL) to prior state-of-the-art methods on the Route-42 benchmark, where the results show the mean and standard deviation over 3 evaluations for each model. TransFuser [21] and NEAT [7] are both imitation learning methods that enhance the driving state representation using attention mechanism, without incorporating historical information. Since NEAT associates the images with the BEV representation using intermediate attention maps, it performs better than Transformer, highlighting the value of BEV-form description. Roach performs on par with NEAT by improving the quality of demonstrations via a well-trained RL expert, albeit lacking attention mechanism. WOR [2] is a model-based RL method learning from pre-recorded driving logs, with comparable performance to NEAT and Roach. InterFuser [24] emerges as the most powerful autonomous driving model in CARLA, outperforming the previous four methods by enhancing safety and interpretability with intermediate features and constrained actions. Our proposed TDH-RL method, with BEV-based environment description and history-

Table 1. Comparison of our TDH-RL with other methods on Route-42 benchmark

Method	DS % ↑	RC % ↑	IS ↑
TransFuser [21]	53.40 ± 4.54	72.18 ± 4.17	0.74 ± 0.04
NEAT [7]	65.10 ± 1.75	79.17 ± 3.25	0.82 ± 0.01
Roach [30]	65.08 ± 0.99	85.16 ± 4.20	0.77 ± 0.02
WOR [2]	67.64 ± 1.26	90.16 ± 3.81	0.75 ± 0.02
InterFuser [24]	91.84 ± 2.17	97.12 ± 1.95	**0.95 ± 0.02**
TDH-RL(Ours)	**93.07 ± 1.58**	**98.31 ± 1.78**	**0.95 ± 0.02**

feedforward state representation, achieves slightly superior performance compared to InterFuser, and pushes the performance ceiling of the Route-42 benchmark without the need for complex feature encoding structures or safety constraints. This showcases the efficacy of our method in advancing the state-of-the-art in autonomous driving research. Moreover, TDH-RL can also be considered an advanced expert for providing high-quality demonstrations in imitation learning (IL)-based methods.

Ablation Study. To investigate the influence of historical feedforward information, we compared our method to a baseline model that did not include the Transformer-based historical behavior encoder. Additionally, we analyzed the influence of historical observations represented by various output features from the Transformer-based behavior encoder.

Addition of History-Feedforward Encoding. Table 2 and Table 3 present the evaluation results of the baseline model and our proposed TDH-RL on the Nocrash and Offline-LB benchmarks, respectively. Both models were trained in an end-to-end manner with exploration-enhanced objectives for optimization. Our TDH-RL consistently outperforms the baseline model across all evaluation settings, indicating that the inclusion of a history-feedforward mechanism can significantly improve driving performance, even in challenging urban scenarios characterized by higher traffic density and more diverse routes. To further compare the learning performance of the two models, Fig. 4 displays the learning curves within 1.4e7 timesteps. The blue lines represent our TDH-RL model, while the red lines represent the baseline model. As depicted in Fig. 4, both models initially perform poorly with low rewards and frequent collisions. Through continuous exploration and optimization, they gradually learned to perform more efficiently and approached convergence after 1e7 timesteps. Regarding the overall change in the learning curves, TDH-RL demonstrates a more stable and efficient learning process compared to the baseline model that only considers current perceptions. Notably, TDH-RL achieves an asymptotic performance improvement of approximately 20%, showcasing enhanced exploration and exploitation capabilities.

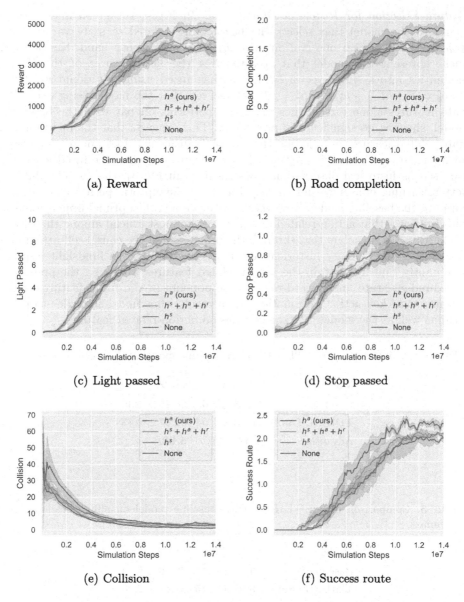

Fig. 4. Learning curves of models on 6 towns, where solid lines and shaded areas show the mean and standard deviation across 3 seeds, respectively, and the lines of each trial are smoothed via exponential weighted moving averaging. There are four models with different types of state encoding: (1) "None": the baseline model only without Transformer-based historical behavior encoder; (2) "h^s": the model using the state-related feature as the historical observation; (3) "$h^s + h^a + h^r$": the model using the fused feature of three modalities to represent the historical observation; (4) "h^a": our proposed TDH-RL model which incorporates the action-related feature for state encoding. Each model is learned end-to-end using the same exploration-enhanced objective for optimization, and is evaluated every 1e4 steps.

Feature Selection for Historical Observation. In Fig. 4, the green lines correspond to the model that selects the last feature related to state variable in $\mathbf{h}_{out}$ as the historical observation $o_t = h_{t-1}^s$. On the other hand, the orange lines represent the model that concatenates the last features of the three modalities into a single vector, which is then fused using additional FC layers $o_t = \text{FC}(h_{t-1}^s + h_{t-1}^a + h_{t-1}^r)$. It is observed that the model using encoded state features alone (green lines in Fig. 4) performs on par with the baseline model, showing no significant improvement. This suggests that the inclusion of state-related features encoded by the causal Transformer does not provide useful information for inferring actions. In the meantime, when the fused features are used as historical observations (orange lines in Fig. 4), the model exhibits better learning efficiency and slightly improved asymptotic performance compared to the baseline model. However, it fails to reach the optimal upper bound. This indicates that action-related features are the most crucial among the three modalities. We speculate that through the processing of the causal self-attention mechanism, the primary historical information with decision hindsight is predominantly captured within the action-related features. However, the presence of redundant state and reward-related components in the fused features may obscure some key action information. As a result, this model struggles to perform as well as the model that solely relies on action-related features.

Table 2. Comparison of our TDH-RL with baseline model on Nocrash benchmark

Model	DS % ↑	SR % ↑
Baseline	92.75 ± 2.63	86.75 ± 4.35
TDH-RL(Ours)	96.23 ± 1.54	92.67 ± 2.47
Advantage	**3.48**	**5.92**

Table 3. Comparison of our TDH-RL with baseline model on offline-Leaderboard benchmark

Model	DS % ↑	SR % ↑
Baseline	84.83 ± 3.17	72.26 ± 5.73
TDH-RL(Ours)	89.02 ± 2.65	81.11 ± 2.20
Advantage	**4.19**	**8.85**

5 Conclusion

We present an end-to-end urban navigation framework with Transformer-based decision hindsight. Following the POMDP-based problem formulation, the historical behavior information is encoded as supplementary observations using autoregressive modeling, implemented through a causal Transformer. By combining the historical observations with current perception, a state representation with enhanced global awareness is constructed to serve as the state input for out RL-based navigation policy. Correspondingly, using the history-feedforward state encoding as upstream support, we establish an end-to-end learning framework based on the exploration-enhanced PPO, aiming to obtain a reliable navigation model with decision hindsight. Through extensive validations on challenging urban navigation tasks, our proposed method outperforms the previous methods on benchmarks, demonstrating significant improvement in learning efficiency and inference reliability. Moving forward, our research will focus on introducing the parameterized motion skills for more informative exploration and accelerated reward signaling. Additionally, we aim to explore efficient human-in-the-loop learning methods to ensure high safety and generalizability when transferring our approach to real-world scenarios.

References

1. Ahmed, M., Abobakr, A., Lim, C.P., Nahavandi, S.: Policy-based reinforcement learning for training autonomous driving agents in urban areas with affordance learning. IEEE Trans. Intell. Transp. Syst. **23**(8), 12562–12571 (2022). https://doi.org/10.1109/TITS.2021.3115235
2. Chen, D., Koltun, V., Krähenbühl, P.: Learning to drive from a world on rails. In: 2021 IEEE/CVF International Conference on Computer Vision, ICCV 2021, Montreal, QC, Canada, 10–17 October 2021, pp. 15570–15579. IEEE (2021). https://doi.org/10.1109/ICCV48922.2021.01530
3. Chen, J., Li, S.E., Tomizuka, M.: Interpretable end-to-end urban autonomous driving with latent deep reinforcement learning. IEEE Trans. Intell. Transp. Syst. **23**(6), 5068–5078 (2022). https://doi.org/10.1109/TITS.2020.3046646
4. Chen, J., Yuan, B., Tomizuka, M.: Model-free deep reinforcement learning for urban autonomous driving. In: 2019 IEEE Intelligent Transportation Systems Conference (ITSC), pp. 2765–2771 (2019). https://doi.org/10.1109/ITSC.2019.8917306
5. Chen, L., et al.: Decision transformer: Reinforcement learning via sequence modeling. In: Ranzato, M., Beygelzimer, A., Dauphin, Y.N., Liang, P., Vaughan, J.W. (eds.) Advances in Neural Information Processing Systems, vol. 34, Annual Conference on Neural Information Processing Systems 2021, NeurIPS 2021(December), pp. 6–14, 2021. Virtual, pp. 15084–15097 (2021). https://proceedings.neurips.cc/paper/2021/hash/7f489f642a0ddb10272b5c31057f0663-Abstract.html
6. Chen, M., Xiao, X., Zhang, W., Gao, X.: Efficient and stable information directed exploration for continuous reinforcement learning. In: ICASSP 2022-2022 IEEE International Conference on Acoustics, Speech and Signal Processing (ICASSP), pp. 4023–4027 (2022). https://doi.org/10.1109/ICASSP43922.2022.9746211
7. Chitta, K., Prakash, A., Geiger, A.: NEAT: neural attention fields for end-to-end autonomous driving. In: 2021 IEEE/CVF International Conference on Computer Vision, ICCV 2021, Montreal, QC, Canada, 10–17 October 2021, pp. 15773–15783. IEEE (2021). https://doi.org/10.1109/ICCV48922.2021.01550
8. Codevilla, F., Santana, E., López, A.M., Gaidon, A.: Exploring the limitations of behavior cloning for autonomous driving. In: 2019 IEEE/CVF International Conference on Computer Vision, ICCV 2019, Seoul, Korea (South), October 27

- November 2, 2019, pp. 9328–9337. IEEE (2019). https://doi.org/10.1109/ICCV. 2019.00942

9. Deshpande, N., Vaufreydaz, D., Spalanzani, A.: Navigation in urban environments amongst pedestrians using multi-objective deep reinforcement learning. In: 2021 IEEE International Intelligent Transportation Systems Conference (ITSC), pp. 923–928 (2021). https://doi.org/10.1109/ITSC48978.2021.9564601

10. Dosovitskiy, A., Ros, G., Codevilla, F., López, A.M., Koltun, V.: CARLA: an open urban driving simulator. In: 1st Annual Conference on Robot Learning, CoRL 2017, Mountain View, California, USA, 13–15 November 2017, Proceedings. Proceedings of Machine Learning Research, vol. 78, pp. 1–16. PMLR (2017). http:// proceedings.mlr.press/v78/dosovitskiy17a.html

11. Fakoor, R., Chaudhari, P., Soatto, S., Smola, A.J.: Meta-q-learning. In: 8th International Conference on Learning Representations, ICLR 2020, Addis Ababa, Ethiopia, 26–30 April 2020. OpenReview.net (2020). https://openreview.net/ forum?id=SJeD3CEFPH

12. Huang, C., et al.: Deductive reinforcement learning for visual autonomous urban driving navigation. IEEE Trans. Neural Netw. Learn. Syst. **32**(12), 5379–5391 (2021). https://doi.org/10.1109/TNNLS.2021.3109284

13. Janner, M., Li, Q., Levine, S.: Offline reinforcement learning as one big sequence modeling problem. In: Ranzato, M., Beygelzimer, A., Dauphin, Y.N., Liang, P., Vaughan, J.W. (eds.) Advances in Neural Information Processing Systems 34: Annual Conference on Neural Information Processing Systems 2021, NeurIPS 2021(December), pp. 6–14, 2021. virtual, pp. 1273–1286 (2021). https://proceedings.neurips.cc/paper/2021/hash/ 099fe6b0b444c23836c4a5d07346082b-Abstract.html

14. Kargar, E., Kyrki, V.: Increasing the efficiency of policy learning for autonomous vehicles by multi-task representation learning. IEEE Trans. Intell. Veh. **7**(3), 701–710 (2022). https://doi.org/10.1109/TIV.2022.3149891

15. Khalil, Y.H., Mouftah, H.T.: Exploiting multi-modal fusion for urban autonomous driving using latent deep reinforcement learning. IEEE Trans. Veh. Technol. **72**(3), 2921–2935 (2023). https://doi.org/10.1109/TVT.2022.3217299

16. Li, W., Luo, H., Lin, Z., Zhang, C., Lu, Z., Ye, D.: A survey on transformers in reinforcement learning. CoRR abs/2301.03044 (2023). https://doi.org/10.48550/ arXiv.2301.03044

17. Liu, H., Huang, Z., Wu, J., Lv, C.: Improved deep reinforcement learning with expert demonstrations for urban autonomous driving. In: 2022 IEEE Intelligent Vehicles Symposium (IV), pp. 921–928 (2022). https://doi.org/10.1109/IV51971. 2022.9827073

18. Loynd, R., Fernandez, R., Celikyilmaz, A., Swaminathan, A., Hausknecht, M.J.: Working memory graphs. In: Proceedings of the 37th International Conference on Machine Learning, ICML 2020, 13–18 July 2020, Virtual Event. Proceedings of Machine Learning Research, vol. 119, pp. 6404–6414. PMLR (2020). http:// proceedings.mlr.press/v119/loynd20a.html

19. Melo, L.C.: Transformers are meta-reinforcement learners. In: Chaudhuri, K., Jegelka, S., Song, L., Szepesvári, C., Niu, G., Sabato, S. (eds.) International Conference on Machine Learning, ICML 2022, 17–23 July 2022, Baltimore, Maryland, USA. Proceedings of Machine Learning Research, vol. 162, pp. 15340–15359. PMLR (2022). https://proceedings.mlr.press/v162/melo22a.html

20. Parisotto, E., et al.: Stabilizing transformers for reinforcement learning. In: Proceedings of the 37th International Conference on Machine Learning, ICML 2020,

13–18 July 2020, Virtual Event. Proceedings of Machine Learning Research, vol. 119, pp. 7487–7498. PMLR (2020). http://proceedings.mlr.press/v119/parisotto20a.html

21. Prakash, A., Chitta, K., Geiger, A.: Multi-modal fusion transformer for end-to-end autonomous driving. In: IEEE Conference on Computer Vision and Pattern Recognition, CVPR 2021, virtual, 19–25 June 2021, pp. 7077–7087. Computer Vision Foundation/IEEE (2021). https://doi.org/10.1109/CVPR46437.2021. 00700, https://openaccess.thecvf.com/content/CVPR2021/html/Prakash_Multi-Modal_Fusion_Transformer_for_End-to-End_Autonomous_Driving_CVPR_2021_paper.html

22. Schulman, J., Moritz, P., Levine, S., Jordan, M.I., Abbeel, P.: High-dimensional continuous control using generalized advantage estimation. In: Bengio, Y., LeCun, Y. (eds.) 4th International Conference on Learning Representations, ICLR 2016, San Juan, Puerto Rico, 2–4 May 2016, Conference Track Proceedings (2016). http://arxiv.org/abs/1506.02438

23. Schwarting, W., Alonso-Mora, J., Rus, D.: Planning and decision-making for autonomous vehicles. Ann. Rev. Control Robot. Autonom. Syst. 1(1), 187–210 (2018). https://doi.org/10.1146/annurev-control-060117-105157

24. Shao, H., Wang, L., Chen, R., Li, H., Liu, Y.: Safety-enhanced autonomous driving using interpretable sensor fusion transformer. In: Liu, K., Kulic, D., Ichnowski, J. (eds.) Conference on Robot Learning, CoRL 2022, 14–18 December 2022, Auckland, New Zealand. Proceedings of Machine Learning Research, vol. 205, pp. 726–737. PMLR (2022). https://proceedings.mlr.press/v205/shao23a.html

25. Vinyals, O., et al.: Grandmaster level in starcraft ii using multi-agent reinforcement learning. Nature 575(7782), 350–354 (2019). https://doi.org/10.1038/s41586-019-1724-z

26. Wang, C., Wang, J., Shen, Y., Zhang, X.: Autonomous navigation of UAVs in large-scale complex environments: a deep reinforcement learning approach. IEEE Trans. Veh. Technol. 68(3), 2124–2136 (2019). https://doi.org/10.1109/TVT.2018. 2890773

27. Wang, P., Chan, C.Y.: Formulation of deep reinforcement learning architecture toward autonomous driving for on-ramp merge. In: 2017 IEEE 20th International Conference on Intelligent Transportation Systems (ITSC), pp. 1–6 (2017). https://doi.org/10.1109/ITSC.2017.8317735

28. Wu, J., Huang, W., de Boer, N., Mo, Y., He, X., Lv, C.: Safe decision-making for lane-change of autonomous vehicles via human demonstration-aided reinforcement learning. In: 2022 IEEE 25th International Conference on Intelligent Transportation Systems (ITSC), pp. 1228–1233 (2022). https://doi.org/10.1109/ITSC55140. 2022.9921872

29. Zambaldi, V.F., et al.: Deep reinforcement learning with relational inductive biases. In: 7th International Conference on Learning Representations, ICLR 2019, New Orleans, LA, USA, 6–9 May 2019. OpenReview.net (2019). https://openreview. net/forum?id=HkxaFoC9KQ

30. Zhang, Z., Liniger, A., Dai, D., Yu, F., Van Gool, L.: End-to-end urban driving by imitating a reinforcement learning coach. In: 2021 IEEE/CVF International Conference on Computer Vision (ICCV), pp. 15202–15212 (2021). https://doi.org/10.1109/ICCV48922.2021.01494

Identifying Self-admitted Technical Debt with Context-Based Ladder Network

Aiyue Gong[1,2], Fumiyo Fukumoto[3(✉)], Panitan Muangkammuen[2],
Jiyi Li[3], and Dongjin Yu[1]

[1] School of Computer Science and Technology, Hangzhou Dianzi University,
Hangzhou 310018, China
{airaa,yudj}@hdu.edu.cn
[2] Integrated Graduate School of Medicine, Engineering, Agricultural Sciences,
Faculty of Engineering, University of Yamanashi, Kofu 400-8511, Japan
g21dts04@yamanashi.ac.jp
[3] Faculty of Engineering, Graduate Faculty of Interdisciplinary Research,
University of Yamanashi, Kofu 400-8511, Japan
{fukumoto,jyli}@yamanashi.ac.jp

Abstract. Technical debt occurs when development teams take actions
to expedite the delivery of a project at the cost of poor code quality and
additional work of later refactoring. The accumulation of technical debt
will make the software fixes prohibitively expensive. As a typical type of
technical debt, Self-Admitted Technical Debt (SATD) is acknowledged
by developers in code comments. Identifying SATD in code comments
can improve code quality. However, manually discerning whether code
comments contain SATD would be expensive and time-consuming. To
solve this problem, we propose a method to apply the Ladder Network
with the pre-training model to identify SATD based on the labeled data
from 10 open source projects and the unlabeled data from another ten
projects. By comparing with the original model of Ladder Network, and
other semi-supervised learning models, the results show that the proposed method performs better in technical debt identification. In addition, the proposed method also achieves better results compared with
supervised learning methods. This shows that our approach can make
better use of unlabeled data to improve classification performance.

Keywords: Self-admitted Technical Debt · Semi-supervised
Learning · Ladder Network

1 Introduction

The concept of technical debt was introduced by Cunningham [1] in 1993. It
describes the immature decisions intentionally or unintentionally made by developers to achieve short-term goals. In the short term, these decisions can solve
the current issues in software, while they undoubtedly pose hidden risks in the
project. Zazworka et al. [2] have pointed out that accumulated technical debt in

B. Luo et al. (Eds.): ICONIP 2023, CCIS 1969, pp. 84–97, 2024.
https://doi.org/10.1007/978-981-99-8184-7_7

software development will have a significant impact in the future. So balancing the resolution approach and the technical debt introduced becomes a crucial consideration for developers.

There are many attempts focused on self-admitted technical debt (SATD), which is mentioned by Potdar and Shihab [3]. SATD intentionally introduced technical debt, which is a branch of technical debt. Developers record technical debt to facilitate repayment by themselves or other developers in the future. To facilitate mutual understanding of the code in the development team, code comments will be written close to the code. Therefore, the code comments have also become a suitable place to record SATD. As software is continuously developed and updated, technical debt is always inevitable. It is necessary to invent methods to identify technical debt.

Since most of the content in code comments is natural language text, identifying technical debt through code comments can also be considered as text classification task. In previous work, many deep learning methods have been applied to SATD identification [4–8]. In 2019, Ren et al. [4] applied convolution neural network (CNN) to extract text features of SATD. Santos et al. [5,6] used word embedding model to preprocess code comment, and then used Long Short-Term Memory (LSTM) model to build a classifier to identify SATD. Yu et al. [7] employed Bi-directional Long Short-Term Memory (BiLSTM) to identify SATD. Li et al. [8] identified SATD through graph neural networks (GNN), and used focal loss to deal with the problem of data imbalance between different classes.

Most of SATD identification approaches have relied on supervised learning methods. However, supervised learning requires a large amount of labeled training data, the cost of obtaining labeled data is high. Different annotators may also have ambiguity about the same comment, which requires multiple people to discuss and reach a conclusion. In previous studies, some unsupervised models were also proposed to identify SATD [3,9,10]. Potdar and Shihab [3] proposed using pattern matching to identify SATD. They summarized 62 patterns by manually reading code comments. However, the limited 62 patterns make it possible to miss SATD when faced with some complex technical debt that needs to be judged by context. Farias et al. [9] extended the pattern content by adding part of speech and code tags to the original matching patterns. Guo et al. [10] proposed a MAT method to identify SATD by investigating the distribution of task tags. The effect even outperforms some supervised learning methods.

Obtaining SATD data with labels requires substantial manual effort, while unlabeled data is easy to obtain. We thus use pre-training models to obtain contextual representation, especially because it is capable to learn rich contextual features from both labeled and unlabeled data to help the model better identify technical debt information. In this paper, we propose self-admitted technical debt with a ladder network (SATD-LN), which combines supervised learning and unsupervised learning. This model is a combination of Ladder network [11,12] and the pre-training model. Ladder Networks can play an important role in unsupervised learning. It introduces noise and reconstruction loss functions at multiple encoding layers and gradually removes noise from input data during training, which enables the model to learn more robust and accurate feature representations. It can help the pre-training language model to better capture noise and variation, thereby improving its generalization ability. For the existing model [12], we add a tanh activation function in the last layer. This modification aims to further enhance the expressiveness and nonlinearity of the model. We expect the model to be able to capture more complex patterns, resulting in a good performance for the self-admitted technical debt identification task. The contributions of our work can be summarized as follows:

(1) We are the first to introduce the semi-supervised learning method into the field of SATD identification, achieving promising results with a small amount of labeled data.
(2) We propose the SATD-LN model for SATD identification, by using both labeled data and unlabeled data to obtain rich contextual features.
(3) We conduct extensive experiments on the gold-standard dataset. Experimental results show that our method is effective in identifying SATD.

2 Related Work

2.1 Ladder Network

Semi-supervised learning is between supervised learning and unsupervised learning. It has the advantage of reducing the cost of manual annotation of the labeled data by combining unlabeled data with a small amount of labeled data to train the model. Ladder Network combines the supervised learning method on the basis of the previous unsupervised learning and applies the improved model to semi-supervised learning [11]. This model utilizes a feed-forward model as an encoder to add noise and uses a decoder to map each layer of the encoder. Finally, the whole model is trained by minimizing the loss function. This method enables the model to learn a more robust feature representation of the sample in the noise, making the model more generalizable. Rasmus et al. also proposed a simplified structure called Γ-Model in the paper [11]. In this structure, most of the decoder structure is omitted, and only the top layer is denoised to simplify the model.

2.2 Pre-training Model

The pre-training model is a framework for training on large general corpora and then fine-tuning specific target tasks. The general semantic representation learned by the model in the pre-training task enables the model to perform well on target tasks even with less data.

Bidirectional Encoder Representation from Transformers (BERT) [13] is a pre-training language model based on Transformers, which can capture semantic information from context by pre-training on large-scale unsupervised data. The pre-training process of BERT consists of two unsupervised tasks. They are Masked Language Modeling and Next Sentence Prediction. The pre-trained

BERT model can be trained on various target tasks through fine-tuning. The rich language representation in the pre-training task can be used to improve the performance and effectiveness of various target tasks.

Table 1. Examples of code comments and whether the technical debt is included.

Code Comment	Type
// TODO: delete the file if it is not a valid file	Debt
// FIXME What's the clean way to add a LogTarget afterward?	Debt
// This is really irritating; we need a way to set stuff	Debt
// The current character is always emitted	No-debt

2.3 Self-admitted Technical Debt

Developers may indicate non-optimal code implementations with code comments and wait to pay off the debt later. Therefore, We refer to such comments as self-admitted technical debt (SATD) [3]. Table 1 gives some examples. In the first and second examples, it is evident that they contain technical debt. Words like "TODO" and "FIXME" are exactly the distinguishing features that developers put in code comments to indicate that "there is technical debt waiting to be dealt with". In contrast, in the third example, there is no such distinctive word or phrase as above. We need to use natural language processing techniques to comprehend the semantic context of the sentences and identify the presence of technical debt.

In current self-admitted technical debt identification, many supervised methods have been applied to this task. Ren et al. [4] applied the convolution neural network (CNN) to extract text features. Santos et al. [5,6] used the Long Short-Term Memory (LSTM) model to identify SATD. And Li et al. [8] identified SATD through graph neural networks (GNN). These methods all rely on large number of labeled data. However, manual annotation of data is expensive and time-consuming. In this paper, we use a combination of unlabeled data and a small amount of labeled data and use semi-supervised learning to obtain better SATD identification results than supervised learning.

3   SATD-LN Model

The SATD-LN structure used in the experiments is illustrated in Fig. 1. Based on the Γ-Model of the Ladder Network, we combined it with the pre-training model. Let N be the number of labeled data, and $\{(\mathbf{x}_1, y_1), \ldots, (\mathbf{x}_N, y_N)\}$ be a set consisting of a pair of the data $\mathbf{x}$ and its label y. Likewise, $\{\mathbf{x}_{N+1}, \ldots, \mathbf{x}_{N+M}\}$ be a set consisting of the number of M unlabeled data. As illustrated in Fig. 1, the network consists of two forward passes, where the upper path in Fig. 1 indicates the noise-adding path, and the lower path refers to the noise-free path. The lower path produces clean $\mathbf{z}$ and $\mathbf{y}$ which are given by:

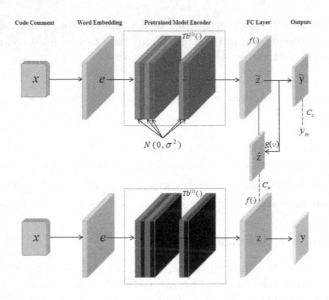

Fig. 1. The SATD-LN framework used in the experiments.

$$\mathbf{z} = f(\mathbf{h}^{(L)}) = N_B(\mathbf{W}\mathbf{h}^{(L)}),$$
$$\mathbf{y} = \phi(\gamma(\mathbf{z} + \beta)),$$
$$\mathbf{h}^{(0)} = \mathbf{e},$$
$$\mathbf{h}^{(l)} = Tb^{(l)}(\mathbf{h}^{(l-1)}),$$

(1)

where $\mathbf{W}$ shows the weight matrix of the linear transformation f, and N_B indicates a batch normalization. $\mathbf{h}^{(l)}$ refers to the state representation of layer l. ϕ refers to an activation function, where β and γ are trainable scaling and bias parameters, respectively. Likewise, $\mathbf{e}$ denotes the input embedding of $\mathbf{x}$ with positional encoding and $Tb^{(l)}$ refers to the transformer block at layer l in the L-layer pre-training language model.

The clean path shares the mappings $Tb^{(l)}$ and f with the noise-adding path. In the upper path of Fig. 1, $\tilde{\mathbf{z}}$ and $\tilde{\mathbf{y}}$ are generated by adding Gaussian noise $\mathbf{n}$ which are defined as:

$$\tilde{\mathbf{z}} = f(\tilde{\mathbf{h}}^{(L)}) + \mathbf{n},$$
$$\tilde{\mathbf{y}} = \phi(\gamma(\tilde{\mathbf{z}} + \beta)),$$
$$\tilde{\mathbf{h}}^{(0)} = \mathbf{e} + \mathbf{n},$$
$$\tilde{\mathbf{h}}^{(l)} = Tb^{(l)}(\tilde{\mathbf{h}}^{(l-1)}) + \mathbf{n}.$$

(2)

Given the input $\mathbf{x}_n$, a supervised cost C_s is calculated as the average negative log-probability of the noisy output $\tilde{\mathbf{y}}$ which matches the target y_n:

Table 2. Statistics on datasets with labeled code comments.

Project	Version	Contributors	SATD	Non-SATD	Sum
Apache Ant	1.7.0	74	131	3,967	4,098
ArgoUML	0.34	87	1,413	8,039	9,452
Columba	1.4	9	204	6,264	6,468
Eclipse	2.4.1	30	2,104	4,286	6,390
Hibernate	3.3.2	8	472	2,496	2,968
JEdit	4.2	57	256	10,066	10,322
JFreeChart	1.0.19	19	209	4,199	4,408
JMeter	2.10	33	374	7,683	8,057
JRuby	1.4.0	328	622	4,275	4,897
SQuirreL	3.0.3	46	284	6,929	7,215
Total	-	691	6,069	58,204	62,275

$$C_s = -\frac{1}{N} \sum_{n=1}^{N} \log P(\tilde{\mathbf{y}} = y_n | \mathbf{x}_n), \tag{3}$$

The unsupervised denoising cost C_u is given by:

$$C_u = \frac{1}{N+M} \sum_{n=1}^{N+M} \frac{\lambda}{d} \| \mathbf{z}_n - N_B(\hat{\mathbf{z}}_n) \|,$$

$$\hat{\mathbf{z}} = g(\tilde{\mathbf{z}}, \mathbf{u}),$$

$$\mathbf{u} = N_B(\tilde{\mathbf{y}}), \tag{4}$$

where λ is a coefficient for unsupervised cost, and d refers to the width of the output layer. g is a liner function with respect to $\tilde{z}$ and consists of its own learnable parameters [11] and reconstructs the denoised $\hat{\mathbf{z}}$ for given $\tilde{\mathbf{z}}$ and $\tilde{\mathbf{y}}$. The final cost C is given by:

$$C = C_s + C_u \tag{5}$$

4 Experimental Setup

4.1 Data and Evaluation Metrics

The labeled data of our experiments derived from the code comments in the dataset created by Maldonado et al. [14]. This dataset collects code comments from 10 open-source projects. These projects are used in different fields, which is representative of the self-admitted technical debt research in code comments. We present the statistical content of the labeled data in Table 2. The unlabeled data part refers to the code comment dataset contributed by Guo et al. [15]. After our processing, the dataset includes 80,702 unlabeled code comments.

We utilize F1-score as evaluation metrics. The F1-score is the harmonic mean of *precision* and *recall*. *Precision* refers to the proportion of the true positive class among all samples predicted by the model as the positive class. *Recall* refers to the proportion of all samples that are truly positive that the model successfully predicts as positive. The F1 score provides a balanced measure that combines both performance metrics. It is calculated using the following equation:

$$\text{F1-score} = \frac{2 * precision * recall}{precision + recall} \tag{6}$$

Table 3. Some hyperparameter settings in the model.

Hyperparameter	Values
Dimensional of Word embedding	768
Dropout	0.5
Learning rate	2e−5
Dropout	0.5
Noise	0.2

4.2   Implementation Details

We need to filter some potentially useless information to improve the detection effect of our model for SATD. We refer to the work of Yu et al. [7] to perform the following data pre-processing operations on the dataset. We remove duplicate comments in the dataset. Then we convert all characters to lowercase and remove some invalid information such as punctuation and URLs. Lastly, we removed code comments that represent revision history with dates.

During the training process, we combined nine projects into one dataset and used the remaining one project as a test set. The combined dataset is split into training and validation sets in a ratio of 8:2. We will randomly select data from the training set and validation set according to different data volumes. Because the data is randomly selected, we conduct multiple random selections and repeated experiments under different parameter conditions in order to eliminate errors. Adam optimizer is used during model training. We trained our model using the Titan RTX graphics card. Some hyperparameters are set in Table 3.

4.3   Baselines

We combined the pre-trained model on the basis of the original model of the Ladder Network for training. We compared the training results of the modified model with those of the original Ladder Network. Moreover, we use three other semi-supervised learning methods to compare with our results.

Virtual Adversarial Training (VAT) [16] - A regularization method that adds perturbations to the model to maintain smoothness.

MixText [17] - A method that first guesses the low-entropy labels of the unlabeled data, and then interpolates the labeled and unlabeled data.

Pseudo Labeling [18] - The model is first trained on labeled data, and it predicts on unlabeled data with pseudo-labels. Then put the unlabeled data into the model as labeled data to continue training.

To verify the effectiveness of the model proposed in this paper, we choose five mainstream supervised learning methods. They are Transformer [19], convolutional neural network (CNN) [20], recurrent neural network (RNN) [21], graph convolutional neural network (GCN) [22] and GNNSI proposed by Li et al. [8]. The reasons for choosing these methods are as follows. Transformer can handle the global relationship and contextual relationship of text. CNN is suitable for processing the local relationship and context relationship of text and has high computational efficiency. RNN is suitable for processing sequence data and capturing temporal dependencies in sequences. GCN can capture the relationship between tokens for information dissemination and aggregation. The GNNSI model is currently the best-performing supervised model in SATD identification.

Table 4. F1-score obtained by SATD-LN with and without tanh activation function.

Project	SATD-LN	Without Tanh	With ReLu
Apache Ant	**0.730**	0.630	0.632
ArgoUML	**0.911**	0.753	0.809
Columba	**0.859**	0.871	0.843
Eclipse	**0.728**	0.728	0.645
Hibernate	**0.877**	0.621	0.811
JEdit	0.778	**0.818**	0.622
JFreeChart	**0.829**	0.703	0.758
JMeter	**0.829**	0.731	0.817
JRuby	**0.849**	0.789	0.708
SQuirreL	**0.755**	0.638	0.769
Average	**0.815**	0.728	0.741
Improved	-	11.95%	9.98%

5 Results and Discussions

5.1 The Results of SATD-LN

Recall that we add a tanh activation function to the existing model [13] to further enhance expressiveness and nonlinearity of the model. We compared SATD-LN with and without the activation function by using the data which randomly chose 400 labeled and 1,000 unlabeled samplings. The results are shown in Table 4.

   Result: As can be seen from Table 4, adding the tanh activation function has improved the effectiveness of our model on the SATD identification task. The average F1 score increased by 11.95%. Compared with other activation functions, tanh, and sigmoid are suitable for classification problems, while activation functions such as ReLu and Leaky ReLU are more suitable for regression problems. In classification problems, sigmoid is suitable for the output layer, while tanh is suitable for the hidden layer, so we choose the tanh activation function here. Tanh activation function introduces a new nonlinear transformation in the last layer of the model. This synergy may lead to improved performance of the model on the self-admitted technical debt identification task.

5.2   The Performance Against the Number of Labeled Data

We examined the performance of SATD-LN and Ladder Network against the number of labeled data. Because we hope that the training of the entire model is simple and does not take too much time so that we can use fewer resources to train a model with better results. We randomly selected 1000 unlabeled data to research the classification performance of SATD-LN and Ladder Network when the amount of labeled data is 100, 200, 400, 600, and 800 respectively. We train for 50 epochs each time and repeat the training five times under different data conditions to obtain the average value. We take the average F1-score of ten projects as the evaluation metric. The experimental results are shown in Fig. 2.

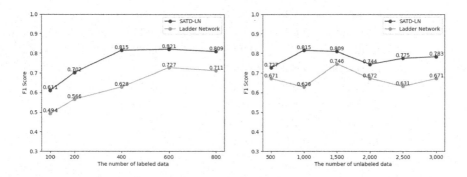

Fig. 2. The left-hand-side of Fig. 2 is the F1-score obtained by two models against the number of labeled data. The right-hand-side of Fig. 2 is the F1-score obtained by two models against the number of unlabeled data.

Result: As can be seen from Fig. 2, when the number of labeled data reaches 400, the F1-score of the SATD-LN model exceeds 0.8. After that, as the amount of labeled data increases, the F1-score stabilizes above 80%. Our SATD-LN model, which has been modified from the original model, has an average increase of 20.84% in the F1-score in the identification of SATD. After the amount of labeled data reaches 400, the reason for the decrease in improvement may be that when first adding labeled data to the model, the model gets a lot of information in the initial stage. However, as the amount of labeled data increases, diminishing returns are observed as the model has already captured most data features.

5.3   The Performance Against the Number of Unlabeled Data

We conducted an experiment to examine the performance of SATD-LN and Ladder Network against the number of unlabeled data. In the previous section of the research, we found that when the labeled data reached 400, Adding labeled data did not significantly improve the classification performance, so we chose 400 labeled data as our fixed value and studied the classification performance of SATD-LN and Ladder Network when the labeled data volume is 500, 1,000, 1,500, 2,000, 2,500, and 3,000, respectively. The results are shown in Fig. 2.

Result: We can see from Fig. 2 that when the amount of labeled data is 400, the performance of the model decreases after the number of unlabeled data exceeds 1,500, and the Ladder Network drops more than SATD-LN. The F1-score of the SATD-LN model has increased by an average of 16.13% in the identification of SATD. This shows that the use of SATD-LN for SATD identification

Table 5. F1-score of our method and three semi-supervised learning methods for identifying SATD on ten projects.

Project	SATD-LN	VAT	MixText	Pseudo Labeling
Apache Ant	**0.730**	0.665	0.596	0.671
ArgoUML	**0.911**	0.882	0.844	0.784
Columba	**0.859**	0.801	0.718	0.775
Eclipse	**0.728**	0.693	0.602	0.706
Hibernate	**0.877**	0.864	0.810	0.858
JEdit	**0.778**	0.694	0.662	0.732
JFreeChart	**0.829**	0.749	0.719	0.779
JMeter	**0.829**	0.779	0.745	0.773
JRuby	**0.849**	0.834	0.760	0.821
SQuirreL	**0.755**	0.708	0.668	0.593
Average	**0.815**	0.767	0.712	0.749
Improved	-	6.26%	14.47%	8.81%

needs to select an appropriate ratio of labeled data and unlabeled data. Increasing the amount of unlabeled data may introduce more noise due to different feature distributions. When the labeled data is 400, unlabeled data exceeding 1500 will affect the model performance.

5.4   The Effectiveness of SATD-LN Against Other Semi-supervised Learning Methods

In this section, we choose three additional semi-supervised learning methods (VAT, MixText and Pseudo Labeling) to compare the results of our model. Four models were trained using 400 labeled data and 1,000 unlabeled data randomly drawn from the dataset. The data is randomly selected five times, and the model training results are averaged. The evaluation metric is the F1 score of the ten projects. The experimental results are shown in Table 5.

Result: It can be seen from Table 5 that SATD-LN is superior to VAT, MixText, and Pseudo Labeling in identifying SATD, and the improvement of F1-score is 6.26%, 14.47%, and 8.81%, respectively. One possible reason is that Ladder Network focuses on hierarchical feature learning methods. Combined with the pre-training model BERT, it can perform multi-layer feature extraction, which helps the model capture features of different layers.

5.5   The Effectiveness of SATD-LN Against Other Supervised Learning Methods

Since the data used by the model in this paper is far less than that used in the related supervised learning model, we refer to the work of Li et al. [8] for the training results of the supervised learning model in this section. We use the same experimental method and settings as in the previous subsection to compare the F1-scores of different models. The experimental results are shown in Table 6.

Result: It can be seen from Table 6 that SATD-LN is superior to GNNSI, Transformer, CNN, RNN, and GCN in identifying SATD, and the improvement of F1-score is 2.39%, 16.95%, 21.86%, 22.59%, and 97.57%, respectively. Our model is better than the average F1-score of GNNSI, which has the best effect before, and our model only needs to use 400 labeled data. We introduce the Ladder Network model structure so that we can reduce the need for labeled data. At the same time, we added the pre-training model. I think the pre-training model has a significant improvement in performance, which can help us extract effective information even on a small number of samples.

Table 6. F1-score of our method and five supervised learning methods in identifying SATD on ten projects.

Project	SATD-LN	GNNSI	Transformer	CNN	RNN	GCN
Apache Ant	**0.730**	0.658	0.528	0.528	0.500	0.240
ArgoUML	**0.911**	0.894	0.853	0.845	0.817	0.558
Columba	0.859	**0.930**	0.785	0.797	0.715	0.438
Eclipse	**0.728**	0.685	0.530	0.444	0.382	0.282
Hibernate	**0.877**	0.866	0.810	0.756	0.792	0.562
JEdit	**0.778**	0.616	0.418	0.464	0.413	0.368
JFreeChart	**0.829**	0.778	0.707	0.597	0.615	0.341
JMeter	0.829	**0.864**	0.811	0.780	0.781	0.481
JRuby	0.849	**0.901**	0.864	0.810	0.789	0.520
SQuirreL	0.755	**0.758**	0.658	0.656	0.635	0.327
Average	**0.815**	0.795	0.696	0.668	0.644	0.412
Improved	-	2.39%	16.95%	21.86%	22.59%	97.57%

6   Conclusions

In this paper, we are the first to introduce semi-supervised learning to SATD identification. We combined the Ladder Network with pre-training models to propose a semi-supervised learning method applied to SATD identification, and the results showed that our method is effective. Our model achieved better results on far less data than supervised learning. In the future, we plan to explore adding more unlabeled data to enhance the performance of identification SATD.

Acknowledgements. We would like to thank anonymous reviewers for their helpful comments and suggestions. This work is supported by Support Center for Advanced Telecommunications Technology Research (SCAT), JKA, Kajima Foundation's Support Program, JSPS KAKENHI (No. 21K12026, 22K12146 and 23H03402), and National Natural Science Foundation of China (No. 62372145).

References

1. Cunningham, W.: The WyCash portfolio management system. ACM SIGPLAN OOPS Messenger **4**(2), 29–30 (1992)
2. Zazworka, N., Shaw, M.A., Shull, F., Seaman, C.: Investigating the impact of design debt on software quality. In: Proceedings of the 2nd Workshop on Managing Technical Debt, pp. 17–23. Association for Computing Machinery, USA (2011)
3. Potdar, A., Shihab, E.: An exploratory study on self-admitted technical debt. In: Proceedings of the 2014 IEEE International Conference on Software Maintenance and Evolution, pp. 91–100. IEEE Computer Society (2014)
4. Ren, X., Xing, Z., Xia, X., Lo, D., Wang, X., Grundy, J.: Neural network-based detection of self-admitted technical debt: from performance to explainability. ACM Trans. Softw. Eng. Methodol. (TOSEM) **4**(2), 1–45 (2019)

5. Santos, R.M., Junior, M.C.R., de Mendonça Neto, M.G.: Self-admitted technical debt classification using LSTM neural network. In: Latifi, S. (ed.) 17th International Conference on Information Technology–New Generations (ITNG 2020). AISC, vol. 1134, pp. 679–685. Springer, Cham (2020). https://doi.org/10.1007/978-3-030-43020-7_93

6. Santos, R., Santos, I., Júnior, M.C., et al.: Long term-short memory neural networks and Word2Vec for self-admitted technical debt detection. In: ICEIS, vol. 2, pp. 157–165 (2020)

7. Yu, D., Wang, L., Chen, X., Chen, J.: Using BiLSTM with attention mechanism to automatically detect self-admitted technical debt. Front. Comp. Sci. **15**(4), 1–12 (2021). https://doi.org/10.1007/s11704-020-9281-z

8. Li, H., Qu, Y., Liu, Y., Chen, R., Ai, J., Guo, S.: Self-admitted technical debt detection by learning its comprehensive semantics via graph neural networks. Softw. Pract. Exp. **52**(10), 2152–2176(2022)

9. de Freitsa Farias, M.A., de Mendonça Neto, M.G., da Silva, A.B., Spínola, R.O.: A contextualized vocabulary model for identifying technical debt on code comments. In: 2015 IEEE 7th International Workshop on Managing Technical Debt, pp. 25–32(2015)

10. Guo, Z., Liu, S., Liu, J., et al.: MAT: a simple yet strong baseline for identifying self-admitted technical debt. arXiv preprint arXiv:1910.13238 (2019)

11. Rasmus, A., Berglund, M., Honkala, M., et al.: Semi-supervised learning with ladder networks. In: Proceedings of the 28th International Conference on Neural Information Processing Systems, pp. 3546–3554. MIT Press, Montreal, Canada (2015)

12. Muangkammuen, P., Fukumoto, F., Li, J., et al.: Exploiting labeled and unlabeled data via transformer fine-tuning for peer-review score prediction. In: Findings of the Association for Computational Linguistics: EMNLP 2022, pp. 2233–2240. Association for Computational Linguistics (2022)

13. Devlin, J., Chang, M.W., Lee, K., et al.: BERT: pre-training of deep bidirectional transformers for language understanding. In: Proceedings of the 2019 Conference of the North American Chapter of the Association for Computational Linguistics: Human Language Technologies, Volume 1 (Long and Short Papers), pp. 4171–4186. Association for Computational Linguistics (2018)

14. da Silva, M., Shihab, E., Tsantalis, N.: Using natural language processing to automatically detect self-admitted technical debt. IEEE Trans. Softw. Eng., 1044–1062(2017)

15. Guo, Z., Liu, S., Liu, J., et al.: How far have we progressed in identifying self-admitted technical debts? A comprehensive empirical study. ACM Trans. Softw. Eng. Methodol. (TOSEM) **30**(4), 1–56 (2021)

16. Miyato, T., Dai, A.M., Goodfellow, I.J.: Adversarial training methods for semi-supervised text classification. In: Proceedings of the 5th International Conference on Learning Representations, ICLR 2017, Toulon, France, pp. 24–26 (2017)

17. Chen, J., Yang, Z., Yang, D.: MixText: linguistically-Informed interpolation of hidden space for semi-supervised text classification. In: Proceedings of the 58th Annual Meeting of the Association for Computational Linguistics, pp. 2147–2157. Association for Computational Linguistics (2020)

18. Rizve, M.N., Duarte, K., Rawat, Y.S., Shah, M.: In defense of pseudo-labeling: an uncertainty-aware pseudo-label selection framework for semi-supervised learning. arXiv preprint arXiv:2101.06329 (2021)

19. Vaswani, A., Shazeer, N., Parmar, N., et al.: Attention is all you need. In: Advances in Neural Information Processing Systems, vol. 30, pp. 5998–6008 (2017)

20. Kim, Y.: Convolutional neural networks for sentence classification. In: Proceedings of the 2014 Conference on Empirical Methods in Natural Language Processing, pp. 1746–1751. Association for Computational Linguistics (2014)
21. Liu, P., Qiu, X., Huang, X.: Recurrent neural network for text classification with multi-task learning. In: Proceedings of the 25th International Joint Conference on Artificial Intelligence, pp. 2873–2879 (2016)
22. Yao, L., Mao, C., Luo, Y.: Graph convolutional networks for text classification. In: Proceedings of the AAAI Conference on Artificial Intelligence, vol. 33, no. 1, pp. 7370–7377 (2019)

NDGR: A Noise Divide and Guided Re-labeling Framework for Distantly Supervised Relation Extraction

Zheyu Shi[1,2], Ying Mao[1,2], Lishun Wang[1,2], Hangcheng Li[1,2], Yong Zhong[1,2], and Xiaolin Qin[1,2(✉)]

[1] Chengdu Institute of Computer Applications, Chinese Academy of Sciences, Chengdu 610041, China
{shizheyu21,maoying19}@mails.ucas.edu.cn, qinxl2001@126.com
[2] University of Chinese Academy of Sciences, Beijing 100049, China

Abstract. Distant supervision (DS) is widely used in relation extraction to reduce the cost of annotation but suffers from noisy instances. Current approaches typically involve selecting reliable instances from the DS-built dataset for model training. However, these approaches often lead to the inclusion of numerous noisy instances or the disregard of a substantial number of valuable instances. In this paper, we propose NDGR, a novel training framework for sentence-level distantly supervised relation extraction. Initially, NDGR partitions the noisy data from the DS-built dataset by employing a Gaussian Mixture Model (GMM) to model the loss distribution. Afterwards, we utilize a guided label generation strategy to generate high-quality pseudo-labels for noisy data. By iteratively executing the processes of noise division and guided label generation, NDGR helps refine the noisy DS-built dataset and enhance the overall performance. Our method has been extensively evaluated on commonly used benchmarks, and the results demonstrate its substantial improvements in both sentence-level evaluation and noise reduction.

Keywords: Distantly Supervised · Relation Extraction · Label Noise

1 Introduction

Relation extraction (RE) is a fundamental task in the field of Information Extraction (IE), which aims at extracting structured relations between named entity pairs from unstructured text. Most existing methods approach this task by employing supervised training of neural networks, requiring a significant amount of manually labeled data. It is widely acknowledged that data annotation is a

Supported by Sichuan Science and Technology Program (2019ZDZX0006, 2020YFQ0056), Science and Technology Service Network Initiative (KFJ-STS-QYZD-2021-21-001), and the Talents by Sichuan provincial Party Committee Organization Department.

laborious and time-consuming task. In order to address this challenge, Mintz et al. [19] introduced Distant Supervision (DS), an approach that automatically annotates textual data by aligning relation facts extracted from knowledge graphs with the unlabeled corpus. Regrettably, this annotation paradigm inevitably leads to a problem of noise. Hence, there is a need to explore de-noise DSRE methods to minimize the impact of noisy instances.

Currently, there exist two primary approaches for reducing noise in DSRE: the bag-level method and the sentence-level method. The bag-level methods [12,16,25,28,29] are based on Multi-Instance Learning (MIL), both the training and testing processes are performed at the bag-level. While bag-level approaches are effective in mitigating the influence of noisy data, they do not assign specific labels to each sentence within the bag. Additionally, these approaches overlook cases where all sentences in the bag are false positive samples [23]. These limitations hinder the application of RE in downstream tasks that necessitate sentence-level relation types. Therefore, over the past few years, there has been a growing interest in sentence-level DSRE methods. Most existing sentence-level DSRE methods [4,8,11,22,31] employ adversarial learning, reinforcement learning, or frequent patterns to filter out noisy data. Although these methods are effective at handling noisy data, they have certain limitations, including reliance on prior knowledge, subjective sample construction, and accessing external data.

In this paper, we propose NDGR, a training framework for sentence-level DSRE. In contrast to previous methods, our method is independent of prior knowledge or external data. It can automatically identify noisy instances and re-label them during training, thereby refining the dataset and enhancing performance. Specifically, NDGR first divides the noisy instances from DS built data by modeling the loss distribution with a GMM. Since noisy instances contain valuable information, we consider them as unlabeled data and employ a guided label generation strategy to produce high-quality pseudo-labels for the purpose of transforming them into training data. Although the above-mentioned method can improve performance, it fails to fully exploit the potential of each component. Hence, we further design an iterative training algorithm to fully refine the DS-built dataset through iterative execution of noise divide and guided label generation.

The main contributions of this paper are as follows:

- We propose the use of the Gaussian Mixture Model to model the data loss distribution for sentence-level distant supervised RE, which effectively separates noisy data from DS-built data.
- We design a guided label generation strategy, with the aim of generating high-quality pseudo labels to avoid the gradual drift problem in the distribution of sentence features.
- We develop NDGR, a sentence-level DSRE training framework, which combines a noise division, guided label generation, and iterative training to refine DS-built data.

- Our proposed method makes a great improvement over previous sentence-level DSRE methods on widely used datasets, not only relation extraction ability but also noise filtering ability.

2 Related Work

To address the issue of insufficient annotated data in relation extraction, Mintz et al. [19] firstly align unlabeled text corpus with structured data to automatically annotate data. While this method has the capability to autonomously label data, it is bound to engender the issue of wrong labeling. To minimize the impact of mislabeled data, Riedel et al. [25] relaxes the basic assumption of DS to the At-Least-One assumption and applied Multi-Instance Learning [10,25] to the task of DSRE. In MIL, all sentences with the same entity pair are put into a bag, and assumes that at least one sentence in the bag expresses the relation. Existing bag-level approaches have mainly focused on mitigating the impact of potentially noisy sentences within the bag. Some methods [9,14,16,28] utilize the attention mechanism to assign different weights to the sentences in the bag. These approaches aim to enhance the impact of accurate sentences while mitigating the influence of erroneous ones. Other ways involve using reinforcement learning or adversarial learning [8,24,26,29] to select clean sentences from the bag to train the model. Nevertheless, recent research [4] indicates that bag-level DSRE methods have a limited effect on sentence-level prediction. Besides, bag-level methods are unable to assign a specific sentence label to each sentence in the bag and disregard the fact that all sentences in the bags are noisy samples.

Thus, sentence-level distantly supervised relation extraction has received increasing attention in recent years. Sentence-level DSRE methods typically employ sampling strategies to filter noisy data. Jia et al. [11] refined the DS dataset by identifying frequently occurring relation patterns. Ma et al. [18] utilized complementary labels to create negative samples and employed negative training to filter out noisy data. Li et al. [15] incorporated a small amount of external reliable data in their training process through meta-learning. Adjusting the loss function [5,7] is also a commonly used method to reduce the impact of noisy samples. Despite the effectiveness of these methods in handling noisy data, they possess certain potential limitations, such as reliance on prior knowledge, subjectivity in sample construction, and access to external data. Different from previous work, our method iteratively executes noise divide and guided label generation to refine the DS-built data, independent of external data and prior knowledge.

3 Methodology

In this section, we introduce NDGR, a novel training framework for sentence-level DSRE. As shown in Fig. 1, our method comprises three main steps: (1) Divide noisy data from the DS-built dataset by modeling the loss distribution with a GMM (Sect. 3.1); (2) Generate high-quality pseudo-labels for unlabeled

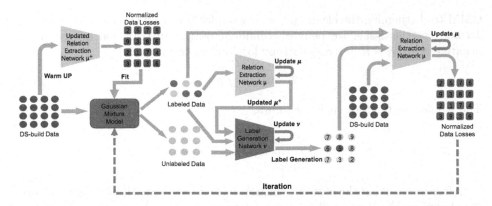

Fig. 1. An overview of the NDGR. There are three main steps: (1) Divide noisy data by employing a GMM to model the distribution of losses; (2) Generate high-quality pseudo labels for unlabeled data by guided label generation strategy; (3) Iterative training to further strengthen performance.

data using a guided label generation strategy (Sect. 3.2); (3) Iterative training based on (1) and (2) to further refine DS-built dataset (Sect. 3.3).

3.1   Noise Division by Loss Modeling

Previous research suggests that deep neural networks exhibit a swifter adaptation to clean data in contrast to noisy data [1], resulting the loss value incurred by clean data is lower than that of noisy data [2]. Hence, we attempt to separate noisy instances from the DS-built dataset by the loss value of each sample. Inspired by Li et al. [13], we aim to get the clean probability of each sample by fitting a Gaussian Mixture Model (GMM) [21] to the sample loss distribution.

Formally, we denote the input DS-built dataset as $D = \{(S, Y\} = \{(s_i, y_i)\}_{i=1}^{N}$, where $y_i \in \{1, ..., C\}$ is the class label for the ith input sentence s_i. In the initial stage of our method, we warm up the Relation Extraction Network (REN) on all DS-built data for a few epochs to get the initial loss distribution. Specifically, for a model with parameters θ, we have:

$$L(\theta) = \sum_{i=1}^{N}(l_i) = \sum_{i=1}^{N} loss(p_i, y_i) \tag{1}$$

$$p_i = M_\theta(s_i) \tag{2}$$

where p_i is the probability distribution of the relation. M_θ consists of the encoder module which converts the input sentence s_i into sentence representation h and the fully connected layer that applies h for classification.

However, we found the network would quickly overfit the noisy data during the warm-up phase, which resulted in most samples having similar normalized loss values close to zero (as shown in Fig. 2(a)). In this case, it's difficult for

GMM to distinguish the clean and noisy samples based on the loss distribution. To address this issue, we penalize confident output distribution by adding a negative entropy $-H$ to cross-entropy loss during training [20].

$$H = -p_i log(p_i) \tag{3}$$

The whole loss function as Eq. 4 shows:

$$L(\theta) = \sum_{i=1}^{N} \{loss(p_i, y_i) + p_i log(p_i)\} \tag{4}$$

After maximizing the entropy, the normalized loss is distributed more evenly (as shown in Fig. 2(b)) and is easier to model by GMM. We then apply the Expectation-Maximum (EM) algorithm to fit a two-component Gaussian Mixture Model [21]. For each sample, we calculate a posterior probability $p(g|l_i)$ as its clean probability c_i, where g is the smaller mean Gaussian component. A threshold Th is established to partition the training data into two separate sets. The samples with c_i greater than Th are assigned to a clean labeled set X, otherwise are assigned to a noisy unlabeled set U.

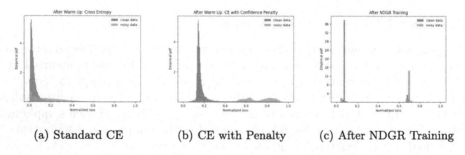

(a) Standard CE   (b) CE with Penalty   (c) After NDGR Training

Fig. 2. Loss distribution when training on Noisy-TACRED. (a) Training using the standard cross-entropy loss function, may lead to overfitting and overly confident predictions. (b) Adding a negative entropy to cross-entropy loss leads to the normalized loss being distributed more evenly. (c) After NDGR training, the clean and noisy data are further separated.

3.2 Guided Label Generation and Exploitation

Following the isolation of noisy data from the dataset, the majority of preceding research employed exclusively clean data for training purposes, thereby overlooking the valuable information embedded in the noisy data. However, proper handling of noisy data can effectively improve the performance of the model. In this section, we introduce the guided label generation strategy, aimed at generating high-quality pseudo-labels for noisy data.

To avoid the gradual drift in the distribution of sentence features caused by the noise present in the generated pseudo labels [17], we construct two networks: the Relation Extraction Network (REN) which is trained to extract relations from unstructured text and the Label Generation Network (LGN) which has the same architecture as REN but is trained separately to generate pseudo labels for the unlabeled data.

To distinguish, we denote the parameters of REN as μ and the parameters of LGN as ν. Using the updated REN as a reference, we let the LGN learn to evaluate the quality of the generated labels. To optimize the ν, we adopt the loss function as follows:

$$L(\nu) = \sum_{i=1}^{N} (loss(M_{\mu^+}(s_i), y_i) + W loss(M_\nu(s_i), y_i)) \tag{5}$$

where $W \in [0, 1]$ is a manually set hyperparameter, μ^+ denotes the parameters of the REN after a gradient update based on the loss function defined in Eq. 4. In this way, we calculate the loss value and update ν using the updated parameters μ^+. This can aid LGN in acquiring a deeper understanding of the training procedure employed by REN. To prevent the accumulation of errors caused by noise in the generated labels during training, LGN is trained on labeled set X.

After optimizing the LGN for a few epochs, we generate pseudo labels for the unlabeled set U. For each sample u_i in U, the generated pseudo label is defined:

$$label = \arg\max(M_{\nu^+}(u_i)) \tag{6}$$

where ν^+ is the updated parameters of the LGN.

The relation associated with the highest probability after softmax is considered as the pseudo-label. We utilize these generated pseudo-labels by amalgamating the re-labeled set R with the labeled set X to form an enhanced dataset. Subsequently, this dataset is employed to retrain REN, leading to performance improvement.

3.3   Iterative Training Algorithm

While dividing noisy data by modeling the loss distribution with a GMM and implementing the guided label generation strategy for re-labeling the noisy data can refine the dataset and enhance performance, it fails to fully exploit the potential of each component. Therefore, we employ iterative training to further enhance performance.

As shown in Fig. 1, for each iteration, we firstly divide the DS-built into labeled data and unlabeled data by modeling the data loss distribution with GMM (before the first iteration, we warm up the model to get the initial normalized loss distribution). Following M epochs of REN training using labeled data, we leverage the updated REN as a reference to optimize LGN. Subsequently, the updated LGN is employed for generating pseudo-labels pertaining to the unlabeled data. By amalgamating the labeled and re-labeled data, a novel refined dataset is formulated. Prior to retraining REN with the refined dataset, a model re-initialization step is executed to avert overfitting. This process ensures that the models are optimized using a high-quality dataset and incorporates randomness, thereby improving the robustness of our method. Finally, we input the origin DS-build data into the updated REN to obtain the new normalized loss distribution for fitting the GMM and perform re-initialization of both the REN and LGN to enter the next iteration. Figure 2(c) shows the loss distribution after NDGR training. As seen, there is a substantial margin in the loss values between most of the noisy data and clean data. Most of the noisy data has been successfully separated, with only an acceptable amount of clean data being misclassified, demonstrating the robust de-noising ability of NDGR.

4   Experiments and Analysis

To evaluate the efficacy of the proposed method, we divided the experiment into two parts and conducted tests on two datasets: (1) The first part is to validate the effectiveness of our proposed method at the sentence-level evaluation. Numerous previous DSRE approaches employ a held-out evaluation, where both the training and test sets are constructed using the DS method. According to the study, Gao et al. [6] suggest that using a held-out test set cannot accurately demonstrate the model's performance. Hence, we utilized a manually-labeled test set to evaluate the model. (2) In the second part, a series of experiments are designed to evaluate the efficacy of each component in NDGR. Since the DS-build dataset cannot label whether this instance is mislabeled, we construct a noisy dataset called Noisy-TACRED from the manually labeled dataset.

4.1   Datasets

We evaluate our method on two widely-used datasets: the NYT dataset and the Noisy-TACRED dataset.

NYT: Riedel et al. [25] constructed this dataset by aligning the New York Times corpus with entity-relationship triples from Freebase. The original training and test sets are both established using the DS method, encompassing noisy data. For a more precise evaluation, we employ the original training set alongside a manually annotated sentence-level test set [11].

Noisy-TACRED: The original TACRED dataset, constructed by Zhang et al. [33], comprises 80% of instances labeled as "NA". The "NA" rate is similar to the NYT dataset which is constructed by the DS method, hence analysis on this dataset is more reliable. To create the Noisy-TACRED dataset, we select noisy instances randomly with a noisy ratio of 30%. For each noisy instance, a noisy label is assigned by randomly selecting a label from a complementary class. The selection probability of a label is determined by its class frequency, this approach helps to preserve the original distribution of data.

4.2   Baseline Models

We compare our method with multiple strong baselines as follows:

PCNN [29]: A *bag-level* method with multi-instance learning to address the wrong label problem.

PCNN+SelATT [16]: A *bag-level* de-noise method uses the attention mechanisms to reduce the impact of noisy data.

PCNN+RA_BAG_ATT [27]: A *bag-level* method that utilizes the inter-bag and intra-bag attention mechanisms to alleviate noisy instances.

CNN+RL_1 [24]: A *bag-level* method that applies reinforcement learning to recognize the false positive samples and then the filtered data reallocated as negative samples.

CNN+RL_2 [4]: A *sentence-level* method which incorporates reinforcement learning to jointly train a RE model for relation extraction and a selector to filter the potential noisy samples.

ARNOR [11]: A *sentence-level* method that selects reliable instances by rewarding high attention scores on specific patterns.

SENT(BiLSTM) [18]: A *sentence-level* DSRE method filters the noisy data by negative training and performs a re-label process to transform the noisy data into useful data.

CNN [30], **BiLSTM** [32] and **BERT** [3] are widely-used models for *sentence-level* relation extraction without denoising method.

4.3   Implementation Details

Our proposed method employs the BiLSTM as the sentence encoder. 50-dimensional GloVe vectors [16] are used as word embeddings during training. Furthermore, we incorporate 50-dimensional randomly initialized position and entity type embeddings throughout all training phases. The hidden size of the BiLSTM is set to 256. It is optimized using the Adam optimizer with a learning rate of 1e−5. The weight parameter W in Eq. 4 is assigned a value of 5e−1.

The hyperparameters are tuned by performing a grid search on the validation set. When training on the NYT dataset, we first warm up the REN using all DS-built data for 4 epochs. We then perform a total of 8 iterations, with each iteration involving training the model for 15 epochs. The data-divide threshold is set to $Th = 0.7$. During the training phase on the Noisy-TACRED, the REN is initially warmed up using all DS-built data for 45 epochs. Following this warm-up phase, the model goes through 8 iterations, where each iteration involves 50 epochs of training. The training procedure utilizes a data-divide threshold Th of 0.6 and sets the learning rate to 5e−4.

4.4   Sentence-Level Evaluation

Table 1 shows the results of our method and other baseline models on sentence-level evaluation. Consistent with previous methods [11,15], we calculate Micro-Precision (Prec.), Micro-Recall (Rec.), and Micro-F1 (F1) to evaluate the effectiveness of our approach. According to the results, we can observe that: (1)

Table 1. Results of our method and other baseline models on sentence-level evaluation. The first part of the table is ordinary RE methods without denoising and the second part of the table is distant RE methods. The results with "*" are bag-level methods and the results with "†" are sentence-level methods.

Method	Dev			Test		
	Prec.	Rec.	F1	Prec.	Rec.	F1
CNN†	38.32	65.22	48.28	37.75	64.54	46.01
PCNN*	36.09	63.66	46.07	36.06	64.86	46.35
BiLSTM†	36.71	66.46	47.29	35.52	67.41	46.53
BERT†	34.78	65.17	45.35	36.19	70.44	47.81
PCNN+SelATT*	46.01	30.43	36.64	45.51	30.03	36.15
PCNN+RA_BAG_ATT*	49.84	46.90	48.33	56.76	50.60	53.50
CNN+RL_1*	37.71	52.66	43.95	39.41	61.61	48.07
CNN+RL_2†	40.00	59.17	47.73	40.23	63.78	49.34
ARNOR†	62.45	58.51	60.36	65.23	56.79	60.90
SENT (BiLSTM)†	66.71	57.27	61.63	71.22	59.75	64.99
NDGR (BiLSTM)	**74.34**	**58.46**	**65.45**	**78.30**	**56.97**	**65.95**

When trained with DS-built data without de-noise, all baseline models performed poorly. Even the highly acclaimed pre-trained language model BERT, renowned for its superior performance in sentence-level relation extraction tasks on clean datasets, demonstrated subpar results. This phenomenon underscores the substantial impact of noisy samples in the dataset on model training, particularly for pre-trained language models, which are prone to overfitting such data. (2) The bag-level methods demonstrate poor performance in sentence-level evaluation, indicating their unsuitability for downstream tasks that require precise sentence labels. Therefore, it is imperative to explore sentence-level methods for DSRE. (3) The proposed NDGR method achieves a significant improvement over previous sentence-level de-noise methods. Our implementation utilizing BiLSTM as the sentence encoder results in a 0.96% improvement in the F1 score compared to SENT. Additionally, it exhibits significantly higher precision while maintaining comparable recall. These outcomes highlight the effectiveness of our strategy for data division and guided label generation.

4.5 Analysing on Noisy-TACRED

In this section, we analyze the efficacy of noise division and label generation process on the Noisy-TACRED dataset.

Table 2. Model performance on clean-TACRED and noisy-TACRED.

	Method	Prec.	Rec.	F1
Clean TACRED	BiLSTM+ATT	67.7	63.2	65.4
	BiLSTM	61.4	61.7	61.5
Noisy TACRED	BiLSTM+ATT	32.8	43.8	37.5
	BiLSTM	37.8	45.5	41.3
	NDGR (BiLSTM)	**86.4**	**43.3**	**57.7**

Evaluation on Noisy-TACRED: We trained in Clean-TACRED and Noisy-TACRED respectively, and the results are shown in Table 2. Comparing the results on two datasets, we can find the performance of baseline models degraded significantly, the F1 value of BiLSTM+ATT decreased by 27.9, while the F1 value of BiLSTM decreased by 20.2. By employing our proposed NDGR on the noisy data, the BiLSTM model demonstrates comparable performance to the model trained on clean data. This finding indicates that our methodology effectively mitigates the impact of incorrectly labeled data in the DS-built dataset.

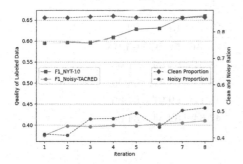

Fig. 3. Experimental Details of Divide Data by Loss Modeling

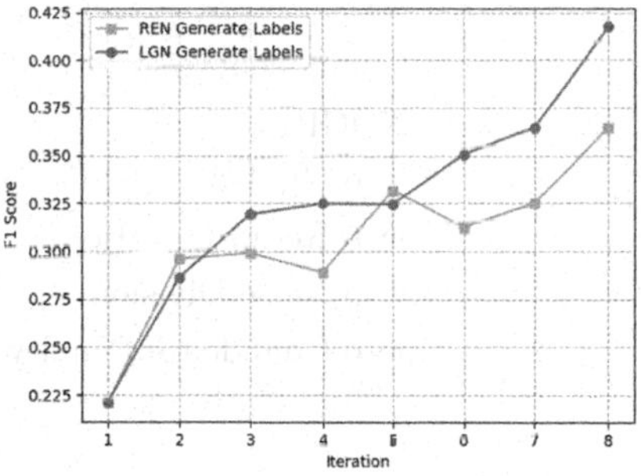

Fig. 4. The quality of the pseudo labels generated by REN and LGN.

Effects of Divide Data by Loss Modeling: As described in Sect. 3.1, we employ GMM to model the loss distribution, and subsequently leverage the posterior probability to partition the noisy data within the dataset constructed by DS. To demonstrate the efficacy of the loss modeling by GMM in distinguishing between clean and noisy data. We first evaluated the quality of the labeled data, we trained the model only with the labeled data and ignored the unlabeled data on both NYT and Noisy-TACRED. Additionally, we compute the ratio of clean data within the labeled set and noisy data within the unlabeled set on Noisy-TACRED.

The results are shown in Fig. 3, we can observe: 1) The F1 score exhibits a gradual increase with each iteration for both the NYT and Noisy-TACRED datasets, suggesting an improvement in the quality of the labeled data. 2) As the iteration progressed, the ratio of clean data in the labeled set and the ratio of noise data in the unlabeled set both increased. The proportion of clean data can reach approximately 85%, while the proportion of noisy data amounts to around 51%. These observations affirm the effective noise filtering capability of GMM-based loss modeling within the dataset constructed using DS.

Effects of Guided Pseudo Label Generation: As described in Sect. 3.2, once the noisy data has been separated from the training dataset, we employ the guided label generation strategy for converting the unlabeled data into valuable training data. To verify the effectiveness of this strategy, we separately used REN and LGN to generate pseudo labels and calculated the F1 score between the generated labels and the original labels of the TACRED, the results are shown in Fig. 4. As seen, the pseudo-labels generated directly by REN exhibit inferior performance compared to those generated by LGN. This demonstrates that the label generation strategy we employ can reduce noise in generating labels and improve performance.

Table 3. Ablation study on NYT dataset

Components	Prec.	Rec.	F1
NDGR	**78.30**	**56.97**	**65.95**
w/o Guided Label Generation	58.31	64.09	61.06
w/o Re-initialization	55.11	63.47	58.99
w/o Noise Division	45.11	69.97	54.85
w/o Confidence Penalty	47.64	62.54	54.08

4.6   Ablation Study

We conduct an ablation study to demonstrate the contribution of each component in our proposed method on the NYT dataset. We specifically assess the performance by removing certain components, including guided label generation, re-initialization, noise divide, and confidence penalty. The results are shown in Table 3, we can observe that: 1) Without the guided label generation strategy, instead of employing LGN for label generation, we employ REN to directly generate labels. The presence of noise in the generated labels results in error accumulation during iterations, causing a progressive drift in the distribution of sentence features, thereby impacting the overall system performance. 2) Re-initialization has a great contribution to the performance. In the label generation phase, despite utilizing labeled data for training REN and LGN with the intention of mitigating the impact of noise, it is inevitable that certain noise data will become incorporated and the models will adapt to them. With re-initialization, REN will initialize the overfitting parameters and retrain on the refined dataset, thus contributing to better performance. 3) Noise division significantly impacts the performance. Without noise division, we use the original DS-built data to optimize the REN and LGN, then generate pseudo-labels for all DS-built data. However, due to the presence of noise during the training process, the quality of these labels diminishes, resulting in inferior performance. Moreover, as the iterations progress, the performance further deteriorates. 4) Confidence penalty contributes a lot to the performance. In the absence of the confidence penalty, most of the normalized loss values are comparable and tend toward zero. This makes it challenging for the GMM to effectively filter out noisy samples based on the distribution of losses. The presence of numerous mislabeled data has significantly affected the subsequent label generation process, resulting in a decline in the quality of the newly constructed training dataset.

5 Conclusion

In this paper, we propose NDGR, a novel sentence-level DSRE training framework that incorporates noise division, guided label generation, and iterative training. Specifically, NDGR first separates noisy data from the training dataset by modeling the data loss distribution with a GMM. Next, we assign pseudo-labels for unlabeled data using a guided label generation strategy to reduce the noise in the generated pseudo-labels. Through iterative execution of noise division and guided label generation, NDGR helps re-fine the noisy DS-built data and enhance the performance. Extensive experiments on widely-used benchmarks have demonstrated that our method has significant improvement in sentence-level relation extraction and de-noise effect.

Acknowledgements. This work was supported by Sichuan Science and Technology Program (2019ZDZX0006, 2020YFQ0056), Science and Technology Service Network Initiative (KFJ-STS-QYZD-2021-21-001), and the Talents by Sichuan provincial Party Committee Organization Department.

References

1. Arpit, D., et al.: A closer look at memorization in deep networks. In: International Conference on Machine Learning, pp. 233–242. PMLR (2017)
2. Chen, P., Liao, B.B., Chen, G., Zhang, S.: Understanding and utilizing deep neural networks trained with noisy labels. In: International Conference on Machine Learning, pp. 1062–1070. PMLR (2019)
3. Devlin, J., Chang, M.W., Lee, K., Toutanova, K.: BERT: pre-training of deep bidirectional transformers for language understanding. In: Proceedings of the 2019 Conference of the North American Chapter of the Association for Computational Linguistics: Human Language Technologies, Volume 1 (Long and Short Papers), Minneapolis, Minnesota, June 2019, pp. 4171–4186. Association for Computational Linguistics (2019). https://doi.org/10.18653/v1/N19-1423. https://aclanthology.org/N19-1423

4. Feng, J., Huang, M., Li, Z., Yang, Y., Zhu, X.: Reinforcement learning for relation classification from noisy data (2018)
5. Fu, B., Peng, Y., Qin, X.: Learning with noisy labels via logit adjustment based on gradient prior method. Appl. Intell. **53**, 24393–24406 (2023). https://doi.org/10.1007/s10489-023-04609-1
6. Gao, T., et al.: Manual evaluation matters: reviewing test protocols of distantly supervised relation extraction. arXiv preprint arXiv:2105.09543 (2021)
7. Ghosh, A., Kumar, H., Sastry, P.S.: Robust loss functions under label noise for deep neural networks. In: Proceedings of the AAAI Conference on Artificial Intelligence, vol. 31 (2017)
8. Han, X., Liu, Z., Sun, M.: Denoising distant supervision for relation extraction via instance-level adversarial training. arXiv preprint arXiv:1805.10959 (2018)
9. Han, X., Liu, Z., Sun, M.: Neural knowledge acquisition via mutual attention between knowledge graph and text. In: Proceedings of the AAAI Conference on Artificial Intelligence, vol. 32 (2018)
10. Hoffmann, R., Zhang, C., Ling, X., Zettlemoyer, L., Weld, D.S.: Knowledge-based weak supervision for information extraction of overlapping relations. In: Proceedings of the 49th Annual Meeting of the Association for Computational Linguistics: Human Language Technologies, Portland, Oregon, USA, June 2011, pp. 541–550. Association for Computational Linguistics (2011). https://aclanthology.org/P11-1055
11. Jia, W., Dai, D., Xiao, X., Wu, H.: ARNOR: attention regularization based noise reduction for distant supervision relation classification. In: Meeting of the Association for Computational Linguistics (2019)
12. Jiang, X., Wang, Q., Li, P., Wang, B.: Relation extraction with multi-instance multi-label convolutional neural networks. In: Proceedings of the 26th International Conference on Computational Linguistics: Technical Papers, COLING 2016, pp. 1471–1480 (2016)
13. Li, J., Socher, R., Hoi, S.C.: DivideMix: learning with noisy labels as semi-supervised learning. arXiv preprint arXiv:2002.07394 (2020)
14. Li, Y., et al.: Self-attention enhanced selective gate with entity-aware embedding for distantly supervised relation extraction. In: Proceedings of the AAAI Conference on Artificial Intelligence, vol. 34, pp. 8269–8276 (2020)
15. Li, Z., et al.: Meta-learning for neural relation classification with distant supervision. In: Proceedings of the 29th ACM International Conference on Information & Knowledge Management, pp. 815–824 (2020)
16. Lin, Y., Shen, S., Liu, Z., Luan, H., Sun, M.: Neural relation extraction with selective attention over instances. In: Proceedings of the 54th Annual Meeting of the Association for Computational Linguistics (Volume 1: Long Papers), pp. 2124–2133 (2016)
17. Liu, S., Davison, A., Johns, E.: Self-supervised generalisation with meta auxiliary learning. In: Advances in Neural Information Processing Systems, vol. 32 (2019)
18. Ma, R., Gui, T., Li, L., Zhang, Q., Zhou, Y., Huang, X.: SENT: sentence-level distant relation extraction via negative training. arXiv preprint arXiv:2106.11566 (2021)
19. Mintz, M., Bills, S., Snow, R., Jurafsky, D.: Distant supervision for relation extraction without labeled data. In: Proceedings of the Joint Conference of the 47th Annual Meeting of the ACL and the 4th International Joint Conference on Natural Language Processing of the AFNLP, pp. 1003–1011 (2009)

20. Pereyra, G., Tucker, G., Chorowski, J., Kaiser, Ł., Hinton, G.: Regularizing neural networks by penalizing confident output distributions. arXiv preprint arXiv:1701.06548 (2017)
21. Permuter, H., Francos, J., Jermyn, I.: A study of Gaussian mixture models of color and texture features for image classification and segmentation. Pattern Recogn. **39**(4), 695–706 (2006)
22. Qin, P., Xu, W., Wang, W.Y.: Robust distant supervision relation extraction via deep reinforcement learning. In: Meeting of the Association for Computational Linguistics (2018)
23. Qin, P., Xu, W., Wang, W.Y.: DSGAN: generative adversarial training for distant supervision relation extraction. arXiv preprint arXiv:1805.09929 (2018)
24. Qin, P., Xu, W., Wang, W.Y.: Robust distant supervision relation extraction via deep reinforcement learning. arXiv preprint arXiv:1805.09927 (2018)
25. Riedel, S., Yao, L., McCallum, A.: Modeling relations and their mentions without labeled text. In: Balcázar, J.L., Bonchi, F., Gionis, A., Sebag, M. (eds.) ECML PKDD 2010. LNCS (LNAI), vol. 6323, pp. 148–163. Springer, Heidelberg (2010). https://doi.org/10.1007/978-3-642-15939-8_10
26. Shang, Y., Huang, H.Y., Mao, X.L., Sun, X., Wei, W.: Are noisy sentences useless for distant supervised relation extraction? In: Proceedings of the AAAI Conference on Artificial Intelligence, vol. 34, pp. 8799–8806 (2020)
27. Ye, Z.X., Ling, Z.H.: Distant supervision relation extraction with intra-bag and inter-bag attentions. arXiv preprint arXiv:1904.00143 (2019)
28. Yuan, Y., et al.: Cross-relation cross-bag attention for distantly-supervised relation extraction. In: Proceedings of the AAAI Conference on Artificial Intelligence, vol. 33, pp. 419–426 (2019)
29. Zeng, D., Liu, K., Chen, Y., Zhao, J.: Distant supervision for relation extraction via piecewise convolutional neural networks. In: Proceedings of the 2015 Conference on Empirical Methods in Natural Language Processing, pp. 1753–1762 (2015)
30. Zeng, D., Liu, K., Lai, S., Zhou, G., Zhao, J.: Relation classification via convolutional deep neural network. In: Proceedings of the 25th International Conference on Computational Linguistics: Technical Papers, COLING 2014, pp. 2335–2344 (2014)
31. Zeng, X., He, S., Liu, K., Zhao, J.: Large scaled relation extraction with reinforcement learning. In: Proceedings of the AAAI Conference on Artificial Intelligence, vol. 32 (2018)
32. Zhang, S., Zheng, D., Hu, X., Yang, M.: Bidirectional long short-term memory networks for relation classification. In: Proceedings of the 29th Pacific Asia Conference on Language, Information and Computation, pp. 73–78 (2015)
33. Zhang, Y., Zhong, V., Chen, D., Angeli, G., Manning, C.D.: Position-aware attention and supervised data improve slot filling. In: Proceedings of the 2017 Conference on Empirical Methods in Natural Language Processing, EMNLP 2017, pp. 35–45 (2017). https://nlp.stanford.edu/pubs/zhang2017tacred.pdf

Customized Anchors Can Better Fit the Target in Siamese Tracking

Shuqi Pan[1], Canlong Zhang[1,2(✉)], Zhixin Li[1,2], Liaojie Hu[3], and Yanfang Deng[1]

[1] Key Lab of Education Blockchain and Intelligent Technology, Ministry of Education, Guangxi Normal University, Guilin, China
lizx@gxnu.edu.cn, dyf@stu.gxnu.edu.cn
[2] Guangxi Key Lab of Multi-source Information Mining & Security, Guangxi Normal University, Guilin, China
[3] The Experimental High School Attached to Beijing Normal University, Beijing, China

Abstract. Most existing siamese trackers rely on some fixed anchors to estimate the scale and aspect ratio for all targets. However, in real tracking, different targets have different sizes and shapes, these predefined anchors are not enough to cover all possible scales and aspect ratios caused by various movement and deformation, so an adaptive scale and aspect ratio estimation method is expected for robust online tracking. In this paper, a customized anchor generation module is first proposed to estimate the shape of the target and generate customized anchors adapted to the target. Then, through an anchor adaptation module, each anchor information is embed into corresponding feature to learn more discriminative features. Finally, We design a Target-aware feature correlation module to reduce the interference of background information. It takes the region of interest of template as variable template and its central subregion as central template, and then performs global and local correlation operations, respectively. Experiments on benchmarks including OTB100, VOT2019, LaSOT, UAV123, and VOT2018 show that our tracker achieves promising performance.

Keywords: Visual tracking · Feature adaptation · Feature correlation

1 Introduction

As a popular computer vision processing technology, target tracking has been applied widely in security monitoring, intelligent transportation, autonomous driving, human-computer interaction, and other fields. Designing robust target tracking algorithms is still challenging due to occlusion, image blurring, deformation and background interference. Siamese tracking is the most popular tracking framework at present. Bertinetto et al. [1] is the first to apply full convolution in Siamese network to achieve correlation matching between the search image and the template image. After that, many improved siamese trackers have been proposed.

© The Author(s), under exclusive license to Springer Nature Singapore Pte Ltd. 2024
B. Luo et al. (Eds.): ICONIP 2023, CCIS 1969, pp. 112–124, 2024.
https://doi.org/10.1007/978-981-99-8184-7_9

SiamRPN [13] introduced a region proposal network (RPN) into siamese networks. RPN mapped anchors with fixed aspect ratios to image features for classification and regression by convolutional networks. In order to handle image data with different scales and aspect ratios, they need to design anchors based on heuristic prior knowledge. However, in real videos, the scale and aspect ratio of the target varies with the target and the camera, so the fixed anchors will reduce the robustness of the tracker. SiamRPN++ [12] employed a fixed-size template to calculate the similarity between template and search frame. However, this fixed feature region may retain a lot of background information or lose a large amount of foreground information when the tracked target has different scales. Therefore, the tracker will be ineffective against background interference.

To address the above issues. Inspired by [19], we propose a customized anchor proposal subnet. The customized anchor generation module generates customized anchors based on the location and shape of the target, adapting to the scale variation of the target. More flexible and versatile. The anchor adaptation module is introduced to embed the semantic information of the customized anchor into the correlation features. It will transform the features at each indivision in the feature map according to the shape of the anchor so that the feature map adapts to the shape of the anchor.

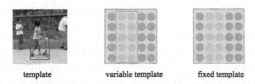

Fig. 1. Ground-truth bounding boxes of the targets are marked by red boxes. The background and foreground features in the template correspond to the blue and gray origin points. The variable template based on the target-aware mechanism is marked by a green box. It acquires the body area of the target under the supervision of the bounding box. The fixed template features cropped out by fixed regions are marked by yellow boxes, which may contain a lot of background information or missing foreground information. (Color figure online)

To reduce the background information in the template, we generate variable templates based on target-aware, as shown in Fig. 1. We designed a Target-aware feature correlation module for calculating correlation of feature. The variable template which can adapt to the target and the search features are subjected to self-attention and cross-attention calculations to obtain global features while fusing depth correlation operations to further capture local spatial features. It effectively suppresses the interference of background information and reduces the semantic ambiguity in the attention mechanism, achieving robust tracking. In summary the contributions of this work are as follows:

– In this work, we propose a siamese tracking network based on customized anchors. Instead of designing fixed scale anchors, customized anchors are gen-

erated from correlation features between template and search frame. It makes the tracker more adaptable to the shape change of the target during tracking.
- We propose a structure of local and non-local feature aggregation for cross-correlation operation, which helps the tracker resist background information interference during tracking.
- The evaluation results on the five datasets show that our tracker significantly improves the performance of the baseline tracker.

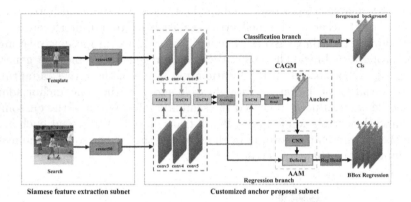

Fig. 2. The network architecture of our framework consists of two main subnets: a siamese feature extraction subnet for feature extraction and a customized anchor proposal subnet. The customized anchor proposal subnet is divided into a customized anchor generation module (CAGM) and an anchor adaptation module (AAM), a classification branch and a regression branch.

2 Proposed Method

2.1 Overall Framework

Figure 2 shows the general framework of our proposed SiamCAFT. We take the template z in the initial frame and the search x in the current frame as the inputs to the network. The features of z and x are extracted by a siamese feature extraction subnet which consisting of two Resnet50 networks sharing weights. The customized anchor proposal subnet consists of a classification branch for classifying and a regression branch for refinement of the corresponding customized anchors.

2.2 Target-Aware Feature Correlation Module

We use the target-aware feature correlation module to calculate the correlation information between search features and template features. Given the template features F_t and the search features F_s. To reduce the interference information of the background in the template, We obtain the variable template F_v through a target-aware mechanism under the supervision of the template bounding box Bt. As shown in Fig. 3, we consider $1 \times 1 \times c$ in the feature map as a token, T_s and T_v denote the set of all feature tokens in F_s and F_v. h_s^i denotes the feature vector in token i, h_v^j denotes the feature vector in token j. Where token $i \in T_s$, token $j \in T_v$. We can simply use the inner product between features vectors as the similarity measure. The attention weights are calculated as follows:

$$\left[w_s^i, w_v^i\right] = Softmax \left(\frac{[K_s, K_v] \cdot q_s^i}{\sqrt{d}}\right) \tag{1}$$

where q_s^i denotes the query vector of token h_s^i, K_s denotes the key matrix corresponding to the search features F_s, and K_v denotes the key matrix corresponding to the variable template feature F_v. The cross-attention weight w_v^i determines the similarity between the search token h_s^i and the variable template region tokens. The self-attention weight w_s^i determines the similarity between the search token h_s^i and all search region tokens. The j-th item ($1 \leq j \leq n$, n is the number of search area tokens) of w_s^i determines the similarity between h_s^i and j-th of T_s.

The cross-attention calculation of F_s and F_v allows us to discriminate the foreground from the background better. F_s perform self-attention calculations to better distinguish the distractors in search image. We represent the self-attention aggregated features and the cross-attention aggregated features as:

$$\begin{aligned} h_{self}^i &= w_s^i \cdot V_s \\ h_{cross}^i &= w_v^i \cdot V_v \end{aligned} \tag{2}$$

where V_s denotes the value matrix corresponding to the search region, V_v denotes the value matrix corresponding to the variable template region.

We can fuse attention aggregation features with the search token h_s^i to obtain a more powerful feature representation empowered by the target information as:

$$\hat{h}^i = ReLU \left(h_s^i || h_{self}^i || h_{cross}^i\right) \tag{3}$$

where $||$ represents vector concatenation. The feature vectors at all locations in the feature map are computed in parallel to generate a global attention features map for subsequent tasks.

To provide more detailed target information to increase the diversity of template features, we obtain the central template F_c by center cropping F_t, by performing DW-Corr [12] with the search features F_s, we further capture the local spatial context and reduce the semantic ambiguity in the attention mechanism as:

$$P_{local} = \varphi [F_s] \star \varphi [F_c] \tag{4}$$

where $\star$ denotes the DW-Corr [12] operation with $\varphi [F_c]$ as the 3×3 convolution kernel, P_{local} indicates local correlation features.

Finally, global attention features are fused with local convolutional features to obtain a more powerful feature representation.

2.3   Customization Anchor Proposal Subnet

We obtain $F_t = \left[F_t^3, F_t^4, F_t^5\right]$, $F_s = \left[F_s^3, F_s^4, F_s^5\right]$ as template features and search features from Resnet50's $conv3$, $conv4$, and $conv5$ layers respectively.

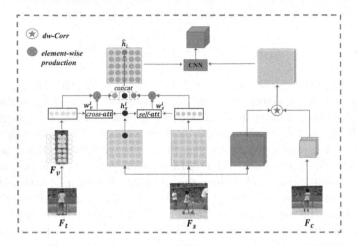

Fig. 3. Illustration of the proposed Target-aware feature correlation module. The global correlation operation is performed with the variable template F_v. Also local correlation operations are performed with central template F_c. Finally the global features and local features are aggregated for implementing downstream tasks.

Customized Anchor Generation Module. We squeeze features F_t, F_s with different depths as well as perceptual fields to obtain $F_{at} = Squeeze(F_t^3, F_t^4, F_t^5)$ and $F_{as} = Squeeze(F_s^3, F_s^4, F_s^5)$. Feed them into TACM to integrate the feature information from the template and search images, and then go through an anchor head to generate the customized anchor adapted to the target. The above calculation can be formulated as:

$$P_{anch}^{w \times h \times 2} = conv_{anch}\left(TACM((F_{at}), (F_{as}))\right) \tag{5}$$

where the two channels of P_{anch} represent a_w and a_h, the anchor head $conv_{anch}$ consists of tow 1×1 convolutional layers. Due to the excessive range of anchor height and width, directly generated customized anchors of width(C_w) and height(C_h) are not stable, we map C_w and C_h by outputting a_w, a_h, and going through the following transformations:

$$C_w = \sigma \cdot s \cdot e^{a_w}$$
$$C_h = \sigma \cdot s \cdot e^{a_h} \tag{6}$$

s is the stride of convolution, which is 8, σ is 8.

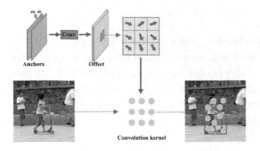

Fig. 4. The sketch of the AAM.

Classification and Regression. In the classification and regression task, we stratify F_t, F_s into TACM stratification to calculate the similarity, and then take its average to obtain $\bar{F} = Avg(TACM(F_s^3, F_t^3), TACM(F_s^4, F_t^4), TACM(F_s^5, F_t^5))$. The $\bar{F}$ is then split into classification and regression branches. Features with different receptive fields have different sensitivities to the shape of the target, enabling the tracker to adapt to targets of different sizes. $\bar{F}$ through a cls head $conv_{cls}$ consisting of tow 1×1 convolution to generate a Cls map:

$$P_{cls}^{w \times h \times 2} = conv_{cls}(\bar{F}) \tag{7}$$

where the two channels of P_{cls} represent the probability that an anchor belongs to foreground or background.

In order to obtain more accurate regression features, we will not directly output the regression results, but use the Anchor adaptation module(AAM) to guide it. AAM can encode and refine the feature of the regression branch based on the shape of the anchor, as show in Fig. 4, each anchor in the anchor map is first mapped into a 3×3 offset kernel after one 1×1 convolution layer, then the offset is embedded into the regular 3×3 convolution kernel (see left image) to adjust kernel (see right image) that can more accurately cover the target region. In other words, through position-shift learning guided by customized anchor, the sample points of the convolutional kernel on feature map can adaptively move to the target region, thereby adapting the regression features to the shape of the anchor, which is beneficial for bounding box regression.

Finally, the feature map adjusted by AAM is then fed into the regression head to form the final regression map. The above calculation can be formulated as:

$$P_{regs}^{w \times h \times 4} = conv_{regs}\left(AAM(\bar{F}, a_w, a_h)\right) \tag{8}$$

where the four channels of P_{regs} represents the distance between the center point and the height and width of the customized anchor and bounding box, and regression head $conv_{regs}$ consists of two 1×1 convolution layers.

2.4   Training

For the classification branch, refer to SiamBAN [3]. We construct two ellipses to divide the positive and negative samples with the help of ground-truth bounding box.(g_x, g_y) denotes the center of the ground truth-bounding box, and (g_w, g_h) denotes the width and height of the ground-truth bounding box. With (g_x, g_y), (g_x, g_y) as the center and $(\frac{g_w}{4}, \frac{g_h}{4})$, $(\frac{g_w}{2}, \frac{g_h}{2})$ as the axis lengths, respectively. We can obtain the ellipses $E1$, $E2$:

$$\frac{(p_i - g_x)^2}{\frac{g_w}{4}} + \frac{(p_j - g_y)^2}{\frac{g_h}{4}} = 1$$

$$\frac{(p_i - g_x)^2}{\frac{g_w}{2}} + \frac{(p_j - g_y)^2}{\frac{g_h}{2}} = 1 \tag{9}$$

If the location (p_i, p_j) of the feature points falls inside the ellipse $E1$, the customized anchors with this location as the center point will be used as positive samples. If the location is outside the ellipse $E2$, these anchors far from the target center will be used as negative samples. The rest are ignored.

For regression branches and predictions of customized anchors, We only train the anchors on the positive sample positions(p_i^o, p_j^o). C_w,C_h denote the shape of the customized anchor. The regression objective can be formulated as:

$$d_x = \frac{g_x - p_i^o}{g_w}, d_w = \ln \frac{g_w}{C_w}$$

$$d_y = \frac{g_y - p_j^o}{g_w}, d_h = \ln \frac{g_h}{C_h} \tag{10}$$

We use a smooth L_1 loss function to optimize the predictions of customized anchors, as:

$$L_{shape} = L_1(C_w - T_w) + L_1(C_h - T_h) \tag{11}$$

We define the loss function for multitasking as follows:

$$L = \lambda_c L_{cls} + \lambda_r L_{regs} + \lambda_s L_{shape} \tag{12}$$

L_{cls} is cross-entropy loss and L_{regs} is *GIOU* loss function. For the hyperparameters $(\lambda_c, \lambda_r, \lambda_s)$, we set them all to 1.

3   Experiments

3.1   Implementation Details

ILSVRC-VID/DET [17], COCO [15], LaSOT [6], GOT-10k [9] datasets are utilized as the training sets. We initialize our backbone networks with the weights pre-trained on ImageNet and the parameters of the last three layers will be used for training. Our network is trained with stochastic gradient descent (SGD) with a minibatch of 32 pairs. We train a total of 20 epochs, using a warmup learning rate of 0.001 to 0.005 in the first 5 epochs and a learning rate exponentially decayed from 0.005 to 0.00005 in the last 15 epochs. In the first 10 epochs, we only train the customized anchor guidance subnet, and fine-tune the backbone network in the last 10 epochs at one-tenth of the current learning rate. Weight decay and momentum are set as 0.0001 and 0.9.

3.2 Datasets and Evaluation Metrics

We evaluate our algorithm on the OTB100, LaSOT, VOT2018, UAV123, and VOT2019 datasets. Specifically. OTB100 [21] is a tracking dataset containing 100 fully annotated sequences, and each sequence in the dataset includes challenging attributes for tracker performance. LaSOT [6] consists of 1400 sequences with a total frame size of over 3.5M. Unlike existing datasets, LaSOT provides both visual bounding box annotations and a rich natural language specification. They measure tracker performance in terms of success rate (Succ).

Both VOT2018 [10] and VOT2019 [11] contain 60 video sequences with various challenging factors. Compared to VOT2018, VOT2019 sequences were replaced by 20%. They use accuracy (A), robustness (R), and expected average overlap (EAO) to evaluate and measure the different trackers. UAV123 [16] is a new aerial video benchmark that contains 123 video sequences taken from low altitude aerial angles. They measure tracking performance in terms of accuracy(A) and precision rate(Pr), and rank precision plots based on the area under the curve.

Table 1. Ablation study for SiamCAFT on VOT2019

CAGM	AAM	TACM	DW-Corr	PW-Corr	EAO↑	A↑	R↓
			✓		0.285	0.599	0.482
✓			✓		0.287	0.597	0.446
✓	✓		✓		0.303	0.598	0.406
✓	✓			✓	0.274	0.596	0.532
✓	✓	✓			0.327	0.604	0.371

3.3 Ablation Experiments

In this section, we perform an ablation study on the VOT2019 [11] dataset to analyze the effectiveness of the CAGM, AAM and TACM. The evaluation results are presented in Table 1. Baseline tracker uses DW-Corr for cross-correlation operation. As shown in Table 1, by introducing CAGM to generate customized anchors, the robustness(R) score improves by 7% compared to the baseline tracker, which indicates that anchors generated according to correlational feature can better describe the target in the tracking task.

The introduction of the AAM to adapt the anchor information improves the robustness score by 15% and the EAO score by 6% compared to the baseline tracker, demonstrating that AAM effectively facilitates anchor localization and regression.

To verify the validity of TACM, we compare it with variants which utilize DW-Corr [12] and PW-Corr [14] as correlation operations. The tracker with TACM improves the robustness score by 8.6% and the eao score by 8% compared to the tracker with DW-Corr. Compared to the tracker with PW-Corr, the robustness score improves by 30% and the EAO score improves by 19%. As show in Fig. 5(a), it is demonstrated that TACM effectively suppresses background information and learns the relationship of feature local spatial context. Overall, SiamCAFT improves the baseline tracker by 14.7%/23% in terms of EAO/R metrics.

3.4   Comparison with State-of-the-art Methods

We compared SiamCAFT with several state-of-the-art trackers on five challenging datasets, including OTB100, VOT2019, UAV123, LaSOT, and VOT2018.

Table 2. Comprehensive comparisons on UAV123 experiments. The best two results are shown in **bold** and *italic*, respectively.

Tracker	SiamFC [1]	SiamRPN [13]	DaSiamRPN [24]	SiamCAR [8]	SiamRPN++ [12]	SiamFC++ [22]	ours
A↑	0.485	0.557	0.569	0.614	0.613	*0.623*	**0.631**
Pr↑	0.693	0.768	0.781	0.760	*0.807*	0.781	**0.829**

UAV123. As shown in Table 2, our method obtains the highest accuracy (A) score and precision (Pr) score among all the methods. Compared with baseline tracker SiamRPN++, accuracy score improves from 0.613 to 0.631, an improvement of 3%, and precision score improves from 0.807 to 0.829, improving by 2.7%.

Table 3. Comprehensive comparisons on OTB100 experiments. The best two results are shown in **bold** and *italic*, respectively.

Tracker	TransT [2]	SiamRPN++ [12]	SiamBAN [3]	AiATrack [7]	DropTrack [20]	SiamCAR [8]	ours
Succ↑	69.4	69.6	69.6	69.6	69.6	*69.8*	**70.6**

OTB100. As shown in Table 3, our tracker achieves a state-of-the-art success (Succ) score of 70.6%, which is the best among the competitors. Both SiamCAR and SiamRPN++ use DW-corr for feature information correlation, which does not suppress background noise well. Therefore the tracking results of SiamCAR and SiamRPN++ are not as good as ours (Table 4).

Table 4. Comprehensive comparisons on VOT2019 experiments. The best two results are shown in **bold** and *italic*, respectively.

Tracker	PACNet [23]	SiamFC++ [22]	ATOM [4]	SiamBAN [3]	SiamRPN++ [12]	SiamFC [1]	ours
A↑	0.573	0.575	*0.603*	0.602	0.599	0.470	**0.604**
R↓	0.401	0.406	0.411	*0.398*	0.482	0.958	**0.371**
EAO↑	*0.300*	0.284	0.292	**0.327**	0.285	0.163	**0.327**

VOT2019. SiamCAFT achieved an EAO score of 0.327 and a robustness (R) score of 0.371, which is the best among the competitors. Specifically, SiamCAFT outperformed the baseline tracker SiamRPN++ by 4.2% in the EAO. And the robustness is better than SiamBAN with the same EAO score by 12.8%.

Table 5. Comprehensive comparisons on LaSOT experiments. The best two results are shown in **bold** and *italic*, respectively.

Tracker	SiamFC [1]	SiamRPN++ [12]	SiamBAN [3]	ATOM [4]	SiamCAR [8]	CGACD [5]	ours
Succ↑	33.6	49.6	51.4	51.5	51.6	*51.8*	**52.7**

LaSOT. As show in Table 5, SiamCAFT achieved a competitive success (Succ) score of 52.7%. SiamCAFT outperformed the baseline tracker SiamRPN++ by a significant 6% in terms of success score due to its powerful target classification capability and accurate bounding box prediction (Table 6).

Table 6. Comprehensive comparisons on VOT2018 experiments. The best two results are shown in **bold** and *italic*, respectively.

Tracker	SiamFC++ [22]	SiamRPN++ [12]	ATOM [4]	ULAST [18]	SiamCAR [8]	SiamBAN [3]	ours
A↑	0.587	**0.604**	0.590	0.571	0.574	*0.597*	**0.604**
R↓	0.183	0.234	0.204	0.286	0.197	*0.178*	**0.145**
EAO↑	0.426	0.417	0.401	0.355	0.423	**0.452**	*0.435*

VOT2018. Our tracker achieves the best result in terms of a robustness (R) score of 0.145. SiamBAN performs better in terms of EAO score due to its unique regression module. It is worth pointing out that SiamCAFT outperforms SiamBan by 18% in terms of robustness metrics. CAGM effectively generates customized anchors for target localization and regression in Fig. 5(a).

(a)   (b)

Fig. 5. (a) Comparison of our SiamCAFT with the state-of-the-art tracker on three challenging sequences of V0T2018. (b) Attention visualization results of the tracker on three sequences of OTB100, where the ground truth is highlighted in red. The first row shows the results derived from SiamCAFT with TACM, while the second row displays the attention results for SiamCAFT with DW-Corr. (Color figure online)

4 Conclusion

In this paper, a robust and effective online tracking method SiamCAFT is proposed, which mainly consists of a customized anchor generation module (CAGM) and an anchor adaptation module (AAM), as well as a target-aware feature correlation module (TACM). Specifically, CAGM estimates the shape of each target to obtain a more representative anchor. AAM embeds the customized anchors information into deeper features to learn more discriminative and valuable features. TACM embeds variable templates and central templates into the search features for feature interaction, fusing locally correlated features and globally correlated features. It effectively suppresses the interference of background information to have robust tracking. Experimental results on five challenging datasets show that SiamCAFT performs well against several state-of-the-art trackers.

Acknowledgements. This work is supported by National Natural Science Foundation of China (Nos. 62266009, 61866004, 62276073, 61966004, 61962007), Guangxi Natural Science Foundation (Nos. 2018GXNSFDA281009, 2019GXNSFDA245018, 2018GXNSFDA294001), Guangxi Collaborative Innovation Center of Multi-source Information Integration and Intelligent Processing, Innovation Project of Guangxi Graduate Education(YCSW2023187), and Guangxi "Bagui Scholar" Teams for Innovation and Research Project.

References

1. Bertinetto, L., Valmadre, J., Henriques, J.F., Vedaldi, A., Torr, P.H.S.: Fully-convolutional Siamese networks for object tracking. In: Hua, G., Jégou, H. (eds.) ECCV 2016. LNCS, vol. 9914, pp. 850–865. Springer, Cham (2016). https://doi.org/10.1007/978-3-319-48881-3_56

2. Chen, X., Yan, B., Zhu, J., Wang, D., Yang, X., Lu, H.: Transformer tracking. In: Proceedings of the IEEE/CVF Conference on Computer Vision and Pattern Recognition, pp. 8126–8135 (2021)
3. Chen, Z., Zhong, B., Li, G., Zhang, S., Ji, R.: Siamese box adaptive network for visual tracking. In: Proceedings of the IEEE/CVF Conference on Computer Vision and Pattern Recognition, pp. 6668–6677 (2020)
4. Danelljan, M., Bhat, G., Khan, F.S., Felsberg, M.: ATOM: accurate tracking by overlap maximization. In: Proceedings of the IEEE/CVF Conference on Computer Vision and Pattern Recognition, pp. 4660–4669 (2019)
5. Du, F., Liu, P., Zhao, W., Tang, X.: Correlation-guided attention for corner detection based visual tracking. In: Proceedings of the IEEE/CVF Conference on Computer Vision and Pattern Recognition, pp. 6836–6845 (2020)
6. Fan, H., et al.: LaSOT: a high-quality benchmark for large-scale single object tracking. In: Proceedings of the IEEE/CVF Conference on Computer Vision and Pattern Recognition, pp. 5374–5383 (2019)
7. Gao, S., Zhou, C., Ma, C., Wang, X., Yuan, J.: AiATrack: attention in attention for transformer visual tracking. In: Avidan, S., Brostow, G., Cissé, M., Farinella, G.M., Hassner, T. (eds.) Computer Vision, ECCV 2022, Part XXII. LNCS, vol. 13682. pp. 146–164. Springer, Cham (2022). https://doi.org/10.1007/978-3-031-20047-2_9
8. Guo, D., Wang, J., Cui, Y., Wang, Z., Chen, S.: SiamCAR: Siamese fully convolutional classification and regression for visual tracking. In: Proceedings of the IEEE/CVF Conference on Computer Vision and Pattern Recognition, pp. 6269–6277 (2020)
9. Huang, L., Zhao, X., Huang, K.: GOT-10k: a large high-diversity benchmark for generic object tracking in the wild. IEEE Trans. Pattern Anal. Mach. Intell. **43**(5), 1562–1577 (2019)
10. Kristan, M., et al.: The sixth visual object tracking VOT2018 challenge results. In: Leal-Taixé, L., Roth, S. (eds.) ECCV 2018. LNCS, vol. 11129, pp. 3–53. Springer, Cham (2019). https://doi.org/10.1007/978-3-030-11009-3_1
11. Kristan, M., et al.: The seventh visual object tracking VOT2019 challenge results. In: Proceedings of the IEEE/CVF International Conference on Computer Vision Workshops (2019)
12. Li, B., Wu, W., Wang, Q., Zhang, F., Xing, J., Yan, J.: SiamRPN++: evolution of Siamese visual tracking with very deep networks. In: Proceedings of the IEEE/CVF Conference on Computer Vision and Pattern Recognition, pp. 4282–4291 (2019)
13. Li, B., Yan, J., Wu, W., Zhu, Z., Hu, X.: High performance visual tracking with Siamese region proposal network. In: Proceedings of the IEEE Conference on Computer Vision and Pattern Recognition, pp. 8971–8980 (2018)
14. Liao, B., Wang, C., Wang, Y., Wang, Y., Yin, J.: PG-Net: pixel to global matching network for visual tracking. In: Vedaldi, A., Bischof, H., Brox, T., Frahm, J.-M. (eds.) ECCV 2020. LNCS, vol. 12367, pp. 429–444. Springer, Cham (2020). https://doi.org/10.1007/978-3-030-58542-6_26
15. Lin, T.-Y., et al.: Microsoft COCO: common objects in context. In: Fleet, D., Pajdla, T., Schiele, B., Tuytelaars, T. (eds.) ECCV 2014. LNCS, vol. 8693, pp. 740–755. Springer, Cham (2014). https://doi.org/10.1007/978-3-319-10602-1_48
16. Mueller, M., Smith, N., Ghanem, B.: A benchmark and simulator for UAV tracking. In: Leibe, B., Matas, J., Sebe, N., Welling, M. (eds.) ECCV 2016. LNCS, vol. 9905, pp. 445–461. Springer, Cham (2016). https://doi.org/10.1007/978-3-319-46448-0_27

17. Russakovsky, O., et al.: ImageNet large scale visual recognition challenge. Int. J. Comput. Vis. **115**(3), 211–252 (2015)
18. Shen, Q., et al.: Unsupervised learning of accurate Siamese tracking. In: Proceedings of the IEEE/CVF Conference on Computer Vision and Pattern Recognition, pp. 8101–8110 (2022)
19. Wang, J., Chen, K., Yang, S., Loy, C.C., Lin, D.: Region proposal by guided anchoring. In: Proceedings of the IEEE/CVF Conference on Computer Vision and Pattern Recognition, pp. 2965–2974 (2019)
20. Wu, Q., Yang, T., Liu, Z., Wu, B., Shan, Y., Chan, A.B.: DropMAE: masked autoencoders with spatial-attention dropout for tracking tasks. In: Proceedings of the IEEE/CVF Conference on Computer Vision and Pattern Recognition, pp. 14561–14571 (2023)
21. Wu, Y., Lim, J., Yang, M.H.: Online object tracking: a benchmark. In: Proceedings of the IEEE Conference on Computer Vision and Pattern Recognition, pp. 2411–2418 (2013)
22. Xu, Y., Wang, Z., Li, Z., Yuan, Y., Yu, G.: SiamFC++: towards robust and accurate visual tracking with target estimation guidelines. In: Proceedings of the AAAI Conference on Artificial Intelligence, vol. 34, pp. 12549–12556 (2020)
23. Zhang, D., Zheng, Z., Jia, R., Li, M.: Visual tracking via hierarchical deep reinforcement learning. In: Proceedings of the AAAI Conference on Artificial Intelligence, vol. 35, pp. 3315–3323 (2021)
24. Zhu, Z., Wang, Q., Li, B., Wu, W., Yan, J., Hu, W.: Distractor-aware Siamese networks for visual object tracking. In: Ferrari, V., Hebert, M., Sminchisescu, C., Weiss, Y. (eds.) ECCV 2018. LNCS, vol. 11213, pp. 103–119. Springer, Cham (2018). https://doi.org/10.1007/978-3-030-01240-3_7

Can We Transfer Noise Patterns?
A Multi-environment Spectrum Analysis
Model Using Generated Cases

Haiwen Du[1,2], Zheng Ju[2], Yu An[2], Honghui Du[2], Dongjie Zhu[3(✉)],
Zhaoshuo Tian[1], Aonghus Lawlor[2], and Ruihai Dong[2(✉)]

[1] School of Astronautics, Harbin Institute of Technology, Harbin, China
[2] Insight Centre for Data Analytics, Dublin, Ireland
ruihai.dong@ucd.ie
[3] School of Computer Science and Technology, Harbin Institute of Technology,
Weihai, China
zhudongjie@hit.edu.cn

Abstract. Spectrum analysis systems in online water quality testing
are designed to detect types and concentrations of pollutants and enable
regulatory agencies to respond promptly to pollution incidents. However,
spectral data-based testing devices suffer from complex noise patterns
when deployed in non-laboratory environments. To make the analysis
model applicable to more environments, we propose a noise patterns
transferring model, which takes the spectrum of standard water samples in different environments as cases and learns the differences in their
noise patterns, thus enabling noise patterns to transfer to unknown samples. Unfortunately, the inevitable sample-level *baseline noise* makes the
model unable to obtain the paired data that only differ in dataset-level
environmental noise. To address the problem, we generate a sample-
to-sample case-base to exclude the interference of sample-level noise
on dataset-level noise learning, enhancing the system's learning performance. Experiments on spectral data with different background noises
demonstrate the good noise-transferring ability of the proposed method
against baseline systems ranging from wavelet denoising, deep neural
networks, and generative models. From this research, we posit that
our method can enhance the performance of DL models by generating
high-quality cases. The source code is made publicly available online at
https://github.com/Magnomic/CNST.

Keywords: 1D data denoising · Noise patterns transferring ·
Spectrum analysis · Signal processing

1 Introduction

Laser-induced fluorescence spectrum (LIFS) analysis is used to interpret spectrum data obtained via laser-induced fluorescence (LIF) spectroscopy in a specific range of wavelengths, frequencies, and energy levels [10]. Its advantages

© The Author(s), under exclusive license to Springer Nature Singapore Pte Ltd. 2024
B. Luo et al. (Eds.): ICONIP 2023, CCIS 1969, pp. 125–139, 2024.
https://doi.org/10.1007/978-981-99-8184-7_10

include the high sampling frequency, high accuracy, and easy deployment, making it a popular water quality monitoring approach [3,27]. However, avoiding the impact of *environmental noise* on the analysis results is challenging because it changes with weather, temperature, and water impurities [16,26].

Denoising is essential in analysing spectral and other one-dimensional (1D) data, as it can improve the accuracy and reliability of subsequent analyses. Traditional methods that rely on mathematical models or filters (e.g., moving average or wavelet transformation) can reduce high-frequency or specific noise patterns while consuming less computing resources [21]. However, they cannot learn complex nonlinear relationships between the noisy and the corresponding clean signal. Deep learning(DL)-based denoising methods use a data-driven approach. They are trained on a large dataset to learn the underlying patterns and relationships between the data [5,13].

Nevertheless, we found limitations in this approach due to the complex noise sources and noise patterns in LIFS. Specifically, we cannot obtain paired clean-noisy data to analyse the noise pattern between two datasets because the noise in spectral data is from both *baseline noise* ξ and *environmental noise* N [1,20]. *Environmental noise* arises from backgrounds, light sources and other stable factors over time, exhibiting differences at the dataset-level. The *baseline noise* originates from random disturbances during the sampling process, such as baseline drift, which is present even in laboratory environments. Different from *environmental noise*, *baseline noise* shows differences at the sample-level, i.e., every sample has different *baseline noise*. However, as both noise sources are low-frequency and superimposed on the signal, extracting only the *environmental noise* from the LIFS to train a noise patterns transferring model is complex. We can only feed the samples with both *environmental noise* and *baseline noise* differences to the noisy pattern learning model, which is inefficient.

In this paper, we propose a method to enhance the performance of denoising models based on generated cases, i.e., we generate paired samples that differs only in *environmental noise*. In turn, we learn the differences to achieve noise pattern transformation. Firstly, the model extracts noise patterns from a dataset-to-dataset (D2D) case-base that consists of two groups of samples from dataset D_T and dataset D_S. Then, we generate a sample-to-sample (S2S) case-base in which all pairs of the sample differ only in going from *environmental noise* N_S to N_T (N_{S-T}). Finally, we train a noise patterns transferring model using the S2S case-base. The highlights are:

1. we develop a noise transfer model to transfer the *environmental noise* pattern N_S of dataset D_S into the *environmental noise* pattern N_T of dataset D_T (also denoising when D_T is *environmental noise*-free);
2. propose a network structure for extracting *environmental noise* N_S, signal X, and *baseline noise* ξ_T from a D2D case S and T, thereby generating a new sample G that only have *environmental noise* pattern differences N_{S-T} with sample T. It is the first work to enhance the learning performance of denoising models by generating new cases; and

3. verify the contribution of our model on 1D LIFS noise patterns transferring tasks. It significantly improves the performance of chemical oxygen demand (COD) parameter analysis model under different noise patterns.

In the following section, we discuss related work and present motivations for this work. Section 3 presents the proposed model that captures noise pattern differences between signals in a different dataset. In Sect. 4, we describe the experimental setup. Then, Sect. 5 presents and analyses the evaluation results. Section 6 concludes the paper and discusses future work.

2 Background and Motivation

2.1 Problems in Traditional Denoising Models

In LIF spectroscopy, noise can arise from various sources, including electronic noise in the sensors, light source fluctuations, and sample matrix variations. It can affect the LIFS analysis process by introducing artefacts, such as spurious peaks or baseline drift, that can obscure or distort spectral features associated with water quality parameters [25]. For example, baseline drift is a noise that can make it challenging to identify and quantify the fluorescence peaks accurately. Similarly, high-frequency noise can introduce spurious peaks in the spectra that can be misinterpreted as actual spectral features [18]. Therefore, avoiding the noise's influence on the analysis result is vital for researchers in analysing LIFS.

Various denoising techniques can mitigate noise in LIFS, such as smoothing algorithms or wavelet transform-based methods [21]. These methods can remove parts of noise while preserving the spectral features of interest. Another popular approaches are machine learning algorithms, such as DNNs [9]. Compared with traditional methods, their adaptability to various noise types and their robustness to complex scenes make DNNs an attractive option for denoising tasks.

Nevertheless, we are facing a dilemma in applying DNNs, which is caused by the *internal noise*, i.e., even in the laboratory environment, the spectral data still carries low-frequency noise because of fuzzy factors such as baseline drift. To make matters worse, the two measurements for the same water sample show different *internal noise* patterns. For datasets D_S and dataset D_T collected in different environments, when we try to transfer the *environmental noise* pattern of a sample S in dataset D_S into the *environmental noise* pattern of dataset D_T, which have the relations in Eq. 1, where S and T are the LIFSs of the same water sample X in D_S and D_T, N_{S-T} is the *environmental noise* difference (dataset-level) between D_S and D_T, ξ_S and ξ_T are the *internal noise* (sample-level) in samples S and T.

$$S = X + N_S + \xi_S$$
$$T = X + N_T + \xi_T \qquad (1)$$
$$N_{S-T} = N_T - N_S$$

We find that it does not satisfy the relationship in Eq. 2 but in Eq. 3. Then, we give examples in Fig. 1 to better represent the terms in noise patterns.

$$T = S - N_{\text{S}-\text{T}} \tag{2}$$

$$T = S - N_{\text{S}-\text{T}} - \xi_S + \xi_T \tag{3}$$

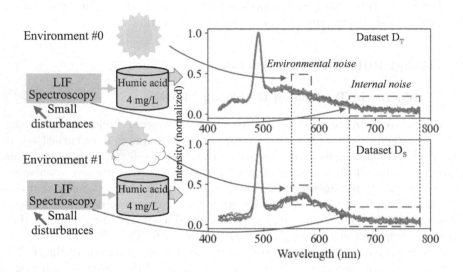

Fig. 1. We provide LIFS examples of two standard solutions in different environments. The blue boxes show that each sample has a different *internal noise* ξ, even from the same dataset. The red boxes show the *environmental noise* N, at the dataset level, are shared by all the samples in the same dataset, respectively. For ease of presentation, all S and T that occur together in the following paper represent samples corresponding to the same water sample.

Unfortunately, the DNN models show bad performance when learning the dataset-level noise pattern transformation from N_{S} to N_{T} because of the existence of sample-level noise ξ_S and ξ_T.

2.2   Studies Related to Noise Pattern Transferring

Since *internal noise* is inevitable in LIFS, we hope all the samples have the same *environmental noise* patterns by removing their *environmental noise* patterns differences. With this, we only need to train the analysis model in one *environmental noise* pattern. Then we can process the data in various environments by transferring their *environmental noise* to the target pattern [2,7].

We refer to the generative methods, which use the LIFS of the same standard solution in different environments as cases. Therefore, noise pattern differences can be learned by analysing these cases and applied to transfer noise patterns of unknown samples. However, directly training a DNN-based noise-transferring model like DnCNN [28] using these cases is less efficient because the *internal noise* is sample-level low-frequency noise and superimposed on the signal and *environmental noise*.

We have two feasible approaches to achieve it. The first is to design a model that does not strictly require pairwise data for training, which CycleGAN [15] represents. CycleGAN uses an adversarial and cycle consistency loss, encouraging the network to learn a consistent mapping between the two domains. Recently, CycleGAN has been used to suppress unknown noise patterns and get satisfactory results in 1D signal processing, such as audio and seismic traces [12].

Alternatively, we create new cases that have *environmental noise* (dataset-level) difference but the same *internal noise* (sample-level). Compared with CycleGAN, it is more explainable and provides a much more guided process. It is challenging to extract only *environmental noise* because *internal noise* is superimposed on it. Although both *internal noise* and *environmental noise* are low-frequency noise, the frequency of the *internal noise* is slightly higher than the *environmental noise*. It means that the curve fluctuations caused by internal noise are in the smaller wavelength range. As the feature maps in deep convolutional neural network (DCNN) can help to generate cases [14] and feature information [23], we try to use a pre-trained DCNN to extract these noise patterns separately. Based on the theory that the deeper the convolutional layer is, the more pooling layers it passes through and the increasing range of its extracted features [24], we propose the hypothesis that deeper convolutional layers can better extract the *environmental noise* N_S and signal X of the samples in datasets D_S, and shallower convolutional layers can better extract the *internal noise* ξ_T of the samples in datasets D_T. We can use them and generate a sample $G = X + N_S + \xi_T$ such that its difference from T is N_{S-T}, as shown in Fig. 2.

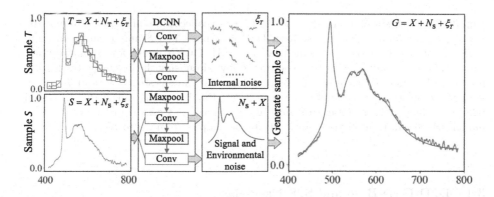

Fig. 2. We provide an example of our hypothesis. Sample S and sample T are fed into the DCNN to get their feature maps on deep and shallow convolutional layers, which correspond to extracting *internal noise* and *environmental noise* (also the signal). Then, we use this information to synthesise G.

3   Noise Patterns Transferring Model

Since the samples in different datasets have both sample-level (noise difference in *internal noise*) and dataset-level (noise difference in *environmental noise*) noise pattern differences, traditional solutions meet difficulties when trying to extract and learn *environmental noise* without the interference of *internal noise*. This section details our generated cases (GC)-based method to achieve noise patterns transferring. The overall workflow of our method is shown in Fig. 3.

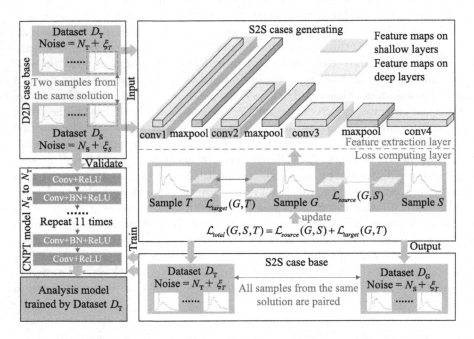

Fig. 3. The workflow of our method. D_S and D_T are collected in different environments. Firstly, we randomly select a case from D2D case-base, i.e., S and T and feed them with an initialised sample G into the pre-trained 1D DCNN model for extracting their feature maps. Secondly, we get the generated G by minimising the loss using the feature maps of G, S and T. Then we get an S2S case that is consisted of G and T, which with the same ξ_T and only differ in N_S and N_T. Finally, we use all of the generated S2S cases, i.e., S2S case-base, to train the curve noise patterns transferring (CNPT) model, which can transfer the noise patterns of samples from N_S to N_T.

3.1   D2D Case-Base and S2S Case-Base

Our system has two case-bases, D2D case-base and S2S case-base. If we want to transfer the *environmental noise* pattern N_S in D_S to the *environmental noise* pattern N_T in D_T, a D2D case is a sample S in D_S and random sample T in D_T that is measured using the same standard solution but under different environments. Although all of the S2S and D2D cases have the same N_{S-T}, D2D cases have different *internal noise* while S2S cases have the same *internal noise*. In other words, an S2S case is a pair of a sample G in D_G and a sample T that has the same *internal noise* ξ_T but differs in N_S and N_T.

Since S2S cases cannot be directly obtained by LIF spectroscopy because of the existence of *internal noise*, to learn the pattern of N_{S-T}, the first step is to generate S2S cases using the D2D case-base that satisfies Eq. 4.

$$G = T + N_{S-T} \tag{4}$$

According to Eq. 1, G can be represented as Eq. 5.

$$\begin{aligned} G &= S - \xi_S + \xi_T \\ &= X + N_S + \xi_T \end{aligned} \tag{5}$$

To this step, we get a S2S case, i.e., G and T, because they only have the difference in N_T and N_S. It enables us to train a model that can learn the pattern of this transformation. The key point, therefore, is how to extract X, N_S and ξ_T from D2D case-base.

3.2   S2S Cases Generating Model

According to the hypothesis in Sect. 2.2, we propose a 'feature extraction-matching' method to generate G in the S2S case. Firstly, we use a pre-trained DCNN-based 1D LIFS analysis model that contains multiple convolutional and pooling layers to extract features. Then, we initialise G and design a loss function that makes the feature of G approximate X, ξ_T, and N_S from the D2D case. Lastly, we get an S2S case that is consisted of G and T by minimising the loss.

As X, N_T and ξ_T are wavelength (x-axis) related but show different signal frequency features. We propose a loss function $\mathcal{L}_{total}$ that consists of source loss $\mathcal{L}_s$ and target loss $\mathcal{L}_t$ in Eq. 6, where α and β are the weights. l_t and l_s are the layers to compute $\mathcal{L}_t$ and $\mathcal{L}_s$.

$$\mathcal{L}_{total}(G, S, T, l_s, l_t) = \alpha \mathcal{L}_s(G, S, l_s) + \beta \mathcal{L}_t(G, T, l_s, l_t) \tag{6}$$

$\mathcal{L}_s$ is used to construct the N_S and X in the generated sample G. As N_S and X are x-axis related and show lower-frequency features, we use deeper feature maps and position-related loss to reconstruct X and N_S from S (see Eq. 7.)

$$\mathcal{L}_s(G, S, l_s) = \left(F_i^{l_s}(G) - F_i^{l_s}(S) \right)^2 \tag{7}$$

$\mathcal{L}_t$ is used to construct the *internal noise* ξ_T of T in G. As we discussed, its frequency is higher than *environmental noise* and signal. We use feature maps on shallow layer l_t and gram matrix (see Eq. 8 and Fig. 4) to control its feature shape and an MSE loss on deep layer feature l_s to control its x-axis position. The reason we introduced the gram matrix is that we cannot synthesise ξ_T with the same loss function structure for T as for S. It would cause the overall loss to converge towards $\mathcal{L}_s$ or $\mathcal{L}_t$, and thus we would not be able to obtain features from both S and T.

$$Gram_{ij}^{l_t}(Y) = \sum_{k \in l_t} F_{ik}^{l_t}(Y) \cdot F_{jk}^{l_t}(Y) \tag{8}$$

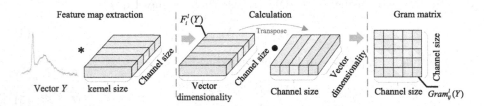

Fig. 4. shows how the gram matrix is calculated. Suppose that we input the sample Y to the DCNN for its convolution layer l; we get the feature matrix with the size of $P_l * M_l$, where P_l is the channel size of l, M_l is the size of feature map in l, the convolution of Y on the ith channel is $F_i^l(Y)$. Then, by multiplying the obtained matrix with its transpose, we obtain the gram matrix of $P_l * P_l$, where $Gram_{i,j}^l(Y)$ represents the correlation of Y on the ith and j^{th} features in convolution layer l.

The overall $\mathcal{L}_t$ is calculated as Eq. 9, where ω_l is the loss weight of l.

$$\mathcal{L}_t(G, T, l_s, l_t) = \sum_i \left(F_i^{l_s}(G) - F_i^{l_s}(T) \right)^2 + \sum_{l \in l_t} \omega_l E_l$$
$$E_l = \frac{1}{4P_l^2 M_l^2} \sum_{i,j} \left(Gram_{ij}^l(G) - Gram_{ij}^l(T) \right)^2 \tag{9}$$

3.3   Noise Patterns Transferring Model

S2S case-base provides noise pattern differences N_{S-T} in sample-to-sample pairs, i.e., G and T. It enables us to train the CNPT model with $CNPT(G) = T$. Similar to denoising tasks, we use the residual learning method, which involves learning the residual mapping between the pairwise samples to obtain the CNPT model that achieves transferring N_S to N_T.

The CNPT model's network structure consists of 17 convolutional layers and rectified linear unit (ReLU) layers, allowing it to capture complex patterns and features. The inputs to CNPT are the generated sample $G = X + N_S + \xi_T$ and the corresponding sample $T = X + N_T + \xi_T$. CNPT is trained to learn the residual mapping between the S2S case $\mathcal{R}(G, \phi) = N_{S-T} = N_T - N_S$.

We use the MSE loss function to train the CNPT to minimise the learned noise pattern residual and the expected residuals. The loss function is shown in Eq. 10, where the Q represents the number of S2S cases we feed the model.

$$\mathcal{L}_{CNPT} = \frac{1}{2Q} \sum_{i=1}^{Q} \| \mathcal{R}(G; \phi) - (G - T) \|^2 \tag{10}$$

Ideally, the trained CNPT model should learn the noise pattern differences between N_T and N_S by feeding S2S cases into it.

4 Environmental Setup

4.1 Model Configurations and Runtime Environments

In this section, we introduce all the hyper-parameters and the configurations in the experiments to guide the researchers in refining the model in different applications and environments.

For the S2S case generating model, the feature extraction model is a four-layer 1D convolutional neural network in which kernel size is 1×7. This pre-trained model is from a COD parameter analysis model, which is trained by 6,000 samples and achieves 92.53% accuracy. The feature maps for calculating $\mathcal{L}_t$ are from conv1 and conv2 layers, while conv3 is used for x-axis-related MSE loss. The ratio of α and β is set to 1: 2e5 and $\omega_l = 0.2 * \omega_{l-1}$. We iteratively execute the L-BFGS function 150 times to minimise the loss.

For the CNPT model, the kernel size is 1×7. The training batch size is 64, and the training epochs are 100. The learning rate we use is 1e−3, and the input data is normalised by the min-max normalisation method before feeding into the neural network. The data for the training-to-validation ratio is 80% to 20%.

The COD parameter analysis model, used for verifying the performance of noise patterns transferring, has the same network structure as the feature extraction model. The training batch size is 128, the epochs are 200, and the learning rate is 3e−4. All the data fed to this model for validation are not used for training or testing the CNPT model.

All the experiments were performed on a Linux server for the hardware environments, with CUDA version 11.8 using GeForce GTX 4090 Graphics Cards. We use the PyTorch framework to write the codes.

4.2 Datasets

The datasets are LIF spectral data of humic acid solution with different noise patterns, open-sourced at [6]. The COD of solutions is measured using laser-induced fluorescence spectroscopy [22], in which a 405 nm semiconductor laser was used as the excitation source. The excited light signals from the water sample pass through a spectroscopic filter system and be focused into the CCD by a lens. The dataset contains three groups of LIFS, which are collected from 10 solutions that the COD parameter ranges from 0 mg/L to 30 mg/L in three different noise patterns D_A, D_B, and D_C. The noise strength of them is ordered as $D_A < D_B << D_C$. Each solution sample is measured 200 times, and all the data are min-max normalised. We provide examples of LIF spectra in Fig. 5.

4.3 Baselines

To verify the performance of our model, we select competitive models with good performance in terms of noise processing as baselines. In other words, they are

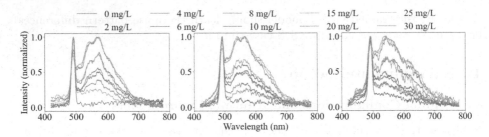

Fig. 5. The LIFS of standard water sample under three noise patterns: The left figure represents a dark room setting (D_A), the central one portrays a natural light condition (D_B), while the right one is from a static pollutant background environment (D_C).

used for transferring noise patterns, not to get spectral analysis results. All of our model's and baselines' outputs are fed into the same COD parameter analysis model that is trained by the samples in D_T, which can check whether the samples that are transferred to the noise pattern N_T can be correctly identified by the analytical model trained with D_T. The baselines are as follows:

Rule-based Models: use wavelets, Fourier transforms, or filtering to reduce the noise of the data. They have the advantage of not requiring paired data and learning processes. For LIFS, we use the 'db8' wavelet to achieve the denoising task [8]. This method trains the classification model using the denoised samples in D_T. Then, the CNPT model learns the noise pattern difference between denoised samples in D_S and D_T.

DNN Models: use end-to-end training process. DnCNN [28] is a state-of-the-art image denoising method that uses residual learning to remove noise from images. In this paper, we implemented a 1D DnCNN model with a 1×7 kernel size to learn the noise pattern differences between D_S and D_T. Since the spectral data in different noise patterns are not paired, we randomly pick samples in D_S and D_T corresponding to the same standard solution as paired data to train it.

Generative Models: learn at the scale of the dataset, which makes the model find noise pattern differences in the adversarial learning process. Cycle-GAN [15] is a deep learning model that can learn to translate images from one domain to another without paired examples. Therefore, CycleGAN is a suitable approach to handle the unpaired LIFS data. We implemented a 1D CycleGAN model, which can transfer the noise pattern of samples from D_S to D_T.

5   Experimental Results

We discuss the experimental results in two parts: 1) check if the gram matrix works well in the S2S cases generating process; 2) test the model's performance by feeding the output of CNPT to a trained COD parameter analysis model.

5.1   Effects on S2S Cases Generating

One of the contributions of our method is to use the unpaired data with different *environmental noise* patterns and *internal noise* to generate paired data that only have *environmental noise* differences. It comes from our S2S cases generation process based on the 'feature extraction-matching' method and the loss on different feature maps. Therefore, this section shows how the sample G is generated in feature maps and gram matrix view. We present the convolution kernel and the gram matrix changes in the training process in Fig. 6.

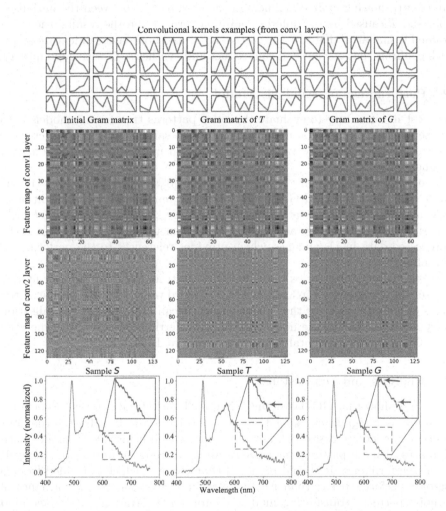

Fig. 6. We show the convolution kernel used to compute the feature map in the top part of the figure. Then, we show the initial, target, and final state of the gram matrix of the signal G in the training progress of the conv1 and conv2 layers. The bottom part shows the corresponding signals T, S, and G in this example.

Since the convolution kernel provides rich patterns for the feature maps, we can also analyse the signal patterns from the feature maps. For example, we find that the feature in which id is 0 has a strong auto-correlation on the conv1 (the value at position 0,0 in the gram matrix), which means the pattern corresponding to the convolution kernel with ID 0 (upper left corner in the convolution kernel example) appears most frequently in the signal.

It matches our expectation since its pattern is similar to the baseline drift, i.e., the initialisation signals and T have large oscillations on a smaller scale (1×5). In contrast, conv2 has a more extensive feature extraction scale (1×25) than conv1, showing more differences in the feature map between the initialised signal and T caused by the different noise patterns. From the results, the G we obtained is similar to that of T in *internal noise* and both layers of feature maps, which confirms that our method can extract and synthesise G with good quality.

5.2   Performance on Analysis Model

The most intuitive way to evaluate the noise patterns transferring models is to analyse their output using a model trained by the samples with the target noise pattern. In this section, we validate the proposed method in comparison with baselines.

The baselines we use are wavelet denoising [8], autoencoder (AE) [4], 1D DnCNN [17], and 1D CycleGAN [11]. They are representatives of traditional, DNN, and adversarial generation methods and are suitable for the target task. Besides, we present the accuracy of the COD parameter analysis model using the unprocessed data in $D_\mathbb{S}$ and $D_\mathbb{T}$, which provides references to the performance of baselines.

We use two training ratios (30% and 50%) to train the noise transfer model, i.e., use only a training set in 3 and 5 of 10 standard water solutions, while all ten groups of data are used for the validation set. It can indicate the generalisation ability and the effect of the noise patterns transferring method with generated cases (GC). The experimental results are shown in Table 1.

5.3   Analysis and Discussions

The results show that GC-based DnCNN CNPT model (GC-DnCNN) performs best and has a significant advantage over the baselines. It validates that our method can generate cases with better quality, which helps CNPT model learn differences in noise patterns between datasets, thereby improving the analysis model's performance. It should be noted that GC-AE, which is trained by S2S case-base performs better than AE. It better indicates that the performance of the noise patterns transferring model is improved by training with generated cases (the S2S case-base). It validates that we can enhance the performance of DL models by generating high-quality case-bases.

The CycleGAN model significantly performs poorly even lower than the unprocessed samples. It may be because the data set is not large enough, which make CycleGAN overfitting [19]. Besides, AE performs badly because it directly

Table 1. The experimental results of our model and baselines. Transfer targets correspond to D_S and D_T. For example, when the transfer target is N_B to N_A, it means $D_S = D_B$ and $D_T = D_A$ and the analysis model is trained by D_A. For the results, D_S and D_T show the accuracy when feeding the unprocessed data in D_S and D_T to the analysis model. In contrast, others feed the data in D_S that are processed by corresponding baselines to the analysis model. Each group of results uses two training ratios: 30% and 50%, which indicates how many standard water solutions we use when training the CNPT model. We also tested the results using an autoencoder (AE) as a CNPT model, which tests if our generated cases can improve the performance of different case-based noise transfer DL models.

Transfer target	Training ratio	D_S	Wavelet	CycleGAN	AE	GC-AE	DnCNN	GC -DnCNN	D_T
N_B to N_A	50%	63.83	74.90	76.75	50.60	57.00	71.50	**82.60**	90.78
	30%		78.40	76.50	53.25	57.30	74.25	**82.58**	
N_C to N_A	50%	33.94	59.30	37.50	37.45	42.85	55.80	**73.55**	90.78
	30%		69.30	39.10	30.70	44.80	62.15	**70.85**	
N_C to N_B	50%	55.11	62.60	66.25	48.15	53.45	58.95	**88.60**	98.78
	30%		73.45	65.25	53.35	57.00	69.45	**85.05**	
N_A to N_B	50%	59.67	49.40	37.25	39.15	49.05	65.35	**94.00**	98.78
	30%		58.20	42.25	39.00	45.30	63.70	**94.95**	
N_A to N_C	50%	29.72	53.65	40.20	23.65	31.35	51.20	**56.25**	88.28
	30%		52.90	46.85	27.20	38.45	53.95	**56.26**	
N_B to N_C	50%	36.44	54.85	49.75	39.45	39.70	52.45	**59.40**	88.28
	30%		60.90	57.90	39.15	41.60	59.85	**63.80**	

smooths the *internal noise* when transfers the position related *environmental noise*. It will also lead to overfitting problems because the *internal noise* can show low-frequency noise patterns that are superimposed on *environmental noise* and signal. The spurious peaks will mislead AE to transfer it to a wrong noise or signal pattern. This phenomenon does not occur in two other baselines, i.e., wavelet and DnCNN, demonstrating the satisfactory results of traditional signal noise processing methods and the DNN-based models' generalisation capability.

6 Conclusions

In this paper, we propose a noise pattern transferring approach for learning noise pattern differences from S2S cases, which is the first work to enhance the learning effect of noise pattern transferring models by generating cases. The most novelty idea is that we generate an S2S case-base using the 'feature extraction-matching' method to enhance the learning ability of the noise pattern transferring model. Experimental results on a COD concentration measurement task show that our GC-based noise pattern transferring model outperforms important baselines ranging from wavelet denoising, DNN, and generative models.

The excellent performance of the generated cases on different CNPT models also validates that our model can be used as a plug-in to the noise processing systems. When existing cases cannot express enough features for noise processing systems, implementing reasonable case-generating methods will contribute to performing the relevant tasks. In future work, we will continue exploring principles and applications of generated cases in noise processing systems.

Acknowledgements. This document is the results of the research project funded by the Science Foundation Ireland (SFI) [SFI/12/RC/2289_P2]; Beijing Dublin International College Fund; China scholarship council Grant 202106120101.

References

1. Abdessamad, E., Saadane, R., El Aroussi, M., Wahbi, M., Hamdoun, A.: Spectrum sensing with an improved energy detection. In: 2014 International Conference on Multimedia Computing and Systems (ICMCS), pp. 895–900. IEEE (2014)
2. An, Y., et al.: Current state and future directions for deep learning based automatic seismic fault interpretation: a systematic review. Earth Sci. Rev. **243**, 104509 (2023)
3. Bukin, O., et al.: New solutions of laser-induced fluorescence for oil pollution monitoring at sea. Photonics **7**, 36 (2020)
4. Chandra, B., Sharma, R.K.: Adaptive noise schedule for denoising autoencoder. In: Loo, C.K., Yap, K.S., Wong, K.W., Teoh, A., Huang, K. (eds.) ICONIP 2014. LNCS, vol. 8834, pp. 535–542. Springer, Cham (2014). https://doi.org/10.1007/978-3-319-12637-1_67
5. Chen, S., et al.: Olive oil classification with laser-induced fluorescence (LIF) spectra using 1-dimensional convolutional neural network and dual convolution structure model. Spectrochim. Acta Part A Mol. Biomol. Spectrosc. **279**, 121418 (2022)
6. Du, H.: Laser-induced fluorescence spectral data of humic acid solution in different noise patterns (2023). https://doi.org/10.21227/7r0c-mf67
7. Du, H., et al.: Disentangling noise patterns from seismic images: noise reduction and style transfer. IEEE Trans. Geosci. Remote Sens. **60**, 1–14 (2022)
8. He, W., Zi, Y., Chen, B., Wang, S., He, Z.: Tunable Q-factor wavelet transform denoising with neighboring coefficients and its application to rotating machinery fault diagnosis. Sci. China Technol. Sci. **56**, 1956–1965 (2013)
9. Hu, F., et al.: Identification of mine water inrush using laser-induced fluorescence spectroscopy combined with one-dimensional convolutional neural network. RSC Adv. **9**(14), 7673–7679 (2019)
10. Hu, F., et al.: Selection of characteristic wavelengths using SPA for laser induced fluorescence spectroscopy of mine water inrush. Spectrochim. Acta Part A Mol. Biomol. Spectrosc. **219**, 367–374 (2019)
11. Kaneko, T., Kameoka, H., Hiramatsu, K., Kashino, K.: Sequence-to-sequence voice conversion with similarity metric learned using generative adversarial networks. In: Interspeech, vol. 2017, pp. 1283–1287 (2017)
12. Kaneko, T., Kameoka, H., Tanaka, K., Hojo, N.: CycleGAN-VC2: improved CycleGAN-based non-parallel voice conversion. In: 2019 IEEE International Conference on Acoustics, Speech and Signal Processing (ICASSP), ICASSP 2019, pp. 6820–6824. IEEE (2019)
13. Kazemzadeh, M., Hisey, C.L., Zargar-Shoshtari, K., Xu, W., Broderick, N.G.: Deep convolutional neural networks as a unified solution for Raman spectroscopy-based classification in biomedical applications. Optics Commun. **510**, 127977 (2022)

14. Kenny, E.M., Keane, M.T.: On generating plausible counterfactual and semi-factual explanations for deep learning. In: Proceedings of the AAAI Conference on Artificial Intelligence, vol. 35, pp. 11575–11585 (2021)

15. Kwon, Y.H., Park, M.G.: Predicting future frames using retrospective cycle GAN. In: Proceedings of the IEEE/CVF Conference on Computer Vision and Pattern Recognition, pp. 1811–1820 (2019)

16. Laurent, G., Woelffel, W., Barret-Vivin, V., Gouillart, E., Bonhomme, C.: Denoising applied to spectroscopies-part i: concept and limits. Appl. Spectrosc. Rev. **54**(7), 602–630 (2019)

17. Liu, Z., Wu, H., Du, H., Luo, Z., Tang, M.: Distributed temperature and curvature sensing based on Raman scattering in few-mode fiber. IEEE Sens. J. **22**(23), 22620–22626 (2022)

18. Loh, W., et al.: Operation of an optical atomic clock with a Brillouin laser subsystem. Nature **588**(7837), 244–249 (2020)

19. Peng, X., et al.: Contour-enhanced CycleGAN framework for style transfer from scenery photos to Chinese landscape paintings. Neural Comput. Appl. **34**(20), 18075–18096 (2022)

20. Santos, G.J.E., Rivera, M., Eiswirth, M., Parmananda, P.: Effects of noise near a homoclinic bifurcation in an electrochemical system. Phys. Rev. E **70**(2), 021103 (2004)

21. Sobolev, I., Babichenko, S.: Application of the wavelet transform for feature extraction in the analysis of hyperspectral laser-induced fluorescence data. Int. J. Remote Sens. **34**(20), 7218–7235 (2013)

22. Tian, Z., et al.: Rapid water quality assessment by micro laser–induced fluorescence spectrometer. In: Advanced Solid State Lasers. Optica Publishing Group (2019). Paper JTh3A.46

23. Turner, J.T., Floyd, M.W., Gupta, K., Oates, T.: NOD-CC: a hybrid CBR-CNN architecture for novel object discovery. In: Bach, K., Marling, C. (eds.) ICCBR 2019. LNCS (LNAI), vol. 11680, pp. 373–387. Springer, Cham (2019). https://doi.org/10.1007/978-3-030-29249-2_25

24. Turner, J.T., Floyd, M.W., Gupta, K.M., Aha, D.W.: Novel object discovery using case-based reasoning and convolutional neural networks. In: Cox, M.T., Funk, P., Begum, S. (eds.) ICCBR 2018. LNCS (LNAI), vol. 11156, pp. 399–414. Springer, Cham (2018). https://doi.org/10.1007/978-3-030-01081-2_27

25. Wang, H., Zhao, Z., Wang, Z., Xu, G., Wang, L.: Independent component analysis-based baseline drift interference suppression of portable spectrometer for optical electronic nose of internet of things. IEEE Trans. Industr. Inf. **16**(4), 2698–2706 (2019)

26. Yang, Z., Albrow-Owen, T., Cai, W., Hasan, T.: Miniaturization of optical spectrometers. Science **371**(6528), eabe0722 (2021)

27. Zacharioudaki, D.E., Fitilis, I., Kotti, M.: Review of fluorescence spectroscopy in environmental quality applications. Molecules **27**(15), 4801 (2022)

28. Zhang, K., Zuo, W., Chen, Y., Meng, D., Zhang, L.: Beyond a Gaussian denoiser: residual learning of deep CNN for image denoising. IEEE Trans. Image Process. **26**(7), 3142–3155 (2017)

Progressive Supervision for Tampering Localization in Document Images

Huiru Shao[1,2], Kaizhu Huang[3], Wei Wang[1], Xiaowei Huang[2], and Qiufeng Wang[1(✉)]

[1] School of Advanced Technology, Xi'an Jiaotong-Liverpool University, Suzhou, China
Huiru.Shao21@student.xjtlu.edu.cn, {Wei.Wang03,Qiufeng.Wang}@xjtlu.edu.cn
[2] University of Liverpool, Liverpool, UK
xiaowei.huang@liverpool.ac.uk
[3] Duke Kunshan University, Suzhou, China
kaizhu.huang@dukekunshan.edu.cn

Abstract. Tampering localization in document images plays an important role in the field of forensic and security, which has made great progress in recent years, however it is far from being solved. In this work, we aim to improve the tampering localization performance by refining both sides of the localization model. On one hand, we propose a multi-view enhancement (MVE) module at the input side, which combines RGB image, noise residual and texture information to obtain more forensic traces for tampering localization. On the other hand, at the output side, we propose both progressive supervision (PS) and detection assistance (DA) modules to enrich more detailed supervision information. Under the progressive supervision, we calculate BCE loss at each scale to extensively explore multi-scale features, which are vital for the tampering localization. To explore the tampering detection model, we adopt a KL loss to align both tampering localization and detection scores in the DA module, benefiting the estimation of global tampered probability. In the experiments, we evaluate the proposed method on the benchmark dataset DocTamper and the results demonstrate its effectiveness.

Keywords: Tampering localization · Document image · Progressive supervision · Multi-view enhancement

1 Introduction

Due to widely used image editing tools with convenient operations (e.g., copy-move, splicing and inpainting), more challenges have been posed to the field of multimedia forensics. Specially, document images usually contain important semantic information such as ID numbers, amount of transactions set in the contract [16,21], therefore it is essential to localize tampered regions in document images for the multimedia forensics.

© The Author(s), under exclusive license to Springer Nature Singapore Pte Ltd. 2024
B. Luo et al. (Eds.): ICONIP 2023, CCIS 1969, pp. 140–151, 2024.
https://doi.org/10.1007/978-981-99-8184-7_11

Tampering localization has attracted much attention recently. In the natural images, there are abundant color changes, visual texture and divergent edge of candidate regions to obtain inconsistencies for tampering localization [5]. However, such information is usually not sufficient in the document images, where the colors of background are pretty similar and the candidate contents for tampering in document images are similar, such as font and size [17]. Therefore, it is challenging to directly apply the achievements of tampering localization in natural images to document images. For document images, some works explored visual forensic features about typography, such as the inconsistency about shapes, weights, angles and serifs of characters [3], which has achieved higher performance for those document images with obvious tempered traces. But most of realistic tampered document images do not remain strong visual tempered traces, resulting lower performance.

It is essential to obtain sufficient forensic trace for the tampering detection/localization [11]. To this end, we refine both the input side and output side of the model in this paper. On one hand, at the input side of the network, we propose a multi-view enhancement (MVE) module to concatenate RGB signals, noise residual and texture information as multi-view forensic clue maps for tampering localization. Different from the previous works using texture [4] that captured handcrafted statistic features designed for specific phenomena or attributes, we leverage strong and comprehensive deep features and apply it in the task of forgery localization regarding as pixel-level classification. Compared to the noise signals [23] which utilized results processed by the SRM (Spatial Rich Model) filter layer, we employ residual of subtraction between RGB images and results processed by the Bayar layer. On the other hand, we propose a progressive supervision and detection assistance loss at the output side of the network, which helps to optimize the model with more sufficient information. In the progressive supervision, we calculate the BCE loss at different scale of images, which can help localize tampered regions in the document images progressively. For the detection assistance loss, we utilize a KL loss to align the tampering localization and detection scores which is from a tampering detection block. As the detection task only predicts the image whether contains tampering or not, it can be regarded as a more accurate task than the localization task. By integrating such KL loss alignment, it helps our model to obtain a global tempered probability. In the experiments, we evaluate the proposed method on the benchmark tampered document image dataset (DocTamper [12]) and achieve promising performance.

The main contributions of this work can be summarized as follows:

- We propose a multi-view enhancement block at the input side of the tampering localization model to enhance the forensic traces, which concatenates RGB signals, noise residual and texture information
- We propose a progressive supervision paradigm to calculate the loss at different scales, which can efficient localize tampered regions in the document images progressively

– We integrate a Kullback-Leibler (KL) loss to refine image-level tampered
probability from a predicted localization mask with detection score in global
view, which promotes the performance of tampering localization.

2 Related Works

2.1 Natural Image Tampering Localization

We detail some works about tampering localization based on natural images
from the perspective of noise which is the kind of invisible forensic trace, such as
noise caused by in-device acquisition process and out-device image processing.
One straightforward idea to highlight this kind of weak signal is to suppress
strong signal–visual image content based on natural images in previous works
through high-pass filters or denoising [6,23], such as the spatial rich models
(SRM) [23]. But different from natural images, for document images, high-pass
filtered results, which are the edge of characters not that of tampered region, have
less forensic information for tampering localization. Guillaro et al. used denoisng
to obtain information of noise recently [6] , However, framework of denoiser is
based on the hypothesis of one predefined noise model which is too ideal to be
realistic. Therefore we abandon denoising to obtain noise and utilize the residual
of subtraction between original image and the result of content suppression layer
to extract background and noise information.

2.2 Document Image Tampering Localization

Texture is a visual feature that reflects homogeneous phenomena in images,
reflecting the surface structure and arrangement attributes of objects with slow
or periodic changes. Different from natural images, texture in document images is
hardly visible especially in the background. When cheap fake occurs, the texture
surrounding tampered regions may be changed especially post operations applied
after tampering. Some works utilized LBP to extract relevant texture images
and use handcrafted features to represent the characteristics about changed texture [4]. We also use LBP, but explore the ability of deep feature extractors in
the field of tampering detection.

In recent years, tampering localization has made promising progress. At first,
as the extension of application in the field of source printer identification, they
focused on the apparent inconsistency of visual content preprocessed by OCR
system. For example, Hardik et al. [7] proposed a set of features for characterizing
text-line-level geometric distortions; Romain et al. [3] tried to seek characters,
words or sentences in a document with font properties different from their sur-
roundings as a clue to detect document forgery. But they all only work well
on clean document images with unified template. With development of deep
learning, some works tried to use general and strong feature extractors. Yu Sun
et al. [15] proposed the model built following the encoder-decoder structure using
a pre-trained residual network as the encoder to extract rich features and generating the localization mask in the decoder. While the dataset used in these two

methods are limited, which can not prove its cross-domain generalization ability in the domain of document tampering detection. Wenbo Xu et al. [20] proposed a two-stream network for grasping forgery traces in the aspects of spatial and frequency information. However, their performance is relatively poor, which can not prove its effectiveness. Yuxin Wang et al. [17] used features extracted by the frameworks of scene text detection and trained on the images tampered by SRNet [18], which is effective in capturing high frequency clues left by generative tampering and could hardly work well on other tampered types and more complex scenarios.

3   Methodology

Figure 1 shows an overview of our method for tampering localization in document images, which mainly contains four blocks: Multi-view enhancement (MVE) block, Backbone network, Detection Assistance (DA) block and progressive supervision (PS) block. Firstly, we obtain multi-view signals from an image processing module for a input image. Next, such multi-view signals are concatenated and input to backbone network to generate multi-scale features. Lastly, we adopt spatio-channel Attention modules on each scale to predict the tampering masks, and calculate the supervision loss progressively at each scale with the corresponding ground-truth masks. Meanwhile, we also utilize a detection module to obtain the detection score, which will be used for the both detection loss and KL loss with the prediction from localization block. To be noted, we only use the first-scale predicted mask (i.e., $Mask^1_{pred}$) for the tampering location in the test stage. In the following, we will give more details of the proposed MVE, DA and PS blocks, while we adopt a widely used light-weight model HRNet as our backbone network [14].

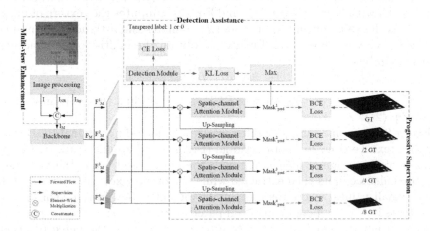

Fig. 1. The overall diagram of the proposed network.

3.1   Multi-view Enhancement

For the tampering localization in document images, it is essential to seek anomaly among all pixels. To this end, human commonly make full use of visual clues by comparing discrepancies and similarities of color and texture, as well as the logic compatibility of suspected regions with their context. In this work, we do not consider logic context for the simplicity. Instead, we focus on the visual clues which is commonly imitated by RGB signals. However, due to monotonous color in document images, the changes of color and texture are more difficult to perceive than those in natural images. Therefore, the network can be confused and over-fit in some visual patterns which are not related to the task of tampering localization. Therefore, besides RGB images, we use the additional operator to expose and highlight odd texture. To overcome this issue, some forgery localization methods in natural images utilize one specific convolutional layer to suppress visual contents and call results of the suppressed layer as noise. Motivated by this, we utilize residuals between original images and outputs of the suppressed layer, and call them as noise residuals, to highlight the inconsistencies in font properties as well as enhancing that in the background which can hardly be seized by the original suppressed layer. We transform original RGB images I to noise residual I_{NR} and LBP maps I_{lbp} in Eq. (1) and Eq. (2) to expose multiple and useful information from different perspectives:

$$I_{NR}(i,j) = \begin{cases} I(i,j) - BLayer(I(i,j)) & I(i,j) > BLayer(I(i,j)) \\ 255 + I(i,j) - BLayer(I(i,j)) & I(i,j) < BLayer(I(i,j)), \end{cases} \quad (1)$$

$$I_{lbp}^p = LBP(I_p) = \sum_{n=0}^{7} 2^n u[I_{q_n} - I_p], \quad (2)$$

where $BLayer$ in Eq. (1) is abbreviation of $BayarLayer$ which is a Convolutional operation [2] with kernel size of $5*5$, and Eq. (2) computes the texture of images where I_p is the value of central pixel and I_{q_n} denotes the n-th neighboring pixel value for a $3*3$ neighborhood. Finally, we obtain the multi-view enhancement forensic map by

$$I_M = Cat(I, I_b, I_{lbp}), \quad (3)$$

where $Cat(\cdot)$ means concatenation of RGB, noise residual (Eq. (1)) and texture maps (Eq. (2)).

Figure 2 shows one sample of different signals, including original RGB image (Fig. 2a), texture (Fig. 2b), noise residual (Fig. 2c) and ground-truth (GT) mask (Fig. 2d) of this original image. Through the suppressed layer, the anomalous texts are enhanced and then extracted while the other pixels are overlooked due to very small values, where there is significant differences in the color and size of the texts in the tampered region compared to that in the authentic region. In this case, it is very easy for network to distinguish between tampered pixels and other pixels. Although inconsistent fonts properties are important visual traces, when this kind of inconsistencies are small, authentic texts are also be remained leading to misjudgement with less forensic traces to be utilized in the noise

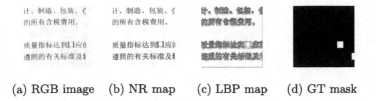

(a) RGB image   (b) NR map   (c) LBP map   (d) GT mask

Fig. 2. The Multi-view enhancement maps.

map. At the same time, if this noise map is used to train a tampering system, it can also be misled in authentic documents where texts are required to have different font weights, colors, and sizes. Therefore we use noise residual to obtain traces remained by inconsistent font properties as well as the forensic traces in the background. The above considerations can be observed by comparing Fig. 2b and Fig. 2c. Absolutely due to the influence of suppressed layer in the calculation process of noise residual, texts and background in the tampered regions for forensics are strengthened compared to the original RGB image as shown in Fig. 2a. Texture is shown to be different between tampered and authentic regions in Fig. 2d after obtaining the texture map by LBP, which gives another kind of forensic clue for tampering localization.

3.2 Progressive Supervision Module

Progressive mechanism has been widely used in the field of computer vision [22], which aims to solve the multi-scale problem. Inspired by these works, we explore the applicability of Progressive Supervision Module (PSM) for tampering localization in document images to supervise divergent sizes of features processed by backbone and attention modules. The predicted mask in each scale will be a coarse advisor for the next scale. The progressive mechanism is reflected in the supervision of predicted tampering masks shown in Fig. 1. The Spatio-Channel Attention Module utilized in each scale has the same structure. Considering that our task is about localization which needs positional information, we utilize the attention module constructing from the perspectives of both channel and spatial position to learn effective feature representation. In Fig. 3, we visualize both features before and after the finest-scale Spatio-Channel Attention Module (SCAM) to indicate its feasibility. Each input size is twice of the size of its coarser Spatio-Channel Attention Module, therefore there is an up-sampling operation before the next scale. We obtain the predicted mask with finest scale as the final results. During the training, the predicted mask will approach the GT mask, which means that the advisor will give more accurate advice to the next scale, and we will finally get trustworthy results in this way. In summary, Eq. (4) and Eq. (5) can describe the process of PSM:

$$M_{pred}^i = SCAM[U[M_{pred}^{i+1}] \odot F_M^i], i = 3, 2, 1, \tag{4}$$

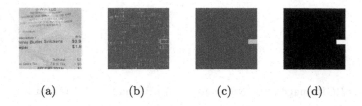

(a)   (b)   (c)   (d)

Fig. 3. Visualization of features before and after SCAM: (a) RGB image, (b) before SCAM, (c) after SCAM, (d) GT mask.

$$L_{loc}^i = BCE(M_{pred}^i, GT_M^i), i = 4, 3, 2, 1, \tag{5}$$

where M_{pred}^{i+1} means the predicted mask from previous coarse scale; $U[.]$ denotes the operation of Up-sampling; $SCAM$ is the process of Spatio-Channel Attention Module; GT_M^i is the groudtruth mask in the i-th scale; $BCE(.)$ represents binary cross-entropy loss to supervise the task of tampering localization.

3.3 Detection Module

The objective of tampering localization is to locate the fine-grained spatial position of tampered regions, which is commonly considered as pixel-level classification. In general, this is more difficult than tampering detection focusing on image-level classification. Obviously, if there are some pixels are predicted to be tampered, the image can be classified into the tampered category. Based on this, the results of tampering detection can be transformed from that of tampering localization. This method for obtaining the results of tampering detection is simple and straightforward but fully depend on the results of tampering localization which are more likely be inaccurate in practice. As observed in our experiments, adding additional detection module can improve the performance of tampering localization. To make full use of global and local information from Backbone module, we utilize multi-scale features from backbone one by one through upsampling coarse scale features to fuse with fine scale features. We use cross-entropy loss (L_{dec}) to supervise this process by

$$L_{dec} = CE(s_d, l_d), \tag{6}$$

where s_d and l_d mean the score of detection module and ground-truth of detection module; $CE(\cdot)$ represents cross-entropy loss.

Tampering detection concentrates on image-level forensic clues, so we regard this is Coarse-Grained task and easier to obtain promising performance than the fine-grained task of tampering localization, which is more likely to tend to be confused in some pixels due to the very low discrimination between tampered and authentic pixels. From another perspective, the aim of tampering localization is consistent with that of tampering detection and we can obtain tampering detection results by analyzing the maximum of pixel-level tampering score in tampering localization maps. Therefore, we use L_{kl} in Eq.(7) to align the distribution of image-level tampering score with that from the detection module, which is beneficial to distinguish tampered and authentic pixels.

$$L_{kl} = KL(s_l, s_d), \tag{7}$$

$$s_l = \log \sum_{i,j} e^{M_{pred}^{1,i,j}}, \tag{8}$$

where $KL(\cdot)$ represents Kullback-Leibler loss, s_l is the image-level score by computing a differentiable approximate maximum function Eq. (8) from pixel-level probabilistic score maps in the localization branch [6,9], where $M_{pred}^{1,i,j}$ represents the probability of pixel (i,j) as tampering pixel in the predicted mask at the finest scale. In summary, the loss of DA module can be represented by

$$L_{DA} = L_{dec} + L_{kl}. \tag{9}$$

Finally the total loss function of our framework is calculated by both progressive supervision (Eq. (5)) and detection assistance loss (Eq. (9)) by

$$L = \sum_{i=1}^{4} L_{loc}^i + L_{DA}. \tag{10}$$

4   Experiments

4.1   Experimental Setting

Datasets: To be our best of knowledge, there are not many public large datasets about document tampering localization. In our experiments, we adopt the largest dataset DocTamper released in 2023 [12], which contains various types of documents, such as contracts, invoices, and sroies, with both Chinese and English. There are three tampered types in the DocTamper: copy-move, splicing and inpainting. In the DocTamper, there are four sub-sets: one training set with 12,000 images, and three test sets containing TestingSet/FCD (First Cross Domain)/SCD (Second Cross Domain) with 30,000/4,000/36,000 images. The TestingSet contains images with the same domain as training set while FCD and SCD images are from another two different other domains.

Metrics: In the evaluation, we consider the tampering localization as binary pixel-level classification and adopt Precision (P), Recall (R) and F-score (F) as the evaluation metric.

Implementation Details: Our model is implemented by PyTorch on GeForce RTX 3090. The input size is set to 256 * 256 for training stage and arbitrary-size images can be utilized in testing stage. We train our model with DocTamper training dataset, optimize it by Adam [13] with a batch size of 10. The initial learning rate is 2e–4 and halved every 5 epochs.

4.2   Ablation Study

To evaluate the usefulness of each component in our proposed multi-view enhancement block containing original RGB image (RGB), texture information (LBP) and noise residual (NR), we conduct ablation study on each component with the same network (baseline + PS (progressive supervision) + Dec (detection assistance module) + KL loss), and the results are shown in Table 1. Since RGB images are commonly used in different tasks, we first only use RGB images. Shown in Table 1, with adding more information, the performance is gradually improved, which is consistent in three testing datasets and the values of each row demonstrate that each kind of information is feasible for tampering localization. Since TestingSet is in the same domain as training set, the gap of results caused by inputting different information is relatively smaller than them in another two cross domain datasets. It indicates that the input with three kinds of multi-view information has the best generalization. From the results in FCD, we observe that without noise residual, the performance is dropped significantly and also in SCD the results with RGB and noise residual are relatively better than them with RGB and LBP, which reflects the superiority of noise residual in generalization compared to LBP.

Table 1. Ablation study on multi-view enhancement. Bold text means the best.

Method	TestingSet			FCD			SCD		
	P	R	F	P	R	F	P	R	F
RGB	63.93	87.63	72.05	43.30	97.87	58.49	57.64	81.27	64.24
RGB + LBP	64.92	93.95	75.54	43.75	**99.72**	59.15	57.87	88.51	68.05
RGB + NR	64.08	93.96	73.67	61.91	99.17	75.75	59.01	89.35	69.06
RGB + LBP + NR	**69.84**	**97.22**	**79.79**	**71.59**	98.72	**82.72**	**65.93**	**90.94**	**74.07**

At the output side of the network, we propose a Progressive Supervision (PS) and detection assistance loss to optimize the model more effectively, where the DA loss (Eq. (9)) contains detection (Dec) based cross-entropy loss (Eq. (6)) and KL loss (Eq. (7)). To evaluate the effectiveness of each supervision loss, we conduct the ablation study on different supervision information, and the results are shown in Table 2, where the baseline method means it does not have PS and DA loss (i.e., only the loss L_{loc}^1 are used). In summary, adding of each loss is useful observed from the results of F-score in three testing datasets. Apart from baseline, the generalization of these frameworks based on FCD is pretty good while the performance in SCD is relatively poor due to that images in SCD are obtained in different environment and the format in each image is not as standard as that in FCD which will influence generalization. Among the six frameworks in Table 2, the value of recall is superior to that of precision obtained by frameworks with progressive supervision while by contrast the value of precision is superior to that of recall obtained by frameworks without progressive supervision. It is more likely to be caused by that progressive supervision makes use of multi-scale information and tend to predict as many positive pixels as possible; the information obtained in frameworks without progressive supervision is not much richer than that in frameworks with progressive supervision, which leads to less predicted positive pixels. Fortunately in our final solution, the value of recall is almost perfect which is very close to 1 and the value of precision is not low. It indicates that almost all pixels labeled as 1 can be judged correctly and though some pixels labeled as 0 are predicted wrongly by our method, the number of these wrongly predicted pixels is acceptable.

Table 2. Ablation study on different supervision information.

Method	TestingSet			FCD			SCD		
	P	R	F	P	R	F	P	R	F
Baseline	80.39	61.15	64.89	85.18	56.85	63.20	79.76	51.43	56.24
Baseline + PS	65.99	92.47	75.23	64.68	98.69	77.29	58.72	87.46	67.62
Baseline + Dec	**84.30**	64.29	69.43	**94.51**	68.44	77.23	**89.09**	57.27	63.17
Baseline + PS + Dec	66.64	96.08	76.02	67.23	98.61	79.49	60.20	88.83	69.49
Baseline + Dec + KL	83.72	71.91	74.14	88.20	79.46	81.24	88.38	63.24	66.71
Baseline + PS + Dec + KL	69.84	**97.22**	**79.79**	71.59	**98.72**	**82.72**	65.93	**90.94**	**74.07**

4.3 Comparison with Other Methods

We evaluate our method with some state-of-the-art natural image manipulation localization methods [5,8,19], one document tampering localization method [12] and two semantic segmentation methods [1,10] with their officially released codes as shown in Table 3. We can see that our proposed method is more promising than most of them and exposes the effectiveness and applicability in document tampering localization, and achieve the state-of-the-art F-score on Testingset and FCD, and second-score on SCD. However, our performance in the metric of precision is lower than that of [12], which made the performance of F-score is not as excellent as recall in our method. Although [12] has balanced values of precision and recall, they misjudge relatively more tampered pixels to be authentic ones. In this work, we misjudge more authentic pixels to be tampered ones while almost all pixels with positive label are predicted correctly. Hence one advantage of our method is that we will save time and narrow down the scope for the subsequent re-judgment process in the future by only considering these predicted positive pixels. From another perspective, the above phenomenon is more likely to be caused by that the amount of forensic clues in different images is uneven. This also indirectly reflects the difficulty and uncertainty of extracting appropriate information and the necessity of converting images into forensic domain in tampering localization. Meanwhile, the performance based on semantic segmentation methods is poor, confirming this argument once again, whose extracted features are more likely to concentrate on some visual patterns, not the forensic traces expected in the task of tampering localization.

Table 3. Comparison with other methods. We use bold and underline to indicate optimal performance and suboptimal performance.

Method	Testingset			FCD			SCD		
	P	R	F	P	R	F	P	R	F
Mantra-Net [19]	12.30	20.40	15.30	17.50	26.10	20.90	12.40	21.80	15.70
MVSS-Net [5]	49.40	38.30	43.10	48.00	38.10	42.40	47.80	36.60	41.40
BEiT-Uper [1]	56.40	45.10	50.10	55.00	43.60	48.70	40.80	39.50	40.20
Swin-Uper [10]	67.10	60.80	63.80	64.20	47.50	54.60	54.10	61.20	57.40
CAT-Net [8]	<u>73.70</u>	66.60	70.00	64.40	48.40	55.30	64.50	61.80	63.10
DocTamper [12]	**81.40**	<u>77.10</u>	<u>79.20</u>	**84.90**	<u>78.60</u>	<u>81.60</u>	**74.50**	<u>76.20</u>	**75.40**
ours	69.84	**97.22**	**79.79**	<u>71.59</u>	**98.72**	**82.72**	<u>65.93</u>	**90.94**	<u>74.07</u>

5 Conclusion

In this paper, we aim to improve the tampering localization in document images by enhancing both input and out sides. At the input side, we propose a multi-view enhancement module to combine three types of signals including RBG channels, noise residual and texture information, which help obtain more strong forensic traces. At the output side, we propose a progressive supervision loss and detection assistance loss to guide the model optimization more effectively. Our progressive supervision calculates the loss progressively at each scale with the corresponding ground-truth masks, which will help to capture the tampered regions. With the help of detection module, we can align tampering localization and detection scores by a KL loss, benefiting the estimation of global tampered probability from localization branch. By the extensive experiments on the benchmark dataset DocTamper, we can see each refinement strategy can improve the localization performance, and the combination of all strategies will boost the performance further, achieving a new state-of-the-art F-score on the testingset of DocTamper.

Acknowledgements. This research was funded by National Natural Science Foundation of China (NSFC) no.62276258, Jiangsu Science and Technology Programme no. BE2020006-4, European Union's Horizon 2020 research and innovation programme no. 956123, and UK EPSRC under projects [EP/T026995/1]

References

1. Bao, H., Dong, L., Piao, S., Wei, F.: BEiT: BERT pre-training of image transformers. In: ICLR (2022)
2. Bayar, B., Stamm, M.C.: Constrained convolutional neural networks: a new approach towards general purpose image manipulation detection. IEEE Trans. Inf. Forensics Secur. **13**, 2691–2706 (2018)

3. Bertrand, R., Terrades, O.R., Gomez-Krämer, P., Franco, P., Ogier, J.M.: A conditional random field model for font forgery detection. In: ICDAR, pp. 576–580 (2015)
4. Cruz, F., Sidere, N., Coustaty, M., d'Andecy, V.P., Ogier, J.M.: Local binary patterns for document forgery detection. In: ICDAR, pp. 1223–1228 (2017)
5. Dong, C., Chen, X., Hu, R., Cao, J., Li, X.: Mvss-net: multi-view multi-scale supervised networks for image manipulation detection. IEEE Trans. Pattern Anal. Mach. Intell. **45**(3), 3539–3553 (2022)
6. Guillaro, F., Cozzolino, D., Sud, A., Dufour, N., Verdoliva, L.: TruFor: leveraging all-round clues for trustworthy image forgery detection and localization. In: CVPR, pp. 20606–20615 (2023)
7. Jain, H., Joshi, S., Gupta, G., Khanna, N.: Passive classification of source printer using text-line-level geometric distortion signatures from scanned images of printed documents. Multimedia Tools Appl. **79**(11–12), 7377–7400 (2020)
8. Kwon, M.J., Nam, S.H., Yu, I.J., Lee, H.K., Kim, C.: Learning jpeg compression artifacts for image manipulation detection and localization. Int. J. Comput. Vision **130**(8), 1875–1895 (2022)
9. Liu, J., Zheng, L.: A smoothing iterative method for the finite minimax problem. J. Comput. Appl. Math. **374**, 112741 (2020)
10. Liu, Z., Hu, H., Lin, Y.E.A.: Swin transformer v2: Scaling up capacity and resolution. In: CVPR, pp. 12009–12019 (2022)
11. Mayer, O., Stamm, M.C.: Forensic similarity for digital images. IEEE Trans. Inf. Forensics Secur. **15**, 1331–1346 (2020)
12. Qu, C., et al.: Towards robust tampered text detection in document image: new dataset and new solution. In: CVPR, pp. 5937–5946 (2023)
13. Sharma, M., Pachori, R., Rajendra, A.: Adam: a method for stochastic optimization. Pattern Recogn. Lett. **94**, 172–179 (2017)
14. Sun, K., Xiao, B., Liu, D., Wang, J.: Deep high-resolution representation learning for human pose estimation. In: CVPR, pp. 5686–5696 (2019)
15. Sun, Y., Ni, R., Zhao, Y.: MFAN: multi-level features attention network for fake certificate image detection. Entropy **24**(1), 118–133 (2022)
16. Verdoliva, L.: Media forensics and deepfakes: an overview. IEEE J. Sel. Top. Sig. Process. **14**(5), 910–932 (2020)
17. Wang, Y., Xie, H., Xing, M., Wang, J., Zhu, S., Zhang, Y.: Detecting tampered scene text in the wild. In: ECCV, pp. 215–232 (2022)
18. Wu, L., et al.: Editing text in the wild. In: ACM MM, pp. 1500–1508 (2019)
19. Wu, Y., AbdAlmageed, W., Natarajan, P.: Mantra-net: manipulation tracing network for detection and localization of image forgeries with anomalous features. In: CVPR, pp. 9543–9552 (2019)
20. Xu, W., et al.: Document images forgery localization using a two-stream network. Int. J. Intell. Syst. **37**(8), 5272–5289 (2022)
21. Yang, Q., Huang, J., Lin, W.: SwapText: image based texts transfer in scenes. In: CVPR, pp. 14700–14709 (2020)
22. Yi, P., Wang, Z., Jiang, K., Jiang, J., Ma, J.: Progressive fusion video super-resolution network via exploiting non-local spatio-temporal correlations. In: ICCV, pp. 3106–3115 (2019)
23. Zhou, P., Han, X., Morariu, V.I., Davis, L.S.: Learning rich features for image manipulation detection. In: CVPR, pp. 1053–1061 (2018)

Multi-granularity Deep Vulnerability Detection Using Graph Neural Networks

Tengxiao Yang[1], Song Lian[1], Qiong Jia[2], Chengyu Hu[1], and Shanqing Guo[1(✉)]

[1] School of Cyber Science and Technology, Shandong University, Jinan, China
{202117071,202137086}@mail.sdu.edu.cn, guoshanqing@sdu.edu.cn
[2] Beijing Institute of Computer Technology and Applications, Beijing, China

Abstract. The significance of vulnerability detection has grown increasingly crucial due to the escalating cybersecurity threats. Investigating automated vulnerability detection techniques to avoid high false positives and false negatives is an important issue in the current software security field. In recent years, there has been a substantial focus on deep learning-based vulnerability detectors, which have achieved remarkable success. To fill the gap in multi-granularity program representation, we propose MulGraVD, a deep learning-based vulnerability detector at the function level. MulGraVD captures the continuity and structure of the programming language by considering information at word, statement, basic block, and function granularity respectively. To overcome the constraint posed by hyperparameter layers in the information aggregation process of graph neural networks, MulGraVD serially passes information from coarse to fine granularity, which facilitates the mining of vulnerability patterns. Our experimental evaluation on FFMPeg+Qemu and ReVeal datasets shows that MulGraVD significantly outperforms existing state-of-the-art methods in terms of precision, recall, and F1 score, with an average improvement of 11.62% in precision, 27.69% in recall, and 19.71% in F1 score.

Keywords: Vulnerabilty detection · Deep learning · Graph neural network

1 Introduction

In recent years, the rapid growth in software systems coupled with their increasing complexity has resulted in a surge of exposed software vulnerabilities (e.g. Buffer Overflow [2], Divide By Zero [5]). To cope with complex software security issues and improve the efficiency and accuracy of vulnerability detection analysis, traditional vulnerability detection methods include static analysis, dynamic analysis, and symbolic execution. Johnson *et al.* [14] proposed that the traditional approach suffers from high false positive or false negative rates. As the volume of open-source vulnerability data continues to increase, machine learning and deep learning based vulnerability detectors have attracted a lot of attention.

Current machine learning based vulnerability detection models require determining software metrics to characterize vulnerabilities. Software metrics include occurrence frequencies [23], imports and function calls [21], etc. The papers on deep learning based vulnerability detection are mainly focused on the source code domain [29], partly in the intermediate representation [16] and binary [15].

Deep learning vulnerability detection that utilizes source code as input encompasses different levels such as file [12], slice [18,19], function [22,29], path [10], and statement [13]. To better understand the structural nature of programs, work in this area has gradually moved from treating programs as sequences of languages [22] to using abstract syntax trees, data flows, control flows and so on to represent vulnerabilities [7,29]. Many approaches such as [13,17] use program dependency graphs in program representation. Zhou *et al.* [29] introduced the ComputedFrom, LastRead, and LastWrite types of edges to identify multiple types of vulnerabilities. We perceive that when using graph neural networks, the problem of being unable to absorb the far-hop rows needed to identify vulnerabilities arises due to the number of hyperparameter layers.

Therefore, in order to address the above issues and capture the continuity and structure of the programming language, we propose a new multi-granularity vulnerability detection model MulGraVD, which is based on Gated Graph Neural Network(GGNN) and Transformer. To prove the effectiveness of our model, we conducted experiments on FFMPeg+Qemu [29] and ReVeal [7] datasets. In summary, this paper makes the following contributions:

- We propose an effective multi-granularity vulnerability detection method MulGraVD, which explicitly characterizes vulnerabilities at the granularity of word, statement, basic block, and function.
- In the design of MulGraVD, the limitation of vulnerability-related rows being inaccessible in information aggregation due to the graph neural network's hyperparameter layers is overcome by progressively incorporating coarser granularity features.
- The results on the FFMPeg+Qemu and ReVeal datasets show that MulGraVD improves accuracy by 6.26%, precision by 11.62%, recall by 27.69%, and F1 score by 19.71% on average compared to the baseline approaches. Each granularity shows its additivity to the model. The word granularity, statement granularity, basic block granularity, and function granularity have 5.90%, 3.49%, 4.46%, and 4.66% improvement on the F1 score respectively.

The remainder of this paper is organized as follows. Section 2 describes our motivation. Section 3 discusses the design of MulGraVD. Section 4 shows the experimental evaluation and results. Section 5 lists the previous related work. Conclusions are provided in Sect. 6.

2   Motivation

2.1   Multi-granularity Program Representation Covering Multiple Vulnerability Types

The function granularity is used to capture program continuity. The remaining three granularities are used to depict the structure of functions. Based on Yamaguchi's [27] analysis, the combined representation of these three granularities can cover NULL Pointer Dereference, Integer Overflows, Resource Leaks, and so on. Table 1 displays the granularity representation coverage on partial vulnerability types detected by Flawfinder and Cppcheck in the FFmpeg+QEMU and ReVeal datasets.

Table 1. Coverage of different granularity representation for modeling partial vulnerability types in FFMPeg+Qemu and ReVeal datasets.

CWE ID	Coverage granularity	% in FFMPeg+Qemu	% in ReVeal
CWE-120 [2]	Word + Sentence	33.58	26.07
CWE-362 [4]	–	2.89	1.93
CWE-190 [3]	Word + Sentence + Block	2.31	2.15
CWE-369 [5]	Word + Sentence	0.02	–
CWE-476 [6]	Word + Sentence + Block	0.81	0.83

2.2   Overcoming the Constraint Posed by Hyperparameter Layers

In graph neural networks, the hyperparameter layers are generally determined based on model performance such as accuracy and so on, independent of the furthest hop of vulnerability-related rows in the entire datasets. If the number of layers is set according to the farthest hop, the problem of oversmoothing [9] will occur, resulting in similar features for all nodes. Figure 1 shows a code snippet of CVE-2012-2793 [1], in which we plot a six-hop data dependency path. When the layer value is greater than or equal to six, s_8 can absorb the information in s_{11} and thus determine the vulnerability.

3   The MulGraVD Framework

This section describes our vulnerability detection model MulGraVD based on GGNN and Transformer. Its framework is shown in Fig. 2, which includes the three sequential components: 1) Multi-granularity Program Representation, which learns program representation from word, statement, basic block, and function granularity. 2) Multi-granularity Information Propagation, where the constraint posed by hyperparameter layers is overcome in the aggregation process. 3) Representation Learning Model, which is trained after extracted features have been resampled.

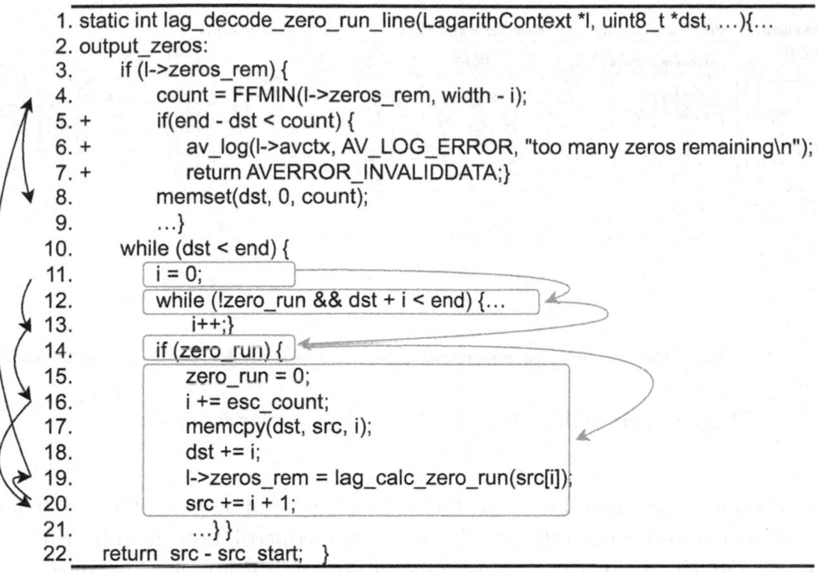

```
1. static int lag_decode_zero_run_line(LagarithContext *l, uint8_t *dst, ...){...
2. output_zeros:
3.     if (l->zeros_rem) {
4.         count = FFMIN(l->zeros_rem, width - i);
5. +       if(end - dst < count) {
6. +           av_log(l->avctx, AV_LOG_ERROR, "too many zeros remaining\n");
7. +           return AVERROR_INVALIDDATA;}
8.         memset(dst, 0, count);
9.         ...}
10.    while (dst < end) {
11.        i = 0;
12.        while (!zero_run && dst + i < end) {...
13.            i++;}
14.        if (zero_run) {
15.            zero_run = 0;
16.            i += esc_count;
17.            memcpy(dst, src, i);
18.            dst += i;
19.            l->zeros_rem = lag_calc_zero_run(src[i]);
20.            src += i + 1;
21.        ...} }
22.    return  src - src_start;  }
```

Fig. 1. CVE-2012-2793 in FFmpeg. The green boxes are the basic blocks and the black lines display a six hop data dependent path. (Color figure online)

3.1   Multi-granularity Program Representation

Word Granularity. The Abstract Syntax Tree (AST) is a tree-like representation of the abstract syntax structure of the source code. AST nodes attributes include location, key, code, location, and so on. GRU is used to capture the sequence of depth-first nodes sequence of the AST subtree of statement s and its initial word sequence to obtain the embedding EMB(s).

Statement Granularity. The statement granularity graph is constructed by data dependencies and control dependencies in the program. If the definition of a variable in one statement can reach its usage in another statement, there is a data dependency between these two statements (indicated by black arrows in Fig. 2 and the relevant variables are marked). Control dependency is a constraint resulting from the control flow of the program (indicated by yellow arrows in Fig. 2). The node features are aggregated and updated using GGNN. It is given by:

$$x_j^t = GRU \left(x_j^{t-1}, \sum_{(i,j) \in E} g\left(x_i^{t-1}\right) \right) \qquad (1)$$

where x_j^t is the embedding of vertex v_j at the tth layer in the statement granularity graph. E is the set of edges and $g(.)$ is a transformation function absorbing the embedding of all neighbors of vertex v_j.

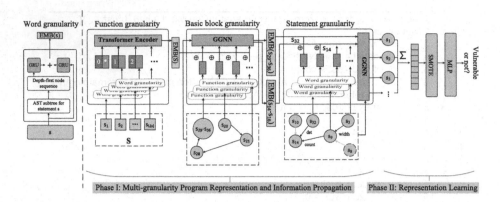

Fig. 2. The Architecture of MulGraVD (Color figure online)

Basic Block Granularity. The basic block granularity is derived from the single-in, single-out principle on the program control flow graph. Each node in this granularity contains at least one program statement. The path options are determined by conditional statements such as if, for, etc. The generation of node feature representation mimics the function granularity and aggregates information through GGNN.

Function Granularity. The function granularity representation allows the programming logic of the source code to be preserved. The sequence of statements is represented as a function embedding EMB(S) using a transformer encoder. In this case, we use ViT's [11] approach to add additional learnable [class] embeddings at the zero position and then use the embedding obtained by the transformer encoder at zero position as the function granularity embedding.

3.2   Multi-granularity Information Propagation

MulGraVD uses the statement granularity graph structure as the skeleton of the final generated graph embedding. It serially passes information from coarse to fine granularity, ultimately generating rich node feature representations for the nodes in the statement granularity graph. Our model maintains the mapping between statements and basic blocks during information propagation, allowing statements to absorb information from their corresponding basic block. The basic block granularity graph also absorbs information from function granularity. We illustrate this process by the generation of initial representation at statement granularity, which is given by:

$$x_j^0 = EMB\left(s_{v_j}\right) \oplus GRU\left(x_{b_j}^{l-1}, \sum_{(b_i, b_j) \in E} g\left(x_{b_i}^{l-1}\right)\right) \tag{2}$$

$$x_{b_j}^0 = EMB\left(b_j\right) \oplus EMB(S) \tag{3}$$

where b_j denotes the basic block corresponding to the statement granularity node v_j. s_{v_j} denotes the statement corresponding to node v_j. l is the layers of the graph neural network at the basic block granularity.

Analysis of Overcoming Constraints Posed by Layers. Taking Fig. 1 as an example, When the layers of basic block granularity are three, the basic block corresponding to $s_{15} - s_{20}$ can absorb the information of s_{11}. It is subsequently concat with the node features of s_{20} in the statement granularity graph. Even if the layers of statement granularity are three, it is still able to aggregate information from s_{11} (node with 6-hop distances in statement granularity graph). If the information of a particular row cannot be aggregated by these two stages, then a coarse representation can be provided by the function granularity. It is important to note that MulGraVD does not make the node features in the statement granularity graph similar, leading to oversmoothing problem. Because we use the concat function to absorb coarser granularity information while preserving the original features.

3.3 Representation Learning Model

We use the representation learning model proposed in the Reveal [7]. The imbalance between the number of vulnerable and non-vulnerable categories is handled by SMOTE [8], and then a multilayer perceptron (MLP) is used for classification. We use triple loss function $\mathcal{L}_{trp}$, including (a) cross-entropy loss $\mathcal{L}_{CE}$, (b) projection loss $\mathcal{L}_p$, and (c) regularization loss $\mathcal{L}_{reg}$.

$$\mathcal{L}_{trp} = \mathcal{L}_{CE} + \alpha * \mathcal{L}_p + \beta * \mathcal{L}_{reg} \tag{4}$$

$$\mathcal{L}_{CE} = -\sum \hat{y} \cdot \log(y) + (1 - \hat{y}) \cdot \log(1 - y) \tag{5}$$

$$\mathcal{L}_p = |D\left(h\left(x_g\right), h\left(x_{same}\right)\right) - D\left(h\left(x_g\right), h\left(x_{diff}\right)\right) + \gamma| \tag{6}$$

$$\mathcal{L}_{reg} = \|h\left(x_g\right)\| + \|h\left(x_{same}\right)\| + \|h\left(x_{diff}\right)\| \tag{7}$$

Where x_g is the embedding representation of the entire graph. $h\left(.\right)$ is a function to obtain the intermediate layer projection. α and β are two hyperparameters indicating the contribution of projection loss and regularization loss respectively. $D(.)$ denotes the cosine distance between the two vectors.

4   Evaluation

To explore the performance of MulGraVD, we answered the following research questions:

4.1   Research Questions

Q1 How does our vulnerability detection method perform in terms of various metrics compared to other deep learning based vulnerability detection methods.
Q2 Whether the feature representation extracted by MulGraVD is sufficiently discriminative.
Q3 How do the four granularities involved in MulGraVD contribute to the performance of the model.
Q4 Whether the model judgment basis (by interpreting the model) is correct.
Q5 How effective is the GGNN compared to other approaches in the field of graph neural networks.

4.2   Datasets

We used two public vulnerability datasets Reveal [7] and FFMPeg+Qemu [29]. Reveal was collected from Linux Debian Kernel and Chromium, containing 18169 methods, of which 1658 are vulnerable. The FFMPeg+Qemu dataset collected by [29] contains 22316 methods, of which 45.02% are vulnerable.

4.3   Experimental Design of Research Questions

Q1 How does our vulnerability detection method perform in terms of various metrics compared to other deep learning based vulnerability detection methods.

Table 2. Accuracy, Precision, Recall and F1 score of vulnerability detection methods on FFMPeg+Qemu and ReVeal datasets

Approach	FFMPeg+Qemu				ReVeal			
	Accuracy	Precision	Recall	F1 score	Accuracy	Precision	Recall	F1 score
Russel *et al.*	57.92	53.81	39.46	45.51	91.23	26.22	12.21	16.62
VulDeePecker	54.10	50.00	21.08	29.66	89.05	18.55	6.13	9.21
SySeVR	51.76	48.07	68.87	56.41	80.66	24.72	40.23	30.13
Devign	61.94	56.48	69.03	62.12	91.81	32.20	29.23	29.23
Reveal	62.04	56.34	75.15	64.32	84.13	33.13	70.16	44.88
IVD$_{ETECT}$	56.98	52.14	70.89	60.08	77.40	24.33	57.23	34.15
MulGraVD	**68.79**	**63.46**	**75.89**	**69.01**	86.89	**39.12**	**72.77**	**50.80**

In Q1, we compare MulGraVD with state-of-the-art deep learning-based vulnerability detection methods: 1) **Russell** *et al.* [22]: features are extracted from the lexed representation of code by CNN or RNN and classified using ensemble classifier. 2) **VulDeePecker** [19]: extract library/API function calls and generate code gadgets, which are later learned by the BLSTM network. 3) **SySeVR** [18]: absorb semantic information introduced by data dependencies and control

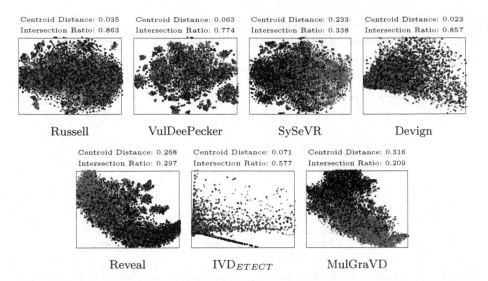

Fig. 3. t-SNE plots, centroid distance, and intersection ratio on FFMPeg+Qemu dataset.

dependencies, and support multiple neural network types. 4) **Devign** [29]: represent the program as a graph structure and perform graph-level prediction. 5) **Reveal** [7]: use GGNN for feature extraction from the code property graph, and use triple loss to optimize the model. 6) **IVD**$_{ETECT}$ [17]: consider statements and surrounding contexts through data and control dependencies, and provide fine-grained interpretation through GNNExplainer [28].

Table 2 summarizes the performance comparison of MulGraVD and other deep learning-based vulnerability detection models. The relative accuracy gain by MulGraVD is averagely 11.33%, at least 6.75% on the FFMPeg+Qemu dataset. The relative gain of precision, recall, and F1 score are 10.65%, 18.48%, and 15.99% on average respectively. On the ReVeal dataset, our average gains on each metric were: precision 12.60%, recall 36.91%, and F1 score 23.43%. The least gains are: precision 5.99%, recall 2.61%, and F1 score 5.92%, respectively.

Q2 Whether the feature representation extracted by MulGraVD is sufficiently discriminative.

We use t-SNE [20] to visualize the learned features and analyze the distinguishability of functions. In addition to t-SNE visualization, we calculate the Euclidean distance between the centers of two clusters and model the two categories as circles and calculate the percentage of the intersection part.

Figure 3 shows the dimensional representation of the features extracted from the FFMPeg+Qemu dataset. In FFMPeg+Qemu dataset, the average Euclidean distance between the two categories centers of our model is the farthest, 0.316. Meanwhile, the Intersection ratio is the smallest, 0.209, which is the best among all models. It is followed by Reveal and again by SySeVR, which is consistent with the phenomenon observed in Fig. 3. In the ReVeal dataset, our method ranks second only to the Reveal model.

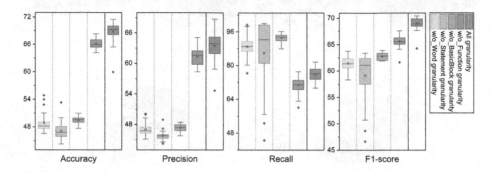

Fig. 4. Performance of models (missing granularity) and full models on FFM-Peg+Qemu dataset.

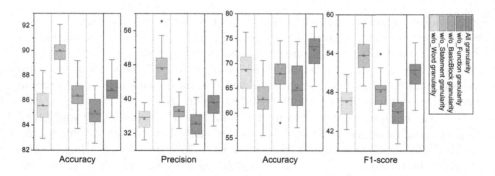

Fig. 5. Performance of models (missing granularity) and full models on ReVeal dataset.

Q3 How do the four granularities involved in MulGraVD contribute to the performance of the model.

For Q3, we conducted a series of experiments to investigate the role of a particular granularity by removing it. The results are shown in Fig. 4 and Fig. 5. Combining the experimental results on both datasets, the model that considers all four granularities simultaneously accounts for four of the eight optimal values. The model that does not consider statements granularity accounts for three. We use the F1 score to measure the gain of different granularities. The average gain of F1 score for word, statement, basic block, and function are 5.90%, 3.49%, 4.46%, and 4.66% respectively. In summary, each granularity has a contribution to our model.

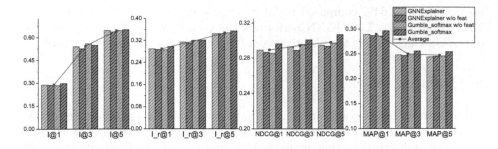

Fig. 6. I@k, I_r@k, NDCG@k and MAP@k calculated by GNNExplainer and its variants

Q4 Whether the model judgment basis (by interpreting the model) is correct.

We use GNNExplainer [28] and its variant using Gumbel-Softmax to interpret the model. The model is iterated 50 times for each data and then sorted in descending order according to the value. In Fig. 6, on the MAP@k metric, the corresponding MAP value decreases as the k value increases. MAP@5 is 0.25, which is about 0.04 different from MAP@1, indicating that the more forward our model is, the more accurate the prediction is. I@5 indicates that the probability that the first five rows given by the explanatory model intersect with the actual vulnerability rows is 64.5%, which indicates that the model largely makes judgments through vulnerability-related rows.

Table 3. The effect of GGNN on MulGraVD

	FFMPeg+Qemu				ReVeal			
	Accuracy	Precision	Recall	F1 score	Accuracy	Precision	Recall	F1 score
GGNN	68.79	63.46	75.89	69.01	86.89	39.12	72.77	50.80
	(2.13)	(2.74)	(3.31)	(1.15)	(1.07)	(2.55)	(3.10)	(2.13)
GAT	59.85	54.59	75.44	63.10	80.29	26.87	65.94	38.08
	(1.31)	(1.64)	(7.46)	(1.87)	(1.64)	(2.21)	(3.91)	(2.13)
GCN	64.16	59.88	67.29	63.09	86.05	36.28	68.69	47.43
	(1.22)	(2.47)	(6.39)	(1.57)	(0.75)	(2.07)	(2.92)	(1.90)

Q5 How effective is the GGNN compared to other approaches in the field of graph neural networks.

We evaluated the effectiveness of graph neural network selection for MulGraVD. The GGNN significantly outperforms the two commonly used graph neural networks, GAT and GCN in Table 3. The number of layers of GAT and GCN was kept consistent with that of GGNN. Our model increased F1 scores by an average of 4.64% compared to GCN, and 5.91% and 12.72% compared to GAT respectively. The GGNN performed the best in all eight metrics for both datasets, which proves the effectiveness of our chosen graph neural network.

5   Related Work

Deep learning based static analysis, dynamic analysis, and symbolic execution are developed to detect vulnerabilities. In terms of static analysis, [18,19] perform vulnerability detection by extracting program slices. Chakraborty *et al.* [7] generates program representations by code property graph and optimizes the model by the triple loss function. In terms of dynamic analysis, She *et al.* [25] uses surrogate neural network models to learn smooth approximations of the program's branching behaviors. She *et al.* [24] uses a Multi-Task Neural Network to guide the mutation process. The deep learning based symbolic execution work [26] learns an approximation of the relationship between program values of interest through neural networks.

6   Conclusion

We introduce a multi-granularity deep learning vulnerability detection model, MulGraVD, which can overcome the constraint posed by hyperparameter layers. It has gained improvements in several evaluation metrics. Future directions include applying the model to detect cross-function vulnerabilities to better suit the potential nature of software source code and vulnerabilities.

References

1. CVE-2012-2793. https://www.cve.org/CVERecord?id=CVE-2012-2793
2. CWE-120: Buffer overflow. https://cwe.mitre.org/data/definitions/120.html
3. CWE-190: Integer overflow. https://cwe.mitre.org/data/definitions/190.html
4. CWE-362: Race condition. https://cwe.mitre.org/data/definitions/362.html
5. CWE-369: Divide by zero. https://cwe.mitre.org/data/definitions/369.html
6. CWE-476: Null pointer dereference. https://cwe.mitre.org/data/definitions/476.html
7. Chakraborty, S., Krishna, R., Ding, Y., Ray, B.: Deep learning based vulnerability detection: are we there yet. IEEE Trans. Software Eng. **48**, 3280–3296 (2021)
8. Chawla, N.V., Bowyer, K.W., Hall, L.O., Kegelmeyer, W.P.: Smote: synthetic minority over-sampling technique. J. Artif. Intell. Res. **16**, 321–357 (2002)
9. Chen, D., Lin, Y., Li, W., Li, P., Zhou, J., Sun, X.: Measuring and relieving the over-smoothing problem for graph neural networks from the topological view. In: Proceedings of the AAAI Conference on Artificial Intelligence, vol. 34, pp. 3438–3445 (2020)
10. Cheng, X., Zhang, G., Wang, H., Sui, Y.: Path-sensitive code embedding via contrastive learning for software vulnerability detection. In: Proceedings of the 31st ACM SIGSOFT International Symposium on Software Testing and Analysis, pp. 519–531 (2022)

11. Dosovitskiy, A., Beyer, L., et al.: An image is worth 16x16 words: transformers for image recognition at scale. arXiv preprint arXiv:2010.11929 (2020)
12. Du, X., et al.: LEOPARD: identifying vulnerable code for vulnerability assessment through program metrics. In: 2019 IEEE/ACM 41st International Conference on Software Engineering (ICSE), pp. 60–71. IEEE (2019)
13. Hin, D., Kan, A., Chen, H., Babar, M.A.: LineVD: statement-level vulnerability detection using graph neural networks. In: Proceedings of the 19th International Conference on Mining Software Repositories, pp. 596–607 (2022)
14. Johnson, B., Song, Y., Murphy-Hill, E., Bowdidge, R.: Why don't software developers use static analysis tools to find bugs? In: 2013 35th International Conference on Software Engineering (ICSE), pp. 672–681. IEEE (2013)
15. Le, T., et al.: Maximal divergence sequential autoencoder for binary software vulnerability detection. In: International Conference on Learning Representations (2019)
16. Li, X., Wang, L., Xin, Y., Yang, Y., Tang, Q., Chen, Y.: Automated software vulnerability detection based on hybrid neural network. Appl. Sci. **11**(7), 3201 (2021)
17. Li, Y., Wang, S., Nguyen, T.N.: Vulnerability detection with fine-grained interpretations. In: Proceedings of the 29th ACM Joint Meeting on European Software Engineering Conference and Symposium on the Foundations of Software Engineering, pp. 292–303 (2021)
18. Li, Z., Zou, D., Xu, S., Jin, H., Zhu, Y., Chen, Z.: SySeVR: a framework for using deep learning to detect software vulnerabilities. IEEE Trans. Dependable Secure Comput. **19**(4), 2244–2258 (2021)
19. Li, Z., et al.: VulDecPecker: a deep learning-based system for vulnerability detection. arXiv preprint arXiv:1801.01681 (2018)
20. Van der Maaten, L., Hinton, G.: Visualizing data using t-SNE. J. Mach. Learn. Res. **9**(11) (2008)
21. Neuhaus, S., Zimmermann, T., Holler, C., Zeller, A.: Predicting vulnerable software components. In: Proceedings of the 14th ACM Conference on Computer and Communications Security, pp. 529–540 (2007)
22. Russell, R., et al.: Automated vulnerability detection in source code using deep representation learning. In: 2018 17th IEEE International Conference on Machine Learning and Applications (ICMLA), pp. 757–762. IEEE (2018)
23. Scandariato, R., Walden, J., Hovsepyan, A., Joosen, W.: Predicting vulnerable software components via text mining. IEEE Trans. Software Eng. **40**(10), 993–1006 (2014)
24. She, D., Krishna, R., Yan, L., Jana, S., Ray, B.: MTFuzz: fuzzing with a multi-task neural network. In: Proceedings of the 28th ACM Joint Meeting on European Software Engineering Conference and Symposium on the Foundations of Software Engineering, pp. 737–749 (2020)
25. She, D., Pei, K., Epstein, D., Yang, J., Ray, B., Jana, S.: Neuzz: efficient fuzzing with neural program smoothing. In: 2019 IEEE Symposium on Security and Privacy (SP), pp. 803–817. IEEE (2019)
26. Shen, S., Shinde, S., Ramesh, S., Roychoudhury, A., Saxena, P.: Neuro-symbolic execution: augmenting symbolic execution with neural constraints. In: NDSS (2019)
27. Yamaguchi, F., Golde, N., Arp, D., Rieck, K.: Modeling and discovering vulnerabilities with code property graphs. In: 2014 IEEE Symposium on Security and Privacy, pp. 590–604. IEEE (2014)

28. Ying, Z., Bourgeois, D., You, J., Zitnik, M., Leskovec, J.: GNNExplainer: generating explanations for graph neural networks. Adv. Neural Inf. Process. Syst. **32** (2019)
29. Zhou, Y., Liu, S., Siow, J., Du, X., Liu, Y.: Devign: Effective vulnerability identification by learning comprehensive program semantics via graph neural networks. In: Advances in Neural Information Processing Systems, vol. 32 (2019)

Rumor Detection with Supervised Graph Contrastive Regularization

Shaohua Li(iD), Weimin Li(✉)(iD), Alex Munyole Luvembe(iD), and Weiqin Tong(iD)

School of Computer Engineering and Science, Shanghai University,
Shanghai 200444, China
{flowingfog,wmli,luvembe,wqtong}@shu.edu.cn

Abstract. The rapid spread of rumors on social networks can significantly impact social stability and people's daily lives. Recently, there has been increasing interest in rumor detection methods based on feedback information generated during user interactions and the propagation structure. However, these methods often face the challenge of limited labeled data. While addressing data dependency issues, graph-based contrastive learning methods struggle to effectively represent different samples of the same class in supervised classification tasks. This paper proposes a novel Supervised Graph Contrastive Regularization (SGCR) approach to tackle these complex scenarios. SGCR leverages label information for supervised contrastive learning and applies simple regularization to the embeddings by considering the variance of each dimension separately. To prevent the collapse problem, sessions belonging to the same class are pulled together in the embedding space, while those from different categories are pushed apart. Experimental results on two real-world datasets demonstrate that our SGCR outperforms baseline methods.

Keywords: Rumor Detection · Contrastive Learning · Graph Neural Network · Gradient Conflict

1 Introduction

The spread of rumors may have the following negative social impacts: first, people may accept deliberately fabricated lies as facts; second, rumors can change the way people respond to legitimate information; finally, the prevalence of rumors has the potential to undermine the credibility of the entire social ecosystem. Given the possible adverse effects of rumors, efficient methods for automatically identifying social media rumors are required.

The dissemination of information in social networks has received much attention from researchers [9,12,22]. The spread of true and false rumors on social media elicited different responses, forming rumor propagation cascades that are helpful for automatic rumor identification. Methods based on user participants utilize the wisdom of the crowd. The opinion of a single user is likely to be wrong,

B. Luo et al. (Eds.): ICONIP 2023, CCIS 1969, pp. 165–176, 2024.
https://doi.org/10.1007/978-981-99-8184-7_13

but combined with the stances and emotions in the feedback of multiple users, we can approach the truth.

Many previous studies have explored the feasibility of using user feedback during propagation to identify rumors in conjunction with community and network research [8,10,19]. With the development of graph neural networks and their excellent performance in structured data processing, graph neural networks [6] have achieved widespread applications in a number of fields, including rumor detection [2,3,7,17,21]. However, most current rumor detection methods using graph neural networks rely on limited labels when updating node representations. Some researchers [20] have attempted to reduce their reliance on label information by employing contrastive learning to discover invariant representations through data augmentation. In these methods, each sample is paired with its augmentation to form positive pairs, while the sample is paired with augmentations of other samples to create negative pairs for contrastive learning. However, in supervised rumor classification tasks, such methods may struggle to effectively handle situations where different samples belong to the same class.

This paper proposes a novel rumor identification method named Supervised Graph Contrastive Regulation (SGCR). SGCR benefits from crowd wisdom and identifies rumors by encoding the rumor propagation tree. SGCR uses contrastive learning to train the encoder and then get graph-level representations with the fusion of multiple pooling techniques. Given the features of input graphs, SGCR first generates contrastive views $G' = X', A'$ and $G'' = X'', A''$ through graph enhancement $\mathcal{A}_{p_X, p_A}$, then the graph encoder applies attention mechanism to aggregate neighborhood information and update node representations in a weighted manner, and then get graph-level representations Z' and Z'' after pooling. Then, Z' is sent to the classified for prediction results. What's more, we make supervised contrastive regularization between Z' and Z'' to mine the invariance characteristics of different rumors, where the similarity between two representations from the same class is maximized, the variance regularization term along the batch dimension and the invariance criterion between each pair of vectors is minimized. Therefore, the main contributions of this paper can be summarized as follows:

- We propose a novel graph neural network architecture to represent rumor information with user feedback and obtain graph-level representations through fusion pooling;
- We offer a supervised contrast regularization method as an auxiliary task for rumor detection, which reduces dependence on labeled data and enables samples of the same class to learn similar representations.
- We conduct experiments on two real-world datasets, and the proposed model outperforms the baseline in most metrics;

2 Preliminary Knowledge

We define rumors as a set of statements $\mathcal{C} = \{C_1, C_2, \cdots, C_n\}$, where each statement C_i contains the source post s_i and corresponding responses r_{i*}, i.e.,

$C_i = \{s_i, r_{i1}, r_{i2}, \cdots, r_{i,n_i}\}$. We represent the graph structure of C_i as $\mathcal{G}_i = \{V_i, E_i\}$, where V_i is the set of nodes and E_i represents the connections between nodes in the graph. The representations of nodes in the graph form a feature matrix X_i. We consider rumor detection with user feedback as a supervised graph classification task. which learns a classifier f from the labeled statements:

$$f : G_i \rightarrow Y_i \tag{1}$$

where Y_i takes one of two categories: false rumor and true rumor.

3   Method

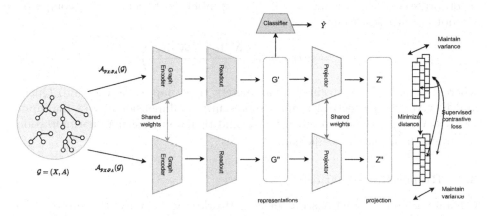

Fig. 1. Architecture of SGCR.

This section proposes a novel rumor detection method that combines supervised contrastive learning and graph regularization, called Supervised Graph Contrastive Regulation (SGCR). As shown in Fig. 1, SGCR uses the method of contrastive learning to train the encoder and pooling layer to mine the essential characteristics of different rumor propagation structures. Given the features of input graphs, SGCR first generates contrastive views $\mathcal{G}' = \{X', A'\}$ and $\mathcal{G}'' = \{X'', A''\}$ through graph enhancement $\mathcal{A}_{p_X, p_A}$, then the graph encoder applies attention mechanism to aggregate neighborhood information and update node representations in a weighted manner, and we can get graph-level representations G' and G'' with readout. G' is then sent to the fully connected classifier to get the prediction results. For supervised contrastive regularization, we adopt a projector f_ϕ that maps the graph-level representations into projections in an embedding space where the supervised contrastive loss will be computed.

3.1   Graph Representation

In SGCR, we consider the propagation structure of a rumor as a homogeneous graph and employ a graph neural network to fuse features of the original claim and user preferences. With the attention mechanism, the graph neural network will pay more attention to neighbors important to a node in the graph.

We aggregate information from neighboring nodes to update node representations with attention mechanism [15,18]:

$$\mathbf{h}'_i = \mathbf{W}_2\mathbf{h}_i + \sum_{j \in \mathcal{N}(i)} \alpha_{i,j}\mathbf{W}_2\mathbf{h}_j, \tag{2}$$

where j represents the neighboring nodes of i, α_{ij} denotes the attention coefficient between node i and node j, and W_2h_j projects h_j to Value vector. We incorporate a self-loop using W_1h_i. The attention coefficient $\alpha_{i,j}$ is computed via dot product attention:

$$\alpha_{i,j} = softmax\left(\frac{(\mathbf{W}_3\mathbf{h}_i)^\top(\mathbf{W}_4\mathbf{h}_j)}{\sqrt{d}}\right) \tag{3}$$

where W_3 and W_4 are trainable parameters that map x_i and x_j into Query and Key vectors respectively, and d represents the dimensionality of the input vectors. We scale the dot product by $\frac{1}{\sqrt{d}}$ and then apply softmax normalization to obtain the attention coefficients.

3.2   Rumor Classification

Combining user interactions on social networks can significantly improve the performance in rumor identification tasks [4,11]. We treat rumor classification with user engagement information as a graph classification task and obtain a graph-level representation of the rumor propagation structure through fusion-pooling:

$$G = FC(Max(\mathbf{H}, \mathcal{G}_i)\|Mean(\mathbf{H}, \mathcal{G}_i)) \tag{4}$$

where the max pooling and mean pooling operations are performed on nodes in graph G_i, and $\|$ connects the graph-level representations from different kinds of pooling methods. Then we take the graph level representations as the input to the classifier and generate the prediction results $\hat{Y}$:

$$\hat{\mathbf{Y}} = Softmax(FC(\mathbf{G})) \tag{5}$$

We take the cross-entropy between the classification outputs of the model $\hat{Y}$ and the ground truth labels Y as the loss function for the classification task:

$$\mathcal{L}_{CE}(\mathbf{Y}, \hat{\mathbf{Y}}) = -(\mathbf{Y} \cdot log(\hat{\mathbf{Y}}) + (1 - \mathbf{Y}) \cdot log(1 - \hat{\mathbf{Y}})) \tag{6}$$

It should be noted that during the testing phase, we cease augmentation for the classification branch to obtain graph-level representations of the original structures and node features through the encoder.

3.3 Supervised Contrastive Learning with Regulation

Some research has attempted to use contrastive learning as an auxiliary task to mine intrinsic data features and reduce dependence on limited labeled data. However, in supervised rumor classification tasks, contrastive learning-based methods [7,17] often struggle to handle cases where different samples from the same class are treated as negative pairs. In this paper, we combine supervised contrastive learning [5] and regularization [1], using supervised contrastive regularization as an auxiliary task to reduce reliance on limited labeled data.

The supervised graph contrastive regularization first transforms the graph with an augmentation operation $\mathcal{A}_{p_X,p_A}(\mathcal{G})$ to obtain two augmented views $\mathcal{G}'$ and $\mathcal{G}''$. We randomly remove edges in G_i with probability p_A and mask node features with probability p_X to generate two different augmented views of $\mathcal{G}_i$, $\mathcal{G}_i' = (X_i', A_i')$ and $\mathcal{G}_i'' = (X_i'', A_i'')$. Use the graph encoder f_θ to encode augmented views into representations $H' = f(X', A')$ and $H'' = f(X'', A'')$. Then we adopt fusion pooling operation as the readout function to process these representations to get graph-level representations G' and G''. The MLP projector f_ϕ further processes the representations into projections $Z' = f_\phi(G')$ and $Z'' = f_\phi(G')$. The supervised contrastive loss is computed between projected graph-level representations Z' and Z''.

$$Z' = \sigma(BN(FC(G'))) \tag{7}$$

where FC stands for a fully connected layer, BN represents batch normalization, and σ denotes the activation function.

The supervised contrastive regularization loss between Z' and Z'' comprises variance, invariance, and supervised contrastive loss terms. We use $Z' = [z_1', \cdots, z_n'] \in \mathbb{R}^{n \times d}$ and $Z'' = [z_1'', \cdots, z_n''] \in \mathbb{R}^{n \times d}$ represents the graph-level representation of the two branches of the Siamese architecture, which consists of m vectors of dimension d. Let Z_i denote the i−th vector in Z, and $Z_{:,j}$ denote the vector consisting of each value of dimension j in all vectors in Z. The variance regularization term v can be defined as a hinge loss along the graph-level representation standard deviation of batches:

$$v(Z) = \frac{1}{d} \sum_{j=1}^{d} max(0, \gamma - \sqrt{Var(Z_{:,j}) + \epsilon}) \tag{8}$$

where γ is the target value hyperparameter representing the standard deviation, ϵ is a small scalar against numerical instability, and $Var(x)$ is the unbiased variance estimator given by:

$$Var(x) = \frac{1}{n-1} \sum_{i=1}^{n} (x_i - \bar{x})^2, \tag{9}$$

where n is the number of x and $\bar{x}$ is the average of x. This criterion will enforce that the variance within the current batch is γ along each dimension, preventing crash solutions where all inputs map to the same vector.

We define the invariance criterion s between Z' and Z' as the mean squared Euclidean distance between each pair of vectors:

$$s(Z, Z') = \frac{1}{m} \sum_i \|Z_i' - Z_i''\|_2^2 \tag{10}$$

For the supervised contrastive criterion sup between Z and Z, we consider intra- and inter-view supervised contrastive losses. Only considering the mean square Euclidean distance between pairs of vectors cannot handle the case where multiple samples belong to the same class. To handle any number of positive pairs belonging to the same class, we consider the InfoNCE loss between anchors and samples of the same class within and between views:

$$\begin{aligned} sup(Z', Z'') &= \sum_{i=1}^{2N} \mathcal{L}_i^{sup} \\ &= \sum_{i=1}^{2N} \frac{-1}{2N_{\tilde{y}_i} - 1} \sum_{j=1}^{2N} \mathbb{1}_{i \neq j} \cdot \mathbb{1}_{\tilde{y}_i = \tilde{y}_j} \cdot log \frac{exp(z_i \cdot z_j / \tau)}{\sum_{k=1}^{2N} \mathbb{1}_{i \neq k} \cdot exp(z_i \cdot z_k \tau)} \end{aligned} \tag{11}$$

where $N_{\tilde{y}_i}$ is the number of mini-batch rumors with the same label y_i as anchor i. Note that for the convenience of calculation, we splice Z' and Z'' to get the feature matrix $[z_1, \cdots, z_{2n}] \in \mathbb{R}^{2n \times d}$. During training, for any i, the encoder is tuned to maximize the numerator of the logarithmic parameter in the equation while minimizing its denominator.

The overall loss function is a weighted average of the invariance, variance, and covariance terms:

$$\mathcal{L}_{REG}(Z', Z'') = \lambda\{sup(Z', Z'')\} + \mu\{v(Z') + v(Z'')\} + \nu s(Z', Z''), \tag{12}$$

where λ, μ, and ν are hyperparameters that control the importance of each term in the loss. In our experiments, we set $\nu = 1$, $\lambda = \mu = 25$, $\epsilon = 0.0001$.

3.4 Joing Training with Gradient Projection

We update the parameters θ based on the cross-entropy loss $\mathcal{CE}$ of the rumor classification task together with supervised contrast regularization loss $\mathcal{L}_{REG}$:

$$\mathcal{L} = \mathcal{L}_{CE} + \alpha \mathcal{L}_{REG} \tag{13}$$

where α is the balanced cross-entropy loss $\mathcal{L}_{CE}$ and supervised regularization loss $\mathcal{L}_{REG}$, in our experiments, we empirically set it to 0.1.

4   Experiments

To verify the effectiveness of SGCR, we compare it with several recently proposed methods, especially those based on rumor propagation paths on two real datasets. In addition, we also compare the performance of different models on the early detection task to understand the real application value of the model. Finally, to clarify the contribution of different components and features to the rumor identification task, we conduct ablation experiments on SGCR.

4.1 Datasets

We conduct experiments on data obtained through FakeNewsNet [16], which consists of true and false news data checked on two fact-checking sites, Politifact and Gossipcop, as well as the social feeds of engaged users on Twitter. The statistics of the dataset are shown in Table 1:

Table 1. Dataset statistics.

	Politifact	Gossipcop
#graphs	314	5464
#true	157	2732
#fake	157	2732
#total nodes	41,050	314,262
#total edges	40,740	308,798
#Avg. nodes per graph	131	58

4.2 Baselines

We compare the proposed model with algorithms proposed in recent years, especially those based on rumor propagation paths. The baseline models compared in our experiments are as follows:

- CSI [14] combines text and user response features, including a capture module that captures the time-varying features of a user's response to a given document, a scoring model that scores user engagement, and an ensemble module.
- GCNFN [13] generalizes convolutional neural networks to graphs, allowing the fusion of various features such as content, user profiles, interactions, and rumor propagation trees.
- GNN-CL [4] applies continuous learning to progressively train GNNs to achieve balanced performance on existing and new datasets, avoiding retraining the model from scratch on the entire data.
- BiGCN [2] proposes a bidirectional graph convolutional network that aggregates the features of adjacent nodes in the rumor propagation tree in both top-down and bottom-up directions.
- GACL [17] uses adversarial training to generate challenging features and takes these features as difficult negative samples in contrastive learning to help the model strengthen feature learning from these difficult samples.

We add two additional baselines that directly classify word2vec and BERT source content embeddings. We employ a 3-layer fully connected network and apply DropOut to prevent overfitting.

Many baseline methods utilize additional information, and to ensure fair comparisons, we implement the baseline algorithms only with features such as news content, user profiles, user preference embeddings, and rumor propagation trees. We apply the same batch size (64) and hidden layer dimension (64) for different methods. In the division of the data set, the training set, validation set, and test set account for 70%, 10%, and 20%, respectively.

4.3   Experimental Results and Analysis

The performance of SGCR and baseline models on two real datasets is shown in Table 2.

Table 2. Prediction Performance on Real World Datasets.

Model	Politifact				Gossipcop			
	Acc.	Prec.	Rec.	F1	Acc.	Prec.	Rec.	F1
Word2Vec	0.8438	0.9224	0.8530	0.8359	0.7706	0.8065	0.7174	0.7696
BERT	0.8594	0.9309	0.8387	0.8551	0.7550	0.7567	0.7482	0.7504
CSI	0.8281	0.8947	0.8293	0.8181	0.9159	0.9290	0.9003	0.9153
GCNFN	0.8594	0.8810	0.9024	0.8458	0.9406	0.9339	0.9498	0.9400
GNN-CL	0.6562	0.8182	0.5000	0.6532	0.9296	0.9611	0.8908	0.9280
BiGCN	0.8906	0.9381	0.8537	0.8843	0.9598	0.9437	0.9596	0.9596
GACL	0.8594	0.8929	0.8065	0.8585	0.9027	0.9194	0.792	0.9010
SGCR	**0.9688**	**1**	**0.9412**	**0.9687**	**0.9799**	**0.9738**	**0.9859**	**0.9797**

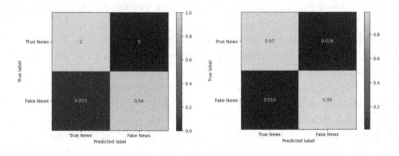

Fig. 2. Confusion matrix on Politifact and Gossipcop.

Both *Word2Vec+MLP* and *Bert+MLP* abandon the rumor propagation structure and only utilize news content features. The performance of these two models is close to several other methods on the smaller Politifact dataset but worse on the larger Gossipcop dataset. We believe this is because rumors in the Gossipcop dataset are more deceptive and cannot be effectively identified by simple text features, requiring the help of crowd wisdom. The *CSI* model

Table 3. Prediction performance on Politifact and Gossipcop.

Model	Politifact		Gossipcop	
	ACC	F1	ACC	F1
$SGCR_{-graph}$	0.8594	0.8551	0.7706	0.7690
$SGCR_{-sup}$	**0.9688**	**0.9687**	0.9781	0.9777
$SGCR_{-reg}$	0.9531	0.9531	0.9744	0.9741
$SGCR$	**0.9688**	**0.9687**	**0.9799**	**0.9796**

considers user participation, but the fixed user representations are spliced with the content embedding to get the score, so the effect is not as good as the model based on message passing. The poor performance of *GNN-CL* on the Politifact dataset is because the user profile contains insufficient information. *BiGCN* extracts more structural information from the Politifact dataset, with fewer but deeper rumor propagation trees, and achieves better results. The *GACL* view improves the generalization ability of the model through supervised contrastive learning with adversarial training, but it has not achieved good results on the Politifact and Gossipcop datasets.

The confusion matrix for SGCR is shown in the Fig. 2. **SGCR** has achieved high accuracy on both categories and outperformed the baseline model across all metrics., especially the smaller Politifact. Under insufficient training, the SGCR makes full use of the class consistency information within and between views to mine the essential characteristics of different classes of rumor propagation structures, alleviates the problem of insufficient labeled data, and achieves a significant improvement compared to the baseline models.

4.4   Ablation Study

We conducted ablation experiments to clarify the contribution of different features and components to the model. The experimental results are shown in Table 3.

- In $SGCR_{-graph}$, we remove user involvement and classify rumor embeddings directly.
- In $SGCR_{-sup}$, we remove the supervised learning in supervised contrastive regularization.
- In $SGCR_{-reg}$, we removed the regularization term and solely employed supervised contrastive learning as an auxiliary task to learn consistent representations for samples of the same class.
- In $SGCR$, we use the full SGCR model.

After removing the user interactions, $SGCR_{-graph}$ is equivalent to the BERT model in the baselines and performs poorly on the GossipCop dataset, which makes it more difficult to distinguish authenticity. After adding user interaction information ($SGCR_{-reg}, SGCR_{-sup}, SGCR$), the performance of the model is significantly improved, especially on the Gossipcop dataset, which is large in scale and difficult to identify solely by rumor content, illustrating the effectiveness of user stances reflected during user interactions on the task of rumor identification. The comparison between $SGCR_{-reg}$ and $SGCR_{-sup}$ shows that contrastive regularization can mine the essential characteristics of the rumor propagation structure and improve the model's effectiveness on data sets that are difficult to identify. $SGCR$ contains all the features and functions and achieves the best scores in each indicator. The comparison of $SGCR$ and $SGCR_{-sup}$ shows that supervised contrastive learning can effectively alleviate the problem of insufficient labels in small-scale datasets.

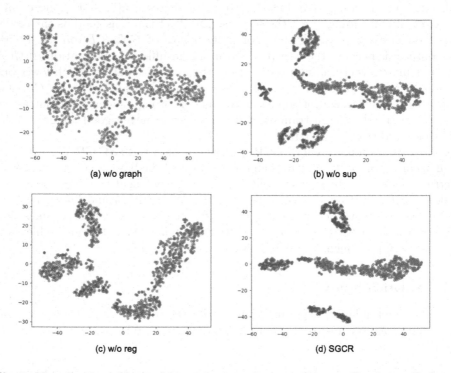

Fig. 3. Visualization of learned latent feature representations on Gossipcop. Red and green dots represent false and real rumors, respectively. (Color figure online)

To understand the impact of various features and functionalities on the final embeddings, we qualitatively visualize the learned latent representations using t-SNE. As shown in Fig. 3, in $SGCR_{-graph}$, which uses only the root node features, the representations of false rumors cannot be distinguished from those of true rumors. In $SGCR_{-sup}$, some false rumors are misclassified as true rumors in the lower cluster due to the lack of supervised information from same-class samples. In $SGCR_{-reg}$, the absence of regularization leads to larger distances between some false rumors, resulting in more misclassifications. $SGCR$, on the other hand, extracts essential features of different types of rumors, enhancing its ability to identify rumors, and the learned representations exhibit clear separation boundaries.

5   Conclusion

This paper proposes a novel GNN-based rumor detection method that incorporates supervised contrastive learning and regularization, called Supervised Graph Contrastive Regulation (SGCR). SGCR takes the embeddings of rumor statements and user interactions involved in rumor propagation as the initial features of nodes, then applies weighted attention to aggregate neighbor features and then adopts fusion pooling to obtain graph-level representations for rumor classification. In addition, we introduce supervised graph contrastive learning as an auxiliary task to constrain the consistency of node representations in the same class. Experimental results on two real-world datasets show that our proposed SGCR model achieves significant improvements over baseline methods.

Acknowledgements. This work was supported in part by the National Key Research and Development Program of China (grant number 2022YFC3302601), the High-Performance Computing Center of Shanghai University, and the Shanghai Engineering Research Center of Intelligent Computing System (2252600) for providing the computing resources.

References

1. Bardes, A., Ponce, J., LeCun, Y.: VICReg: variance-invariance-covariance regularization for self-supervised learning (2021). https://doi.org/10.48550/ARXIV.2105.04906, https://arxiv.org/abs/2105.04906
2. Bian, T., et al.: Rumor detection on social media with bi-directional graph convolutional networks (2020)
3. Dou, Y., Shu, K., Xia, C., Yu, P.S., Sun, L.: User preference-aware fake news detection. In: Proceedings of the 44nd International ACM SIGIR Conference on Research and Development in Information Retrieval (2021)
4. Han, Y., Karunasekera, S., Leckie, C.: Graph neural networks with continual learning for fake news detection from social media. CoRR abs/2007.03316 (2020)
5. Khosla, P., et al.: Supervised contrastive learning. Adv. Neural. Inf. Process. Syst. **33**, 18661–18673 (2020)
6. Kipf, T.N., Welling, M.: Semi-supervised classification with graph convolutional networks. arXiv preprint arXiv:1609.02907 (2016)
7. Li, S., Li, W., Luvembe, A.M., Tong, W.: Graph Contrastive Learning With Feature Augmentation for Rumor Detection. IEEE Transactions on Computational Social Systems, pp. 1–10 (2023). https://doi.org/10.1109/TCSS.2023.3269303, conference Name: IEEE Transactions on Computational Social Systems
8. Li, W., Guo, C., Liu, Y., Zhou, X., Jin, Q., Xin, M.: Rumor source localization in social networks based on infection potential energy. Inf. Sci. **634**, 172–188 (2023)
9. Li, W., Zhong, K., Wang, J., Chen, D.: A dynamic algorithm based on cohesive entropy for influence maximization in social networks. Expert Syst. Appl. **169**, 114207 (2021)

10. Li, W., et al.: Evolutionary community discovery in dynamic social networks via resistance distance. Expert Syst. Appl. **171**, 114536 (2021)
11. Lu, Y.J., Li, C.T.: GCAN: graph-aware co-attention networks for explainable fake news detection on social media. In: Proceedings of the 58th Annual Meeting of the Association for Computational Linguistics, pp. 505–514. Association for Computational Linguistics, Online (2020)
12. Ma, J., Gao, W., Wong, K.F.: Rumor detection on twitter with tree-structured recursive neural networks. In: Proceedings of the 56th Annual Meeting of the Association for Computational Linguistics (Volume 1: Long Papers), pp. 1980–1989. Association for Computational Linguistics, Melbourne, Australia (2018)
13. Monti, F., Frasca, F., Eynard, D., Mannion, D., Bronstein, M.M.: Fake news detection on social media using geometric deep learning (2019)
14. Ruchansky, N., Seo, S., Liu, Y.: CSI: a hybrid deep model for fake news detection. In: Proceedings of the 2017 ACM on Conference on Information and Knowledge Management, pp. 797–806. CIKM '17, Association for Computing Machinery, New York, NY, USA (2017)
15. Shi, Y., Huang, Z., Feng, S., Zhong, H., Wang, W., Sun, Y.: Masked label prediction: Unified message passing model for semi-supervised classification. arXiv preprint arXiv:2009.03509 (2020)
16. Shu, K., Mahudeswaran, D., Wang, S., Lee, D., Liu, H.: Fakenewsnet: a data repository with news content, social context and dynamic information for studying fake news on social media. arXiv preprint arXiv:1809.01286 (2018)
17. Sun, T., Qian, Z., Dong, S., Li, P., Zhu, Q.: Rumor detection on social media with graph adversarial contrastive learning. In: Proceedings of the ACM Web Conference 2022, pp. 2789–2797. WWW '22, Association for Computing Machinery, New York, NY, USA (2022)
18. Vaswani, A., et al.: Attention is all you need. In: Advances in Neural Information Processing Systems 30 (2017)
19. Xu, Y., Zhuang, Z., Li, W., Zhou, X.: Effective community division based on improved spectral clustering. Neurocomputing **279**, 54–62 (2018)
20. Yang, X., Lyu, Y., Tian, T., Liu, Y., Liu, Y., Zhang, X.: Rumor detection on social media with graph structured adversarial learning. In: Proceedings of the Twenty-ninth International Conference On International Joint Conferences on Artificial Intelligence, pp. 1417–1423 (2021)
21. Zhang, C., Li, W., Wei, D., Liu, Y., Li, Z.: Network dynamic GCN influence maximization algorithm with leader fake labeling mechanism. IEEE Transactions on Computational Social Systems, pp. 1–9 (2022). https://doi.org/10.1109/TCSS.2022.3193583
22. Zhou, X., Li, S., Li, Z., Li, W.: Information diffusion across cyber-physical-social systems in smart city: a survey. Neurocomputing **444**, 203–213 (2021)

A Meta Learning-Based Training Algorithm for Robust Dialogue Generation

Ziyi Hu and Yiping Song[✉]

National University of Defense Technology, Changsha, China
songyiping@nudt.edu.cn

Abstract. There are many low-resource scenarios in the field of dialogue generation, such as medical diagnosis. Many dialogue generation models in these scenarios are usually unstable due to the lack of training corpus. As one of the most popular training algorithms in recent years, meta learning has achieved remarkable results. MAML in meta learning can find a fast adaptive initialization parameter for a series of low resource tasks, which is often used to solve the problem of low resource, and has achieved excellent performance in image classification tasks. However, in the field of text generation, such as dialogue generation, because of the large vocabulary, long sequence and large number of parameters involved in text generation, the effect of MAML is unstable. Therefore, this paper proposes a high robust text generation training framework based on meta learning for the dialogue generation task. By identifying the significant information in the model parameters, the optimizer can train the important parameters more concentrated in the limited data in the bi-level optimization. Experiments show that our method has a good performance on the BLEU scores on the six single-domain datasets.

Keywords: Low resource · Dialogue generation · Robustness

1 Introduction

Nowadays, more and more text generation tasks are based on low-resource scenarios. In the case of insufficient training corpus, many text generation models are prone to overfitting during training, and the model has poor robustness and cannot adapt well to specific tasks.

Recently, a rapidly developing training algorithm in the low-resource field is meta learning. As a kind of low-resource training model, meta learning has achieved remarkable results in solving the shortage of training corpus. Meta learning, literally translated as learning to learn, specifically means that with the training of existing data, the model learns significant features of the data, to achieve the purpose of quickly adapting to new tasks. At present, the research on meta learning can be roughly divided into five independent directions: the method based on metric learning, initialization methods with strong generalization, the method based on optimizer, the method based on additional external storage and the method based on data augmentation. Among these, more

B. Luo et al. (Eds.): ICONIP 2023, CCIS 1969, pp. 177–190, 2024.
https://doi.org/10.1007/978-981-99-8184-7_14

researches are based on initialization methods with strong generalization, and the best performance is Model-Agnostic Meta-Learning algorithm (MAML) [1], which has also become the backbone of the meta-learning research field.

At present, there have been many improvements based on MAML [2–5] . However, the performance of MAML in some tasks is still not stable enough. Once some links are wrong or the input data is destroyed during the training process, the trained model will have problems such as poor generalization ability and poor robustness. In order to solve this problem, this paper works under the training framework of MAML and focuses on the dialogue generation task. By using Fisher information, parameters that contribute more to model training in optimization are identified, so that the optimizer can use more limited data to train these parameters to improve the robustness of the whole model. The main contributions of this paper are as follows:

- In this paper, a low-resource neural network training algorithm with high robustness via meta learning is proposed for dialogue generation tasks;
- Fisher information is introduced to identify the important parameters of the model, and the optimal mask is formulated;
- Experiments on six task domains show that the trained model has strong robustness.

2   Related Work

From the existing research work, how to improve the robustness or stability of text models can be studied from three levels - sample level, semantic level and model level.

From the perspective of samples, the methods to improve the robustness of the model can be explored from the selection of sampling methods, the construction of datasets, the construction of adversarial samples, the modification of input data and so on. These methods can be used to make full use of limited data during the training of the model and enhance the stability and robustness of the model. [6] generated pseudo-OOD samples by using in-distribution (IND) data, introduced a new regularization loss and fine-tuned the model, which improved the robustness of the model towards OOD data. Based on "Machine reading comprehension models are vulnerable and not robust to adversarial examples", [7] proposed Stable and Contrastive Question Answering (SCQA) to improve invariance of model representation to alleviate the robustness issue. Increasing search steps in adversarial training can significantly improve robustness, but a too large number of search steps can hurt accuracy. Therefore, [8] proposed friendly adversarial data augmentation (FADA) to generate friendly adversarial data and performed adversarial training on FADA to obtain stronger robustness via fewer steps. Specifically, they recovered all the adversarial counterpart to the sentences word by word and named these data with only one word recovered as friendly adversarial data, which in their experiments can improve model robustness without hurting accuracy. To address the issue of low-resource target data, [9] proposed a unique task transferability measure named Normalized Negative Conditional Entropy (NNCE), through which the source datasets can be fully used.

It is relatively difficult to solve the problem of model robustness from the semantic level, but actually, the model is more prone to become unstable because of semantic problems. [10] argued for semantic robustness, which can improve performance compared to models robustness in the classical sense, such as the adversarial robustness. [11] measured the proximity, connectedness and stability of the faithfulness of explanations by counterfactuals, analyzed the role that latent input representations play in robustness and reliability – the importance of latent representation in the counterfactual search strategy. [12] design a novel prompt-based history modeling approach in Conversational Question Answering, which is simple, easy-to-plug into practically any model, and shows strong robustness in all kinds of settings.

When training, the model can obtain strong robustness by performing certain operations on the model or training process, such as adding disturbance constraints, changing the loss function or activation function, adding external modules, and performing regularization processing. [13] used the Fisher information to calculate the Fisher mask and further obtained the sparse perturbation, which can not only reduce the computation cost of optimization potentially, but also make the optimizer focus on the optimization of the important parameters. [14] proposed a memory imitation meta-learning (MemIML) method that enhanced the model's reliance on support sets for task adaptation and improved the robustness of the model. [15] outlines potential directions in enhancing distributional robustness of LLMs to mitigate a performance drop under distribution shift, proposed an accurate model generalization measure and methodologies for robustness estimation of both generative and discriminative LLMs. [16] proposed a weakly supervised neural model in low resources, which did not require relevance annotations, instead it was trained on samples extracted from translation corpora as weak supervision. Based on Multilingual neural machine translation (Multi-NMT) with one encoder-decoder model, [17] proposed a compact and language-sensitive method for multilingual translation and found that this method is more effective in low-resource and zero-shot translation scenarios.

3   Method

The work in this paper is based on text generation tasks, so the basic model is a seq2seq text generation model –the domain adaptive dialogue generation model (DAML). Since MAML is model-agnostic, it can be adapted to various scenarios. The following content will introduce how Fisher information is skillfully integrated with the DAML model.

3.1   FDAML

The method proposed in this paper is to improve the training of the DAML model by introducing Fisher information in the outer optimization stage without

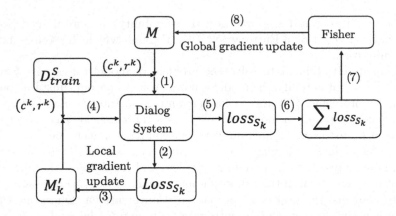

Fig. 1. The Training Process of FDAML. M represents the original model for each cycle and M'_k represents the temporary model. Step (1) is to input the data extracted from D^S_{train} into the dialogue system; Step (2) is to calculate the loss $Loss_{S_k}$ through each input and the generated system response; Step (3) is to update the temporary model through local gradient descent; Step (4) is to input (c^k, r^k) into the temporary model; Step (5) is to calculate the new loss $loss_{S_k}$; Step (6) is to sum the losses corresponding to each data and use the sum as the loss of the outer layer optimization; Step (7) is to calculate the Fischer information of parameters and identify important parameters; Step (8) is to update the original model through global gradient descent. In the next cycle, the original model is replaced with the updated model, and these steps are repeated to train the model.

changing the main framework of the training of the DAML model. The improved model is named FDAML, of which the training process is shown in Fig. 1.

This paper regards the objective function L_{meta} of the outer optimization phase as the objective function of the mask variable $\mathbf{m}$:

$$L_{meta} = L_{meta}(\mathbf{m}) \tag{1}$$

then uses the second-order Taylor expansion at $\mathbf{m} = \mathbf{1}$ to approximate the objective function :

$$L_{meta}(\mathbf{m}) \approx L_{meta}(\mathbf{1}) - \mathbf{g}^T(\mathbf{1} - \mathbf{m}) + \frac{1}{2}(\mathbf{1} - \mathbf{m})^T \mathbf{H}(\mathbf{1} - \mathbf{m}) \tag{2}$$

where $\mathbf{g} = E(\frac{\partial}{\partial \mathbf{m}} L_{meta}(\mathbf{1}))$, $\mathbf{H} = E(\frac{\partial^2}{\partial \mathbf{m}^2} L_{meta}(\mathbf{1}))$.

It is easy to know that when the model has converged to a local minimum, the first derivative g of the objective function L_{meta} with respect to the mask variable is close to 0 [18], that is, the second term in Formula (2) is approximately 0, so Formula (3) is obtained :

$$L_{meta}(\mathbf{m}) \approx L_{meta}(\mathbf{1}) + \frac{1}{2}(\mathbf{1} - \mathbf{m})^T \mathbf{H}(\mathbf{1} - \mathbf{m}) \tag{3}$$

Since $L_{meta}(1)$ is a constant relative to the mask variable, the objective of the outer optimization phase can be written as :

$$\underset{m}{\operatorname{argmin}} L_{\text{meta}}\left(\boldsymbol{m}\right) \approx \underset{m}{\operatorname{argmin}}(\boldsymbol{1}-\boldsymbol{m})^T \mathbf{H}(\boldsymbol{1}-\boldsymbol{m}). \tag{4}$$

It can be seen from Formula (4) that the optimal mask is determined by the Hessian matrix. However, since the accurate calculation of the Hessian matrix is almost impossible here, this paper uses the empirical Fisher information matrix of the mask variable to approximate the Hessian matrix :

$$\hat{I}(\theta) = \frac{1}{N} \sum_{j=1}^{N} \left(\frac{\partial}{\partial m} L_{\text{meta}}\left(r_j^k, 1\right)\right) \cdot \left(\frac{\partial}{\partial m} L_{\text{meta}}\left(r_j^k, 1\right)\right)^T \tag{5}$$

In this paper, the optimal mask variable is calculated by Formula (5), so as to select the important parameters of the model. However, in the actual calculation process, it is difficult to use the complete Fisher information matrix to solve the optimization problem. Therefore, in order to simplify the calculation, this paper assumes that the Fisher information matrix is a diagonal matrix.

Algorithm 1. FDAML

Input: dataset in Source Domain D_{train}^S; α, β;
1: Randomly initialize model M
2: **while** not done **do**
3:      **for** S_k in Source Domain **do**
4:          Sample data c^k from D_{train}^S
5:          Evaluate L_{S_k}
6:          $M_k' \leftarrow M_k' - \alpha * \nabla L_{S_k}$
7:          Evaluate l_{S_k}
8:      **end for**
9:      Evaluate $L_{\text{meta}} = \sum l_{S_k}$
10:     Evaluate Empirical Fisher $\hat{I}(\theta)$ by Formula 5
11:     $m_1 \leftarrow ArgTopk(\hat{I}(\theta), (1-s) \cdot |\theta|)$
12:     $m_0 \leftarrow ArgTopk(-\hat{I}(\theta), s \cdot |\theta|)$
13:     Update mask m by merging $m = m_1 \cup m_0$
14:     $M \leftarrow M - \beta * (\nabla L_{meta} \odot m)$
15: **end while**

The algorithm of model training is shown in Algorithm 1, steps 1–8 normally perform inner layer optimization training and calculate the optimization loss. Then, by calculating the Fisher information by Formula (5) and comparing its size, the parameters with larger importance scores are found and the mask is formulated :

$$m_1 \leftarrow ArgTopk(\hat{I}(\theta), (1-s) \cdot |\theta|) \tag{6}$$

$$m_0 \leftarrow ArgTopk(-\hat{I}(\theta), s \cdot |\theta|) \tag{7}$$

$$m = m_1 \cup m_0 \tag{8}$$

where the $ArgTopk(a, k)$ function returns the first k maximum elements in a, and returns the minimum element when a is negative; s represents the sparse ratio, and its value is obtained by directly calculating the network parameters of each layer. $|\theta|$ represents the total number of parameters ; the m_1 and m_0 represent the set of elements 1 and 0 in the mask m, respectively. The m is a sparse binary code, where the element 1 corresponds to an important parameter.

After formulating the mask, the mask is applied to the corresponding gradient, so that the optimizer can focus on the important parameters, and then updates the model :

$$M \leftarrow M - \beta * (\nabla L_{meta} \odot m) \tag{9}$$

In this way, the model finishes a complete training, and training in this way repeatedly, the model will be constantly optimized, so that it can quickly adapt to different tasks.

3.2   Fisher Information

Fisher information is a method for measuring the amount of information about the unknown parameters of an observable random variable with respect to its distribution. It is defined as the variance of the Score function, as shown in Formula (10), and the Score function is the first-order derivative of the log-likelihood function, as shown in Formula (11) :

$$I(\theta) = Var(S(\theta)) \tag{10}$$

$$S(\theta) = \frac{\partial \ln L(\theta)}{\partial \theta} \tag{11}$$

where $L(\theta)$ represents the likelihood function.

The expectation of the Score function is 0, as shown in Formula (12) :

$$E(S(\theta)) = 0 \tag{12}$$

Then it can be obtained from Formula (10) and (12) :

$$I(\theta) = Var(S(\theta)) = E(S^2(\theta)) - E^2(S(\theta)) \tag{13}$$

$$I(\theta) = E(S^2(\theta)) \tag{14}$$

Formula (14) is a commonly used calculation formula for Fisher information.

In general, assuming that θ represents the parameter vector, and the distribution of the observation X is $P(x|\theta)$, then the Fisher information matrix of the parameter is :

$$I(\boldsymbol{\theta}) = E[\nabla \ln P(x|\boldsymbol{\theta})(\nabla \ln P(x|\boldsymbol{\theta}))^T] \tag{15}$$

However, calculating the expectation of the likelihood function composed of the observation distribution directly is difficult. The empirical distribution can be used to approximate the expected value in the overall distribution. The empirical Fisher formula is :

$$\hat{I}(\theta) = \frac{1}{N} \sum_{i=1}^{N} P\left(x_i \mid \theta\right) \cdot \nabla \ln P(x \mid \boldsymbol{\theta}) \cdot \nabla \ln P(x \mid \boldsymbol{\theta})^T \tag{16}$$

where $X = \{x_1, \cdots, x_N\}$ is the sample data . Generally, the empirical Fisher information formula is used more frequently.

In the second-order optimization method, the Fisher information matrix is often used to replace the Hessian matrix. The relationship between the two is : the Fisher information matrix is equal to the negative expectation of the Hessian Metrics of the logarithmic likelihood function :

$$I(\theta) = -E(H_{ln(P(x|\theta))}) \tag{17}$$

3.3   DAML

The basic idea of the Model-Agnostic Meta-Learning algorithm is to fix the model structure and find a suitable initialization for the model parameters through the existing data, so that the model can quickly adapt to different tasks through a small amount of data and achieve good performance on the data set. MAML is essentially a two-layer optimization algorithm. It first trains the model based on all tasks and performs inner optimization. Then go back to the initial position, summarize the loss of all tasks, find a direction that can take into account multiple tasks to use gradient descent and perform outer optimization.

The DAML model based on meta learning [19] is an end-to-end trainable dialogue system model. It applies MAML to the dialogue domain, so that it can find an initialization method that can accurately and quickly adapt to the unknown new domain in the case of a small amount of data. Therefore, the model can effectively adapt to the dialogue task in the new domain with only a small number of training samples.

The DAML model can be seen as a combination of MAML and Sequicity [20]. It applies MAML to the Sequicity, so as to find a suitable initialization for the model parameters during training. Therefore, the model structure of DAML is shown in Fig. 2.

There are two optimization processes in the training of the DAML model. First, for each task domain, a certain size of data sampling is performed. Then the extracted training data (c^k, r^k) are input into the sequence model to obtain the corresponding generated system response. For all task domains, cross entropy is used as the loss function :

$$L_{S_k}\left(M, c^k, r^k\right) = \sum_{j=1}^{N} r_j^k \log P_M\left(r_j^k\right) \tag{18}$$

In the training, a one-step gradient descent is used for each task domain, and a temporary model M_k' is updated. Then, based on the updated temporary

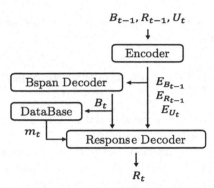

Fig. 2. Model Structure of DAML [19]. In the structure, B_t represents the hidden state of the dialogue at time t. R_t and U_t respectively represent the machine's response at time t and the user's answer at time t. m_t is a simple label that helps generate the response, representing the state of the next step in the conversation. The model takes the dialogue context $c = \{B_{t-1}, R_{t-1}, U_t\}$ as input, and the generated sentence R_t as output.

model, the same training data (c^k, r^k) is used to calculate the new loss l_{S_k} of each M'_k. Then, the updated loss values l_{S_k} on all task domains are summed to obtain L_{meta}, and L_{meta} is set as the objective function of the outer optimization stage:

$$L_{\text{meta}} = \sum_{S_k} l_{S_k} \left(M'_k, c^k, r^k \right) \qquad (19)$$

Finally, the model is updated by minimizing the objective function L_{meta}.

4   Experiments

4.1   Datasets

To compare with the original DAML model, this paper uses the SimDial datasets [19], which has six dialogue domains: restaurant, weather, bus, movie, restaurant-slot and restaurant-style, where restaurant-slot data has the same slot type and sentence generation templates as the restaurant but a different slot vocabulary and restaurant-style has the same slots but different natural language generation (NLG) templates compared to the restaurant.

In this paper, restaurant, weather and bus are chosen as source domains. For each source domain, there are 900, 100, 500 conversations for training, validation and testing correspondingly. The rest three domains are for evaluation, which are considered as target domains. 9 dialogues (1% of source domain) are generated for each domain's adaptation training. For testing, 500 dialogues for each target model are used. Movie is chosen to be the new target domain for evaluation. In this paper, two dialogue domains, restaurant-slot and restaurant-style, are used to evaluate the performance of the model on the task domain with fewer common features with the task source domain.

4.2   Metrics

In this paper, two metrics, BLEU score and entity F1 score, are selected to evaluate the performance of the model. They are the most important and persuasive metrics used in [1] has exhaustively demonstrated the fast adaptation speed to new tasks of the MAML.

BLEU. BLEU (bilingual evaluation understudy) is an auxiliary tool for bilingual translation quality assessment. The BLEU score is originally a tool used to evaluate the quality of machine translation. It judges the quality of sentences generated by machine translation by comparing the similarity between machine translation sentences and professional manual translation sentences. The more similar, the better the quality. In this experiment, the BLEU score is used to evaluate the similarity between the response sentences generated by the model and the real response. The similarity is higher and the performance of the model is better.

Entity F1 Score. Entity F1 score is an indicator to measure the accuracy of the binary classification model, taking into account the accuracy and recall rate of the model. The F1 score can be said to be a weighted average of the accuracy and recall rate. Its maximum value is 1 and the minimum value is 0. The closer the value is to 1, the better the model performance is. For each dialogue data, this paper uses the F1 score to compare the similarity between the key slot information of the response generated by the model and the real response.

4.3   Implementation Details

For all experiments, this paper uses a pre-trained Glove word embedding with a dimension of 50. The optimizer used are Adam and SGD. The learning rate of Adam optimizer is set to 0.003, and the learning rate of SGD is different in different experiments. In this experiment, the SGD optimizer is only used in the outer optimization stage of the initial training stage of the model, and the Adam optimizer is used in other stages. If the validation loss increases during training, the learning rate is reduced to half. The number of data samples captured by a training time is set to 32, and the probability of dropout is set to 0.5.

In the experiment, the total number of training times in the initial training stage is set to 200, and the number of training times in the fine-tuning stage is set to 30. In the fine-tuning stage, the model selected in this paper is: the model obtained after 100 initial training. We tested the 30 models obtained by fine-tuning, and then selected the best results from the obtained test results. In order to prevent the possible loss of important parameters, a strategy is used in the experiment: when the number of valid parameters in a layer of parameters is less than 1000, the parameters are all considered important parameters by default, so the Fisher information is not used in this case.

5   Results

5.1   Main Results

Table 1 shows the BLEU scores and F1 scores of the original DAML model and the FDAML model on six task domains.

Table 1. Performance comparison of the two models.

	Bus		Restaurant		Weather	
	DAML	FDAML1	DAML	FDAML1	DAML	FDAML1
F1	0.9269	0.9257	0.9094	0.9088	0.7967	**0.8058**
BLEU	0.5277	**0.5435**	0.5515	0.5414	0.3670	**0.3927**
	Movie		Restaurant-style		Restaurant-slot	
	DAML	FDAML1	DAML	FDAML1	DAML	FDAML1
F1	0.6702	**0.7304**	0.8503	0.7880	0.8161	**0.8466**
BLEU	0.3979	0.3749	0.4129	**0.4380**	0.4000	**0.4096**

It can be seen from Table 1 that the scores of the FDAML are mostly higher than the original DAML model, which shows that the FDAML model performs better than the DAML model. Specifically, the two scores corresponding to the FDAML model on the Bus task domain are higher than the scores of the DAML model. At the same time, in some other task domains such as the Restaurant domain, the scores of the FDAML model in this paper are not as high as the DAML model. It may be because the training datasets homologous to the test datasets are used during training, so that the model has 'captured' the main data features of the source domain after training. Therefore, the use of Fisher information to identify salient features is not obvious. Nevertheless, the two scores of the FDAML model on the two task domains of Restaurant and Weather are not much different from the DAML model, which also shows that the FDAML model has reached the level of the DAML model. Therefore, the first conclusion can be drawn: the performance of the FDAML model on some test datasets in the same domain as the training datasets can reach the level of the DAML model, even better than the DAML model.

By comparing the data in the Table 1, it can be seen that the BLEU scores and F1 scores corresponding to the FDAML model in the three target domains are mostly higher than the DAML model, which also shows that the overall performance of the model in the three target domains is better than the DAML model. Therefore, the second conclusion can also be drawn: on the new task domain datasets or on the datasets which share few features with the training task domain, using Fisher information can effectively improve the model performance and enhance the robustness of the model. In summary, the method of using Fisher information to improve the robustness of model is effective.

Table 2. Model performance comparison of using two timings when the learning rate is 0.03.

	Bus		Restaurant		Weather	
	FDAML1	FDAML2	FDAML1	FDAML2	FDAML1	FDAML2
F1	**0.6301**	0.6275	0.9040	0.9260	**0.8161**	0.7876
BLEU	0.3301	0.3500	**0.5552**	0.5329	**0.4042**	0.2773
	Movie		Restaurant-style		Restaurant-slot	
	FDAML1	FDAML2	FDAML1	FDAML2	FDAML1	FDAML2
F1	**0.6769**	0.6568	0.7527	0.8448	**0.8581**	0.8422
BLEU	**0.3971**	0.3949	**0.4510**	0.4410	0.4214	0.4320

Table 3. Model performance comparison of using two timings when the learning rate is 0.005.

	Bus		Restaurant		Weather	
	FDAML1	FDAML2	FDAML1	FDAML2	FDAML1	FDAML2
F1	0.5890	0.5903	0.9221	0.9274	**0.8373**	0.8119
BLEU	0.2679	0.2892	0.5537	0.5576	**0.3703**	0.3206
	Movie		Restaurant-style		Restaurant-slot	
	FDAML1	FDAML2	FDAML1	FDAML2	FDAML1	FDAML2
F1	**0.6701**	0.6509	**0.7679**	0.7466	**0.8703**	0.8469
BLEU	**0.3917**	0.3831	0.4069	0.4561	**0.4291**	0.4195

5.2   The Timing of Using Fisher

In addition to the main comparative experiments, this paper also makes some preliminary studies on the timing of using Fisher information, and makes several sets of comparative experiments.

On the timing of the use of Fisher information, a preliminary comparative experiment was conducted under the conditions of SGD learning rates of 0.03 and 0.005, and two timings of the use of Fisher information were set : when the number of training times was 0 and when the number of training times was 50, before that, the model did not use Fisher information during the training process, and after that, the model began to use Fisher information until the end of the training. In this paper, the model is fine-tuned on six task domains, and then the model is tested to obtain BLEU scores and F1 scores. The results of the two groups of comparative experiments are shown in Table 2 and Table 3.

In the tables, the FDAML1 and FDAML2 models represent the models trained using Fisher information from the training times of 50 and 0, respectively. By comparing the data in Table 2 and Table 3, it can be seen that the two scores corresponding to the FDAML model when the number of training times is 50 are mostly higher than the FDAML model when the number of training times is 0. Therefore, a third conclusion can be drawn: using Fisher information to train the model after 50 times of model training is more effective in improving the performance of the model than using Fisher information to train the model at the beginning. A conjecture is made: it may be because the calculation of Fisher information depends on the gradient of the previous round of training parameters, so the previous round of parameters directly affect the reliability of Fisher information. Since the number of training times is 0, Fisher information is used to screen important parameters, which will cause deviations in the training of models. The main reason is that the parameters involved in the training in the first round of training are randomly initialized, which does not contain information that is particularly related to the training task domain. The Fisher information calculated using this set of parameters is random and cannot accurately filter out important parameters. After 50 rounds of training, the Fisher information is calculated to ensure the accuracy of the calculation results, and the Fisher information can be used to extract information related to the task domain from the parameters that have been trained to guide the training of the model.

This result is also consistent with the theoretical part of this paper that Fisher information matrix is used to approximate the Hessian matrix. Specifically, when the objective function of the mask variable is simplified, a conclusion is used: when the model converges to a local minimum, the first derivative of the objective function with respect to the mask variable is approximately 0. The timing of using Fisher information when the number of training times is 50 is selected because during the experiment, by observing the validation loss value of model during training, it is preliminarily judged that when the number of training times is 50, the training of the model has begun to converge, and the objective function is close to the local minimum.

6   Conclusion and Future Work

In order to improve the robustness of text models under low-resource scenarios, this paper proposes a high robust text generation training framework based on meta learning for the dialogue generation task. In the outer optimization stage of model training, the important parameters are identified by calculating and comparing Fisher information, so that the optimizer can focus on the important parameters. In this paper, the evaluation is carried out on six task domains, and the experimental results show that using Fisher information to improve the training of the model can reduce the amount of calculation and improve the robustness of the model while the model training reaches or exceeds the original level.

Due to the limitation of capacity and time, there are still many shortcomings in the research, but these shortcomings also provide goals and directions for future research:

- Consider sparse ratio and stratification;
- Integrate multiple rounds of Fisher information;
- Introduce Fisher information in the inner optimization phase;

– Consider the non-diagonal case of Fischer information matrix.

Acknowledgements. This paper is supported by National Natural Science Foundation of China (NSFC Grant No. 62106275 and No. 62306330) and Natural Science Foundation of Hunan Province (Grant No. 2022JJ40558).

References

1. Finn, C., Abbeel, P., Levine, S.: Model-agnostic meta-learning for fast adaptation of deep networks. In: 34th International Conference on Machine Learning, ICML 2017, vol. 5, pp. 3933–3943. International Machine Learning Society (IMLS) (2017). https://arxiv.org/abs/1703.03400
2. Lux, F., Vu N, T.: Language-agnostic meta-learning for low-resource text-to-speech with articulatory features. In: Proceedings of the Annual Meeting of the Association for Computational Linguistics 2022, vol. 1, pp. 6858–6868. Association for Computational Linguistics (ACL) (2022). https://arxiv.org/abs/2203.03191
3. Gu, J., Wang, Y. Chen, Y.: Meta-learning for low-resource neural machine translation. In: Proceedings of the 2018 Conference on Empirical Methods in Natural Language Processing, EMNLP 2018, pp. 3622–3631. Association for Computational Linguistics (2018). https://arxiv.org/abs/1808.08437
4. Abbas, M., Xiao, Q., Chen, L.: Sharp-MAML: sharpness-aware model-agnostic meta learning (2022)
5. Song, X., Gao W, Yang, Y.: ES-MAML: simple hessian-free meta learning (2019)
6. Sundararaman, D., Mehta, N., Carin, L.: Pseudo-OOD training for robust language models (2022)
7. Yu, H., Wen, L., Meng, H.: Learning invariant representation improves robustness for MRC models. In: Findings of the Association for Computational Linguistics: EMNLP 2022, pp. 3306–3314. Association for Computational Linguistics (2022). https://arxiv.org/abs/1906.04547
8. Zhu, B., Gu, Z., Wang, L.: Improving robustness of language models from a geometry-aware perspective, pp. 3115–3125 (2022)
9. Chen, Z., Kim, J., Bhakta, R.: Leveraging task transferability to meta-learning for clinical section classification with limited data. In: Proceedings of the Annual Meeting of the Association for Computational Linguistics, vol. 1, pp. 6690–6702. Association for Computational Linguistics (2022). https://aclanthology.org/2022.acl-long.461/
10. Malfa E, La., Kwiatkowska, M.: The king is naked: on the notion of robustness for natural language processing. In: Proceedings of the 36th AAAI Conference on Artificial Intelligence, AAAI 2022 vol. 3(6), 11047–11057 (2021)
11. El Zini, J., Awad, M.: Beyond model interpretability: on the faithfulness and adversarial robustness of contrastive textual explanations. In: Findings of the Association for Computational Linguistics: EMNLP 2022, pp. 1391–1402. Association for Computational Linguistics (2022). https://arxiv.org/abs/2210.08902
12. Gekhman, Z., Oved, N., Keller, O.: On the robustness of dialogue history representation in conversational question answering: a comprehensive study and a new prompt-based method (2022)
13. Zhong, Q., Ding, L., Shen, L.: Improving sharpness-aware minimization with fisher mask for better generalization on language models. In: Findings of the Association for Computational Linguistics: EMNLP 2022, pp. 4093–4114. Association for Computational Linguistics (2022). https://arxiv.org/abs/2210.05497

14. Zhao, Y., Tian, Z., Yao, H.: Improving meta-learning for low-resource text classification and generation via memory imitation. In: Proceedings of the Annual Meeting of the Association for Computational Linguistics, vol. 1, pp. 583–595. Association for Computational Linguistics (2022). https://arxiv.org/abs/2203.11670

15. Štefánik, M.: Methods for Estimating and improving robustness of language models. In: NAACL 2022–2022 Conference of the North American Chapter of the Association for Computational Linguistics: Human Language Technologies, Proceedings of the Student Research Workshop, pp. 44–51. Association for Computational Linguistics (2022). https://arxiv.org/abs/2206.08446

16. Zhao, L., Zbib, R., Jiang, Z.: Weakly supervised attentional model for low resource ad-hoc cross-lingual information retrieval. In: Proceedings of the 2nd Workshop on Deep Learning Approaches for Low-Resource Natural Language Processing, pp. 259–264. Association for Computational Linguistics (2021). https://aclanthology.org/D19-6129/

17. Wang, Y., Zhou, L., Zhang, J.: A compact and language-sensitive multilingual translation method. In: 57th Annual Meeting of the Association for Computational Linguistics, Proceedings of the Conference, pp. 1213–1223. Association for Computational Linguistics (2020). https://aclanthology.org/P19-1117/

18. Cun Y, Le., Denker J, S., Solla S, A.: Optimal Brain Damage. In: Advances in Neural Information Processing Systems 2, pp. 598–605, 1990. https://www.semanticscholar.org/paper/Optimal-Brain-Damage-LeCun-Denker/e7297db245c3feb1897720b173a59fe7e36babb7

19. Qian, K., Yu, Z.: Domain adaptive dialog generation via meta learning. In: 57th Annual Meeting of the Association for Computational Linguistics, Proceedings of the Conference, pp. 2639–2649. Association for Computational Linguistics (2019). https://arxiv.org/abs/1906.03520

20. Lei, W., Jin, X., Ren, Z.: Sequicity: simplifying task-oriented dialogue systems with single sequence-to-sequence architectures. In: 56th Annual Meeting of the Association for Computational Linguistics, Proceedings of the Conference (Long Papers), vol. 1, pp. 1437–1447. Association for Computational Linguistics (2018). https://aclanthology.org/P18-1133/

Effects of Brightness and Class-Unbalanced Dataset on CNN Model Selection and Image Classification Considering Autonomous Driving

Zhumakhan Nazir, Vladislav Yarovenko, and Jurn-Gyu Park[✉]

School of Engineering and Digital Sciences,
Nazarbayev University, Astana, Kazakhstan
{zhumakhan.nazir,vladislav.yarovenko,jurn.park}@nu.edu.kz

Abstract. In addition to an approach of combining machine learning (ML) enhanced models and convolutional neural networks (CNNs) for adaptive CNN model selection, a thorough investigation study of the effects of 1) image brightness and 2) class-balanced/-unbalanced datasets is needed, considering image classification (and object detection) for autonomous driving in significantly different daytime and nighttime settings. In this empirical study, we comprehensively investigate the effects of these two main issues on CNN performance by using the ImageNet dataset, predictive models (*premodel*), and CNN models. Based on the experimental results and analysis, we reveal non-trivial pitfalls (up to 58% difference in top-1 accuracy in different class-balance datasets) and opportunities in classification accuracy by changing brightness levels and class-balance ratios in datasets.

Keywords: Image Feature Extraction · Balanced Dataset · Interpretable Models · CNNs

1 Introduction

Object detection is a process of predicting locations and classes (image classification) of objects in an input image. Nowadays, utilizing CNNs is the most common approach for solving this problem [21]. However, even current state-of-the-art CNNs cannot perfectly classify all images; and deeper CNNs result in more inference time and energy consumption [16], although they can improve the accuracy. Moreover, as one of the key applications in real-time, CNNs have been used in autonomous driving [1,2] to efficiently detect and classify objects such as cars, pedestrians, traffic lights etc.; but it requires both high classification accuracy and low inference time.

One solution is to build supervised machine learning (ML) based predictive models (*premodel*) using features of an input image to predict the fastest CNN model that can classify the given image correctly (e.g., *adaptive CNN model*

B. Luo et al. (Eds.): ICONIP 2023, CCIS 1969, pp. 191–203, 2024.
https://doi.org/10.1007/978-981-99-8184-7_15

selection [16]). In our preliminary work [23], we also observe that logistic regression (LR) based *premodel* is an interpretable and accurate model which achieves overall high test accuracy compared to other supervised ML models. (Note that some models are not interpretable, although they could be more accurate). And the *premodel* results clearly present the effects of class-balanced/-unbalanced datasets. Another observation is that brightness is one of most important features [23] in the *premodel*, therefore we start from the intuition that images with higher brightness values have a higher chance of being successfully classified by CNNs, while darker images may be classified less successfully.

Practically, the brightness problem is encountered in autonomous driving very often [20], as the accuracy of CNNs may be worsened significantly during nighttime or extreme weather conditions such as snow and rain. In this paper, with regard to brightness, two different pre-processing methods are studied: 1) categorizing images that have original brightness values, and 2) the brightness of images was reduced manually. Both simulate how camera perceives objects in those severe conditions. Therefore, by studying different brightness groups and analyzing their effects on CNN accuracy the brightness effects can be tested.

Moreover, *premodel's* performance is affected by the class imbalance in training datasets. According to our previous observations [23], more than 80% of the images in the ImageNet 2012 Validation Set [5] can be successfully classified by MobileNet_v1 [9], while 17% cannot be classified by any of MobileNet_v1 [9], Inception [10] or ResNet [8]. Therefore, during the *premodel* training phase to identify CNN that can classify images correctly, it may develop bias towards more images being classifiable (i.e., using unbalanced dataset). This may affect performance of the *premodel* and need to investigate more thoroughly for the the effects of sizes and balance of datasets.

Our paper makes the following contributions:

1. Investigate how the brightness of images can affect the *premodel* and CNN accuracy, and compare different brightness types.
2. Compare the accuracy of CNN models, and test adaptive CNN model selection using the *premodel*.
3. Analyze the effects of size, balance, and type of datasets on the classification accuracy comprehensively.

2   Motivation

As a motivating example (Fig. 1a), to investigate how the different brightness affects the *premodel* and CNNs, we generate the copies of ImageNet dataset, with the reduced brightness levels: 75%, 50%, 25%, and 10%. Figure 1b illustrates the initial test using an unbalanced dataset with randomly chosen 1000 (1k) images. (Full description of an Unbalanced dataset will be given in Methodology section.) The results show that the accuracy drop can be noticed with the brightness decrease. However, the degree of decrease is specific for every CNN: with a 90% brightness decrease (i.e., the top-right 10% in Fig. 1b), Inception_v4 experiences

(a) Images with 100%, 75%, 50%, 25%, and 10% brightness levels

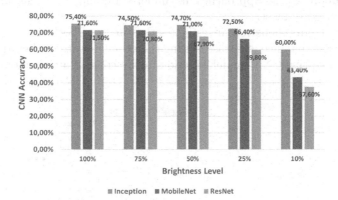

(b) Top-1 CNN Accuracy on Unbalanced Dataset with 1k Images.

Fig. 1. Motivating Example: Correlation between Image Brightness Level and CNN Classification Accuracy.

only a 15% accuracy drop, while ResNet_v1_152 decreases almost by 35% in the same brightness level. Moreover, we are motivated to investigate comprehensively the effects of 1) dataset size, 2) class-balanced/-unbalanced datasets, 3) a specific type of dataset (e.g., car-related dataset) and 4) different types of CNN networks.

3 Related Work

For adaptive CNN model selection, Marco et al. [16] collected features of images using the feature detectors and descriptors from the ImageNet, and built a pre-model which predicts a best CNN model in terms of Top-1 accuracy and inference time among MobileNet [9], Inception [10] and ResNet [8]. They also conducted an extensive investigation to select the most important seven features for *premodel*.

From the perspective of the effect of the *brightness* feature on CNN performance, Xu et al. [22] proposed an adaptive brightness adjustment model, considering the importance of brightness in object detection; and Rodriguez et al. [19] observed that decreasing global illumination levels can degrade classification accuracy. Additionally, Castillo et al. [3] improved the classification performance of cold steel weapons by proposing a preprocessing algorithm based on the brightness; by brightening or darkening input images, their model "strengthens"

CNN's robustness to light conditions. Lastly, Kandel et al. [11] also discussed brightness as a possible augmentation technique by testing brightness values in the range 0%-200%. Their results have shown that current state-of-the-art CNNs perform the best when the brightness is the closest to the original 100% value.

In terms of class-unbalanced datasets, there are several works on the effects of class unbalance distribution in datasets [6,12–15,17]. The main results show that data preprocessing approaches like under-sampling [13–15,17] are effective with unbalanced dataset.

Our work comprehensively investigates the effects of both the *brightness* feature and class-imbalance of datasets using the popular state-of-the-art ImageNet dataset for adaptive CNN model selection in the context autonomous driving.

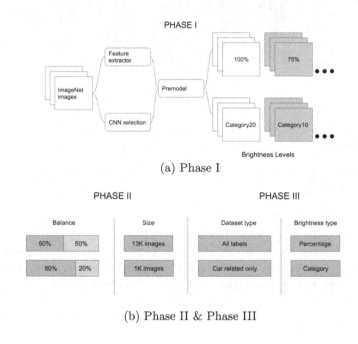

(a) Phase I

(b) Phase II & Phase III

Fig. 2. An Overview of our Methodology.

4   Methodology

Our methodology can be split into three different phases (Fig. 2): 1) Preparation of *premodel* and Data collection of brightness datasets, 2) Dataset size and Class-balance, and 3) Dataset property and Effects of brightness types.

Phase 1: The first phase focuses on creating a predictive model (*premodel*) that can select an optimal CNN for image classification based on the features of an input image. For this, we select the ImageNet 2012 Validation Dataset [5] and calculate the optimal CNN for each image in it. The CNN is considered to be optimal if it can correctly predict the image's label in the shortest amount of time. Next, the features (*brightness* and *hue*) of each image are extracted and recorded. By doing this, we collected a 50k dataset, which has image features as independent variables and an optimal CNN as the dependent/target variable. This dataset was split into training and test datasets, and the Logistic Regression (LR) based predictive model was selected as an accurate and interpretable model among the four ML estimators of Decision Tree, Logistic Regression, SVC Classifier, and MLP Classifier.

Brightness and hue values of all images from ImageNet dataset are extracted and fed to *premodel* which predicts CNN that most efficient on a given image. For instance, if the class is predicted to be "MobileNet", the Mobilenet_v1 CNN is used for the classification. If the class is "Failed", either Inception_v4 or Resnet_v1_152 will be used. One of these two CNNs will be selected beforehand by comparing their performance. The selected network for "Failed" case will be used in all the following datasets, so it is important to conduct this test first.

After that, copies of the ImageNet dataset with reduced brightness are generated. We arbitrarily chose 75%, 50%, 25%, and 10% as brightness levels, creating five different datasets. The brightness is changed using Pillow [4] library in Python. The algorithm uses Formula 1, where *out* is the brightness resulting image, α is the selected coefficient (1 for 100%, 0.75 for 75%, etc.), in_1 is the original image, and in_0 is the pure black image [7].

$$out = (1 - \alpha) * in_0 + \alpha * in_1 \qquad (1)$$

In parallel, the original images of ImageNet dataset were split into 20 categories (2.5k images per category) based on their brightness values. This creates datasets with different brightness levels (referred as *category* dataset) which was not changed artificially, unlike the method described above (referred as *percentage* dataset in the paper). Finally, we choose the five categories which have the most similar average brightness values, compared to the five *category* datasets. For the effects of different types of brightness, which will be covered in the Phase III, using the collected dataset in this phase.

Phase 2: The second phase focuses on the dataset properties, such as class balance and the number of samples (i.e., dataset size). We select only images that belong to the "MobileNet" and "Failed" classes. Image has "MobileNet" class if optimal CNN for it is MobileNet_v1, and "Failed" if all used CNNs fail to classify it. "Inception" and "ResNet" classes were discarded to make the difference between classes more substantial and simplified, representing "MobileNet" and "Failed" as the easiest and the hardest images to classify respectively. Also, the "Inception" and "ResNet" classes had a very small proportion in the dataset, which would result in even smaller balanced datasets.

Since only 6.5k images from ImageNet belong to the "Failed" class, the largest balanced dataset with 50% "MobileNet" and 50% "Failed" classes can have only 13k images, by adopting the under-sampling technique. To make a

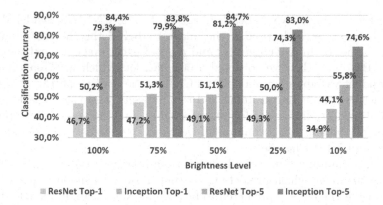

Fig. 3. Comparison of Inception and ResNet accuracy on ImageNet (a "Failed" Unbalanced-1k dataset) with different *percentage* brightness, without *premodel*.

fair comparison between balanced and unbalanced datasets, the unbalanced dataset was made with the same 13k images too, but with an 80%/20% of "MobileNet"/"Failed" ratio. Moreover, smaller size of these datasets with only 1k images were created as well to evaluate the effects of size of datasets.

Phase 3: Lastly, the third phase focuses on the image properties where we created datasets with only car-related labels from ImageNet, such as taxi, minivan, street sign, traffic light, etc. Using these images, Balanced and Unbalanced versions with 0.3k and 0.8k images respectively were created as well. As a result, the 4 datasets each having five different brightness levels were created. Finally, the difference between images with reduced brightness (*percentage* perspective) and images with originally low brightness (*category* perspective) are studied as well.

To properly compare and evaluate the results, the Top-1 and Top-5 accuracy will be recorded for all the datasets with different brightness levels. Based on these results, we evaluate the effects of brightness, dataset size and image properties on classification accuracy.

5   Experimental Results

5.1   Experimental Setup

The initial CNN inference was performed using three different networks pretrained on ImageNet with the Tensorflow library: MobileNet_v1 [9], which is the smallest and least accurate network, Inception_v4 [10] as the most accurate one, and ResNet_v1_152 [8]. The Scikit-learn [18] library is a machine learning Python library, which was used to build the Logistic Regression-based premodel. The brightness manipulation techniques were performed with the help of the Pillow library [4].

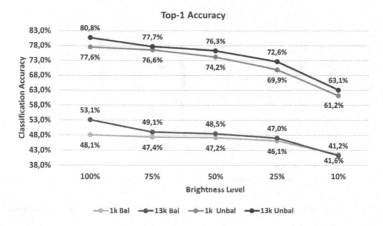

Fig. 4. Classification Accuracy of 1k and 13k Balanced/Unbalanced datasets with different percentage brightness and *Premodel*.

5.2 Results and Analysis

Preliminary Results in Phase I: If the majority of images in all datasets belong to the "MobileNet" class, Mobilenet_v1 should definitely be used as a simple network. However, if the predicted class is "Failed", we can still attempt to classify it using either Inception_v4 or Resnet_v1_152. In order to make a choice between these two, we compare the performance of each network individually on a "Failed" Unbalanced-1k dataset in Fig. 3. (Note that the Top-1 and Top-5 accuracies are lower than general ImageNet results because the "Failed" Unbalanced-1k dataset are used.) In this case, *premodel* is not used yet, as we want to compare the performance of these two CNNs individually to use it for "Failed" cases. On average, Inception_v4 shows a 4% improvement in Top-1 accuracy for higher brightness values (100%, 75%, and 50%), and almost 10% improvements in one-tenth of the initial brightness (i.e., 10%). Similarly, Top-5 accuracy is also better, ranging from a 4% difference with higher brightness values, and more than 10% with lower ones. Considering these results, the Inception_v4 can be selected as the complex CNN. It is used when the image's predicted class is "Failed", and the CNN combination of Mobilenet_v1 and Inception_v4 will be used in all the following tests.

Dataset Size and Class-Balance Results in Phase II: In order to comprehensively investigate any possible effects that are not related to brightness, we check how dataset size and class-balance can affect CNN classification results.

Figure 4 shows the Top-1 classification accuracy of Balanced and Unbalanced datasets with 1k and 13k images. The accuracy is obtained with MobileNet and Inception while selecting CNN for each image with a predictive model. The 13k dataset shows up to 3% better Top-1 accuracy across different brightness levels. The difference between accuracies of 1k and 13k datasets decrease as we decrease

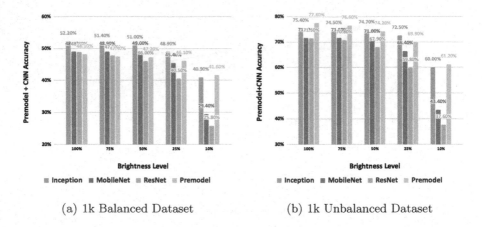

<table>
<tr><td>(a) 1k Balanced Dataset</td><td>(b) 1k Unbalanced Dataset</td></tr>
</table>

Fig. 5. Classification Accuracy of Balanced and Unbalanced 1k Datasets with *Premodel* + CNNs.

the brightness level, especially for the Balanced datasets. Balanced dataset with 13k images and 10% of the brightness has even lower Top-1 accuracy, compared to the balanced 1k dataset. However, this difference is only 0.4%, and considering that 10% brightness is an almost pitch-black image, this is not a critical deviation, and the overall trend remains the same. These results show that a small training dataset can negatively affect the predictive model and its accuracy, which results in lower classification accuracy for 1k datasets. Having larger datasets for the training and testing the predictive model improves proper CNNs selection accuracy, thus resulting in higher image classification accuracy and more reliable results.

The comparison of class-balance effects can be seen in Fig. 4 (with 1k and 13k datasets) and Fig. 5 which shows the classification accuracy of Balanced and Unbalanced datasets with 1k images. On average, compared to the Balanced datasets, all the Unbalanced ones show 25–28% higher Top-1 accuracy respectively. There could be the possible reasons for that, the first one is related to the characteristics of the datasets: Every Balanced dataset has 50% of "Failed" class, which means that none of the used CNNs could predict the correct class, having very complicated images. Therefore, the maximum possible accuracy would be 50% and 80% for Balanced and Unbalanced datasets respectively. In short, Balanced datasets, in which images to be easily classified were removed through the under-sampling, which decreased the accuracy.

It is also important to mention how dataset balance affects the *premodel* itself. If the *premodel* is tested on an Unbalanced dataset, it is more likely to ignore the class with fewer images. For instance, *premodel*, trained on an Unbalanced dataset, will predict mainly "MobileNet" for test images, while ignoring the "Failed" class.

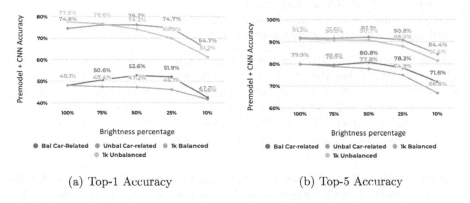

(a) Top-1 Accuracy   (b) Top-5 Accuracy

Fig. 6. Classification Accuracy of 1k and Car-related Balanced and Unbalanced Datasets with *Premodel*.

Table 1. Features and Values that Result in Higher Classification Accuracy.

Feature	Higher Acc	Lower Acc
Avg. Perc. Brightness	$\leq$109.2	>109.2
Contrast	$\geq$65.1	<65.1
Edge Length	$\geq$154.6	<154.6

Dataset Property Results in Phase III: To check how this setup would perform in Autonomous Driving, the car-related dataset was created by selecting images with vehicles in it from ImageNet and split into Balanced and Unbalanced versions. Figure 6 compares each version of the Car-related dataset with the 1k dataset since they are the closest in terms of size. On average, Car-related datasets show higher accuracy on lower brightness levels. To understand the reasons, we analyzed these datasets based on 3 features: average perceived brightness, contrast, and edge length. The first two are global features, and the edge length is a local one. However, it is one of the most important features of the Logistic Regression-based premodel, so we analyzed it as well. The comparison shows following results:

1. Car-related datasets have slightly lower average brightness. In our approach, we scale the brightness by factors (0.75, 0.5, etc.) instead of subtracting a fixed numerical value. Because of that, images with smaller initial brightness experience a smaller decrease, when scaling by a factor. That's one of the reasons why car-related images are more prone to a decrease in brightness.
2. Images in car-related datasets have higher contrast values. This makes the image clearer, and objects more distinguishable. That's why even when the brightness is decreased, the objects can still be found and classified.

3. Images in car-related datasets also have more short edges. While the edge length should not affect the classification accuracy directly, more edges may indicate images with higher quality, and result in slightly better accuracy.

Table 1 shows that datasets with lower brightness (less than 109.2), and higher contrast (higher than 65.1) and edge length (higher than 154.6) values, result in higher *premodel* classification accuracy. We believe that this is the reason why car-related datasets have higher accuracy. Additionally, the accuracy could be slightly affected by the size difference. However, this difference is not that significant, and should not affect the results severely.

Brightness Results in Phase III: Splitting the dataset based on the brightness value resulted in 20 categories with 2.5k images each. Each category has a class distribution, that is the closest to a 1k Unbalanced dataset. To fairly compare the effects of the two brightness approaches both the categories and 1k Unbalanced datasets need to have a similar mean brightness value. For each version of a dataset (100%, 75%, 50%, 25% and 10%) we select the closest category. Table 2 shows which category was selected for each percentage brightness level. Since 25% and 10% result in very low brightness, category 1, which is the darkest one, had to be split even further. The darkest 200 images were selected for

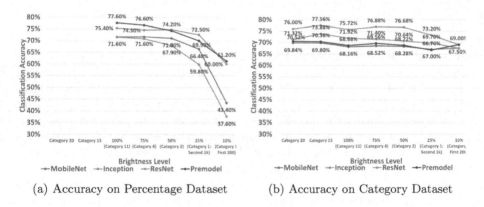

(a) Accuracy on Percentage Dataset   (b) Accuracy on Category Dataset

Fig. 7. Classification Accuracy of *Percentage* and *Category* datasets with *Premodel* + CNNs.

Table 2. Correspondence of Brightness Categories to 1k Unbalanced Dataset.

Percentage	Category #	Mean Brightness
100%	11	117.14
75%	4	87.52
50%	2	58.33
25%	1 (1000–2000)	28.91
10%	1 (1–200)	11.26

10% and 1000–2000 images for 25%. After that, the classification accuracy was calculated for 1k Unbalanced and for its corresponding Category-based dataset with and without the premodel.

Comparing premodel results of Figs. 7a and 7b, it can be seen that *category* accuracy does not change significantly when there is a smaller brightness decrease, between 100% and 50% relative brightness levels. This happens because *category* images were not changed manually. While their average brightness becomes low, this mostly results in a darker background, while the object is still visible. On the other hand, in *percentage* datasets, we reduce the brightness of the whole image, which makes the object less visible.

Accuracy results in Fig. 7 confirm that *category* datasets stay relatively similar, but *percentage* datasets decrease significantly. Additionally, for *percentage* datasets, the results of each CNN individually are worse than the results with the *premodel*. For *category* datasets, *premodel* shows intermediate results which are better than Mobilenet_v1, but worse than Inception_v4.

6 Conclusion and Future Work

In this paper, we comprehensively investigated, measured, and summarized important factors of CNN image classification and adaptive model selection in terms of the class-balance and brightness levels of datasets. First of all, the Inception_v4 shows better classification accuracy than Resnet_v1_152 with all the brightness levels. Using it in the combination with Mobilenet_v1 results in a good balance between accuracy and inference time. Larger datasets have slightly higher accuracy, which is caused by the differences in premodel, while the classification itself is not affected. Unbalanced and Car-Related datasets have slightly higher accuracy than 1k datasets. This is caused by the difference in feature values, such as brightness, contrast, and edge length. Lowering the brightness negatively affects accuracy, with a 90% decrease in brightness resulting in a 12–16% decrease in accuracy. Finally, manually reducing brightness by 90% results in lower accuracy than images that originally had the same brightness level. These effects can be slightly negated by using predictive model.

Our ongoing work includes the extension of autonomous driving-related image datasets and the implementation of multiple object detection to implement this work in self-driving cars, as well as the use of newer CNNs to improve both the classification accuracy and the inference time.

Acknowledgements. This work is supported by the Nazarbayev University (NU), Kazakhstan, under FDCRGP grant 021220FD0851.

References

1. Agnihotri, A., Saraf, P., Bapnad, K.R.: A convolutional neural network approach towards self-driving cars. In: 2019 IEEE 16th India Council International Conference (INDICON), pp. 1–4. IEEE (2019)
2. Bojarski, M., et al.: End to end learning for self-driving cars. arXiv preprint arXiv:1604.07316 (2016)
3. Castillo, A., Tabik, S., Pérez, F., Olmos, R., Herrera, F.: Brightness guided preprocessing for automatic cold steel weapon detection in surveillance videos with deep learning. Neurocomputing **330**, 151–161 (2019)
4. Clark, A.: Pillow (PIL fork) documentation (2015). https://buildmedia. readthedocs.org/media/pdf/pillow/latest/pillow.pdf
5. Deng, J., Dong, W., Socher, R., Li, L.J., Li, K., Fei-Fei, L.: Imagenet: a large-scale hierarchical image database. In: 2009 IEEE Conference on Computer Vision and Pattern Recognition, pp. 248–255. IEEE (2009)
6. García, V., Sánchez, J.S., Mollineda, R.A.: On the effectiveness of preprocessing methods when dealing with different levels of class imbalance. Knowl. Based Syst. **25**(1), 13–21 (2012)
7. Haeberli, P., Voorhies, D.: Image processing by linear interpolation and extrapolation. IRIS Univ. Mag. **28**, 8–9 (1994)
8. He, K., Zhang, X., Ren, S., Sun, J.: Deep residual learning for image recognition. CoRR abs/1512.03385 (2015). http://arxiv.org/abs/1512.03385
9. Howard, A.G., et al.: Mobilenets: efficient convolutional neural networks for mobile vision applications. arXiv preprint arXiv:1704.04861 (2017)
10. Ioffe, S., Szegedy, C.: Batch normalization: Accelerating deep network training by reducing internal covariate shift. In: International Conference on Machine Learning, pp. 448–456. PMLR (2015)
11. Kandel, I., Castelli, M., Manzoni, L.: Brightness as an augmentation technique for image classification. Emerg. Sci. J. **6**(4), 881–892 (2022)
12. Kulkarni, A., Chong, D., Batarseh, F.A.: Foundations of data imbalance and solutions for a data democracy. In: Data Democracy, pp. 83–106. Elsevier (2020)
13. Laza, R., Pavón, R., Reboiro-Jato, M., Fdez-Riverola, F.: Evaluating the effect of unbalanced data in biomedical document classification. J. Integr. Bioinform. **8**(3), 105–117 (2011)
14. Li, Y., et al.: Overcoming classifier imbalance for long-tail object detection with balanced group Softmax. In: Proceedings of the IEEE/CVF Conference on Computer Vision and Pattern Recognition, pp. 10991–11000 (2020)
15. Longadge, R., Dongre, S.: Class imbalance problem in data mining review. arXiv preprint arXiv:1305.1707 (2013)
16. Marco, V.S., Taylor, B., Wang, Z., Elkhatib, Y.: Optimizing deep learning inference on embedded systems through adaptive model selection. ACM. Trans. Embed. Comput. Syst. (TECS) **19**(1), 1–28 (2020)
17. Mountassir, A., Benbrahim, H., Berrada, I.: An empirical study to address the problem of unbalanced data sets in sentiment classification. In: 2012 IEEE International Conference on Systems, Man, and Cybernetics (SMC), pp. 3298–3303. IEEE (2012)
18. Pedregosa, F., et al.: Scikit-learn: machine learning in python. J. Mach. Learn. Res. **12**, 2825–2830 (2011)

19. Rodríguez-Rodríguez, J.A., Molina-Cabello, M.A., Benítez-Rochel, R., López-Rubio, E.: The effect of noise and brightness on convolutional deep neural networks. In: Del Bimbo, A., et al. (eds.) ICPR 2021. LNCS, vol. 12666, pp. 639–654. Springer, Cham (2021). https://doi.org/10.1007/978-3-030-68780-9_49
20. Tian, Y., Pei, K., Jana, S., Ray, B.: Deeptest: automated testing of deep-neural-network-driven autonomous cars. In: Proceedings of the 40th International Conference on Software Engineering, pp. 303–314 (2018)
21. Valueva, M.V., Nagornov, N., Lyakhov, P.A., Valuev, G.V., Chervyakov, N.I.: Application of the residue number system to reduce hardware costs of the convolutional neural network implementation. Math. Comput. Simul. **177**, 232–243 (2020)
22. Xu, N., Huo, C., Pan, C.: Adaptive brightness learning for active object recognition. In: ICASSP 2019–2019 IEEE International Conference on Acoustics, Speech and Signal Processing (ICASSP), pp. 2162–2166. IEEE (2019)
23. Yarovenko, V., Park, J.G., Lee, M.H.: Re-thinking pitfalls of premodel building for adaptive CNNs model selection on imagenet. In: 2022 4th International Conference on Advances in Computer Technology, Information Science and Communications (CTISC), pp. 1–6. IEEE (2022)

HANCaps: A Two-Channel Deep Learning Framework for Fake News Detection in Thai

Krishanu Maity[1], Shaubhik Bhattacharya[1], Salisa Phosit[2],
Sawarod Kongsamlit[2], Sriparna Saha[1], and Kitsuchart Pasupa[2,3](✉) (iD)

[1] Department of Computer Science and Engineering,
Indian Institute of Technology Patna, Patna 801103, India
{krishanu_2021cs19,shaubhik_2111cs19,sriparna}@iitp.ac.in
[2] School of Information Technology, King Mongkut's Institute of Technology
Ladkrabang, Bangkok 10520, Thailand
{63070242,63070245,kitsuchart}@it.kmitl.ac.th
[3] AI Governance Clinic, Electronic Transactions Development Agency (ETDA),
Bangkok 10310, Thailand

Abstract. The rapid advancement of internet technology, widespread smartphone usage, and the rise of social media platforms have drastically transformed the global communication landscape. These developments have resulted in both positive and negative consequences. On the one hand, they have facilitated the dissemination of information, connecting individuals across vast distances and fostering diverse perspectives. On the other hand, the ease of access to online platforms has led to the proliferation of misinformation, often in the form of fake news. Detecting and combatting fake news has become crucial to mitigate its adverse effects on society. This paper presents an investigation into fake news detection in the Thai language. It addresses current limitations in this domain by proposing a novel two-channel deep learning model named HANCaps, which integrates BERT and FastText embeddings with a hierarchical attention network and capsule network. The HANCaps model utilizes the BERT language model as one channel input, while the other channel incorporates pre-trained FastText embeddings. The proposed model undergoes evaluation using a benchmark Thai fake news dataset, and extensive experimentation demonstrates that HANCaps outperforms state-of-the-art methods by up to 3.28% in terms of F1 score, showcasing its superior performance.

Keywords: Fake News · Thai · Hierarchical Attention Network · Capsule Network

1 Introduction

With the advent of the internet, the world has witnessed a revolutionary shift in communication patterns. The proliferation of smartphones and the ubiquity

B. Luo et al. (Eds.): ICONIP 2023, CCIS 1969, pp. 204–215, 2024.
https://doi.org/10.1007/978-981-99-8184-7_16

of social media platforms have played pivotal roles in shaping the way information is disseminated and consumed [16]. This digital transformation has empowered individuals to participate actively in the news ecosystem, enabling them to share, comment, and contribute to online content. According to Statista, as of January 2023, there are approximately 5.18 billion active internet users worldwide, with 4.8 billion being active social media users[1]. This exponential growth in online connectivity has resulted in an unprecedented amount of information being shared and accessed globally. Consequently, the flow of information has become more decentralized, challenging traditional gatekeeping mechanisms and introducing a vast array of perspectives. As of August 2018, approximately two-thirds (68%) of Americans rely on social media platforms as their primary source of news[2].

The democratization of information has inadvertently led to the spread of misinformation and fake news. Fake news refers to deliberately fabricated or manipulated information presented as factual news [1]. It often aims to mislead, manipulate public opinion, or achieve various socio-political objectives. The consequences of fake news can be severe, ranging from undermining trust in legitimate sources of information to influencing elections, exacerbating social tensions, and even inciting violence.

Significant efforts have been made in the research community to address the challenge of fake news detection. Many studies [3,11,12,15] have focused on the English language, benefiting from large datasets and resources. Various approaches, including machine learning techniques, natural language processing (NLP) algorithms, and deep learning models, have been applied to analyze textual content and identify patterns indicative of fake news. Several existing works have achieved promising results in English fake news detection.

While extensive research has been conducted on fake news detection in English, the study of low-resource languages such as Thai remains limited, as described in the next section. This presents a significant challenge, as these languages often lack sufficient labeled data and specialized resources. Consequently, existing models may not perform optimally when applied to the Thai language. Addressing this limitation requires the development of robust and tailored approaches that consider the linguistic characteristics and contextual nuances specific to Thai.

This paper introduces a novel two-channel deep learning framework called HANCaps, which effectively detects fake news in the Thai language. HANCaps harnesses the power of BERT, a pre-trained language model, along with hierarchical attention network (HAN) and capsule networks, to capture the hierarchical relationships inherent in textual features. The model's first channel utilizes the BERT language model, while the second channel incorporates pre-trained FastText embeddings. The proposed framework is evaluated using a benchmark Thai fake news dataset, and extensive experimentation demonstrates HANCaps'

---

[1] https://www.statista.com/statistics/617136/digital-population-worldwide/.

[2] https://www.pewresearch.org/journalism/2018/09/10/news-use-across-social-media-platforms-2018/.

significant superiority over state-of-the-art (SOTA) methods. By leveraging the diverse tags available in the LimeSoda dataset [10] and contextual cues present in the Thai language, our model strives to enhance the accuracy and effectiveness of fake news detection, contributing to advancing this crucial field.

2 Related Works

The first study related to fake news in the Thai language is [2]. They conducted a detection of misinformation from Twitter texts by extracting tweet features and testing them using conventional machine learning techniques, i.e., Support Vector Machines (SVM), Naïve Bayes (NB), and Multilayer Perceptron (MLP). However, the content of the texts was not considered in this study. Following that, there was an attempt to detect unreliable medical articles on Thai websites [14]. They extracted article features from websites in conjunction with content features, TF-IDF, and Bag-of-Words, extracted from some selecting keywords. These features were then tested using conventional machine Learning techniques, including XGBoost, Decision Trees, SVM, Logistic Regression, and k-Nearest Neighbors (kNN). Kaothanthong et al. [6] classified the headline types of articles as clickbait or non-clickbait, as clickbait articles tend to be associated with fake news. They introduced the use of Headline2Vec, a feature derived from the last layer of a Convolutional Neural Network (CNN), and compared it with basic features such as n-Grams and TF-IDF. The features were tested using SVM, NB, and MLP. Meesad [8] presented a framework for detecting fake news, classifying news into three categories: real, fake, and suspicious. They utilized NLP techniques to extract features from the content. Experimental results showed that Long Short-Term Memory (LSTM) achieved the best performance among the other algorithms.

Due to the COVID-19 pandemic, numerous instances of fake news have emerged. In response, Mookdarsanit and Mookdarsanit [9] attempted to develop a system for detecting COVID-19-related misinformation in the Thai language. Since there is a lack of fake Thai-language news datasets specifically related to COVID-19, they adapted a model from the COVID-19 news open datasets, which were translated into Thai. The model was tested using data crawled from Thai websites. Additionally, they employed the feature-shifting technique to increase the number of Thai-language samples for model training. Experimental results demonstrated that ULMFiT outperformed other deep learning models, e.g., BERT and GPT. Subsequently, Payoungkhamdee et al. [10] created a dataset called "LimeSoda", focusing on fake news in the health domain. They evaluated the dataset using deep learning models, including Bidirectional LSTM (Bi-LSTM) with attention, BERT with a linear model, and WangChanBERTa with a linear model. Among them, BERT combined with a linear model yielded the best results. Furthermore, they attempted to understand how the models made decisions by analyzing token-level annotations and attention weights in Recurrent Neural Network-based models or using an embedding layer for transformer-based models. The findings suggest that while machine learning models provide explanations that differ somewhat from human judgments, there are common patterns in how humans and machines categorize words, indicating shared lexical interpretations.

3 Methodology

This section presents the methodology used for detecting fake news in Thai. We introduce a two-channel HAN-based deep neural network model, *HANCaps*, specifically designed for this purpose.

3.1 Proposed *HANCaps* Model

The proposed *HANCaps* model incorporates two distinct channels to capture different aspects of input sentences. The overall architecture of our proposed *HANCaps* model is illustrated in Fig. 1. The first channel employs BERT [4] followed by HAN and capsule network, while the second channel uses pre-trained FastText [5] embedding followed by HAN and capsule network. Given K input sentences where a sentence $S = \{w_1, w_2, \ldots, w_n\}$ comprising n words, both channels process the input using a series of operations as follow.

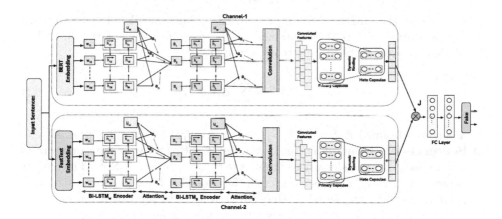

Fig. 1. *HANCaps* architecture

Channel-1.

BERT Embedding. The BERT model processes the input sentence S and generates a sequence output $E_B \in \mathbb{R}^{n \times d}$ of dimensions $max\_sequence\_length \times 768$. The output E_B is then fed through the HAN operates by incorporating attention mechanisms at different levels of the document hierarchy, allowing it to capture local and global dependencies.

Hierarchical Attention Network. We enhanced our model configuration to a hierarchical form to represent context-rich data samples. First, we used Bi-LSTM as a word-level encoder to compute sentence representation.

(i) **Bi-LSTM:** To enhance the contextual representation of the input sequence, we have integrated a Bi-LSTM layer that takes the embedding vector E_B generated by the BERT model as its input. This layer can capture contextual information in both forward and backward directions by processing the input sequence in both directions. At each time step, the hidden state h_t of the Bi-LSTM is obtained by concatenating the hidden state of the forward LSTM $\overrightarrow{h_t}$ and the hidden state of the backward LSTM $\overleftarrow{h_t}$. Consequently, the output of the Bi-LSTM layer is a sequence of hidden states H_e that includes all the hidden states of the input sequence. This representation can be expressed as $H_e = [h_1, h_2, h_3, \ldots, h_n]$.

(ii) **Attention Layer:** We incorporate a word attention layer after the Bi-LSTM layer, which allows the model to focus selectively on important words in the sentence. Formally, given the hidden state h_i of the Bi-LSTM at time step i and the weight vector u_a, the attention score α_i for the i-th word is computed as

$$\alpha_i = \frac{\exp(u_a^T h_i)}{\sum_{j=1}^{n} \exp(u_a^T h_j)}, \tag{1}$$

where n is the length of the input sentence. The word-label (wl) sentence representation S^{wl} is then obtained as the weighted sum of the Bi-LSTM hidden states multiplied by attention weight,

$$S^{wl} = \sum_{i=1}^{n} \alpha_i h_i. \tag{2}$$

Thus, we obtained $E_X^{wl} = [S_1^{wl}, S_2^{wl}, S_3^{wl}, \ldots, S_L^{wl}]$ for an input post X.

(iii) **Sentence-label Encoder:** Next, we apply the Bi-LSTM in the same way as a sentence-label encoder where input is $[S_1^{wl}, S_2^{wl}, S_3^{wl}, \ldots, S_L^{wl}]$. The output generated by Bi-LSTM passes through the attention layer to get the attention score α_i in the sentence label. Here, we simply multiply this attention score with Bi-LSTM hidden states output to get the sentence-label representation E_X^{sl} without performing the weighted sum operation as in the next layer, we apply CNN, which requires 2D input.

CNN [7]. Let the output feature map of this CNN layer be denoted as F. The element-wise dot product is performed between the E_X^{sl} and different filters c_i of size $h \times d$ in the CNN layer. This produces the feature map f_i corresponding to a particular n-gram's presence in the input sentence. The dimension of F is given by $(n - h + 1) \times k$, where h is the filter size and k is the number of filters used. Therefore, F is a collection of k feature maps obtained by sliding the filters over the entire input sequence,

$$F = [\mathbf{F}_1, \mathbf{F}_2, \mathbf{F}_3, \ldots, \mathbf{F}_k]. \tag{3}$$

In our proposed *HANCaps* model, instead of applying a pooling operation to the feature maps, we have used a capsule network [13] to retain the special features often lost during pooling.

Primary Capsule Layer. The capsule network's first layer combines the convolutional features produced by the CNN and creates primary capsules that represent each element in the feature maps using a group of neurons, thereby preserving local word order and semantic representations of words as instantiation parameters rather than scalar values. To generate a set of capsules, denoted as $p_i \in \mathbb{R}^l$, a kernel, denoted as K_i, is applied over the feature maps F, where l is the number of neurons in a capsule. Within the main capsule layer, a channel C_i consisting of a collection of capsules p_i is defined as follows:

$$C_i = \imath(F * K_i + b_i), \tag{4}$$

where $\imath$ refers to a non-linear activation function known as the squashing function, and b_i is a bias term.

Dynamic Routing between Capsules. In this layer, each capsule in the previous layer sends its output vector to all capsules in the next layer. The coupling coefficient between capsule i in the previous layer and capsule j in the next layer, denoted as $c_{i,j}$, is determined by a softmax function over all capsules in the next layer, and is calculated as follows:

$$p_{i,j} = \frac{\exp(b_{i,j})}{\sum_k \exp(b_{i,k})}, \tag{5}$$

where $b_{i,j}$ is the log prior probability that capsule i should be coupled with capsule j. The output of each capsule in the next layer is then calculated as a weighted sum of the predictions from all capsules in the previous layer, weighted by the coupling coefficients:

$$s_j = \sum_i p_{i,j}\hat{a}_{j|i} \text{ and } \hat{a}_{j|i} = W_{ij}a_i, \tag{6}$$

where $\hat{a}_{j|i}$ is the prediction vector of capsule i for the presence of an entity of class j and is defined as the dot product between the output vector of capsule i and a transformation matrix $W_{i,j}$, which learns to represent the instantiation parameters between capsule i and class j. Finally, the output vector of each capsule j is passed through a non-linear squashing function to ensure that its length is between 0 and 1.

Classification Capsule Layer. The final layer of the proposed capsule network is the classification capsule layer, which consists of k capsules with 16-dimensional instantiated parameters. In this layer, each capsule is dedicated to identifying a specific type of hate speech. The hate capsules are generated by routing the previous layer's output to the final layer. The output of the hate capsule layer is a flattened 1D vector of dimension ($k \times 16$). This vector is concatenated with the features generated by Channel-2.

Channel-2. We keep the same architecture as mentioned in Channel-1. The only distinction lies in the choice of embedding generation strategy. While Channel-1 employed BERT for generating input token embeddings, in Channel-2, we utilized FastText for this purpose. This allowed us to examine the impact of different embedding techniques within the same framework.

Fully Connected (FC) Layers. The concatenated outputs of Channel-1 and Channel-2 form a combined representation, denoted as J, for the input sentence X. Subsequently, this representation J is fed into FC layers, consisting of FC_1 with 200 neurons and FC_2 with 100 neurons. Finally, a softmax output layer is applied to predict the probabilities of the sample belonging to the target classes.

3.2 Loss Function

For the purpose of parameter optimization and back-propagation of loss, the categorical cross-entropy loss function $L_{CE}(\hat{Y}, Y)$ has been utilized in this study. It is defined as follows:

$$L_{CE}(\hat{Y}, Y) = -\frac{1}{N} \sum_{j=1}^{M} \sum_{i=1}^{N} Y_i^j log(\hat{Y}_i^j), \tag{7}$$

where $\hat{Y}_i^j$ represents the predicted label and Y_i^j represents the true label. The term N denotes the number of tweets in the dataset, while M represents the number of classes.

4 Experimental Setup

4.1 LimeSoda Dataset

In this research, the LimeSoda dataset [10] was utilized for the purpose of fake news detection in the Thai language. This dataset comprises a total of 7,191 documents sourced from various platforms such as official healthcare departments, official news sources, article websites, web boards, e-commerce sites, and social media platforms, all within the healthcare domain. Each document in the dataset is assigned one of the following classifications: fact, fake, or undefined. Notably, the dataset also includes token-level annotations that facilitate the validation of classifier decisions. These annotations encompass five high-level tags: misleading headline, imposter, fabrication, false connection, and misleading content.

To better utilize the tags mentioned in [10], we have created a zipped sentence where each word is succeeded by its corresponding tag as mentioned in Fig. 2. To investigate the impact of incorporating tags in the fake news detection task, we conducted experiments using two input variations: input sentence only and input sentence with tags (referred to as '+Tags'). By comparing these two settings, we aimed to demonstrate the influence of tag inclusion on the effectiveness of fake news detection.

Input Sentence	'โค', 'วิด', '19', 'แพ้', 'กระเทียม', 'ต้ม', 'แล้ว', 'จิบ', 'บ่อยๆ'
Translation	COVID-19 is susceptible to garlic. Boil it and drink it frequently.
Tags	'Fb-Refer', 'Fb-Refer', 'Fb-Refer', 'Fb-Refer', 'Fb-Refer', '', '', '', ''
Input+Tags	โค Fb-Refer วิด Fb-Refer 19 Fb-Refer แพ้ Fb-Refer กระเทียม Fb-Refer ต้ม '' แล้ว '' จิบ '' บ่อยๆ ''

Fig. 2. A sentence from a fake news sample in the LimeSoda dataset is provided. Here, the "Fb-Refer" tag, representing fabrication, signifies the presence of a regular prelude leading to a fabricated reference.

4.2 Experimental Settings

All our experiments are performed on a machine with an AMD EPYC 7552 48-Core Processor, 512 GB DDR4 RAM, and 5x Nvidia Ampere A100 GPUs totaling 200 GB of graphics memory. To prepare for the experiments, we partitioned the dataset into testing, validation, and training sets, with ratios of 10%, 10%, and 80%, respectively. The models were trained ten times with different random splits to ensure robustness, and the average performance was reported. Various network configurations were tested, and we achieved the best results with a batch size of 16, a learning rate of $1e-4$, and 30 epochs. All models were implemented using Scikit-Learn and PyTorch.

In the baseline setup, we included four commonly used machine learning models: Naïve Bayes, SVM, and Random Forest, using various embedding techniques. For machine learning baselines, we utilized the 768-dimensional pooled output of multilingual BERT (mBERT), which was pre-trained in 104 different languages, including Thai. For FastText embedding, we employed a pre-trained Thai FastText model to extract the embedding of each token and computed the average to obtain a 300-dimensional vector representing the entire sentence.

We established various variants of single-channel and double-channel deep learning baselines by varying the input embedding models followed by different deep learning models such as CNN, Bi-LSTM, HAN, and Capsule network. In the case of single-channel baselines, we first passed the input tweet through BERT or FastText to generate a 2D embedding matrix (E_m) of dimension ($max\_sequence\_length \times d$), where $d = 300$ for FastText and $d = 768$ for BERT. We then passed this E_m through different deep learning models as follows:

(i) HAN: The input embedding E_m is passed through word-label encoder (Bi-LSTM$_w$) followed by a sentence-label encoder (Bi-LSTM$_s$). The weighted sum of the Bi-$LSTM_s$ hidden states multiplied by attention weight is fed into an FC layer (100 neurons) followed by a softmax layer for prediction.

(ii) HAN+Capsule: The input embedding E_m was fed into HAN. The Bi-LSTM$_s$ hidden states output multiplied by attention weight is fed into a 1D CNN with 64 window size two filters. The convoluted feature was then transferred via the capsule network, and the hatred capsule layer's output was flattened and routed through an FC layer. Finally, for the final prediction, a softmax layer was used.

(iii) HAN+CNN: Here, E_m was passed through HAN followed by a 1D CNN with 64 filters of window size 2. We then performed Average Pooling on convoluted features followed by a softmax output layer.

(iv) Bi-LSTM: Input embedding E_m went through a Bi-LSTM layer with 128 hidden states, followed by an FC layer with 100 neurons, and concluded with a softmax layer for the final prediction.

5   Results and Discussion

Table 1 showcases evaluation outcomes for our model, *HANCaps*, and baselines regarding accuracy, precision, recall, and macro F1 score, yielding the subsequent insights: (i) SVM consistently outperforms the other machine learning baselines in terms of F1 score, achieving the best F1 score of 73.89% when combined with both BERT+Fasttext embeddings. (ii) Our proposed model *HANCaps* significantly outperforms the best machine learning baseline (BERT+Fasttext+SVM) with an improved F1 score of 20.57%. (iii) In terms of single-channel deep learning baselines, HAN+Caps network outperforms Bi-LSTM and HAN+CNN with both BERT and Fasttext embedding. BERT+HAN+Capsule with Tags achieved the best F1 score of 89.50% among the single-channel-based deep learning baselines, surpassing BERT+Fasttext+SVM by 15.61% in F1 score. This finding supports the efficacy of deep learning models over machine learning models for hate speech detection in noisy social media data. (iv) The singular results of Channel-1 (BERT+HAN+Caps) and Channel-2 (FastText+HAN+Caps) are 89.50% and 86.46% in terms of F1 score, respectively. However, combining both channels achieves an F1 score of 94.46%, indicating the efficiency of combining BERT and FastText embeddings for handling noisy text. (v) An additional noteworthy observation is that concatenating the associated tag with each word in the input post (represented by +Tags) consistently improves the F1 score. This contrasts the finding from [10], which states that machine learning models provide explanations that differ from human judgments in this dataset, suggesting that the tag is not helpful. However, in our case, utilizing the tag can guide the model and consistently improve the overall performance. (vi) We evaluated other variants of the proposed model and concluded that *HANCaps* (BERT+HAN+Caps, Fasttext+HAN+Caps) achieved the best performance with an F1 score of 94.46, significantly outperforming all the baselines. (vii) When comparing BERT with FastText, we observe that BERT embedded with any deep learning models always performs better than FastText. A similar trend is also observed in the case of machine learning baselines, except for Random Forest. This observation indicates the advantage of the transformer-based pre-trained language model XLNet over FastText in terms of efficient embedding generation of noisy social media text data.

We analyzed prediction errors for fake news by randomly selecting the utilized tag and the non-utilized tag model results from one out of ten trials. The utilized tag model misclassified 6.0% (61/1015), while the non-utilized tag model misclassified 7.0% (71/1015).

Table 1. Results of different baselines and proposed frameworks for Fake news detection in Thai

Embedding	Model	Precision	Recall	F1	Accuracy
	Machine learning baselines				
	Naïve Bayes	64.35	52.58	54.85	51.86
BERT	SVM	61.36	70.62	64.28	69.84
	Random forest	59.63	67.28	61.47	67.18
	Naïve Bayes	48.35	42.58	46.65	47.23
Fasttext	SVM	61.36	70.62	64.28	66.78
	Random forest	63.55	57.28	61.75	63.32
	Naïve Bayes	63.75	55.36	57.23	55.36
BERT+Fasttext	SVM	**76.36**	**72.85**	**73.89**	**72.86**
	Random forest	75.92	71.04	72.42	71.04
	SOTA				
-	BERT & Linear	90.76	89.63	91.18	91.14
	Deep Learning baselines				
	Single channel				
	HAN+Caps	88.39	87.38	88.38	88.38
	HAN	86.51	85.48	86.47	86.48
	HAN+CNN	85.25	84.36	86.15	85.23
BERT	Bi-LSTM	82.02	82.35	82.07	82.09
	HAN+Caps (+Tags)	**89.57**	**88.48**	**89.50**	**89.51**
	HAN (+Tags)	87.21	87.20	86.52	87.18
	HAN+CNN (+Tags)	86.99	85.41	87.13	86.94
	Bi-LSTM (+Tags)	83.17	84.78	84.90	84.90
	HAN+Caps	85.19	84.74	85.72	85.87
	HAN+CNN	83.74	82.25	84.68	83.72
	HAN	82.34	82.11	82.15	82.15
Fasttext	Bi-LSTM	84.12	85.64	84.56	84.87
	HAN+Caps (+Tags)	**86.40**	**85.35**	**86.46**	**86.58**
	HAN+CNN (+Tags)	85.01	83.99	84.72	84.87
	HAN (+Tags)	85.82	85.76	85.76	85.77
	Bi-LSTM (+Tags)	85.94	86.83	86.11	86.33
	Two channel				
BERT+HAN+Caps, Fasttext+HAN		**93.25**	**93.24**	**93.23**	**93.24**
BERT+HAN, Fasttext+HAN+Caps		91.16	91.14	91.14	91.14
BERT+HAN, Fasttext+HAN		89.41	89.27	89.37	89.37
BERT+HAN+Caps, Fasttext+HAN (+Tags)		94.30	94.29	94.29	94.29
BERT+HAN, Fasttext+HAN+Caps (+Tags)		92.16	92.12	92.13	92.12
BERT+HAN, Fasttext+HAN (+Tags)		90.56	90.48	90.50	90.48
	Proposed Model (HANCaps)				
BERT+HAN+Caps, Fasttext+HAN+Caps		**93.27**	**93.19**	**93.24**	**93.37**
BERT+HAN+Caps, Fasttext+HAN+Caps (+Tags)		**93.17**	**94.58**	**94.46**	**94.48**

The errors observed in the non-utilized tag model can be summarized as follows: (i) Challenges in accurately predicting fact news in 25.4% (18/71), with 94.4% (17/18) referred to external organizations (Imposter tag) and 66.7% (12/18) contain clickbait words (Title Clickbait tag). (ii) Lack of ability to distinguish fake news nature in 22.5% (16/71), with 87.5% (14/16) consisting of fabricated content using common fact news words (Fabrication tag).

The utilized tag model achieved 98.6% (70/71) misclassified by the non-utilized tag model. However, false predictions still occurred in 6.0% (61/1015) of the messages. The errors observed in the utilized tag model can be summarized as follows: (i) Misclassifying fact news as fake news in 14.75% (9/61), with 88.9% (8/9) referred to external organizations and medical personnel (Imposter tag), 44.4% (4/9) using persuasive phrases (Misleading tag), and 33.3% (3/9) using attention-grabbing words (Title Clickbait tag) (ii) Misclassifying fake news as fact news in 19.67% (12/61), with 75% (9/12) involve exaggeration and fabricated sources (Fabrication tag), 41.7% (5/12) referred to external organizations, medical personnel, and external unreliable sources (Imposter tag), and 50% (6/12) misclassified the news contain with only one tag.

The most common misclassified tags for fake news were Fabrication and Imposter, while fact news most commonly misclassified were Misleading, Clickbait, and Imposter tags.

6   Conclusion and Future Work

This paper presents HANCaps, a novel two-channel deep learning framework designed for detecting fake news in the Thai language. HANCaps leverages the integration of BERT and FastText embeddings with HAN and capsule networks to capture the hierarchical relationships embedded within textual features. Through extensive experimentation on a benchmark Thai fake news dataset, HANCaps demonstrates remarkable performance, surpassing existing SOTA methods by up to 3.28% in F1 score. By harnessing diverse tags and employing different embedding strategies, our model effectively enhances the accuracy of fake news detection. One limitation of this study is the utilization of token tags as input. The accurate detection of these tags plays a crucial role in achieving explainability, which is one aspect that we plan to address in our future work.

One limitation of this study is the utilization of token tags as input. The accurate detection of these tags plays a crucial role in achieving explainability, which is one aspect that we plan to address in our future work.

Acknowledgements. This work was supported by the Ministry of External Affairs (MEA) and the Department of Science & Technology (DST), India, under the ASEAN-India Collaborative R&D Scheme.

References

1. Allcott, H., Gentzkow, M.: Social media and fake news in the 2016 election. J. Econ. Perspect. **31**(2), 211–236 (2017)
2. Aphiwongsophon, S., Chongstitvatana, P.: Detecting fake news with machine learning method. In: Proceedings of the 15th International Conference on Electrical Engineering/Electronics, Computer, Telecommunications and Information Technology (ECTI-CON), pp. 528–531 (2018)
3. Castillo, C., Mendoza, M., Poblete, B.: Fake news detection: a deep learning approach. ACM Trans. Web **13**(3), 1–28 (2019)
4. Devlin, J., Chang, M., Lee, K., Toutanova, K.: BERT: pre-training of deep bidirectional transformers for language understanding. CoRR abs/1810.04805 (2018)
5. Grave, E., Bojanowski, P., Gupta, P., Joulin, A., Mikolov, T.: Learning word vectors for 157 languages. CoRR abs/1802.06893 (2018)
6. Kaothanthong, N., Kongyoung, S., Theeramunkong, T.: Headline2Vec: a CNN-based feature for Thai clickbait headlines classification. Int. Sci. J. Eng. Technol. **5**(1), 20–31 (2021)
7. Kim, Y.: Convolutional neural networks for sentence classification. In: Moschitti, A., Pang, B., Daelemans, W. (eds.) Proceedings of the 2014 Conference on Empirical Methods in Natural Language Processing, EMNLP 2014, 25–29 October 2014, Doha, Qatar, A meeting of SIGDAT, a Special Interest Group of the ACL, pp. 1746–1751. ACL (2014)
8. Meesad, P.: Thai fake news detection based on information retrieval, natural language processing and machine learning. SN Comput. Sci. **2**(6), 425 (2021)
9. Mookdarsanit, P., Mookdarsanit, L.: The COVID-19 fake news detection in Thai social texts. Bull. Electr. Eng. Inf. **10**(2), 988–998 (2021)
10. Payoungkhamdee, P., et al.: LimeSoda: dataset for fake news detection in healthcare domain. In: Proceedings of the 16th International Joint Symposium on Artificial Intelligence and Natural Language Processing (iSAI-NLP), pp. 1–6 (2021). https://doi.org/10.1109/iSAI-NLP54397.2021.9678187
11. Potthast, M., Kiesel, J., Reinartz, K., Bevendorff, J., Stein, B., Hagen, M.: A stylometric inquiry into hyperpartisan and fake news. In: Proceedings of the 2017 ACM on Conference on Information and Knowledge Management, pp. 1567–1576. ACM (2017)
12. Ruchansky, N., Seo, S., Liu, Y.: CsiNet: towards a more robust fake news detection framework. In: Proceedings of the 26th International Conference on World Wide Web, pp. 797–806. International World Wide Web Conferences Steering Committee (2017)
13. Sabour, S., Frosst, N., Hinton, G.E.: Dynamic routing between capsules. arXiv preprint arXiv:1710.09829 (2017)
14. Saengkhunthod, C., Kerdnoonwong, P., Atchariyachanvanich, K.: Detection of unreliable medical articles on Thai websites. In: Proceedings of the 13th International Conference on Knowledge and Smart Technology (KST), pp. 102–107 (2021)
15. Shahi, G.K., Vaibhav, G., Tiwari, A., Mishra, V., Bansal, S.: Deep learning models for fake news detection: A comparative study. In: Proceedings of the 9th International Conference on Software and Computer Applications, pp. 56–60 (2020)
16. Shu, K., Sliva, A., Wang, S., Tang, J., Liu, H.: Fake news detection on social media: a data mining perspective. ACM SIGKDD Explor. Newsl **19**(1), 22–36 (2017)

Pre-trained Financial Model for Price Movement Forecasting

Chenyou Fan[1][✉][iD], Tianqi Pang[1], and Aimin Huang[2][iD]

[1] South China Normal University, Guangzhou, China
fanchenyou@scnu.edu.cn
[2] Hangzhou Higgs Asset Management Co., Ltd., Hangzhou, China
huangaimin@higgsasset.com

Abstract. We propose the Pre-trained Financial Model (PFM) for price move-ment forecasting, which is critical in the automated trading systems in the Stock and Futures markets. Inspired by recent successes of pre-trained large language models in tackling NLP tasks, our PFM adopts a pretraining-and-finetuning strat-egy for obtaining capable models that are adapted to various downstream price-forecasting tasks. During the pre-training stage, we train a sequence prediction backbone with multi-task learning by adopting both a supervised learning objec-tive and an unsupervised regularization target. Our approach differs from the common masked language modeling (MLM) used in NLP studies. We develop a per-step target variable generation strategy for eliciting future predictions from the transformer encoder-decoder architecture. We verify our pre-trained model on various practical downstream forecasting tasks, including lagged movement regression, movement direction classification, and selective trading with best performing stocks. Specifically, during the fine-tuning stage, we retain the pre-trained encoder and replace the decoder with specific downstream task decoders. We then perform supervised task-specific target generation learning as the fine-tuning process. Through extensive numerical studies and analysis, we demon-strate that our fine-tuned financial model can achieve a 5–15% improvement over downstream regression and classification tasks and over 40% in selective trading task.

Keywords: Financial analysis · Price forecasting · Model Pre-training · Down-stream Task Fine-tuning · Learning-to-Rank · Contrastive Learning

1 Introduction

Price Movement Forecasting (PMF) of commodities, stocks, futures, energies and car-bon credits has attracted increasing attention as an Artificial Intelligence topic [9, 10, 22–26]. PMF technique can assist individual investors to anticipate market risks and minimize losses, and aid policy makers in countering market volatility to promote social welfare and stability. Recent advancements in data-driven models for data analysis have

Supplementary Information The online version contains supplementary material available at https://doi.org/10.1007/978-981-99-8184-7_17.

led to accurate probability mass function (PMF), which is crucial for automated decision making in financial systems. This study aims to develop a comprehensive price movement forecasting framework using a pre-training-and-fine-tuning approach. This approach enhances model capacity through pre-training and adapts to different downstream tasks through light-weight fine-tuning.

Prior research [11,22–24] has primarily approached the task of forecasting price movements as either a regression or trend classification task, aimed at predicting directional changes in stock prices. However, these models are specifically tailored to a single task and are not readily adaptable to other tasks.

Motivated by recent achievements in NLP tasks, we propose a two-stage modeling approach for PMF that includes pre-training a general sequential prediction backbone model and fine-tuning it on various downstream tasks. Our approach involves leveraging a shared transformer encoder model to sequentially encode the entire historical data, without masking any intermediate steps. To train the model, we adopt a multi-task learning approach that employs both supervised and unsupervised objectives. The supervised objective involves minimizing per-step regression error across all future steps. The unsupervised objective ensures the internal consistencies of the predicted future sequence by maximizing similarity between the encoded history and decoded future sequences for a specific asset while minimizing it for two different assets.

We aim at developing a proficient pre-trained financial model (PFM) and applying it to accomplish downstream tasks related to stock and futures trading, such as classifying price movement direction, forecasting price movements, and detecting top-performing stocks. To accomplish each task, we design a customized fine-tuning technique, such as implementing quantile regression for long-term prediction, focal loss for classifying uneven directions, and a unique learning-to-rank methodology for identifying the highest emerging and declining stocks based on specific criteria.

Our contributions can be summarized as follows.

- Our study introduces a novel pre-training-and-fine-tuning approach for modeling financial data, and demonstrates its effectiveness across multiple down-stream tasks.
- In the pre-training stage, we designed a multi-task learning objective with supervised regression and unsupervised contrastive learning for pattern consistency. In the fine-tuning stage, we developed an effective task-aware decoding method which enables quick adaptation to realistic tasks.
- We curated a realistic high-frequency stock dataset containing a billion timestamps of complete Limited Order Book information from the year 2021–2022. We will make this dataset and our code openly available to support future research.
- Experimental results show that our approach outperforms state-of-the-art approaches in regression deviation, classification accuracy, and top-K selection hit rate.

2   Related Work

Price Movement Forecasting (PMF) has gained research interests recently as a typical temporal reasoning task in machine learning and artificial intelligence. From a task formulation perspective, SMF is either formulated as a regression task [5,17,24] or a classification task [9,22,23]. The former task aims to produce numerical price predictions, while the latter one often predicts the movement directions, such as neutral, rising, or falling.

Time-Series Forecasting studies how to effectively learn from past observations and predict the future. The Holt-Winters [13,21] and ARIMA [2] are classical methods. Recent deep-learning-based studies [22–26] focus on formulating the sequence's dynamics with LSTMs [10,11,20] and Transformers [17,23].

Pre-trained Language Models (PLMs) such as BERT [8] and GPTs [3] have achieved great successes in natural language understanding tasks. However, applying proper pre-training to financial tasks remains unknown due to substantial differences between language data and financial sequential data.

Our Contributions. In this study, we propose a pre-training and fine-tuning framework for forecasting price movement. Firstly, we suggest a multi-objective pre-training task with supervised and unsupervised targets. Then, we fine-tune the models on various down-stream tasks to demonstrate their performance advantages.

3   Task Description and Data Features

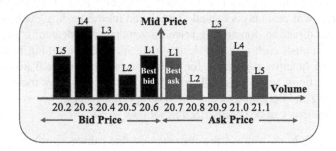

Fig. 1. Details of the limit order book (LOB) at each step t. The "Best ask" refers to the lowest ask price, and "Best bid" refers to the highest bid price.

In this study, we examine a market consisting of N common stocks and M market indexes. Our historical observations comprise the past T_h steps, which we denote as $T_h = \{1, 2, ..., T_h\}$, while our future observations consist of the subsequent T_f steps, denoted as $T_f = \{1, ..., T_f\}$. Specifically, our objective is to utilize the past T_h steps to accurately forecast the subsequent T_f steps. The duration of each step is contingent on the frequency of the data collection scheme. For instance, we utilize our self-collected high-frequency data, which entailed a time interval of 3 s per step.

We begin by elaborating limit order books (LOBs) [12,15,26], which are publicly accessible and maintained by stock exchanges, to represent market information. At each step, a LOB contains multiple ask and bid orders, as depicted in Fig. 1. The volume of v_a is associated with the best ask order price, denoted as $a(t)$, while the best bid order price, denoted as $b(t)$, has a volume of $v_b(t)$. To construct numerical features from the LOBs, we adopt the practice of previous research [4,26] and give them details below.

Mid Price is the logarithm of the middle of the best ask and bid prices, i.e., $p(t) = \log(0.5 * (a(t) + b(t)))$. **Bid-Ask Spread** measures the difference between the ask price and the bid price, i.e., $sp(t) = a(t) - b(t)$, which is mostly positive. **Spread Return** measures the ratio of spread price and mid price, i.e., $r(t) = sp(t)/p(t)$.

Price Movement is the difference between two mid prices of a w-step interval, i.e.,

$$\Delta p(t, w) = p(t) - p(t - w), \tag{1}$$

where window size w controls the time-scale that we are interested in predicting. We will simplify the notation to $\Delta p(t)$ if $w = 1$ in the following equations.

Volume-Weighted Price Movement weighs the average of the bid and ask prices (a.k.a mid prices) by their inverse volumes and subtracts the mid prices, i.e.,

$$\Delta p_v(t) = \frac{b(t) * v_a(t) + a(t) * v_b(t)}{v_a(t) + v_b(t) + \epsilon} - p(t). \tag{2}$$

Signed Transaction Volume quantifies the normalized difference of trade volumes between buying and selling transactions, i.e.,

$$sv(t) = \frac{v_{buy}(t) - v_{sell}(t)}{v_{buy}(t) + v_{sell}(t) + \epsilon}. \tag{3}$$

Bid-Ask Volume Imbalance quantifies the difference between the number of shares at the bid and the ask, i.e.,

$$vi(t) = \frac{v_b(t) - v_a(t)}{v_a(t) + v_b(t) + \epsilon}. \tag{4}$$

We concatenate these factors to build the raw input feature time-series. Then we apply standard **positional embedding** (PE) based on sine and cosine functions to indicate the time step in the trading period per day. Subsequently, we project the feature series onto a fixed dimension d_{model} to obtain the input historical features $L^h = [l_1; \ldots; l_{T_h}] \in \mathbb{R}^{T_h \times d_{model}}$, which represent the LOB time-series features.

At future T_f steps, we project only the positional embeddings to a dimension of d_{model} to generate the matrix $L^f = [e_1; \ldots; e_{T_f}] \in \mathbb{R}^{T_f \times d_{model}}$. The reason behind this is that the LOB attributes are inaccessible during the upcoming time steps.

We focus on predicting accurately the price movements in future steps and their variants, such as the ranking or quantile class of price movements, in downstream tasks.

4   Model Design

Following previous work [9], we develop a Transformer-based encoder-decoder architecture to perform Price Movement Forecasting (PMF) using our designed pretraining-and-finetuning scheme.

We follow the Transformer attention design [19] which converts input features to queries Q, keys K, and values V first, then performing *dot-product attention* as $\sigma(QK^T/\sqrt{d_k})V$ to obtain the history-encoded output feature. Two types of attention modules are used as building blocks.

The *Self-Attention* projects an input feature X (e.g., historical LOB features) onto Q, K, and V, followed by standard dot-product attention. The *Cross-Attention* projects two different input features X_1 onto Q and X_2 onto K and V, respectively, followed by dot-product attention. The cross-attended X_1 and X_2 can be historical and future features, respectively, to incorporate encoded history into future outputs.

4.1 History Encoder

We design a shared Transformer-based encoder for learning to abstract historical temporal information of a stock at both pre-training and fine-tuning stages. As shown in Fig. 2-A, we feed the encoder with the historical LOB features of common stocks, denoted as L^h, as well as the market indexes, denoted as L^I, at each aligned step.

For each stock $i \in \mathcal{N}$, we apply the *self-attention* mechanism to attend to all historical steps in bi-directions, encoding their temporal patterns from the input L_i^h to the hidden feature $H_i^{St} \in \mathbb{R}^{T_h \times \text{dmodel}}$ as $H_i^{St} \leftarrow$ Self-Attention(X_i^S). We also apply self-attention to each market index m and concatenate them into a single feature that encodes the general trend of the market, denoted as $H^{Ind} \in \mathbb{R}^{T_h \times (d_{\text{model}} \times M)}$.

To incorporate the market-level dynamics into individual stock movement predictions, we further design a *cross-attention* mechanism which attends to each stock feature H_i^{St} and the index H^{Ind}, followed by a residual connection, as shown below:

$$H_i^{enc} \leftarrow H_i^{St} \oplus \text{Cross-Attention}(H_i^{St}, H^{Ind}) , \tag{5}$$

where $\oplus$ represents element-wise addition which adds the cross-attended market information to individual stock feature H_i^{St}.

Hereby, the hidden output H_i^{enc} of the encoder effectively incorporates both the internal dynamics of each individual stock and the global context of the entire market.

4.2 Future Decoder

We further designed a Transformer-based decoder to decode the encoded H_i^{enc} for future price movements, as shown in Fig. 2-B.

We adopted the teacher-forcing strategy, which provides the true movement Δpt of the previous step t in the $t + 1$ prediction. Let the future shifted-right targets be $\Delta p' = \{\Delta p_t\}$ for $t \in T_h + \{0, 1, .., T_f - 1\}$, in which Δp_0 is the last observed movement. Then we trained the model to predict Δp for $t \in T_h + \{1, .., T_f\}$ at each future step, as shown in Fig. 2-B.

Concretely, we applied self-attention on the inputs for every stock $i \in \mathcal{N}$ and then cross-attending to the encoded history H_i^{enc} to obtain the decoded future features as:

$$\begin{aligned} G_i &\leftarrow \text{Self-Attention}(\Delta p_i') , \\ H_i^{dec} &\leftarrow G_i \oplus \text{Cross-Attention}(G_i, H_i^{enc}) . \end{aligned} \tag{6}$$

Finally we apply a fully-connected (Fc) layer to predict future movements for each stock i as $\Delta \hat{p}_i \leftarrow \text{Fc}(H_i^{dec})$. We will utilize H_i^{dec} and H_i^{enc} to perform unsupervised contrastive learning, which will be discussed in next section.

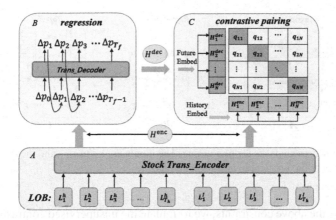

Fig. 2. Pre-training a transformer-based encoder with LOB features using two objectives: 1) unsupervised learning on history-future pairing, and 2) supervised learning on future price changes. The subscripts of $L_t^{T_h}$ and $L_t^{T_I}$ represent historical data and indicator data, respectively, from time nodes 1 to T_h.

5   Model Pre-training

We present our pre-training paradigm for developing a robust model for forecasting stock movements. Our multi-task learning pipeline includes both supervised and unsupervised learning objectives.

Supervised-Learning Task - Regression. At future steps, We use the teacher-forcing paradigm, as described in Sect. 4.2, to predict future movements at step $t \in T_f$ by shifting the input sequence to the right, as shown in Fig. 2-B. Let N be the number of stocks in a batch and $\Delta \hat{p}_t^{(n)}$ be the estimated movement for n-th stock at step t. We minimize the Mean Square Error (MSE) as:

$$L^{reg} = \frac{1}{N} \sum_{n=1}^{N} \frac{1}{T_f} \sum_{t=1}^{T_f} (\Delta \hat{p}_t^{(n)} - \Delta p_t^{(n)})^2 . \tag{7}$$

Unsupervised Task - Contrastive Pairing. We enforce prediction consistency between historical and future steps by designing a contrastive history-future pairing task, as illustrated in Fig. 2-C. Let the encoded history be H^{enc} (Eq. 5) and the decoded future be H^{dec} (Eq. 6). We compute the dot product as the similarity measure for each stock pair (i, j), such that $q_{ij} = H_i^{enc} \cdot H_j^{dec}$. Let $Q = q_{ij}$ for $i, j \in 1, ..., M$, as shown in Fig. 2(C), where we highlight each diagonal element q_{ii} that measures the pattern of the history and future of the same stock.

Normally, the internal dynamics of a single stock should be consistent over a short period, i.e., patterns in the split history and future of the same stock should be more aligned than those of two different stocks. Equivalently, the diagonal elements of Q should be larger than other elements in the same row and column, such that:

$$q_{i,i} > q_{i,j} \wedge q_{i,i} > q_{k,i}, \quad \forall j, k \neq i . \tag{8}$$

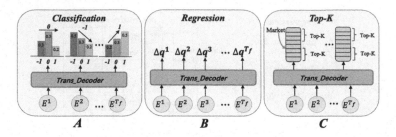

Fig. 3. Three downstream tasks for fine-tuning. The inputs are e_1 to e_{T_f} and the decoders implement the classification, prediction, and sorting tasks.

To impose the above constraints, we formulate the unsupervised contrastive learning objective as an auxiliary classification task as:

$$L^{pair}(Q) = -\frac{1}{2N}\sum_{i=1}^{N}\left(\log\frac{e^{q_{ii}}}{\sum_{j=1}^{N}e^{q_{ij}}} + \log\frac{e^{q_{ii}}}{\sum_{k=1}^{N}e^{q_{ki}}}\right),\qquad(9)$$

in which q_{ii} is maximized as a logit. As Q depends on feature embeddings from the model outputs, we can optimize the model by minimizing L^{chip} using standard SGD.
Multi-task Pre-training Objective. We provide the multi-task pre-training objective by jointly minimizing the MSE regression loss in Eq. (7) and contrastive pairing loss in Eq. (9) as $L^{pre} = L^{reg} + \alpha L^{pair}$ in which α is the scaling factor.

6   Fine-Tuning on Downstream Tasks

In this section, we consider fine-tuning our pre-trained backbone model for three downstream tasks of future forecasting.

In fine-tuning, we follow a realistic practice by making predictions on a subset of future timestamps with a fixed interval that is relatively longer. We consider a set of K future time-steps $\mathcal{S} = \{\Delta t, 2\Delta t, \ldots, K\Delta t\}$, where the interval between each step is $\Delta t = 0.5$-minute. We compute the cumulative price movement over Δt.

Note that the Δt of 0.5-minute is equivalent to 10-steps of time during pre-training, as our data collection rate is 3-seconds per-step.

6.1   Lagged Regression (LR)

We consider a critical task of predicting the lagged price movement over the K future steps $\mathcal{S}$, with a lag of Δt steps, as follows:

$$Y = \{\Delta p(t) = p(t) - p(t - \Delta t),\ \ \forall t \in \mathcal{S}\},\qquad(10)$$

in which $p(t)$ is the mid-price at step t. Predicting lagged movement is a common task that guides realistic decision-making such as when to buy, hold, or sell. Figure 3-B illustrates the fine-tuning task with a designed decoder which decodes the positional encoding e_t at each step t for sequential predictions.

To fine-tune the LR decoder robustly, we perform Quantile Regression (QR) [16] to estimate the distribution of the target price movement. Specifically, we choose a set of quantiles $\mathcal{Q}= [0.2,...,0.5,...,0.8]$ and estimate the predicted movement $\hat{y}^q$ for each quantile q using the asymmetric quantile loss function:

$$\ell^{qr} = \frac{1}{|\mathcal{Q}|} \sum_{q \in \mathcal{Q}} \left(q|y - \hat{y}^q|^+ + (1 - q)|\hat{y}^q - y|^+ \right), \tag{11}$$

where $|\cdot|^+$ truncates values to non-negative. During inference, we estimate the median values of the lagged price movements $\hat{y}^{0.5}$.

6.2 Directional Classification (DC)

We next consider predicting the movement directions for future steps by formulating a 3-way classification task that includes neutral, rising, or falling directions. This is a standard forecasting task in recent works [11,22,24].

Specifically, we collect the cumulative movements considered in Lagged Regression Eq. (10), and divide the range into three intervals as classes: class -1 denotes movement within $(-\infty, -0.1\%)$ as falling, class 0 denotes $[-0.1\%, 0.1\%]$ for neutrality, while class 1 denotes $(0.1\%, \infty)$ for rising.

As shown in Fig. 3-A, we train the DC decoder to produce a 3-way probability distribution, and we utilize the *focal loss* [18] as the training objective to robustly estimate the movement distribution and tackle the class-imbalance issue. We evaluate the 3-way directional classification accuracy over future steps S.

6.3 Top-K Selection (TKS) of Stocks

We now explore a rarely studied task which predicts the top-K performing (and underperforming) stocks for buying (and short-selling). This Top-K selection task can help trading programs to identify quickly rising (and falling) stocks and generate profits.

To evaluate our model's predictions, we use percentages of 5%, 10%, 20% and 30% for the top-rising and falling stocks, and we measure the hit rate as the accuracy metric.

Directly optimizing the above TKS goals with classical machine learning methods is challenging. Therefore, we have designed a differentiable sorting technique (Diff-Sort) to establish the correct ordering of all stocks and train our model to be rank-aware. We will demonstrate that Diff-Sort can effectively improve the hit rates for the TKS task.

Diff-Sort performs a proxy distribution assignment task [1,7]. It estimates the sorting permutation of a non-sorted list by iteratively minimizing the difference between the intermediate permuted list and a pre-defined anchor list of increasing order.

Let $\boldsymbol{y}^t = [y_1^t, ..., y_M^t]^\top \in \mathcal{R}^M$ be a list of movement for M *unique* stocks in the market at the same step t (omitted for simplicity). We aim to search for an $M \times M$ permutation matrix $\boldsymbol{P}^* = \{p_{ij}\}_{i,j=1}^M$ which sorts $\boldsymbol{y}$ into $\boldsymbol{y}'$ in increasing order as:

$$\boldsymbol{y}' = \boldsymbol{P}^*\boldsymbol{y} = [y_1', ..., y_M']^\top, \ s.t. \ y_u' < y_v', \forall u < v. \tag{12}$$

To obtain $\boldsymbol{P}^*$, we introduce the following optimization steps. Let an M-element increasing list be $\boldsymbol{w} = \begin{bmatrix} 1, 2, \cdots, M \end{bmatrix}^\top$, serving as anchor points for pairing with sorted price movements. We build a cost matrix C to measure the difference between movement y_j and anchor point w_i as follows:

$$C = \{c_{ij} = (w_i - y_j)^2\}_{i,j=1}^M \in \mathcal{R}^{M \times M}. \tag{13}$$

We formulate the movement ranking task as a differentiable learning objective of finding the optimal permutation P^* as follows:

$$\hat{P}^* = \arg\min \langle P, C \rangle - \lambda H(P),$$
$$s.t. \ P \geq 0, \ P1_M = 1_M, \ P^\top 1_M = 1_M, \tag{14}$$

in which $H(P) = -\sum_{i,j} P_{i,j} \log P_{i,j}$ is an entropy term; $\langle P, C \rangle$ is the Frobenius product that produces the total cost of permutation P.

Ignoring the entropy term, it's easy to check that Eq. (14) minimizes when y_j and w_i have the same rank in y and w. This is equivalent to sorting y_j to its correct i-th rank by P^*. Thus, the ranking over all-stock movement y is:

$$\hat{R}(y) = \{\hat{r}_i\}_{i=1}^M = \hat{P}^{*\top} w. \tag{15}$$

Problem (14) is convex [6], so we can solve it in iterative and differentiable way. In the training stage, we compare the estimated ranks $\hat{R}(y)$ with actual ranks $R(y) = \{r_i\}_{i=1}^M$ which can be obtained by sorting the actual stock price movements.

We design the movement ranking loss L^{dsir} to penalize inconsistent pairwise orders in a supervised manner such that:

$$L^{rank} = \frac{1}{M^2} \sum_{i=1}^M \sum_{j=1}^M \max(0, -(r_i - r_j)(\hat{r}_i - \hat{r}_j)). \tag{16}$$

In testing, we evaluate the standard top-K hit rate (accuracy) by comparing the predictions with the actual top rising and falling stocks, respectively. We evaluate each $K = \text{round}(r \cdot M)$ with $r = 5\%, 10\%, 20\%$ and 30% over M stocks at future steps.

7 Experiment

We evaluate our Pre-trained Financial Model (PFM) framework using self-collected benchmark datasets and compare it with state-of-the-art methods.

We collected a large-scale dataset called **STAR-22** from the Shanghai Stock Exchange Sci-Tech innovAtion boaRd (SSE STAR MARKET). This dataset includes 4800 stocks from 2021 to 2022, with high-frequency LOB features collected every 3 s, which amounts to 20 steps per minute and 1200 steps per hour.

To improve the evaluation of our methods, we employed three **rolling-based dataset splits** for STAR-22. These splits consist of the test sets for (1) Mar. and April (a downtrending market), (2) Jun. and July (an uptrending market), and (3) Aug. and Sept. of 2022 (a fluctuating market), respectively, as shown in Fig. 4. We excluded May from the splits as the market is closed for Labor Day for the first five days of the month. For each split, all data preceding the test set is further divided for training and validation.

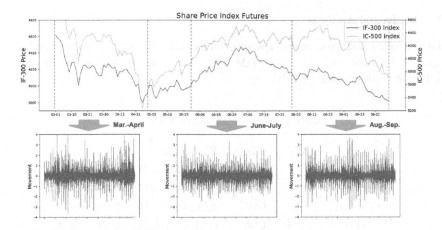

Fig. 4. The top figure shows the market indices (IF-300 and IC-500) on three rolling splits. The bottom figures show corresponding price movement averaged over all stocks and all steps in a 4-hour trading hour.

7.1 Our Methods and Baselines

In our proposed pretraining-and-finetuning scheme, we utilize all the training data in the split during the pre-training stage. For fine-tuning in the downstream task, we adopt two strategies for utilizing the data and compare their performance, as follows.

PFM (Pre-trained Financial Model) is our proposed architecture, which employs a pre-trained encoder and fine-tuned decoder, as explained in Sects. 5 and 6. PFM utilizes the full training data for down-stream task fine-tuning, which covers more than one year of training data. **PFM-FS** (PFM-FewShot) uses the same model architecture as PFM but fine-tunes with only two months of data preceding the testing split. PFM-FS has the advantage of converging much faster due to the utilization of much less data (2 months versus one year).

We compare our methods with existing baselines as follows. **LightGBM** [14] is a gradient boosting tree with a multi-class learning objective used to predict the directional class of future steps. **ALSTM** [11] uses LSTM model with temporal attention and adversarial training to predict price sequences. **TFT** [17] combines LSTM and Transformer for input encoding and modifies the attention as interpretable self-attention layers. **DTML** [23] aggregates all stocks by combining their features as the market context. Instead, we use market indexes as context which is robust to single stock variation and much faster for single-stock inference.

We evaluate the down-stream 3-way Directional Classification (DC) task with standard accuracy, and we evaluate the Lagged Regression (LR) task using the **MAD** (mean absolute deviation) and **R-Squared (R2)** metrics. where $\Delta\hat{p}$ is the predicted movement, Δp is the true movement, and $\Delta\bar{p}$ is the average movement within a day.

For the Top-K Selection (TKS) task, we define the metric **hit rate** (HR) as the ratio of true predicted top stocks to K.

7.2 Results and Analysis

In the following down-stream tasks, we predict the next four future steps with an interval of $\Delta t = 0.5$ minutes, i.e., $\mathcal{S}_f = \Delta t, 2\Delta t, 3\Delta t, 4\Delta t$, which correspond to the next 0.5, 1.0, 1.5, and 2.0 min, respectively.

Table 1. Lagged Regression (LR) results.

Method	MSE ↓	MAD ↓	R2 ↑ ($\times 10^{-2}$)				
			Avg	$1\Delta t$	$2\Delta t$	$3\Delta t$	$4\Delta t$
Linear	1.72	0.68	0.46	1.82	0.15	−0.04	−0.11
LSTM [11]	1.72	0.68	0.51	1.98	0.16	−0.03	−0.07
TFT [17]	1.64	0.66	0.72	2.56	0.26	0.03	0.00
DTML [23]	1.63	0.64	0.78	2.72	0.30	0.07	0.01
PFM-FS (ours)	1.53	0.66	0.83	2.91	0.32	0.08	**0.02**
PFM (ours)	**1.51**	**0.65**	**0.90**	**3.17**	**0.34**	**0.09**	0.01

Lagged Regression Results. In Table 1, we present the Lagged Regression results discussed in Sect. 6.1. We use (↑) to indicate that *the larger the better*, and (↓) to show that *the smaller the better*. We have observed the following trends.

Our fine-tuned model PFM outperforms the best non-finetuned baseline method (DTML) by 7.4% in MSE (1.51 vs. 1.63) and 15.4% in R2 (0.90 vs. 0.78).

Our fine-tuned model PFM-FS, despite having few-shot fine-tuning data, still out-performs the best non-finetuned baseline DTML by 6.0% in MSE and 6.4% in R2.

PFM-FS under-performs PFM by 8.4% in R2 (0.83 vs. 0.90), while using less than 1/6 of the full training data. R2 drops quickly over future in 0.5, 1.0, 1.5, and 2.0 min. E.g., R2 drops from 3.17 (0.5-min) to 0.34 (1.0-min), indicating the volatility and difficulty in predicting long-term future.

Table 2. Directional Classification results in percentage.

Method	mAP ↑	mAcc ↑	Step-Acc ↑			
			$1\Delta t$	$2\Delta t$	$3\Delta t$	$4\Delta t$
Linear	40.1	39.9	43.9	39.1	38.7	37.9
LightGBM [14]	43.1	42.5	45.8	43.2	41.0	39.7
LSTM [11]	41.5	41.3	43.1	42.2	40.7	39.2
TFT [17]	42.0	42.2	44.5	42.8	41.5	40.1
DTML [23]	44.1	44.0	46.0	45.6	43.4	41.9
PFM-FS (ours)	45.3	45.4	47.2	46.9	44.3	43.2
PFM (ours)	**47.2**	**47.1**	**48.9**	**48.1**	**46.2**	**45.3**

Directional Classification Results. In Table 2, we show the Directional Classification results, with the mean average precision (mAP) and mean step-wise accuracy (mAcc) over four future steps. We observe the following trends:

Our PFM outperforms the best baseline DTML by around 7.0% in both mAP and mAcc, indicating the effectiveness of pretraining-and-finetuning scheme.

PFM-FS also outperforms the best baseline DTML by around 3.0% in both mAP and mAcc.

PFM-FS under-performs PFM by only 4%, using less than 1/6 of the full training data. The DC accuracy slightly decreases with increasing time interval from 0.5 to 2 min, by about 8% (from 48.9% to 45.3%), which is less sensitive to the length of the future compared with the regression task. The reason for this is that DC predicts the trend of movement, which could be more consistent than directly regressing the accurate movement rate.

Table 3. Top-K Selection Hit Rate (%). The larger is the better.

Method	Top-K Rising Hit Rate				Top-K Falling Hit Rate				mHR ↑
	5%	10%	20%	30%	5%	10%	20%	30%	
Linear	0.17	3.00	21.8	48.5	0.15	2.17	14.9	37.3	6.88
LSTM [11]	0.19	4.22	28.4	39.1	0.11	5.30	16.7	39.1	7.66
TFT [17]	0.32	4.15	30.8	41.1	0.22	6.13	24.6	42.8	8.74
DTML [23]	0.25	5.37	32.0	49.1	0.19	5.72	29.6	50.8	9.92
PFM-Reg (no top-K tuning)	0.33	10.7	33.4	54.1	0.28	10.3	26.8	53.1	11.5
PFM-FS (ours)	0.56	11.2	39.9	67.2	0.69	11.9	38.7	57.4	13.9
PFM (ours)	0.73	13.8	41.2	71.0	0.92	15.6	44.3	73.8	**16.1**

Top-K Selection Results. In Table 3, we show Top-K Selection (TKS) hit rate of top $\mathcal{R} = \{5\%, 10\%, 20\%, 30\%\}$ of the total number of stocks in the market.

Our PFM and few-shot version PFM-FS are fine-tuned with the Diff-Sort task to improve their TKS capacities, as discussed in Sect. 6.3. We also compare with PFM-Reg, which is the exact regression model fine-tuned with the Lagged Regression (LR) task without being rank-aware. During inference, all models produce movement values, and we sort them to select the top K stocks.

To measure the overall performance, we develop a weighted mean Hit Rate (mHR) metric, which aggregates all top ratios for both rising and falling cases, as follows:

$$\text{mHR} = \left(\sum_{r_i \in \mathcal{R}} \frac{0.5 * (a_{r_i} + b_{r_i})}{r_i} \right) \Big/ \left(\sum_{r_i \in \mathcal{R}} \frac{1}{r_i} \right), \qquad (17)$$

in which a_{r_i} and b_{r_i} are rising and falling hit rate for top ratio r_i, respectively. The importance of each top ratio is weighted by the inverse of r_i, i.e., the accuracy of a small r_i (e.g., 5%) has a greater impact on mHR than a larger one. Based on our analysis, we observe the following trends.

Our PFM outperforms the best baseline method DTML by around 62.3% in mHR (16.1 vs. 9.92) , while our few-shot version PFM-FS also outperforms DTML by 40.1% (13.9 vs. 9.92), as shown in last column of Table 3.

PFM-FS under-performs PFM by 17.5% in mHR (13.9 vs. 16.1), likely due to the limited amount of fine-tuning data.

PFM-Reg significantly under-performs PFM-FS and PFM by 20.9% (11.5 vs. 13.9) and 40% in mHR (11.5 vs. 16.1), respectively. The difference between PFM-Reg and PFM is that PFM utilizes our proposed rank-aware Dff-Sort to effectively establish the correct order of predicted movements over the market.

Zero-Shot Comparison. We further conducted an experiment to verify the zero-shot capacity of our fine-tuned models. To this end, we collected 800 new stocks from the China board of growth enterprises (ChiNext) with the same testing periods as STAR-22. Surprisingly, PFM-FS outperforms PFM in both tasks, by about 6.5% in MAD (0.43 vs. 0.46) and 4.3% in mAcc (0.47 vs. 0.45). This is likely due to the fact that the zero-shot prediction is more closely related to the most recent data trend, which is better captured by PFM-FS as it uses only the fresh history of the past two months. In contrast, PFM explores movement patterns of a longer history of training stocks, which may not be applicable to other stocks due to the zero-shot setting.

8   Conclusion

In this study, we introduce a novel financial model that uses a pre-training-and-fine-tuning pipeline for accurately predicting stock movements. Our methodology involves integrating supervised regression and unsupervised contrastive learning objectives to pre-train a Transformer-based encoder-decoder model. This model captures both the unique stock movement patterns and whole market contexts. We demonstrate that our proposed approach significantly outperforms existing methods, as evidenced by a range of evaluation metrics for realistic downstream tasks.

Acknowledgments. This work is supported by the National Natural Science Foundation of China, Project 62106156, and Starting Fund of South China Normal University.

References

1. Blondel, M., Teboul, O., Berthet, Q., Djolonga, J.: Fast differentiable sorting and ranking. In: ICML (2020)
2. Box, G.E.P., Jenkins, G.M.: Some recent advances in forecasting and control. J. Roy. Statist. Soc. (1968)
3. Brown, T.B., et al.: Language models are few-shot learners. arXiv preprint arXiv:2005.14165 (2020)
4. Cartea, A., Donnelly, R., Jaimungal, S.: Enhancing trading strategies with order book signals. Appl. Math. Financ. **25**(1), 1–35 (2018)
5. Castoe, M.: Predicting stock market price direction with uncertainty using quantile regression forest (2020)
6. Cuturi, M.: Sinkhorn distances: lightspeed computation of optimal transport. In: NeurIPS, pp. 2292–2300 (2013)

7. Cuturi, M., Teboul, O., Vert, J.P.: Differentiable ranking and sorting using optimal transport. In: NeurIPS (2019)
8. Devlin, J., Chang, M.W., Lee, K., Toutanova, K.: Bert: pre-training of deep bidirectional transformers for language understanding. arXiv preprint arXiv:1810.04805 (2018)
9. Fan, C., Lu, H., Huang, A.: A novel differentiable rank learning method towards stock movement quantile forecasting. In: European Conference on Artificial Intelligence (2023)
10. Fan, C., et al.: Multi-horizon time series forecasting with temporal attention learning. In: SIGKDD (2019)
11. Feng, F., Chen, H., He, X., Ding, J., Sun, M., Chua, T.S.: Enhancing stock movement prediction with adversarial training. In: IJCAI (2019)
12. Gould, M.D., Porter, M.A., Williams, S., McDonald, M., Fenn, D.J., Howison, S.D.: Limit order books. Quant. Financ. **13**(11), 1709–1742 (2013)
13. Holt, C.C.: Forecasting seasonals and trends by exponentially weighted moving averages. Int. J. Forecast. (2004)
14. Ke, G., et al.: Lightgbm: a highly efficient gradient boosting decision tree. In: NeurIPS (2017)
15. Kearns, M., Kulesza, A., Nevmyvaka, Y.: Empirical limitations on high-frequency trading profitability. J. Trading **5**(4), 50–62 (2010)
16. Koenker, R., Bassett, Jr., G.: Regression quantiles. Econometrica: J. Economet. Soc. 33–50 (1978)
17. Lim, B., Arık, S.Ö., Loeff, N., Pfister, T.: Temporal fusion transformers for interpretable multi-horizon time series forecasting. Int. J. Forecast. (2021)
18. Lin, T.Y., Goyal, P., Girshick, R., He, K., Dollár, P.: Focal loss for dense object detection. In: ICCV (2017)
19. Vaswani, A., et al.: Attention is all you need. In: NIPS (2017)
20. Wen, R., Torkkola, K., Narayanaswamy, B., Madeka, D.: A multi-horizon quantile recurrent forecaster. arXiv preprint arXiv:1711.11053 (2017)
21. Winters, P.R.: Forecasting sales by exponentially weighted moving averages. Manag. Sci. (1960)
22. Xu, Y., Cohen, S.B.: Stock movement prediction from tweets and historical prices. In: ACL (2018)
23. Yoo, J., Soun, Y., Park, Y., Kang, U.: Accurate multivariate stock movement prediction via data-axis transformer with multi-level contexts. In: SIGKDD (2021)
24. Zhang, L., Aggarwal, C., Qi, G.J.: Stock price prediction via discovering multi-frequency trading patterns. In: SIGKDD (2017)
25. Zhang, Z., Zohren, S.: Multi-horizon forecasting for limit order books: novel deep learning approaches and hardware acceleration using intelligent processing units. arXiv preprint arXiv:2105.10430 (2021)
26. Zhang, Z., Zohren, S., Roberts, S.: Deeplob: deep convolutional neural networks for limit order books. In: IEEE Transactions on Signal Processing (2018)

Impulsion of Movie's Content-Based Factors in Multi-modal Movie Recommendation System

Prabir Mondal[1,2]($\boxtimes$) (iD), Pulkit Kapoor[2], Siddharth Singh[3], Sriparna Saha[2](iD), Naoyuki Onoe[4], and Brijraj Singh[4]

[1] National Institue of Technology Patna, Patna, India
[2] Indian Institute of Technology Patna, Patna, India
prabirm.phd22.cs@nitp.ac.in , sriparna@iitp.ac.in
[3] Indian Institute of Technology Jodhpur, Jodhpur, India
[4] Sony Research India, Bengaluru, India

Abstract. Nowadays the Recommendation System, a subclass of information filtering system does not require any introduction, and the movie recommendation system plays an important role in the streaming platform where a huge number of movies are required to be analyzed before showcasing a perfectly matched subset of them to its users.

The existing works in this domain focus only on the output and consider the model's input similar for all users. But actually, the movie embedding input vector varies on a user basis. A user's perception of a movie depends on the movie's genre as well as its meta information (story, director, and cast). To formulate the fact, we have introduced two scores, (i) *TextLike_score (TL_score)* and (ii) *GenreLike_score (GL_score)*. Our proposed Cross-Attention-based Model outperforms the SOTA (state-of-the-art) by leveraging the effect of the scores and satisfying our factual notion.

In this paper, we have evaluated our model's performance over two different datasets, (i) *MovieLens-100K(ML-100K)* and (ii) *MFVCD-7K*. Regarding multi-modality, the audio-video information of movies' are used and textual information has been employed for score calculation. Finally, it is experimentally proved that the Cross-Attention-based multi-modal movie recommendation system with the proposed *Meta_score* successfully covers all the analytical queries supporting the purpose of the experiment.

Keywords: *TL_score* · *GL_score* · Multihead Cross Attention · Movie Recommendation System · *MFVCD-7K dataset* · *MovieLens-100K dataset*

1 Introduction

Understanding the users' preferences is highly demanding and the popular research topic in the AI/ML field for now. The recommendation system is one

B. Luo et al. (Eds.): ICONIP 2023, CCIS 1969, pp. 230–242, 2024.
https://doi.org/10.1007/978-981-99-8184-7_18

of the best solutions for this real-time problem of the digital world. Personal viewing devices such as Fire TV, Apple TV, etc., and streaming services in OTT platforms such as SonyLiv, Amazon Prime, Netflix, etc. have necessitated a more focused movie series recommendation to match user preferences.

In recent works, the user-movie embeddings generated from the text, audio, or video are taken as the model's input and the corresponding rating value for the pair is predicted. But in our reality, the user prefers those movies that belong to the user's preferred genre, director, cast, and storylines. These preferences are varied from user to user and it is difficult to establish a single-line parameterized function implicitly by the traditional neural network feature extraction technique. Both, the preferences over genres as well as textual information (storylines, director, and cast) are equally important to analytically orient the vector space of user-movie embeddings. And it is also noted that individually, neither of them can help to understand the complete choice of a user. Considering the factual notion, we have introduced two scores, *(i)* **TextLike_score (TL_score)** and *(ii)* **GenreLike_score (GL_score)**. The *TL_score* is the parameter that quantifies how much the user will like the textual information of the movie and *GL_score* is the movie's genre preference score for the user.

Unlike the traditional approach, we have not passed the user and movie embeddings directly to the model. The proposed *Meta_score* (combination of *TL_score* and *GL_score*) based weight is applied to the movie-user embeddings and transformed the vectors into proper magnitude before passing to the rating prediction model.

We have proposed a Cross-Attention-based rating prediction model that takes the audio and video embeddings of movies with their corresponding user embedding as inputs. A self-attention-based fusion technique preceded by multi-head cross-attention is employed to fuse the two different modalities in the proposed model. To evaluate our model, we have used two different datasets, *(i)* **MovieLens-100K (ML-100K)** and *(ii)* **MFVCD-7K**. After a thorough ablation study, experimentally it has been proved that the proposed work outperforms the state-of-the-art and supports our concerns about real-time scenarios.

2   Related Works

Works on Recommendation are basically of three different types, *(i)* *Collaborative Filtering, (ii) Content-Based*, and *(iii) Hybrid*. The collaborative approach in recommendation is common in finding user-user, item-item, or user-item similarity. Authors in [12] used the collaborative approach by considering the implicit feedback of other users while recommending movies to a user. A movie recommendation system by content-based approach had been developed in the work of [19] and it used the genre content-wise correlations among the *MovieLens* movies for its model. The content-based movie recommendation system proposed in [17] uses movies' textual information.

Most of the works in the literature used textual information while trying to predict preferences. In [10], authors analyzed the tweets to understand the users' sentiments and the current trends while recommending movies to users. The power of deep learning has boosted the analyzing scope of multi-modal data rigorously and authors in [4] introduced deep neural network-based trusted filter (DNN-filter) in predicting the rating values for cold-start users. A deep learning-based model for movie recommendation has also been introduced by authors in [15] where textual information is used in the analysis. Multimedia content-based information is effective in the movie recommendation system proposed by authors in [5].

A method in [13,16,18] tried to predict the ratings of the extended multi-modal *MovieLens* dataset. Work proposed by [9] determined the preference of users by tracking their effects or emotions by its ARTIST model. It uses different sensors for capturing human reaction signals and tries to predict the preferences towards the items. By introducing the Graph Attention Network [21] in the proposed model, authors in [3] tried to handle the cold-start problem [11] over the multi-modal *MovieLens* dataset. Similarly in [7,14], the graph attention is used in uniforming the user's as well as the item's auxiliary information for its RS.

3   Problem Statement

The proposed model takes the multi-modal information of a movie along with the user embedding as input and as output, it predicts the rating value as a preference score for the user-movie pair.

$$\hat{r}_{mu} = \sigma([[\bar{a}_m.\bar{u}], [\bar{v}_m.\bar{u}]]) \tag{1}$$

Unlike the conventional approach, we do not generate the user embedding just by simply taking the average of all the movies' embedding watched by the user. Based on the users' preferences over movies' content, the weight is calculated and the weighted average of movie embeddings is used for generating the user embedding. The input modal information(audio or video) is also combined with the weight before passing to the model rather than directly feeding the raw embedding to it. Equation 1 represents the generic functions of the proposed model where $\hat{r}_{mu}$ is the predicted rating value for the user(u) and movie(m) pair. $\bar{a}_m$ and $\bar{v}_m$ are the movie's respective audio and video weighted embeddings. $\bar{u}$ is the user embedding generated from the weighted average of his watched movies' embeddings and [.] represents the crossed attention-based fusion between two embeddings with activation function σ. [,] is the concatenation technique in the equation.

4   Dataset

For the evaluation of the proposed model, we have used two different datasets, *(i) multi-modal* **MovieLens 100K dataset** [16] and *(ii)* **MFVCD-7K** [6].

4.1   Multi-modal *MovieLens 100K (ML-100K)*

The authors in [16] developed this multi-modal dataset by adding the audio-video information in the text-based original *MovieLens 100K* dataset[1]. It contains the audio-video information of 1,494 Hollywood movie trailers, 943 users, and their 100K ratings. Along with that, textual information like the movie's summary, director, and cast are also used in the proposed methodology to analyze the content of the movie.

4.2   *MFVCD-7K* dataset

The Multifaceted Video Clip Dataset (*MFVCD*) contains the embeddings of video clips of movies linked to *MovieLens 20M*[2] dataset. The *MFVCD-7K* contains the audio video embeddings of 6,877 video clips of 796 unique movies and movies' genre information. In our experiment, for audio embeddings, we have used the i-vector (with GMM-128 and tvDim = 200), and embedding from the fc7 layer of AlexNet has been used as video embedding.

5   Methodology

5.1   Embedding Generation

The generation of embedding for audio and video is required for the *ML-100K* dataset only because the *MFVCD-7K* already had the embeddings. For generating the meta score (a combination of Text-score and *GL_score*), we also embedded textual information like the movie's summary, director's name, and cast. Only the multi-modal dataset *ML-100K* [16] has this textual information and that is why we have calculated the *GL_score* for only the *MFVCD-7K* dataset.

Video Embedding. Here are the dataset-wise video embedding techniques:
(a) **For *MFVCD-7K* Dataset:** The *MFVCD-7K* dataset already contains the embedding of video clips of every movie. In our experiment, we have taken the fc7-AlexNet video clip embeddings of every movie clip and averaged them to form a single vector that represents the movie's video embedding of dimension 4096.
(b) **For *ML-100K* Dataset:** The video embedding generation for *ML-100K* is a two-step approach, *(i) Keyframe extraction* and *(ii) Video Embedder*.

(i) Keyframe Extraction: In the proposed approach, the keyframe extraction is very straightforward. 16 frames from all the frames of every movie trailer are extracted by following the systematic sampling and treated as keyframes of the movie trailer. The systematic sampling is done in such a way that the informative frames (not the blank or only text-containing frame) would be collected

---

[1] https://grouplens.org/datasets/movielens/100k/.
[2] https://grouplens.org/datasets/movielens/20m/.

as keyframes. In this way, 16 informative frames covering the entire trailer span are taken and passed to the next module of genre prediction.

(ii) Video Embedder: The Video Embedder is highly valuable for classifying movie genres and generating genre-specific video embeddings. We utilize the advanced TimeSformer model [2] to extract video embeddings from *ML-100K* movie trailers, incorporating keyframes extraction and seamlessly obtaining genre-tailored video embeddings.

We utilized transfer learning in our method by using a pre-trained TimeSformer model, originally trained on the K-600 dataset for video genre classification. To create our video embedder, we made adjustments: removing TimeSformer's classification layers, introducing a 512-neuron embedding layer, and adding an 18-neuron classification layer for multi-label genre classification.

We fine-tuned the TimeSformer model by training its final layers with our movie trailers' keyframes. The resulting 512-dimensional video embedding represents the trailer after training, specifically for *ML-100K*.

Audio Embedding. Similar to generating video embeddings, we obtained 200-dimensional audio embeddings by averaging i-vectors with GMM-128 and tvDim=200 from the *MFVCD-7K* dataset. To process the *ML-100K* dataset, the movie trailer's audio (.wav format) is converted into a 512-dimensional vector using the feature extractor wav2vec2 [1].

To generate task-specific audio embeddings for genres, we utilized a dense layer embedder comprising three dense layers with 512 neurons each, employing Tanh activation. The classification layer, with 18 neurons activated by Sigmoid, produces an 18-dimensional output for genre prediction. The Embedding Layer of the embedder extracts a 512-dimensional audio embedding from the input obtained through wav2vec2.

Text Embedding. The *ML-100K* dataset is the only one with text data, including movie summaries, director names, and cast names. By applying word2vec [8], we obtained a 600-dimensional summary vector, a 600-dimensional cast name vector, and a 300-dimensional director name vector. These were concatenated to form a text embedding of size 1500 for each movie in the dataset.

Due to the lack of mentioned textual information in *MFVCD-7K*, no textual embedding has been generated for its movies.

Content-Based Score Calculation. A user always likes to watch those movies that belong to his preferred genres, are directed by specific directors, have a favorable storyline, and are cast by his favorite actors. Following this notion, we proposed a technique to calculate the *Meta_score* of a user-movie pair that maps the concerned factors of a movie for a user.

The *Meta_score* is the average of the *Text_score* and *GenreLike_score*. Text_score represents the user's liking for textual movie information, while GenreLike_score measures genre preference for user-movie pairs.

To assess movie scores, user-rated movies are divided into five categories based on their rating values. The datasets, *ML-100K* and *MFVCD-7K* include ratings from 0.5 to 5 and encompass 18 genres. Group_1 (Gr._1) comprises movies rated between 0.5 and 1.5, while the other groups are Gr._2: {2 to 2.5}, Gr._3: {3}, Gr._4: {3.5 to 4}, and Gr._5: {4.5 to 5}. These groups are utilized to calculate *Text_score*, *GenreLike_score*, and *Meta_score*.

(i) *TextLike_score (TL_score)*:

To create a user's preference for movie text, 1500-dimensional text embeddings of movies are grouped and aggregated (Eq. 2). Here $t_u^{(Gr)_i}$ is the textual embedding of movies from Group_i for the user u. $i \in \{1, 2, 3, 4, 5\}$ in Eq. 2, 3, 4, 5, 6, 7.

$$T_u^{(Gr)_i} = \sum t_u^{(Gr)_i} \tag{2}$$

When a new movie is considered to recommend to a user, by following Eq. 3, the group-wise text-preference score for the user-movie pair is measured. In this Equation, $dot(,)$ represents the dot product operation between two vectors and t_m is the text embedding of the new candidate movie that would be recommended.

$$t\_score_{um}^{(Gr)_i} = dot(T_u^{(Gr)_i}, t_m) \tag{3}$$

And finally, in Eq. 4, the group-wise textual information preference scores for the user-movie pair are aggregated to evaluate the TextLike_score (*TL_score*). In this equation $mul(,)$ is the scalar multiplication operation and $R_{(Gr)_i}$ is the highest rating value of ith group. So, $R^{(Gr)_1} = 1.5, R^{(Gr)_2} = 2.5, R^{(Gr)_3} = 3, R^{(Gr)_4} = 4$ and $R^{(Gr)_5} = 5$.

$$TL\_score_{um} = \sum mul(log(R^{(Gr)_i}), t\_score_{um}^{(Gr)_i}) \tag{4}$$

(ii) *GenreLike_score (GL_score)*:

Like *TL_score*, the *GenreLike_score(GL_score)* is evaluated for a user-movie pair (um) as follows:

$$G_u^{(Gr)_i} = \sum g_u^{(Gr)_i} \tag{5}$$

$$g\_score_{um}^{(Gr)_i} = dot(G_u^{(Gr)_i}, g_m) \tag{6}$$

$$GL\_score_{um} = \sum mul(log(R^{(Gr)_i}), g\_score_{um}^{(Gr)_i}) \tag{7}$$

Eq. 5, 6, 7 which are as same as Eq. 2, 3, 4. The only difference is the embeddings. Here movies' multi-hot 18-dimensional genre embeddings (g_m) are grouped into five different groups $(G_u^{(Gr)_i})$ based on the rating value range.

(iii) *Meta_score*:

As shown in Eq. 8, the *Meta_score* is the average score of *TL_score* and *GL_score* for the user-movie pair (um). Note that the *ML-100K* dataset contains the textual information and so we have *Meta_score* for this dataset only. For the *MFVCD-7K* dataset we will have only *GL_score*.

$$Meta\_score_{um} = 0.5 * (TL\_score_{um} + GL\_score_{um}) \tag{8}$$

Movie and User Embedding. In the conventional approaches, the generated candidate movie embedding (e_m) is directly passed to the recommendation model along with user embedding for predicting the preference score for the user-movie pair.

$$\bar{e}_{um} = Meta\_score_{um} * e_m \tag{9}$$

$$\bar{e}_u = \frac{1}{N}[\sum \bar{e}_{um}^{(Gr)_j}] \tag{10}$$

In the proposed model, the movie embedding is adjusted with the corresponding user-movie pair (um) $Meta\_score$. Equation 9 generates the adjusted movie embedding $(\bar{e}_{um})$ for user u.

In the proposed methodology, the user embedding is not simply the average of the watched movie embeddings whose rating values are above a certain threshold. As shown in Eq. 10, the user embedding is the average of movie embeddings adjusted in Eq. 9. Note that for generating the user embedding, only the movies belonging to Gr_j groups are considered, where $j \in \{3,4,5\}$ and N is the total number of movies of these groups.

Main Model Architecture: To understand user-movie preferences and predict ratings, cross-attention techniques are essential for establishing a connection between them.

The main model integrated the attention mechanism [20] to consider user-movie preferences. Equations (Eq. 11) and (Eq. 12) were used to generate contextual vectors from (Q, K, V) triplets.

The model incorporates two attention mechanisms: cross-attention and self-attention in the Fusion block as shown in Fig. 1. The architecture (shown in Fig. 1) first performs cross-attention between user-movies pair (um), generating attention-based vectors $\hat{U}$ and $\hat{M}$. These vectors are then concatenated and passed to the self-attention layer in the fusion block. The resulting output undergoes feed-forward and dense layers to predict preferences. The attention blocks employ a multi-head mechanism with four heads.

$$Attention\_weight = Softmax(\frac{Q.K^T}{\sqrt{d_k}}) \tag{11}$$

$$Context\_vector = Attention\_weight * V \tag{12}$$

where d_k is the dimension of the Key vectors (K). The input consists of two vectors: the user embedding $(\bar{e}_u)$ and the modality embedding $(\bar{e}_{um})$. The modality embedding, derived from audio or video of movies, undergoes $Meta\_score$ transformation (Eq. 8) before being fed into the model.

6   Results and Discussion

In the whole experiment, we try to cover four analytical questions to evaluate our model's effectiveness. The research questions can be outlined as follows,

RQ1: Is multi-modality or bi-modality effective in a movie recommendation system?

RQ2: Which content-based weight is better, *TL_score* or *GL_score*?

RQ3: Does the content-based weight play a positive role in generating movie and user embeddings?

RQ4: How does the model perform if Genre embedding or Text embedding is used instead of audio, video, or multi-modal embedding?

The following is the analysis of the findings,

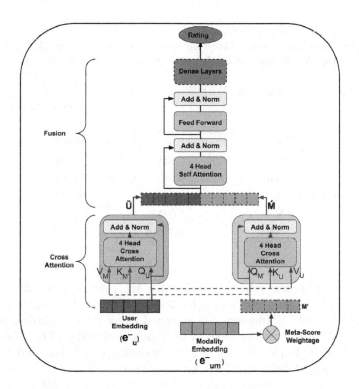

Fig. 1. Cross-Attention-Based Proposed Main Model.

Experimental Setup: In this experiment, three different models have been used (i) the Audio embedding generator, (ii) the Video embedding generator, and (iii) the user-movie preference score predictor (the main model). We evaluated our main model five times and reported the average score in the result tables.

Audio-Video Embedding Generation : We have our audio-video generator for *ML-100K* dataset as the *MFVCD-7K* already provides the embeddings. The audio-video embedding generator actually tries to predict the 18-dimensional genre distribution of the input movie. For this dataset, we have considered 1,285 movies for training and 209 for testing.

For *ML-100K* audio embedding, the extracted audio file (.wav extension) of the movie trailer is converted into a 512-dimensional vector using wav2vec2 [1] and then passed to the audio embedder.

The tanh and Sigmoid activation functions have been used in hidden layers and output layers, respectively, with Categorical Cross Entropy as the loss function. Here the model has been trained using 100 epochs. 94.13% and 91.56% are the training testing accuracy, respectively. After the training, the embedding from the second last layer is taken as the corresponding audio embedding of the movie for the downstream task.

Table 1. Modality and Content-based score wise RMSE results of different datasets.**TL:** *TextLike_score*, **GL:** *GenreLike_score*, **Meta:** *Meta_score*, **A:** *Audio modality*, **V:** *Video modalidy*, **All:** *Audio+Video modality*

Dataset	Modality	With Score(RMSE)			Without Score (RMSE)
		TL	GL	Meta	
ML-100K	**A**	1.0043	1.0022	1.0006	0.9493
	V	1.022	1.021	1.0189	0.9936
	All	0.9515	0.985	**0.9285**	0.9451
MFVCD-7K	**A**	-	0.753	-	0.7341
	V	-	1.0275	-	1.0116
	All	-	**0.7486**	-	0.7447

While generating the video embedding for *ML-100K*, the classification layer of TimeSformer is taken out and a hidden layer (we will treat as the video-embedding-layer) of 512-dimension is attached along with 18-dimensional genre prediction output layer as present in the audio embedder. The TimeSformer [2] takes 96 or 16 frames for the action recognition task but due to making our proposed model light and fast, we have used 16 frames from each movie trailer. The video embedder generation model is trained with 100 epochs utilizing cross-entropy loss. Finally, we gain 90.62% test accuracy and extract the 512-dimensional embedding from the second last layer (the video-embedding layer) for the downstream task.

Main Model Setting: To train our main model (the cross-attention-based model), we have used 150 epochs, the MSE loss function, and the Adam optimizer. For this experiment, we split the dataset in an 80:20 training-testing split and the model performance has been reported in terms of RMSE and mAP@K (with K = 3, 10, 50).

Results: From the Table 1, the three queries (*RQ1, RQ2 and RQ3*) are covered. The table shows the model performance in the RMSE scale for two different datasets. It is clear from the table that the RMSE score is lowest (that is technically good and has been highlighted for making it noticeable) when the bimodality (audio+ video) of the movie is considered. Hence the multi-modality outperforms the unimodality (audio-video modality separately) and covers *RQ1*. Regarding the query *RQ2*, we can compare the *TL_score* and *GL_score* for the *ML-100K* dataset only because the other dataset doesn't have the textual information to form the *TL_score*.

Table 2. Model's performance with only Text, only Genre, only Meta, and Meta+All modality-based embeddings

	ML-100K	MFVCD-7K
Only Text Embedding	1.049	-
Only Genre Embedding	1.1152	0.9581
Only Meta Embedding	0.9942	-
Meta+All modality	**0.9285**	**0.74911**

In the same table, it is clear that none of the scores are superior to another but they affect the model's performance positively when they are combined and form the *Meta_score*. So it is proved that the content-based score is really effective for movie-user embedding and hence query *RQ3* is justified well.

Table 2 shows the ablation studies to deal with *RQ4*. It presents how the model would perform if the user's preference is predicted by analyzing the movie's genre or textual embedding only. For this experiment, we prepared movie and user embeddings in three different ways: *(i) Only Text Embedding, (ii) Only Genre Embedding, and (iii) Only Meta Embedding*. For *(i) Only Text Embedding*, we have considered the movie's 1500-dimensional textual embedding as the movie's embedding and generated the corresponding user's embedding. For *(ii) Only Genre Embedding* the 18-dimensional multi-hot genre embedding is considered as movie embedding and *(iii) Only Meta Embedding* is formed by concatenating the *Only Text Embedding and Only Genere Embedding*. With their different user embeddings, the model is tested on these different movie embeddings and it is found that none of them outperforms the result of *Meta+All modality*. *Meta+All modality* is the combination of Audio-Video embedding with *Meta_score*. Hence our consideration of Meta+All modality is justified and strong enough to defend *RQ4*.

In Table 3(b), the comparison with SOTA has been tabulated and it is experimentally checked that the proposed method has the best RMSE value.

7 Error Analysis

From Table 1: *(i) Individual Score effect:* In this table it is prominent that for the *ML-100K* dataset, neither *TL_score* nor *GL_score* helps the model individually when multi-modality is concerned. But the *Meta_core* has a noticeable impact on it. It supports our notion that the user's preference for a movie relies on both, the movie's genre as well as the movie's meta information (story, director, cast) simultaneously.

(ii) Without Score Model's performance: For unimodality (Audio or Video), without considering any content-based score (neither TL nor GL score) the model shows better results. But for the Meta+All combination, the model really shows the effect of both, multi-modality as well as *Meta_score*. Hence it is justified that to formulate the user's preference, the combination of the movie's genre-text joined information with audio-video information is crucial.

Table 3. (a) mAP over different datasets when the model uses *Meta_score* (and *GL-score* for *MFVCD-7K* dataset) with Audio+Video modality. (b) SOTA Vs. Proposed Model

<table>
<tr><td colspan="4" align="center">(a)</td><td colspan="3" align="center">(b)</td></tr>
<tr><td></td><td>mAP@3</td><td>mAP@10</td><td>mAP@50</td><td>Sl. No.</td><td>Techniques</td><td>RMSE</td></tr>
<tr><td>**ML-100K**</td><td>0.1461</td><td>0.1463</td><td>0.1468</td><td>1</td><td>Siamese [16]</td><td>1.028</td></tr>
<tr><td>**MFVCD-7K**</td><td>0.5278</td><td>0.5286</td><td>0.5287</td><td>2</td><td>GCN [13]</td><td>0.973</td></tr>
<tr><td></td><td></td><td></td><td></td><td>3</td><td>Proposed Model</td><td>**0.9285**</td></tr>
</table>

(iii) No TL_score for MFVCD-7K: For *MFVCD-7K* dataset, the result for *GL_score*+All combination is as same as *Without-score* result. Only the *GL_score* is not sufficient to help the model in preference prediction. Adding the movies' textual information to this dataset and leveraging the *Meta_score*'s effect might improve the model's performance.

From Table 3 (a): *(i) Low mAP@K for ML-100K :* We have evaluated the mAP@K in three different cut-off values, $K = 3, 10, and 100$. For the *ML-100K* dataset, it is seen that the values are quite small. This is because of the high sparsity of the user-movie interaction matrix. But there are 13,800 users in *MFVCD-7K*, and we have chosen 3,000 users randomly for our experiment. For 3,000 users the interaction matrix is quite dense and hence the mAP@K value is comparatively high.

8 Conclusion

The proposed multi-modal movie recommendation system takes the movie's audio-video information with user embedding as input and predicts the preference score for the user-movie pair. The movie's genre, as well as its meta information (the story, director, and cast), are really considering factors for a user when he chooses or rejects a movie. The effect of the *Meta_score* over the cross-attention-based model supports our notion. To evaluate the model's performance, two datasets, *ML-100K* and *MFVCD-7K* have been employed. And it is reported that our model is strong enough to beat the SOTA in both cases.

In the future, we would like to extend the proposed model in a multitask setting by solving some auxiliary tasks like movie genre prediction, and user age/gender prediction along with rating prediction of user-movie pair as the primary task.

Acknowledgement. The authors gratefully acknowledge the support from Sony Research India for conducting this research.

References

1. Baevski, A., Zhou, Y., Mohamed, A., Auli, M.: wav2vec 2.0: A framework for self-supervised learning of speech representations. In: Advances in Neural Information Processing Systems 33, pp. 12449–12460 (2020)
2. Bertasius, G., Wang, H., Torresani, L.: Is space-time attention all you need for video understanding? In: ICML, vol. 2, p. 4 (2021)
3. Chakder, D., Mondal, P., Raj, S., Saha, S., Ghosh, A., Onoe, N.: Graph network based approaches for multi-modal movie recommendation system. In: 2022 IEEE International Conference on Systems, Man, and Cybernetics (SMC), pp. 409–414. IEEE (2022)
4. Choudhury, S.S., Mohanty, S.N., Jagadev, A.K.: Multimodal trust based recommender system with machine learning approaches for movie recommendation. Int. J. Inf. Technol. **13**(2), 475–482 (2021). https://doi.org/10.1007/s41870-020-00553-2
5. Deldjoo, Y.: Enhancing video recommendation using multimedia content. Spec. Top. Inf. Technol. pp. 77–89 (2020). https://doi.org/10.1007/978-3-030-32094-2_6
6. Deldjoo, Y., Schedl, M.: Retrieving relevant and diverse movie clips using the mfvcd-7k multifaceted video clip dataset. In: 2019 International Conference on Content-Based Multimedia Indexing (CBMI), pp. 1–4. IEEE (2019)
7. Feng, C., Liu, Z., Lin, S., Quek, T.Q.: Attention-based graph convolutional network for recommendation system. In: ICASSP 2019–2019 IEEE International Conference on Acoustics, Speech and Signal Processing (ICASSP), pp. 7560–7564. IEEE (2019)
8. Goldberg, Y., Levy, O.: word2vec explained: deriving Mikolov et al'.s negative-sampling word-embedding method. arXiv preprint arXiv:1402.3722 (2014)
9. Kaklauskas, A., et al.: An affect-based multimodal video recommendation system. Stud. Inf. Control **25**(1), 6 (2016)
10. Kumar, S., De, K., Roy, P.P.: Movie recommendation system using sentiment analysis from microblogging data. IEEE Trans. Comput. Soc. Syst. **7**(4), 915–923 (2020)
11. Lam, X.N., Vu, T., Le, T.D., Duong, A.D.: Addressing cold-start problem in recommendation systems. In: Proceedings of the 2nd International Conference on Ubiquitous Information Management and Communication, pp. 208–211 (2008)
12. Lavanya, R., Bharathi, B.: Movie recommendation system to solve data sparsity using collaborative filtering approach. Trans. Asian Low-Resour. Lang. Inf. Process. **20**(5), 1–14 (2021)

13. Mondal, P., Chakder, D., Raj, S., Saha, S., Onoe, N.: Graph convolutional neural network for multimodal movie recommendation. In: Proceedings of the 38th ACM/SIGAPP Symposium on Applied Computing, pp. 1633–1640 (2023)
14. Mondal, P., Kapoor, P., Singh, S., Saha, S., Singh, J.P., Onoe, N.: Task-specific and graph convolutional network based multi-modal movie recommendation system in Indian setting. Procedia Comput. Sci. **222**, 591–600 (2023)
15. Mu, Y., Wu, Y.: Multimodal movie recommendation system using deep learning. Mathematics **11**(4), 895 (2023)
16. Pingali, S., Mondal, P., Chakder, D., Saha, S., Ghosh, A.: Towards developing a multi-modal video recommendation system. In: 2022 International Joint Conference on Neural Networks (IJCNN), pp. 1–8. IEEE (2022)
17. Pradeep, N., Rao Mangalore, K., Rajpal, B., Prasad, N., Shastri, R.: Content based movie recommendation system. Int. J. Res. Ind. Eng. **9**(4), 337–348 (2020)
18. Raj, S., Mondal, P., Chakder, D., Saha, S., Onoe, N.: A multi-modal multi-task based approach for movie recommendation. In: 2023 International Joint Conference on Neural Networks (IJCNN), pp. 1–8. IEEE (2023)
19. Reddy, S.R.S., Nalluri, S., Kunisetti, S., Ashok, S., Venkatesh, B.: Content-based movie recommendation system using genre correlation. In: Satapathy, S.C., Bhateja, V., Das, S. (eds.) Smart Intelligent Computing and Applications. SIST, vol. 105, pp. 391–397. Springer, Singapore (2019). https://doi.org/10.1007/978-981-13-1927-3_42
20. Vaswani, A., et al.: Attention is all you need. In: Advances in Neural Information Processing Systems 30 (2017)
21. Velickovic, P., Cucurull, G., Casanova, A., Romero, A., Lio, P., Bengio, Y., et al.: Graph attention networks. Stat **1050**(20), 10–48550 (2017)

Improving Transferbility of Adversarial Attack on Face Recognition with Feature Attention

Chongyi Lv[1] and Zhichao Lian[2(✉)]

[1] School of Computer Science and Engineering, Nanjing University of Science and Technology, Nanjing, China
creeperlv@icloud.com
[2] School of Cyberspace Security, Nanjing University of Science and Technology, Wuxi, China
lzcts@163.com

Abstract. The development of deep convolutional neural networks has greatly improved the face recognition (FR) technique that has been used in many applications, which raises concerns about the fact that deep convolutional neural networks (DCNNs) are vulunerable to adversarial examples. In this work, we explore the robustness of FR models based on the transferbility of adversarial examples. Due to overfitting on the specific architecture of a surrogate model, adversarial examples tend to exhibit poor transferability. On the contrary, we propose to use diverse feature representation of a surrogate model to enhance transferbility. First, we argue that compared with adversarial examples generated by modelling output features of deep face models, the examples generated by modelling internal features have stronger transferbility. After that, we propose to leverage attention of the surrogate model to a pre-determined intermediate layer to seek the key features that different deep face models may share, which avoids overfitting on the surrogate model and narrows the gap between surrogate model and target model. In addition, in order to further enhance the black-box attack success rate, a multi-layer attack strategy is proposed, which enables the algorithm to generate perturbations guided by features with model's general interest. Extensive experiments on four deep face models show the effectiveness of our method.

Keywords: Adversarial Attack · Transferbility · Face Recognition · Feature Attention

1 Introduction

In recent years, with the rapid development of DCNNs, the accuracy of FR technology improves a lot, resulting FR systems penetrating into all aspects of people's life, such as phone unlocking, payment, security passing and so on.

Nevertheless, researchers found that DCNNs are vulnerable, which means that in test time, after adding carefully designed perturbations on original inputs, these images, called adversarial examples, can easily confuse DCNNs and obtain wrong recognition results. As extension of DCNNs, FR models also suffer from such attacks. For instance,

B. Luo et al. (Eds.): ICONIP 2023, CCIS 1969, pp. 243–253, 2024.
https://doi.org/10.1007/978-981-99-8184-7_19

Komkov et al. [1] found that only putting a piece of paper on the hat can deceive state-of-the-art FR models. Such phenomenon has raised serious concerns about the security of FR applications.

Lots of previous works have demonstrated the vulnerability of deep face models [2, 3], but large parts of them assumed that the attacker has full access to the target model, which is classified as white-box attacks in adversarial attack. White-box attacks use the corresponding examples to adjust perturbations added on the clean examples when the parameters and internal structure of the target model are known. In contrast, black-box attacks generate examples only through the relationship between the input and output of the network. Query-based methods in black-box settings need to interact with target model through large amounts of queries. However, in real-world attack situations, the attacker typically has no rights to get corresponding outputs many times. Goodfellow et al. [4] argued that the reason for the existence of adversarial perturbation is the linear behavior of high-dimensional space of DCNNs, which also provided a reasonable explanation for the transferbility of adversarial examples, meaning that there is still some effect on attacking another unknown model using the examples trained against a white-box model. A lot of works have been proposed to enhance the transferbility of adversarial examples against object classification task, such as using loss-preserving transformation [5–7] and increasing the diversity of surrogate models [8, 9]. Through extensive experiments, the authors in [10] confirmed that above techniques focusing on generating $L_p norm$ constrained adversarial examples against object classification task can be beneficial for improving the transferbility of adversarial examples againtst face recognition models. Considering the distinctiveness of deep face model, Zhong and Deng [9] proposed feature iterative attack method(FIM) based on features extracted by the deep face model, and dropout face attacking network(DFANet) to increase the variability of surrogate models. However, the transferbility gained from the above methods are still unsatisfactory because it is hard for face adversarial examples to escape from local optima.

To improve the transferbility of adversarial examples on FR, this paper explores the effectiveness of deep face model's internal features. In summary, the contributions of our work are three-fold as follows:

- We propose an internal feature iterative attack method(IFIM), which models internal features instead of the output features of target model to generate more transferable adversarial examples. Further, to standardize the optimization direction, we leverage the attention of the surrogate model to a pre-determined intermediate layer to seek features that different deep face models may share, avoiding overfitting on the surrogate model.
- We propose a novel multi-layer attack strategy. Layers at different intermediate depths tend to grasp different features. So fusing multi-layer features together further increase black-box attack success rate.
- Extensive experiments on four mainstream deep face models demonstrate the effectiveness of our method.

2 Related Works

2.1 Adversarial Attack

Adversarial attack [10] is first found in object classification task. After adding perturbations to the original input, DCNNs will output wrong answer with high confidence. Large numbers of attack methods emerged, which provided a research basis for the security of DCNNs.

According to whether or not the attacker has access to the internal structure and parameters of target model, adversarial attack is classified as white-box attacks and black-box attacks. The most classic one in white-box attacks is fast gradient sign method(FGSM) proposed by Goodfellow et al. [4], which constructs an adversarial example by performing gradient update along the direction of the gradient sign at each pixel in one step. Goodfellow also proposed a linear hypothesis that the reason for the existence of adversarial perturbation is the linear behavior of high-dimensional space of deep neural network, which also provided a reasonable explanation for the transferbility of adversarial examples. Many works have improved the FGSM attack performances. Kurakin et al. [11] extended single-step FGSM to iterative fast gradient sign method(I-FGSM) that was strengthened by Dong et al. [12] and Lin et al. [5] in combination with Momentum and Nesterov Momentum, which is known as momentum iterative fast gradient sign method(MI-FGSM) and nesterov iterative fast gradient sign method(NI-FGSM), respectively. Diverse inputs iterative fast gradient sign method (DI^2-FGSM) [6] is another transfer-based attack based on FGSM. DI^2-FGSM performs loss-preserving transformation for input images, including random resizing and padding. This diversified input improves the transferbility of adversarial examples.

In recent years, new transfer-based attack methods emerge in endlessly. Not in the spatial domain, Long et al. [22] focused on the difference between surrogate model and target model and proposed spectrum saliency map, which depicts model characteristics in the frequency domain, improving the transferbility by reducing the difference between spectrum saliency maps. Li et al. [23] found that the noise with property of regional homogeneity is easier to transfer to other models, and the region division scheme designed in the paper also brings good transferbility. Sharma et al. [24] analyzed the composition of perturbation from the perspective of frequency domain, and pointed out the effectiveness of low-frequency part of perturbation in enhancing the transferbility. Several papers [20, 21] have proposed feature space attack methods. Instead of misleading the final classification layer, these methods attacks internal features of surrogate model. These methods will disrupt the feature space to create more transferable examples. However, most methods only work in the non-target attack task that only needs to mislead the model without specifying the specific class of the attack and ignore the effectiveness of feature attention.

2.2 Adversarial Attack on FR

FR [13, 14] has revealed overwhelming verification performance, which is applied in many safety-critical applications ranging from phone unlocking to airport certification. With the development of adversarial attacks, however, not only do we need to understand the operation process of face recognition terminal, but also understand the adversarial robustness of face recognition technology.

Unlike object classification task, current mainstream face recognition models do not attach a specific label to the input face, but extract the feature and then judges the face label according to the feature similarity. Therefore, on the basis of I-FGSM, Zhong et al. [9] modified the label loss to feature loss and FIM is proposed. Further, in order to increase the diversity of surrogate models, Zhong et al. modified the intermediate feature output of surrogate model based on the dropout technique to achieve ensemble-like effects. Komkov and Petiushko [1] let people aware of the vulnerability of face recognition models in the physical domain. The method, dubbed AdvHat, only used a piece of paper to confuse the state-of-the-art public face recognition model ArcFace@ms1m-refine-v2 [13]. AdvHat draws on the idea of expectation over transformation(EOT) [25], takes the possible transformation of the patch placed on the face into account and embeds the possible transformation into the training process to enhance the adversarial patch's robustness in the physical domain. EOT is widely used in physical attack, such as Pautov et al. [26]. Based on EOT, they studied the possible impact of different adversarial patch shapes and positions under the same attack scenario as AdvHat, including nose, forehead, wearable glasses, etc. Yin et al. [15] proposed an imperceptible and transferable attack method on face recognition, called adv-makeup. Specifically, it combines the idea of ensemble attack and meta-learning to narrow the gap between surrogate model and target model. Besides, adv-makeup leverage face makeup to enhance the imperceptibility of attacks, which in physical domain is realized by tattoo stickers. There are a lot of research based on the above two, including improving the adversarial patch robustness in the physical domain [16], reducing chromatic aberration of the sticker caused by the printers [17], and improving physical attack performance [18]. Nevertheless, there are few studies on the transferbility of face adversarial examples and much of them ignore the effectiveness of modelling internal features and feature attention of surrogate model in enhancing transferbility.

3   Methodology

In this section, after brief descriptions of adversarial attacks on deep face model, we propose internal feature iterative attack method (IFIM) in Sect. 3.2. Afterwards, motivated by ensemble attack, we propose to combine feature maps and feature attention of multiple intermediate layers of the deep face model together to further enhance the transferbility in Sect. 3.3.

3.1   Attacks Against Deep Face Model

Assume a FR model $F(x) : X \rightarrow R^d$, where x denotes an input face image and R^d denotes the normalized feature representation vector. Let x^s and x^t be two face images. The FR model first calculates the similarity D_F between two feature vectors. The calculation formula of similarity can be expressed as:

$$D_F\left(x^s, x^t\right) = F(x^s) - F\left(x^t\right)_2^2 \tag{1}$$

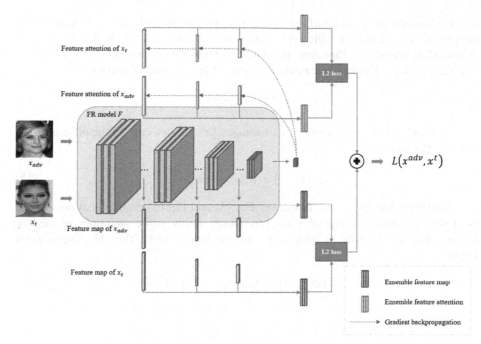

Fig. 1. Overall network architecture

We hope to generate an adversarial example $x^{adv} = x + \epsilon$, where ϵ is the maximum perturbation magnitude. An adversary solves the following optimization problem to generate the adversarial example on the target face model:

$$\arg\min_{x^{adv}} L(x^{adv}, x^t), s.t. x - x_2^{adv} \leq \epsilon \qquad (2)$$

where L is the adversarial loss objective. We use $L = -D_F$ for dodging attack and $L = D_F$ for impersonation attack, which is also named as FIM [9]. In this paper, we mainly focus on the impersonation attack on FR because impersonation attack often possesses smaller search scope [8].

Among existing works on improving the transferbility of adversarial examples, some focus on the feature space [19, 20]. It is easy to extend feature space skills to the attacks on the deep face model. Let l be the intermediate layer of the target model and F_l denotes the intermediate feature map of the target model at the l th layer. Then, the optimization objective is

$$L(x^{adv}, x^t) = F_l(x^{adv}) - F_l(x^t)_2^2 \qquad (3)$$

We denote this algorithm as internal feature iterative attack method(IFIM).

3.2 Feature Space Attack with Feature Attention Loss

We observed that even combining advanced techniques, transfer-based attack performance is still unsatisfactory. Motivated by the fact that feature maps extracted by different architecture models under the same face input often have high similarity, we plan to introduce feature attention of the surrogate model to internal features to standardize the optimization direction, which is helpful to seek the key features that different deep face models may share. Therefore, it is vital to obtain the attention of the model to the intermediate layer features by calculating forward feature map derivation α_l:

$$\alpha_l(x) = \frac{\partial F(x)}{\partial F_l(x)} \tag{4}$$

Therefore, when given a pair of face images $\{x^s, x^t\}$, we can get the differences in feature attention:

$$\Delta_l(x^s, x^t) = \alpha_l(x^s) - \alpha_l(x^t)_2^2 \tag{5}$$

Combining the above two objectives Eq. (3) and Eq. (5) ensures that adversarial examples preserve the original white-box attack ability while undermining the key features that are likely to be adopted by diverse face models. The objective is given below:

$$L(x^{adv}, x^t) = F_l(x^{adv}) - F_l(x^t)_2^2 \\ + \alpha_l(x^{adv}) - \alpha_l(x^t)_2^2 \tag{6}$$

3.3   Ensemble Feature Space Attack with Feature Attention Loss

Motivated by ensemble attack, we propose to generate perturbations based on the feature maps and feature attention of multiple intermediate layers of the surrogate model. Considering the different sizes and channel numbers of feature maps and feature attention at different layers, we first resize the feature maps and feature attention of other layers as of the l th layer, and then concate the feature maps and feature attention, which will be substituted into Eq. (6) to calculate the final loss. Feature fusion can be expressed as(take three layers as example):

$$F_l^{ens}(x^t) = concat(DS(F_{l-m}(x^t)) + F_l(x^t) + UP(F_{l+n}(x^t))) \tag{7}$$

$$\alpha_l^{ens}(x^t) = concat\left(DS\left(\frac{\partial F(x^t)}{\partial F_{l-m}(x^t)}\right) + \frac{\partial F(x^t)}{\partial F_l(x^t)} + UP\left(\frac{\partial F(x^t)}{\partial F_{l+n}(x^t)}\right)\right) \tag{8}$$

where $l - m, l + n$ denote the layers of other two selected feature maps and DS, UP denote resize operation by downsampling and upsampling, respectively. Specially, we use nearest interpolation to adjust the output of different feature layers. So the final ensemble-on-features-based-approach loss is

$$L(x^{adv}, x^t) = F_l^{ens}(x^{adv}) - F_l^{ens}(x^t)_2^2 + \alpha_l^{ens}(x^{adv}) - \alpha_l^{ens}(x^t)_2^2 \tag{9}$$

In this paper, we use I-FGSM to solve above optimization problem. Figure 1 provides a visual overview of the proposed method. Advanced techniques to improve the transferbility can be combined, such as MI-FGSM, DI^2-FGSM and etc. Algorithm 1 summarizes our ensembled algorithm with MI-FGSM.

4 Experiments

In this section, we present the experimental results to demonstrate the effectiveness of the proposed method.

4.1 Experiment Setup

Dataset. The evaluation dataset is LFW and CelebA, which are commonly used to verify the performance of FR models. Specially, for each dataset, we randomly select 1000 pairs of negative face image, which means that the images from the same pair are from different identities and this setting follows [9, 10].

FR Models. To better validate the improvements brought by out method, we select 4 face recognition models, including Facenet, IR152, IRSE50 and MobileFace(MF), as same as [15, 18].

Baseline Attacks. We choose FIM [9], MI-FGSM [12], DI²-FGSM [6] and DFANet [9] as the baseline of attack methods. To make the abbreviation unambiguous, here we use the first character to denote the corresponding combination of methods, except that

Table 1. The success rates of black-box attacks on four FR models, focusing on the effectiveness of feature attention loss term and feature ensemble method. The first column denotes the surrogate model. The 3rd to the 10th columns in the second row denote the target model.

Model	Method	LFW				CelebA			
		Facenet	IR152	IRSE50	MF	Facenet	IR152	IRSE50	MF
Facenet	IFMD [6]	100.0	13.1	38.8	26.5	99.5	12.3	27.9	20.4
	IFMDA [9]	99.7	16.6	39.6	28.8	99.3	13.0	29.6	22.7
	IFMDA-FA	100.0	18.2	41.2	29.5	100.0	14.6	31.3	24.0
	EIFMDA-FA	100.0	**19.7**	**43.5**	**31.6**	100.0	**16.1**	**32.6**	**26.5**
IR152	IFMD [6]	22.3	100.0	90.2	46.1	19.2	100.0	80.1	40.1
	IFMDA [9]	23.7	100.0	91.1	47.5	21.5	100.0	81.7	42.6
	IFMDA-FA	24.2	100.0	93.1	49.6	22.8	100.0	83.8	44.6
	EIFMDA-FA	**27.8**	100.0	**93.3**	**49.7**	**24.1**	100.0	**84.1**	**46.1**
IRSE50	IFMD [6]	23.0	59.8	100.0	96.1	21.4	54.2	100.0	93.2
	IFMDA [9]	25.5	64.2	100.0	97.2	23.0	56.5	100.0	93.4
	IFMDA-FA	28.0	65.1	100.0	98.0	25.3	58.3	100.0	93.9
	EIFMDA-FA	**29.1**	**73.1**	100.0	**98.7**	**25.8**	**63.0**	100.0	**94.6**
MF	IFMD [6]	19.1	19.3	97.7	100.0	18.3	19.3	92.1	100.0
	IFMDA [9]	21.9	22.1	98.2	100.0	20.2	21.9	93.0	100.0
	IFMDA-FA	26.1	27.0	98.5	100.0	23.0	23.3	93.6	100.0
	EIFMDA-FA	**27.1**	**27.7**	**98.7**	100.0	**23.2**	**25.0**	**93.8**	100.0

FMDA denotes the combination of FIM, MI-FGSM, DI^2-FGSM and DFANet. Adding an 'I' before the attack method denotes internal feature space attack. Adding an 'E' before the attack method denotes ensemble attack in Sect. 3.3.

Parameter Settings. The maximum perturbation ϵ is set to 10. Following [9], the iteration N is set to 1000. The learning rate (step size) lr is set to 2/255. For MI-FGSM, we set the decay factor $\mu = 1.0$; for DI^2-FGSM, the transformation probability is set to 1.0 to get the best performance; for DFANet, the drop rate is set to 0.1. For feature space attack, wo choose the same layer, i.e. repeat_3 for Facenet, last layer of 47th block for IR152, last layer of 22th block for IRSE50 and conv_6_dw for MF.

Metrics. ASR(attack success rate) [21] is calculated to compare the performance of different attack methods.

$$ASR = \frac{\sum_i^N sim(F(x_i^s),F(x_i^t)) \geq \delta}{N} \times 100\% \tag{10}$$

Only when the similarity of two face images exceeds a certain threshold, the attack is considered successful, which will be set as the value at 0.001 FAR [15] for each FR model.

4.2 Experimental Results

In this section, we compare our method with the existing state-of-the-art methods. Our method increases black-box attack success rate by a large margin while preserving the white-box attack ability. The results of different attack methods on the two datasets LFW and CelebA are shown in Table 1. It can be seen from the table that our method still maintains a 100% success rate for white-box attacks while the white-box attack performance of IFMD and IFMDA reduced slightly when taking Facenet as the surrogate model. Besides, although IFMDA outperforms IFMD in the black-box settings, our proposed IFMDA-FA in this paper beats the two benchmark methods in all cases by a significant margin. Specially, when MF was set as the surrogate model, IFMDA-FA increased IFMD's attack success rate by an average of 21.21%. In addition, compared with single-layer attack as IFMDA-FA, the multi-layer attack strategy in this paper as EIFMDA-FA also improves the transferbility of face adversarial examples. Taking IRSE50 as a surrogate model, the attack performance has improved by an average of two percentage points, indicating that fusing multi-layer features can further boost the transferability of adversarial examples.

4.3 Experiments for Feature Space Attack

In this section, we show the experimental results of feature space and output space on FR of black-box adversarial attack. We use the output features and internal features of the deep face model to generate adversarial examples to compare the attack performance. For the attack method of the output features, we choose FIM and FMD, then the feature space attack method is an improvement of the two methods, i.e. IFIM and IFMD. The experimental results on the LFW dataset and two FR models are shown in Table 2. The attack performance of feature space is generally better than that of output features under the same constraints, especially when IR152 is selected as the surrogate model, which shows the effectiveness of feature space attack in avoiding overfitting on the deep face models. In fact, there is still a certain similarity in the feature representations of different FR models, which also explains the improvement of the transferbility of face adversarial examples.

Table 2. The success rates of black-box attacks, focusing on evaluating the effectiveness of feature space attack.

	Method	Facenet	IR152	IRSE50	MF
Facenet	FIM [9]	100.0	3.6	9.2	3.5
	FMD [6]	100.0	10.9	25.3	11.8
	IFIM	100.0	6.7	19.3	10.1
	IFMD	100.0	13.1	38.8	26.5
IR152	FIM [9]	3.0	100.0	30.0	5.7
	FMD [6]	10.3	100.0	45.3	17.0
	IFIM	9.5	100.0	74.2	29.5
	IFMD	22.3	100.0	90.2	46.1

4.4 Ablation Study

For the method proposed in this paper, one factor that greatly affects the final result is the l th layer that is determined to obtain the feature map and feature attention. Therefore, we tested different intermediate layers of the FR models to compare the attack performance. We use the IFMDA-FA to generate adversarial examples on the 31-49th layers of IR152's body block and the 6-14th layers of Facenet. The final results are shown in the Fig. 2.

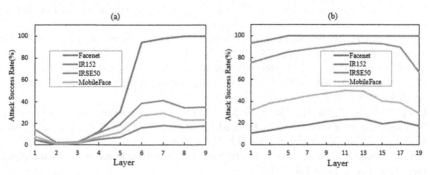

Fig. 2. The success rates under different layer choice. Layers from Facenet(a) and IR152(b) are selected to generate adversarial examples against four FR models. The experiment was conducted on the LFW dataset.

As can be seen from the figure, the choice of feature layer plays an important role in affecting the attack success rate, and different model architecture tend to have inconsistent choice for the best transferbility layer, which is hard to decide and often requires extensive experiments. But it is easy to conclude that examples generated by intermediate layers are more transferable than that by shallow and deep layers. The results in Fig. 2(a) also demonstrate that features extracted by deep layers, which contains more abstract information due to the proximity to the output, i.e. semantic concepts, bring stronger white-box attack ability. Moreover, shallow layers, including more original input pixel information and having smaller receptive fields, are even not suitable for transferable attack.

5 Conclusion

In this paper, we study the transferbility of adversarial examples generated against deep face models. We improve the thinking to shift the focus from output features of the model to internal features, which greatly improves the transferbility. In addition, motivated by the certain similarity of feature maps of different models, we propose to use feature attention to standardize the optimization direction, avoiding overfitting on the surrogate model. Finally, the idea of ensemble attack prompts us to carry out multi-layer feature attack, and we propose to use feature fusion to further enhance the transferility. Extensive experiments show the effectiveness of our method.

References

1. Komkov, S., Petiushko,A.: Advhat: real-world adversarial attack on arcface face id system. In: IEEE Conference on Pattern Recognition (ICPR), pp. 819–826 (2021)
2. Sharif, M., Bhagavatula, S., Bauer, L., Reiter, M.K.: Accessorize to a crime: Real and stealthy attacks on state-of-the-art face recognition. In: Proceedings of the 2016 ACM Sigsac Conference on Computer and Communications Security, pp. 1528–1540 (2016)
3. Yang, L., Song, Q., Wu, Y.: Attacks on state-of-the-art face recognition using attentional adversarial attack generative network. Multimedia Tools Appl. **80**(1), 855–875 (2021)
4. Goodfellow, I.J., Shlens, J., Szegedy, C.: Explaining and harnessing adversarial examples. In: ICLR (2014)
5. Lin, J., Song, C., He, K., Wang, L., Hopcroft, J.E.: Nesterov accelerated gradient and scale invariance for adversarial attacks. In: ICLR (2020)
6. Xie, C., et al.: Improving transferability of adversarial examples with input diversity. In: Proceedings of the IEEE/CVF Conference on Computer Vision and Pattern Recognition, pp. 2730–2739 (2019)
7. Dong, Y., Pang, T., Su, H., Zhu, J.: Evading defenses to transferable adversarial examples by translation-invariant attacks. In: Proceedings of the IEEE/CVF Conference on Computer Vision and Pattern Recognition, pp. 4312–4321 (2019)
8. Liu, Y., Chen, X., Liu, C., Song, D.: Delving into transferable adversarial examples and black-box attacks. In: ICLR (2017)
9. Zhong, Y., Deng, W.: Towards transferable adversarial attack against deep face recognition. IEEE Trans. Inf. Forensics Secur. **16**, 1452–1466 (2020)

10. Xiao, Z., et al.: Improving transferability of adversarial patches on face recognition with generative models. In: Proceedings of the IEEE/CVF Conference on Computer Vision and Pattern Recognition, pp. 11845–11854 (2021)
11. Kurakin, A., Goodfellow, I.J., Bengio, S.: Adversarial examples in the physical world. In: ICLR Workshop, pp. 99–112 (2018)
12. Dong, Y., et al.: Boosting adversarial attacks with momentum. In: Proceedings of the IEEE Conference on Computer Vision and Pattern Recognition, pp. 9185–9193 (2018)
13. Deng, J., Guo, J., Yang, J., Xue, N., Kotsia, I., Zafeiriou, S.: Arcface: additive angular margin loss for deep face recognition. In: Proceedings of the IEEE/CVF Conference on Computer Vision and Pattern Recognition, pp. 4690–4699 (2019)
14. Schroff, F., Kalenichenko, D., Philbin, J.: Facenet: a unified embedding for face recognition and clustering. In: Proceedings of the IEEE Conference on Computer Vision and Pattern Recognition, pp. 815–823 (2015)
15. Yin, B., et al.: Adv-makeup: a new imperceptible and transferable attack on face recognition. In: International Joint Conferences on Artificial Intelligence (2021)
16. Zheng, X., Fan, Y., Wu, B., Zhang, Y., Wang, J., Pan, S.: Robust physical-world attacks on face recognition. Pattern Recogn. **133**, 109009 (2023)
17. Xu, K., et al.: Adversarial t-shirt! evading person detectors in a physical world. In: Proceeding of the European Conference on Computer Vision, vol. 5, pp. 665–681 (2020)
18. Hu, S., et al.: Protecting facial privacy: generating adversarial identity masks via style-robust makeup transfer. In: Proceedings of the IEEE/CVF Conference on Computer Vision and Pattern Recognition, pp. 15014–15023 (2022)
19. Inkawhich, N., et al.: Transferable perturbations of deep feature distributions. In: ICLR (2020)
20. Inkawhich, N., et al.: Feature space perturbations yield more transferable adversarial examples. In: Proceedings of the IEEE/CVF Conference on Computer Vision and Pattern Recognition, pp. 7066–7074 (2019)
21. Deb, D., Zhang, J., Jain, A.K.: Advfaces: adversarial face synthesis. In: IJCB, pp. 1–10 (2020)
22. Long, Y., Zhang, Q., Zeng, B., et al.: Frequency domain model augmentation for adversarial attack. In: Avidan, S., Brostow, G., Cissé, M., Farinella, G.M., Hassner, T. (eds.) Computer Vision–ECCV 2022: 17th European Conference, Tel Aviv, Israel, 23–27 October 2022, Proceedings, Part IV, vol. 13664, pp. 549–566. Springer, Cham (2022). https://doi.org/10.1007/978-3-031-19772-7_32
23. Li, Y., Bai, S., Xie, C., Liao, Z., Shen, X., Yuille, A.: Regional homogeneity: towards learning transferable universal adversarial perturbations against defenses. In: Vedaldi, A., Bischof, H., Brox, T., Frahm, J.-M. (eds.) ECCV 2020. LNCS, vol. 12356, pp. 795–813. Springer, Cham (2020). https://doi.org/10.1007/978-3-030-58621-8_46
24. Sharma, Y., Ding, G.W., Brubaker, M.: On the effectiveness of low frequency perturbations. arXiv preprint arXiv:1903.00073 (2019)
25. Athalye, A., Engstrom, L., Ilyas, A., et al.: Synthesizing robust adversarial examples. In: International Conference on Machine Learning, pp. 284–293. PMLR (2018)
26. Pautov, M., Melnikov, G., Kaziakhmedov, E., et al.: On adversarial patches: real-world attack on arcface-100 face recognition system. In: 2019 International Multi-Conference on Engineering, Computer and Information Sciences (SIBIRCON), pp. 0391–0396. IEEE (2019)

Dendritic Neural Regression Model Trained by Chicken Swarm Optimization Algorithm for Bank Customer Churn Prediction

Qi Wang[1] , Haiyan Zhang[1], Junkai Ji[2] , Cheng Tang[3] , and Yajiao Tang[1(✉)]

[1] College of Economics, Central South University of Forestry and Technology,
Changsha 410004, China
t20060857@csuft.edu.cn
[2] National Engineering Laboratory for Big Data System Computing Technology, Shenzhen
University, Shenzhen 518060, China
[3] Faculty of Information Science and Electrical Engineering, Kyushu University,
Fukuoka 819-0395, Japan

Abstract. Recently, banks are constantly facing the problem of customers churning. Customer churn not only leads to a decline in bank funds and profits but also reduces its credit capacity and affects the bank's operational management. As an important component of Customer Relationship Management, predicting customer churn has been increasingly urgent. Inspired by biological neurons, we build up a dendritic neural regression model (DNRM) with four layers, namely the synaptic layer, the dendritic layer, the membrane layer, and the soma layer for bank customer churn prediction. To pursue better prediction performance in this experiment, the Chicken Swarm Optimization (CSO) algorithm is defined as the training algorithm of DNRM. With the ability to balance exploration and exploitation, CSO is implemented to optimize and improve the accuracy of the DNRM. In this paper, we propose a novel dendritic neural regression model called CSO-DNRM for churn prediction, and the experimental results are based on a benchmark dataset from Kaggle. Compared with other algorithms and models, our proposed model obtains the highest accuracy of 92.27% and convergence speed in customer churn prediction. Due to the novel bionic algorithms and the pruning function of the model, it is evident that our proposed model has advantages in accuracy and computational speed in the field of customer churn prediction and can be widely applied in commercial bank customer relationship management.

Keywords: Dendritic Neural Regression Model · Chicken Swarm Optimization · Customer churn prediction

1 Introduction

Currently, banks are constantly facing the challenge of retaining customers because customers can freely choose to purchase or abandon goods and services that companies provide. Modern banks rely on selling products and services to customers for a living,

B. Luo et al. (Eds.): ICONIP 2023, CCIS 1969, pp. 254–265, 2024.
https://doi.org/10.1007/978-981-99-8184-7_20

which makes customer retention crucial for the healthy development of banks [1, 2]. As an important component of China's economic system, the banking industry plays a crucial role as a financial intermediary in economic development. Recently, as the domestic economy has emerged from the pandemic and gradually recovered, the market environment in the banking industry has improved compared to the period of the pandemic. However, customer churn in banks remains an important issue affecting their credit business and profitability. Due to the continuous loss of customers, financial institutions not only find it difficult to conduct in-depth analysis of rapidly changing customer groups but also have almost no time and cost to interact with a large number of specific customers. Moreover, the non-sharing of customer information between banks makes it difficult for small and medium-sized banks to reduce the cost of retaining existing users. In recent years, attracting new customers costs more than five times more than retaining old ones.

Severe customer churn not only leads to a decline in bank funds and profits but also reduces its credit lending capacity and affects the bank's operational management. With the increasingly fierce market competition, more and more financial institutions value customer management as their most important core means. To some extent, the competition for customer resources also represents the competition for market share. CRM (Customer Relationship Management) is an integrated approach to customer acquisition and retention that uses business intelligence to increase customer value [3]. As an important component of CRM, predicting customer churn has been increasingly urgent. Customer churn is a term that implies customers stop purchasing the goods or services provided by companies for some reason, which has become a significant problem for modern banks [4, 5]. Accurate customer churn prediction helps banks determine whether customers will continue their relationship with banks and convince valuable customers to stay [6].

Recently, more and more scholars have begun to explore churn prediction models with different algorithms. Artificial intelligence is gradually being adopted for customer churn prediction. De Bock, et al. have proposed an ensemble classifier based on the Bagging and Random Subspace Method and implemented generalized additive models (GAMs) as base classifiers. Then they compare GAMensPlus with three ensemble classifiers (Bagging, RSM, and Random Forests) and two individual classifiers (logistic regression, and GAM) and find that GAMensPlus obtains strong classification performance in accuracy and AUC [7]. To identify the churn customers and select features, Ullah, Irfan, et al. implement clustering techniques to propose a churn prediction model. It shows that the model trained by random forest performs well with an accuracy of 88.63% [8]. Inspired by RNNs and CNNs, the bidirectional long short-term memory convolutional neural network (BiLSTM-CNN) model is proposed to improve the prediction accuracy of customer churn rate. The results show that the BiLSTM-CNN model can further improve the prediction performance [9]. Besides, SVM is presented to predict in customer churn and researchers use balancing techniques such as under-sampling, over-sampling, and SMOTE to balance the data [1, 10]. Tsai, et al. used two hybrid models: ANN combined with ANN and SOM combined with ANN for churn prediction. The experimental results show that the two hybrid models outperform the single neural network model in prediction accuracy [11]. Lalwani, Praveen, et al. proposed a methodology consisting of six phases, namely data pre-processing, feature analysis, feature selection, prediction process, K-fold cross-validation, and evaluation process. The results show that Adaboost and XGboost Classifier give the highest accuracy of 81.71% and 80.8% respectively [12]. Other algorithms like random forests [2, 4], decision trees [13], MLP, and the Evolutionary Data Mining Algorithm [14] are also presented to train their churn prediction models. These viewpoints and research findings are enlightening for our research, helping us improve and generate a better model to address customer churn issues. However, most of the above models are complex and difficult to implement on hardware. Therefore, we will propose a relatively simple yet high-precision model to help banks retain customers while reducing costs and improving their profitability.

In financial prediction and classification problems, Wang, Hong, et al. [15] point out that the accuracy obtained by a single best neural network model is sometimes better than that of multiple neural network models. Luo, Xudong, et al. [16] propose a decision tree (DT)-based method for initializing a dendritic neuron model (DNM) and find that the DT-initialized DNM is significantly superior to traditional classification methods such as the original DNM. Tang, Yajiao, et al. [17] propose an evolutionary pruning neural network (EPNN) model trained by the adaptive differential evolution algorithm (JADE) for financial bankruptcy prediction. The results show that the EPNN outperforms the MLP and the pruning neural network model in terms of accuracy and convergence speed. Therefore, we apply a single neural network model in the problem of customer churn prediction in this paper.

To improve the accuracy, robustness, and running speed of the model, we proposed a dendritic neural regression model trained by chicken swarm optimization algorithm (DNRM-CSO) for bank customer churn prediction. Then we compared the following algorithms, namely Back Propagation (BP), Multi-Layer Perceptron (MLP), Brain Storming Optimization (BSO), Support Vector Machine (SVM), and Differential Evolution (DE) with our model. Ultimately, accuracy and convergence rate are implemented as performance metrics to compare the outcomes of different algorithms. After analyzing the results, we found that the proposed model (DNRM-CSO) obtains the best performance in accuracy and convergence speed using the benchmark dataset. Our main contributions are clarified as follows: (1) DNRM-CSO can achieve over 90% accuracy in predicting bank customer churn model, which outperforms many current related studies. (2) Compared to many complex hybrid models, DNRM-CSO has the advantage of being simple, easy to implement, and having a fast computational speed (3) DNRM-CSO can be widely applied to financial prediction and classification problems for different financial institutions. In addition to large banks, small and medium-sized banks can still adopt our model because it can be easily implemented on hardware without reducing its accuracy, which can better save customer management costs for small and medium-sized banks.

The rest of the paper is organized as follows. In Sect. 2, we describe the dendritic neural regression model in detail. In Sect. 3, we explain the methodological underpinnings of CSO and other algorithms. The experimental results will be presented in Sect. 4. Finally, some concluding remarks are given in Sect. 5.

2 Proposed Model

In artificial neural network models, the dendrites of neurons are usually modeled as simple linear structures, and the dendrites weigh the input information and transmit it to the output of the neurons. However, many studies have shown that the function of the dendritic structure of biological neurons is far more complex than simple linear adder. The dendritic structures in biological neurons contain strong local nonlinearity, and synaptic inputs can have nonlinear effects on their neighboring synapses. This feature greatly enhances the role of local nonlinear components in the output of neurons, giving neural networks stronger information processing capabilities.

Inspired by biological neurons, a dendritic neural regression model is built up in this paper with four layers, namely the synaptic layer, the dendritic layer, the membrane layer, and the soma layer. The morphological structure is illustrated in Fig. 1. For the process of the model, the synaptic layer first passes the original signals to every branch. The signals will be multiplied, then the results will be compared with its branch threshold. Thanks to the multiplication operation, many useless dendrites will be removed to improve the calculation speed of the model. Next all dendritic layers will be transferred to the membrane layer by the same sum operation. The soma layer is regarded as the output layer, so the results from the membrane layer will be passed to the soma layer through a sigmoid function.

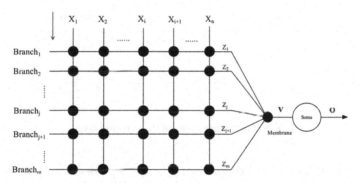

Fig. 1. The architecture of our proposed model.

The synaptic layer is the bridge between other neurons and dendrites, which plays a role in receiving and conducting signals. The function in the synaptic layer is as follows:

$$Y_{ij} = \frac{1}{1 + e^{-k(w_{ij}X_i - q_{ij})}}, \tag{1}$$

where j and X_i denote the j_{th} synaptic layer ($j = 1, 2, 3 \ldots J$) and the i_{th} input signal ($i = 1, 2, 3 \ldots I$) respectively, k is a positive constant, and w_{ij} and q_{ij} are parameters that need to be trained. θ_{ij} is used to describe the threshold of a synaptic layer, which is calculated by the following formula:

$$\theta_{ij} = \frac{q_{ij}}{w_{ij}}, \tag{2}$$

258   Q. Wang et al.

Neural pulses are achieved using specific modes of specific ions. When ions are transported to the receptor, the potential of the receptor changes and determines the excitability or inhibitory characteristics of the synapse. So the threshold θ_{ij} determines the four types of synaptic connections, which are presented in Fig. 2. For direct connection, $0 < q_{ij} < w_{ij}$: for example, $q_{ij} = 0.5$ and $w_{ij} = 1.0$, results in a direct connection. In a word, regardless of the input values x_i, the outputs Y_{ij} always approximate the inputs, namely $Y_{ij} \in (0, 1)$. For inverse connection, $w_{ij} < q_{ij} < 0$: for example, $q_{ij} = -0.5$ and $w_{ij} = -1.0$. Briefly, no matter the values of x_i in $[0, 1]$, Y_{ij} will receive an inverse signal triggered by the input.

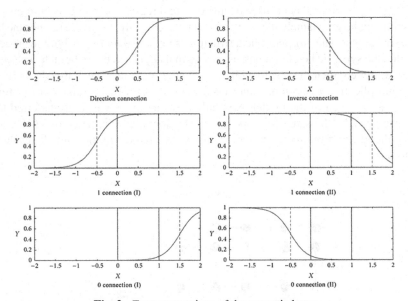

Fig. 2. Four connections of the synaptic layer.

There are two states in the 1 connection: $q_{ij} < 0 < w_{ij}$, for example, $q_{ij} = -0.5$ and $w_{ij} = 1.0$; $q_{ij} < w_{ij} < 0$, e.g., $q_{ij} = -1.5$ and $w_{ij} = -1.0$. In this case, regardless of the input value of x_i, the output value Y_{ij} always approaches 1. There also exist two states in the 0 connection: $0 < w_{ij} < q_{ij}$, e.g., $q_{ij} = 1.5$ and $w_{ij} = 1.0$; $w_{ij} < 0 < q_{ij}$, e.g., $q_{ij} = 0.5$ and $w_{ij} = -1.0$. In this case, regardless of the input value of x_i, the output Y_{ij} always approaches 0. The four types of connections can improve computational speed and accuracy by removing unnecessary synapses and dendrites. For instance, due to multiplication operations, when the calculation result is 0, the entire dendrite will be removed; when the result is 1, the single synapse can also be ignored.

The dendritic layer uses multiplication operations to process the results received from the synaptic layer. A dendrite layer stands for the typical nonlinear interaction of synaptic signals on each branch of dendrites. The whole process can be described as:

$$Z_j = \prod_{i=1}^{I} Y_{ij}. \tag{3}$$

where Y_{ij} represents the output value generated by the j_{th} synapse layer after receiving the i_{th} input signal. Z_j denotes the j_{th} dendrite layer.

The membrane layer connects the dendritic layer and the soma layer. We use sum operations to collect signals processed by the dendritic layer. The corresponding equation is as follows:

$$V = \sum_{j=1}^{J} Z_j. \tag{4}$$

Last, the output from the membrane layer will be transferred to the soma layer. Then the obtained signals will be displayed through a sigmoid function:

$$O = \frac{1}{1 + e^{-k_{soma}(V - \theta_{soma})}}. \tag{5}$$

3 Learning Algorithms

To improve the accuracy of the model, the chicken swarm optimization (CSO) algorithm is used in our DNRM model to train the parameters, namely the weights and thresholds of the synaptic layer. CSO is an optimization algorithm that has high operating efficiency and fast convergence speed [18]. Chicken swarm optimization is a novel bionic swarm intelligence algorithm, which achieves the global optimization effect by simulating the foraging behavior of chickens with hierarchical order [19]. In the CSO algorithm, the whole chicken swarm is divided into several groups based on the size of the fitness value. The flocks of chickens follow the roosters for food, while the chicks forage around the hens. The foraging ability, from strong to weak, is in order of roosters, hens, and chicks. In this hierarchical order, different chicken flocks cooperate and continuously update their positions based on their respective movement patterns until the best foraging position is found. This foraging method not only enhances the local optimization ability of the algorithm but also enhances its global optimization ability.

In a problem of D dimension, we assume N is the population size of the chicken swarm. RN, HN, CN, and MN indicate the size of the roosters, the hens, the chicks, and the mother hens, respectively. On the $j_{th}(j = 1, 2...N)$ search space, the position of $i_{th}(i = 1, 2...N)$ chicken at the t_{th} iteration can be described as $x_{i,j}^t$, and the roosters with better fitness values have priority for food access. So the process can be updated using the following formula:

$$x_{i,j}^{t+1} = x_{i,j}^t * (1 + Randn(0, \sigma^2)) \tag{6}$$

$$\sigma^2 = \begin{cases} 1, & \text{if } f_i \leq f_k, \\ exp(\frac{(f_k - f_i)}{|f_i| + \varepsilon}), & \text{otherwise}, \end{cases} \quad k \in [1, RN], k \neq i. \tag{7}$$

where $Randn(0, \sigma^2)$ is a Gaussian distribution with mean 0 and standard deviation σ^2. ε is a positive constant and $k(k = 1, 2...RN, k \neq i)$ is randomly selected from the roosters group. f stands for the fitness value.

The hens can follow their group-mate roosters to search for food. Therefore, the position update can be calculated as:

$$x_{i,j}^{t+1} = x_{i,j}^t + S1 * Rand * (x_{r1,j}^t - x_{i,j}^t) + S2 * Rand * (x_{r2,j}^t - x_{i,j}^t). \tag{8}$$

$$S1 = exp((f_i - f_{r1})/(abs(f_i) + \varepsilon)). \tag{9}$$

$$S2 = exp((f_{r2} - f_i)). \tag{10}$$

where $Rand$ is a uniform random number in $[0, 1]$, r_1 and r_2 represent for the random rooster and the chicken (rooster or hen) respectively. $(r_1, r_2 = 1, 2...N, r_1 \neq r_2)$ The chicks move around their mother to forage for food, and their corresponding position update equation is as follows:

$$x_{i,j}^{t+1} = x_{i,j}^t + FL * (x_{m,j}^t - x_{i,j}^t.) \tag{11}$$

where $x_{m,j}^t (m \in [1, N])$ is the position of the i_{th} chick's mother, and $FL(FL \in [0, 2])$ is a parameter that means that the chick will follow its mother to find food.

CSO algorithms are mainly constructed by simulating the social behavior of chickens. It can achieve faster convergence speed and higher accuracy in our proposed model with fewer parameters. The initial parameter settings of the DNRM-CSO model are shown in Table 1. The procedure of the CSO algorithm is presented in Fig. 3.

Table 1. The initial parameter settings of our model

Parameter Names	Values
Constant k	5
Data Divide Rate	0.75
Running Times	30
Learning Rate	0.1
Population Size	50
Maximum Iteration	1000

In this paper, we compared other algorithms and models with our proposed model DNRM-CSO. In addition to the CSO algorithm, we also applied algorithms such as BSO, DE, and BP to the DNRM model to demonstrate that our model can achieve higher accuracy and faster convergence and computational speed compared to other evolutionary algorithms and traditional models. The BSO algorithm updates the solution based on the original solution of the population and then finds the optimal solution in the target space. Its main operational processes include initialization, clustering, disrupting class centers, and generating new solutions. Moreover, we also adopted the differential evolution (DE) algorithm to train the parameters. The DE algorithm is a population-based evolutionary algorithm that simulates the process of individual cooperation and competition within a population. It has the advantages of simple principles and fewer parameters that need to be controlled. Its basic process includes initialization, mutation, crossover, and selection. In addition to the algorithms mentioned above, we also compared the accuracy of MLP and SVM with our proposed model in the field of churn prediction. MLP is a popular artificial neural network model applied for prediction, which implements a gradient descent method to adjust the related parameters. SVM uses nonlinear mapping to transform the master data into higher dimensions. As a convex function is required in the optimization process of the SVM, the global minima can be achieved. It can adapt to various kernel functions, such as linear, polynomial, and RBF [20]. We use "SVMlinear", "SVMrbf", "SVMpolynomial" and "SVMsigmoid" to represent four functions used in SVM.

Algorithm 1	Pseudo code of the DNRM-CSO
1:	Initialize the DNRM and the related parameters,
2:	Initialize the population of the chicken swarm, set optimal parameters,
3:	Calculate and rank the chickens' fitness values, re-establish the hierarchy order of the chicken swarm,
4:	**repeat**
5:	Update the positions of roosters by Eq. (6),
6:	Update the positions of hens and chicks by Eq. (8),
7:	Update the positions of chicks by Eq. (11),
8:	Recalculate the fitness value, update the best for each chicken,
9:	**until** The terminal criterion is satisfied, output the optimal results,
10:	The optimal results are sent to the DNRM for training and verification of test data.

4 Application to Bank Customer Churn Prediction

For the statistics used in this paper, we have adopted a benchmark dataset from Kaggle (https://www.kaggle.com/code/psycon/bank-customer-eda-churn/input) to evaluate the performance of our model. The original dataset is composed of more than 10,000 instances based on 23 attributes. These characteristic values include the customer's age, gender, education level, income, marital status, credit limit, etc. We transform the categorical attributes into numerical ones because some learning algorithms require that each sample is expressed as a real number vector. For example, the "Gender" attribute has two labels, namely "M" and "F". And we changed them into "1" and "0" respectively. The first column of data is the client number, which is irrelevant to this study. Therefore, we removed this attribute value, and the same applies to the last two columns of data. In this paper, the state of customer churn is expressed through the binary variable, where "1" represents the "Existing Customer" and "0" denotes the "Attrited Customer". For the missing values in the dataset, we replace them with the average values of the attribute.

The phenomenon of data imbalance always exists in the prediction of bank customer churn. In addition, the huge amount of data and attributes will also make computers overly burdened. There are 8500 existing customers and 1627 lost customers in this dataset, which is considered a data imbalance.

To address this issue and simplify the running process, we randomly select an equal amount of existing customer data and lost customer data before each program run, which accounts for approximately 20% of the total data. We normalize the attributes by the following formula:

$$x_{nomalized} = \frac{x - x_{min}}{x_{max} - x_{min}}, \tag{12}$$

We adapt the classification accuracy rate, sensitivity, and specificity as our performance metrics. Their equations are given below:

$$AccuracyRate = \frac{TN + TP}{TN + TP + FN + FP}\%, \tag{13}$$

$$Sensitivity = \frac{TP}{TP + FN}\%, \tag{14}$$

$$Specificity = \frac{TN}{TN + FP}\%, \tag{15}$$

where TN, TP, FN, and FP represent true positive, true negative, false positive, and false negative respectively. Sensitivity and specificity measure the proportion of true positives to all positives and the proportion of true negatives to all negatives separately. Moreover, the mean squared error is implemented to compare the convergence of different algorithms and models, which is measured by the following equation:

$$MSE = \frac{1}{R} \sum_{a=1}^{R} \left[\frac{1}{N} \sum_{b=1}^{N} (E_{ab} - O_{ab})^2 \right]. \tag{16}$$

where E_{ab} and O_{ab} stand for the predicted output and the actual output respectively. R and N are the running times and the number of instances applied for training.

In this section, we compare DNRM-CSO with other different algorithms and models including BSO, DE, four types of SVM, and BP after processing the original data. For a fair comparison, all the models are equipped with the same parameters and learning rates. As a nonparametric test, the Wilcoxon rank-sum test has been adopted in this paper to detect the significant difference between these algorithms and models. This test allows us to claim that the observed result differences are statistically significant and not simply caused by random splitting effects. Table 2 shows the churn prediction performances of DNRM-CSO and other different algorithms.

As shown in Table 2, our proposed DNRM-CSO model obtains a testing accuracy of 92.27%, which is higher than other algorithms and models. Ranked secondly, SVMrbf obtained an accuracy of 90.34%, which is higher than the 87.77% obtained by MLP and 83.99% obtained by the BSO algorithm. Compared with BSO, DE, MLP, SVM, and BP, our proposed model DNRM-CSO achieves better performance in convergence speed, and the corresponding results are presented in Fig. 4. As it is observed, DNRM-CSO converges very quickly and nearly achieves the best convergence performance at the 70th iteration. It is also obvious that CSO also obtains the highest convergence rate.

Table 2. Experimental results of different algorithms and models

Algorithms/Models	Prediction Accuracy Rate (%)	Sensitivity	Specificity	Wilcoxon rank-sum test (*p-value*)
DNRM-CSO	**92.27**	0.8088	0.9718	N/A
BSO	83.99	0.7647	0.8310	$6.64e^{-06}$
DE	75.54	0	0.9965	$1.73e^{-06}$
MLP	87.77	0.8676	0.8838	0.5846
SVMlinear	85.54	0.8787	0.8768	0.0012
SVMrbf	90.34	0.9007	0.8768	$5.28e^{-05}$
SVMpolynomial	85.54	0.6324	0.8944	$1.73e^{-06}$
SVMsigmoid	85.71	0.8824	0.8486	0.0519
BP	51.08	0	1	$1.72e^{-06}$

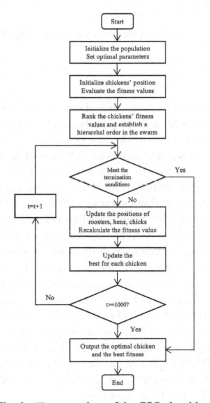

Fig. 3. The procedure of the CSO algorithm.

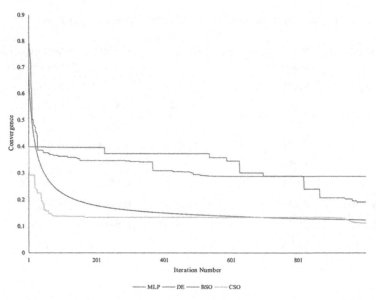

Fig. 4. The convergence curve of the benchmark data using different algorithms.

5 Conclusion

Recently, banks have faced serious customer churn issues. Severe customer churn not only leads to a decline in bank funds and profits but also reduces its credit lending capacity and affects the bank's business model and operational management. To solve the above problems, improve efficiency, and reduce the cost of bank customer relationship management, we introduce a dendritic neural regression model trained by the chicken swarm optimization algorithm, namely DNRM-CSO in this paper. Chicken swarm optimization is an algorithm that has high operating efficiency and fast convergence speed. In contrast with MLP, BP and other algorithms like BSO, DE, and SVM, our proposed model DNRM-CSO performs best in prediction accuracy and convergence speed, which will provide an effective prediction approach for bank customer churn. Besides, due to the novel bionic algorithms and the pruning function of the model, it is evident that DNRM-CSO has advantages in accuracy and computational speed and can be widely applied in commercial bank customer relationship management. Our model also has the following advantages: First, DNRM-CSO can achieve over 90% accuracy in predicting bank customer churn model, which outperforms many current related studies. Second, our model has the advantage of being simple, easy to implement, and having a fast computational speed. Last, DNRM-CSO settles the churn prediction issue efficiently and has universality for different financial institutions. More algorithms and data clustering methods will be analyzed in our future research.

Acknowledgements. This research was supported by the Social Science Fund Project of Hunan Province, China (Grant No. 20YBA260), the Natural Science Fund Project of Changsha, China (Grant No. kq2202297), and Key Project of Hunan Provincial Department of Education: Research on Credit Risk Management of Supply Chain Finance Based on Adaptive Dendritic Neural Network Model (Project No. 22A0178).

References

1. Zhao, X., et al.: Customer churn prediction based on feature clustering and nonparallel support vector machine. Int. J. Inf. Technol. Decis. Mak. **13**(05): 1013–1027 (2014)
2. de Lima Lemos, R.A., Silva, T.C., Tabak, B.M.: Propension to customer churn in a financial institution: a machine learning approach. Neural Comput. Appl. **34**(14): 11751–11768 (2022)
3. Alizadeh, M., et al.: Development of a customer churn model for banking industry based on hard and soft data fusion. IEEE Access 11, 29759–29768 (2023)
4. Xie, Y., et al.: Customer churn prediction using improved balanced random forests. Exp. Syst. Appl. **36**(3), 5445–5449 (2009)
5. Coşer, A., et al.: Propensity to churn in banking: what makes customers close the relationship with a bank? Econ. Comput. Econ. Cybernet. Stud. Res. **54**(2) (2020)
6. Ali, Ö.G., Arıtürk, U.: Dynamic churn prediction framework with more effective use of rare event data: the case of private banking. Exp. Syst. Appl. **41**(17), 7889–7903 (2014)
7. De Bock, K.W., Van den Poel, Dirk.: Reconciling performance and interpretability in customer churn prediction using ensemble learning based on generalized additive models. Exp. Syst. Appl. **39**(8), 6816–6826 (2012)
8. Ullah, I., et al.: A churn prediction model using random forest: analysis of machine learning techniques for churn prediction and factor identification in telecom sector. IEEE Access **7**, 60134–60149 (2019)
9. Liu, Y., et al.: Intelligent prediction of customer churn with a fused attentional deep learning model. Mathematics **10**(24), 4733 (2022)
10. Farquad, M.A.H., Ravi, V., Bapi Raju, S.: Churn prediction using comprehensible support vector machine: an analytical CRM application. Appl. Soft Comput. **19**, 31–40 (2014)
11. Tsai, C.-F., Lu, Y.-H.: Customer churn prediction by hybrid neural networks. Exp. Syst. Appl. **36**(10), 12547–12553 (2009)
12. Lalwani, P., et al.: Customer churn prediction system: a machine learning approach. Computing 1–24 (2022)
13. Höppner, S., et al.: Profit driven decision trees for churn prediction. Eur. J. Oper. Res. **284**(3), 920–933 (2020)
14. Huang, B., Kechadi, M.T., Buckley, B.: Customer churn prediction in telecommunications. Exp. Syst. Appl. **39**(1), 1414–1425 (2012)
15. Wang, H., Xu, Q., Zhou, L.: Large unbalanced credit scoring using lasso-logistic regression ensemble. PLoS ONE **10**(2), e0117844 (2015)
16. Luo, X., et al.: Decision-tree-initialized dendritic neuron model for fast and accurate data classification. IEEE Trans. Neural Netw. Learn. Syst. **33**(9), 4173–4183 (2021)
17. Tang, Y., et al.: A differential evolution-oriented pruning neural network model for bankruptcy prediction. Complexity **2019**, 1–21 (2019)
18. Zhang, Y., et al.: An improved OIF Elman neural network based on CSO algorithm and its applications. Comput. Commun. **171**, 148–156 (2021)
19. Meng, X., et al.: A new bio-inspired algorithm: chicken swarm optimization. In: Advances in Swarm Intelligence: 5th International Conference, ICSI 2014, Hefei, 17–20 October 2014, Proceedings, Part I 5. Springer (2014)
20. Tang, Y., et al.: A survey on machine learning models for financial time series forecasting. Neurocomputing **512**, 363–380 (2022)

BERT-LBIA: A BERT-Based Late Bidirectional Interaction Attention Model for Legal Case Retrieval

Binxia Yang, Junlin Zhu, Xudong Luo(✉), and Xinrui Zhang

Guangxi Key Lab of Multi-Source Information Mining & Security,
School of Computer Science and Engineering, Guangxi Normal University,
Guilin 541004, China
luoxd@mailbox.gxnu.edu.cn

Abstract. Most legal case retrieval methods rely on pre-trained language models like BERT, which can be slow and inaccurate. Alternatively, representation-based models provide quick responses but may not be the most accurate. To address these issues, our paper proposes a BERT-based late bidirectional interaction attention model for similar legal case retrieval. We use a dual BERT model as our backbone network to obtain feature representations of a query and its case candidates. Then, we develop a bidirectional interaction attention network to generate deep interactive attention signals between the query and its corresponding case candidates. Our experiments show that our model is faster and more accurate than existing retrieval models.

Keywords: Legal case retrieval · BERT · Representation-based · Attention mechanism

1 Introduction

In recent years, the rapid digitisation of legal cases and advancements in artificial intelligence (AI) have led to significant progress in the field of LCR. The ability to efficiently retrieve relevant legal cases has become a crucial task in judicial intelligence [9], enabling legal professionals to access precedents [40], analyse legal arguments [15,37], and make informed decisions [30,31,41].

Traditionally, Legal Case Retrieval (LCR) has relied on manual processes and keyword-based searches. However, these methods often suffer from limitations such as incomplete or imprecise search results, making it challenging to find the most relevant cases. To address these limitations, researchers have turned to machine learning and natural language processing techniques to develop automated approaches for LCR.

In particular, similar LCR can be perceived as a type of document retrieval, and numerous document retrieval methodologies based on Pre-trained Language Models (PLMs) have been developed utilizing interaction-based fine-tuning architectures. For instance, Ding et al. [7] and Li et al. [14] select the

ⓒ The Author(s), under exclusive license to Springer Nature Singapore Pte Ltd. 2024
B. Luo et al. (Eds.): ICONIP 2023, CCIS 1969, pp. 266–282, 2024.
https://doi.org/10.1007/978-981-99-8184-7_21

most relevant key blocks from extensive document candidates, form shorter document candidates, and then use BERT [5] to learn the joint representation of the query and critical blocks. The degree of similarity between a query and a document is calculated based on these fine-tuned BERT representations. Similarly, Shao et al. [26] divide a query and its document candidates into multiple paragraphs, input these into BERT for fine-tuning, and then calculate the similarity between two legal cases based on paragraph-level semantic relations. Lastly, Xiao et al. [33] developed a PLM known as Lawformer, which is fine-tuned using hundreds of tokens from a query and thousands of tokens from document candidates, subsequently used to calculate the similarity degree between a query and each document candidate. However, these interaction-based approaches require concatenating the query with the candidate documents and then feeding them into the whole model for fine-tuning. Therefore, although these methods may offer high accuracy, they operate slowly, thereby limiting their application in scenarios with many case candidates.

Representation-based fine-tuning architectures have been explored as alternatives to interaction-based ones, leveraging dual BERT models as encoders for both a query and its case candidate documents. Reimers et al. [23] introduced Sentence-BERT, which utilities two BERT models sharing parameters to model queries and candidate sentences separately, learning meaningful sentence embeddings and significantly reducing retrieval time through pre-caching of candidate sentence embeddings. However, their method's prediction accuracy may need to be improved due to a discrepancy between the classification objective function used for training and the cosine similarity computation performed during prediction. On the other hand, Samuel et al. [10] proposed the Poly-encoder, a transformer architecture that learns global self-attention features rather than token-level ones. While efficient in information retrieval, the absence of deep bidirectional interaction information generation due to the one-sided consideration of information in its dual-encoder system could impact the method's accuracy.

To harness the benefits of both interaction-based and representation-based information retrieval methods and address their disadvantages, in this paper, we propose a hybrid LCR model based on BERT, Attention, and Feedforward Neural Network. BERT is a state-of-the-art PLM that has demonstrated remarkable performance in various natural language processing tasks [28]. By leveraging BERT's contextualised word embeddings and powerful representation learning capabilities, we aim to enhance the accuracy and efficiency of LCR.

Specifically, our model processes a query and its associated case candidates through different encoders to produce feature representations. Acknowledging that independent encoding does not yield optimal interaction information between the query and case candidates, we have developed a bidirectional interaction network that fosters profound attention between them. This network uses a Q2C Attention module and a Self Attention module to incorporate information from the query case to its case candidate and vice versa. Subsequently, the system calculates the similarity degree between the query and each case candidate, ranking them to pinpoint the most analogous one. Our experimental

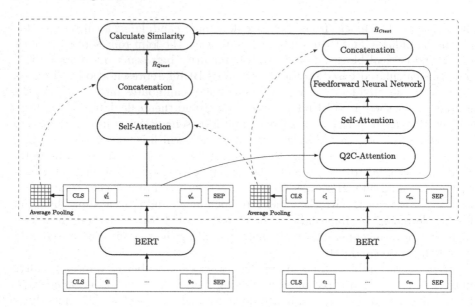

Fig. 1. The overall architecture of our model

results underscore the remarkable potential of representation-based fine-tuning architectures. By crafting a suitable signalling interaction network based on the dual BERT model for a query and its case candidates, we achieved performance nearly akin to the interactive model while preserving the swift retrieval benefits of the representation model. Our proposed model outperforms existing models in similar case retrieval.

The key contribution of our proposed model lies in its ability to integrate the strengths of interaction-based and representation-based approaches for LCR. Interaction-based models typically offer high accuracy by capturing the semantic interactions between a query and candidate documents. On the other hand, representation-based models excel in fast retrieval by generating efficient document representations. By incorporating a late bidirectional interaction attention mechanism into the BERT framework, our model enables deep interactive attention signals between a query and its case candidates while maintaining the advantages of fast retrieval. In summary, our model offers a novel approach to LCR by combining the power of BERT's contextualised representations, bidirectional interaction attention, and fast retrieval capabilities.

The rest of the paper is organized as follows. Section 2 details our model. Section 3 experimentally analysis our model compared with several baseline methods. Section 4 compares our work with the related work. Finally, Sect. 5 concludes this paper and suggests future work.

2   Model

In this section, we will present the definition of our LCR task and explain the details of our model for the task.

2.1 Task Definition

In the legal case retrieval task, the objective of a given query case is to retrieve case candidates that are relevant to the query from a set of case candidates. Formally, given a case query q and a set of case candidates $C = \{c_1, ..., c_m\}$, the task is to retrieve all the cases relevant to q, i.e.,

$$R = \{r_k \mid k = 1, \cdots, m; r_k \in C \land Relevance(r_k, q)\}, \tag{1}$$

where $Relevance(r_k, q)$ denotes that r_k is a case relevant to q (in at least one respect [17]).

2.2 Model Architecture

Figure 1 shows the architecture of our model. The core of our model is to add a late bidirectional interaction attention network to the dual BERT architecture that addresses the inherent defects of the dual BERT. The intuition for this idea is that the dual BERT architecture encodes the query and candidate descriptions separately and thus cannot generate interactions to produce high-quality representations. Our model contains the following three main components.

1) *Query and Candidate Encoder.* We use a dual-encoder structure similar to most previous work [10,12,23], which can obtain excellent retrieval efficiency. Specifically, we use two BERT models as encoders for query and case candidates to encode them separately to generate text representations. This structure can cache the case candidate representations during inference, improving the inference speed. Formally, we construct a query case as a string $Qtext = \{CLS, q_1, \cdots, q_n, SEP\}$, and a case candidate to form another string $Ctext = \{CLS, c_1, \cdots, c_m, SEP\}$, where n, m are the lengths of query and case candidates, respectively, and CLS and SEP are two special tokens in BERT. We then input $Qtext$ and $Ctext$ into the BERT model to obtain the text vector matrices $E_{Qtext} \in \mathbb{R}^{n \times d}$ and $E_{Ctext} \in \mathbb{R}^{n \times d}$ for the query and case candidate, respectively, where $d = 768$ is the dimension of the BERT output.

2) *Q2C Attention Module.* For the sequence matrices E_{Qtext} and E_{Ctext}, we learn the interactions between them by the Query-to-Candidate (Q2C) Attention network. This module computes the mutual attention between each pair of words in the two sequences so that the two sequences interact globally. Similar to [42], the Q2C Attention module uses the standard attention structure, which can fuse word features from query sequences into candidate sequences. Formally, let

$$E_{Qtext} = \{e_q^i \mid i = 1, \cdots, n\}, \tag{2}$$
$$E_{Ctext} = \{e_c^j \mid j = 1, \cdots, m\}, \tag{3}$$

then the similarities between each word in the query sequence and each word in the candidate sequence is calculated by matrix multiplication ($\cdot$ operation), i.e.,

$$R = E_{Ctext} \cdot E_{Qtext}^T, \tag{4}$$

which is a real-number matrix of $n \times m$. Sequentially, we normalize each row and column of R by softmax function to obtain two attention weight matrices $\bar{R}_q \in \mathbb{R}^{n \times m}$ and $\bar{R}_c \in \mathbb{R}^{n \times m}$, respectively:

$$\bar{R}_q = \text{softmax}_{row}(R), \tag{5}$$
$$\bar{R}_c = \text{softmax}_{col}(R). \tag{6}$$

Next, we calculate two attention similarity matrices $\bar{A}_q$ and $\bar{A}_c$ through $\bar{R}_q$ and $\bar{R}_c$ as follows:

$$\bar{A}_q = \bar{R}_q \cdot E_{Qtext}, \tag{7}$$
$$\bar{A}_c = \bar{R}_q \cdot \bar{R}_c^T \cdot E_{Ctext}. \tag{8}$$

Finally, we obtain the final candidate sequence representation fused with query sequence features through the concatenation operation:

$$S_{Ctext} = CAT(E_{Ctext}, \bar{A}_q, E_{Ctext} \odot \bar{A}_q, E_{Ctext} \odot \bar{A}_c), \tag{9}$$

where $\odot$ denotes element-wise multiplication and $S_{Ctext} \in \mathbb{R}^{n \times 4d}$.

3) *Self-Attention Module.* For the sequence matrix E_{Qtext} output by query encoder, we use a self-attention mechanism [36] to map E_{Qtext} into a fixed-size embedding.[1] Before computing the self-attention of a query sequence, we need to integrate the candidate sequence features into E_{Qtext} to obtain the initial query representation:

$$\bar{E}_{Qtext} = \{\bar{e}_q^i \mid i = 1, \cdots, n\}, \tag{10}$$

where

$$\bar{e}_q^i = e_q^i \odot C_{avg}, \tag{11}$$

$$C_{avg} = \frac{1}{m} \sum_{j=1}^{m} e_c^j, \tag{12}$$

where m is the number of the words in the text. Thus, from the initial query representation matrix $\bar{E}_{Qtext}$, we have:

$$H_{Qtext} = \sum_{i=1}^{n} \beta_i \odot e_q^i, \tag{13}$$

where

$$\beta = \text{softmax}(W_2(\tanh(W_1\bar{E}_{Qtext}^T + b_1)) + b_2), \tag{14}$$

where $W_1, W_2 \in \mathbb{R}^{d \times d}$ are two trainable weight matrices, and b_1 and b_2 are biasing. Finally, we obtain the pooled representation of query cases $H_{Qtext} \in \mathbb{R}^d$.

---

[1] Self-attention mechanism computes linear combinations to assign higher weights to the more important words in a sequence. Therefore, it can extract the important word features in a sequence to represent the sequence properly.

Similarly, for the sequence matrix S_{Ctext} output from the Q2C Attention module, we use the same way to obtain the pooled representation of case candidates $H_{Ctext} \in \mathbb{R}^{4d}$ and then input H_{Ctext} into a Feedforward Neural Network (FFN) to map the high-dimensional vector $(4d)$ mapped into a low-dimensional vector (d):

$$\hat{H}_{Ctext} = \text{FFN}(H_{Ctext}) \in \mathbb{R}^d. \tag{15}$$

Finally, we obtain the final representation of the query and case candidates through the concatenation operation:

$$\bar{H}_{Qtext} = \text{CAT}(Q_{avg}, H_{Qtext}), \tag{16}$$

$$\bar{H}_{Ctext} = \text{CAT}(C_{avg}, \hat{H}_{Ctext}), \tag{17}$$

where

$$Q_{avg} = \frac{1}{n} \sum_{i=1}^{n} e_q^i. \tag{18}$$

2.3   Train and Test

Let $Query = \{q_i \mid i = 1, \cdots, N\}$ be the set of all query cases, and N denotes the number of query cases, $Candidate = \{C_i^j \mid i = 1, \cdots, N; j = 1, \cdots, M\}$ be the set of all case candidates, and M denotes the number of case candidates corresponding to a query case. Then the training set contains $N \times M$ samples, where each sample is a query-candidate pair (i.e., $\langle q, c \rangle$). Then, we divide the $N \times M$ sample pairs into two parts: all positive sample pair sets Ω_{pos} and all negative sample pair sets Ω_{neg}. For any positive sample pair $\langle q_i, C_i^j \rangle \in \Omega_{pos}$ and negative sample pair $\langle q_k, C_k^l \rangle \in \Omega_{neg}$, we want to obtain

$$\cos(q_i, C_i^j) > \cos(q_k, C_k^l). \tag{19}$$

where $\cos(\cdot)$ denotes the cosine similarity between a query and case candidates. In other words, we want the similarity scores of positive sample pairs to exceed those of negative sample pairs. Therefore, we can use the following objective function:

$$\mathcal{L} = \ln \left(1 + \sum_{sim(q_i, C_i^j) > sim(q_k, C_k^l)} e^{\gamma(\cos(q_k, C_k^l) - \cos(q_i, C_i^j))} \right), \tag{20}$$

where $\gamma > 0$ is an adjustable hyperparameter and $sim(\cdot)$ denotes the golden similarity between query and case candidates.

At test time, we first calculate the cosine similarity score Y_{cos} between a query and the candidate sequence, then ranks them according to the Y_{cos} from highest to lowest.

3   Experiments

This section will evaluate our model experimentally.

3.1   Datasets and Evaluation Metrics

In the evaluation experiments, we use the LeCaRD dataset, which is developed specifically for legal case retrieval tasks in [17]. The dataset contains 107 query cases and 10,718 case candidates, with each query case corresponding to 100 case candidates. It also includes an additional 18 case candidates. Each sample in the LeCaRD dataset is in the form of a triple $\langle q, c, r \rangle$, where q signifies the query case, c is the case candidate, and $r \in \{0, 1, 2, 3\}$ denotes the similarity score between the query q and case candidate c (the higher the value of r, the greater the similarity between q and c). We split the LeCaRD dataset into training, validation, and test sets at 6:2:2.

Legal case retrieval is to retrieve the top-k most relevant cases from the pool of case candidates. So, we choose Precision (P@k), Mean Average Precision (MAP), and Normalized Discounted Cumulative Gain (NDCG@k) as evaluation metrics to evaluate the performance of a model.

3.2   Baseline Models

We compare our model with some well-known baseline models for Information Retrieval (IR), including:

- TF-IDF [18]: It is a statistical measure dedicated to IR used to assess a word's relevance to a document in a document collection.
- BM25 [25]: It is a standard baseline commonly used for IR to estimate the relevance of a document to a given query.
- Sentence-BERT (SBERT) [23]: It is a representation-based fine-tuned architecture model that uses siamese or triplet network structures to generate semantically meaningful sentence embedding vectors. Moreover, it can pull the sentence embedding distance of semantically similar closer so that it computes the similarity of two sentences.
- OpenCLaP [38]: It is a PLM customized to the Chinese legal domain. It performs well in legal case retrieval because it uses 6.63 million Chinese criminal law documents, pre-trained based on BERT-base [5].
- Lawformer [33]: It is a PLM based on the Longformer [2] structure, which combines local sliding window attention and global task-driven global attention to capture long-range dependencies. Thus, it can handle long Chinese legal documents with thousands of tokens.
- BERT-PLI [26]: It captures paragraph-level semantic relations via BERT and then aggregates these paragraphs to generate interactions and compute correlations between two cases. It has proven to have outstanding performance in legal case retrieval tasks.

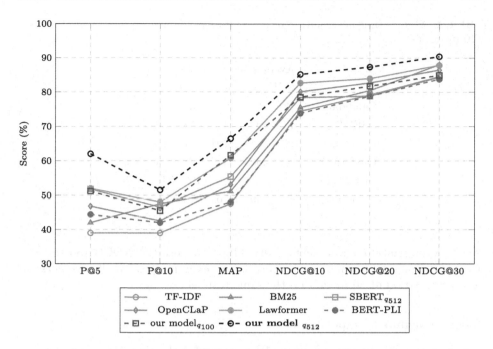

Fig. 2. Comparison of our model and other baseline models on LeCaRD dataset. q_{100} and q_{512} indicate that the maximum input length of the query encoder is 100 and 512, respectively. All results are the average of 10 runs on different random number seeds, except for TF-IDF and BM25.

Among the above baselines, TF-IDF and BM25 are classical IR models; Sentence-BERT is a representation-based fine-tuning architecture model; and OpenCLaP, Lawformer, and BERT-PLI are interaction-based fine-tuning architecture models dedicated to the legal domain.

3.3 Experiment Setting and Implementation

We choose the BERT model proposed in [4] as our backbone and use the deep learning framework PyTorch [21] to implement our model. Specifically, we use the BERTAdam optimiser to train our model with an initial learning rate set to 1e-5 and a batch size of 16. In addition, we set the maximum input length to 512 for both query and candidate encoders, and input sequences less than 512 is padded with [PAD] token.

For classical IR models TF-IDF and BM25, we use the Gensim package to implement it with default parameters. For all neural network baseline models, we set them to the default parameters in the original paper or implementation and train them on our training set. For example, for Lawformer, due to the limitation of GPU resources, we set the maximum input length of the query and case candidates to 300 and 721, respectively, and select all tokens of the query case to perform the global attention mechanism. Besides, we train them in the

Table 1. Average time in seconds to retrieve k case candidates on the LeCaRD dataset (10 retrievals).

| | Average scoring time (s) | | | | | |
	CPU			GPU		
Candidates	1k	5k	10k	1k	5k	10k
SBERT	14	19	24	3	4	5
OpenCLaP	172	814	1701	11	43	84
Lawformer	966	4669	9472	38	116	213
BERT-PLI	-	-	-	109	537	1075
our model$_{q512}$	41	149	276	4	6	8

same experimental environment (Tesla A100) and use the NDCG@30 score of the validation set to achieve early stops for all experiments. Finally, we perform a fair evaluation of all models on the same test set.

3.4   Benchmark Experiment

We show the performance of our model compared with other baseline models on the LeCaRD test set in Fig. 2, regarding metrics such as Precision (*i.e.*, P@5, P@10), Mean Average Precision (*i.e.*, MAP) and Ranking (*i.e.*, NDCG@10, NDCG@20, NDCG@30). From Fig. 2, we can see that in our model, when we set the maximum input length of the query encoder to 100, the model's performance is inferior to that of the interaction-based model. In other words, if our model and the interaction-based architecture model input the same information, then our model is weaker than the interaction-based model in terms of interaction on query and case candidates. However, when we set the maximum input length of the query encoder in our model to 512, its performance improved significantly, achieving the best scores on P@5, P@10, MAP, NDCG@10, NDCG@20, and NDCG@30, which are 10%, 5%, 11%, 6.79%, 8.49%, and 6.06% respectively higher than the representation-based model, and 10%, 3.5%, 5.63%, 3.5%, 3.39% and 2.54% respectively higher than the interaction-based best baseline model.

Table 1 shows the average time it takes for the deep learning models to retrieve k case candidates on the LeCaRD dataset. From the table, we find that the retrieval time of our model is much shorter than that of the interaction-based model, and the time difference becomes more evident as the number of case candidates increases.

3.5   Ablation Experiment

To investigate the effectiveness of the different components used in our model on the model performance, we conducted a series of ablation experiments. Figure 3 shows the results of the ablation experiment. From the figure, we can see:

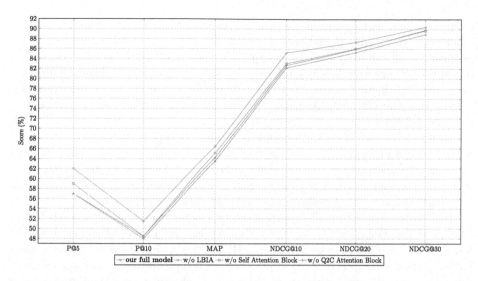

Fig. 3. Ablation tests on the test set. The w/o indicates the removal of a component from our model. For example, w/o LBIA denotes removing the late bidirectional interaction attention network from our model. Other cases are similar.

- Without the late bidirectional interaction attention network, there is a significant degradation in performance compared to the whole model.
- Without the self-attention module, the model cannot generate self-attention and could not fuse features from case candidates. Eventually, its performance is also degraded relative to the whole model.
- Without the Q2C attention module, the model cannot generate enough interaction information in the query and case candidates, and its performance also suffers.

Therefore, the late bidirectional interaction attention network is valuable and practical for the interaction between queries and case candidates.

4   Related Work

The section will compare our work with the related work to show how our work advances state of the art in the field.

4.1   Long Document Retrieval

In LCR tasks, the adjudicative documents involved usually contain long textual content (on average more than 6,000 words in length). Therefore, a LCR actually is a long document retrieval task. Long document retrieval has always been a difficult and challenging task in the field of information retrieval, so attracts many researchers. For example, in 2019, Zhu et al. [42] proposed a neural network model for long document retrieval in the healthcare domain, which uses a deep attention mechanism at the word, sentence, and document levels to allow efficient retrieval over documents of different lengths. In 2020, Arnold et al. [1] proposed a distributed document representation method based on contextual discourse vectors for efficient retrieval of answers from long medical documents. However, the document lengths that their method can handle are around 1000 words, and the efficiency and effectiveness of these methods decreases as the document length increases. The long documents in the legal field are more than six times as long as they are, making it difficult to apply these retrieval methods directly to long legal document retrieval. Moreover, none of them take into account information such as the specific structure and legal expertise in the adjudication documents. However, in this paper we proposed a model for retrieving long legal documents, considering the futures of such documents.

Some researchers have also proposed long document retrieval methods in the legal domain. For example, in 2020, Shao et al. [26] proposed a model of this kind, called BERT-PLI. The model uses BERT to capture semantic relationships within paragraphs and then determines the relevance between two cases by combining paragraph-level interactions. To make it relevant to the legal domain, the model has been fine-tuned using a small-scale case law entailment dataset. However, our model are trained on a much larger dataset and our experiment show our model outperform it significantly. In 2022, Nguyen et al. [19] proposed a general model using deep neural networks with attention mechanisms for statute law document retrieval. The model consists of two hierarchical architectures (Attentive CNN and Paraformer) with sparse attention to representing long sentences and articles. However, their model is not special for LCR, but ours does.

4.2 Neural Information Retrieval

Neural IR (NIR) is the process of using deep neural networks to learn how to match queries and candidates. Currently, some researchers have proposed many NIR methods for applications in various fields, such as legal IR, cross-language retrieval, question answering, and adversarial IR. These methods can usually divide into two types: Interaction-based networks and Representation-based networks.

Interaction-based networks use the dynamic interactions between queries and documents to capture relevance signals and improve information retrieval by explicitly modelling the dependencies and relationships between them.

Therefore, many researchers have developed advanced interactive IR methods on the base of this kind of network structure. For example, in 2020, Wu et al. [32] proposed a context-aware Passage-level Cumulative Gain (PCG) approach, aggregating passages' relevance scores. Then a query and its candidate paragraph are merged into the BERT model for learning their interaction information. In 2020, Jiang et al. [11], and in 2021, Yu et al. [34] used BERT in a cross-linguistic task to learn the representation of interactive information. However, the weakness of the interaction-based network is that it cannot learn the embedding representation of queries and candidate texts individually. Therefore, for an input query, it needs to recompute each document representation in the candidate set, which is an unbearable computational effort in large-scale data retrieval.

Representation-based networks leverage deep learning techniques to learn meaningful and semantically rich representations of textual data, enabling automated feature extraction, enhanced semantic understanding, and adaptability for improved information retrieval performance.

So, also many researchers study the methods of this kind. For example, in 2013, Huang et al. [8] proposed a deep structured semantic model, which uses a multilayer feedforward neural network to learn the semantic representation and calculate the similarity by cosine similarity. In 2014, Shen et al. [27] improved the model structure using convolutional neural networks based on DSSM. In 2016, palangi et al. [20], on the other hand, used LSTM encoders to generate embedded representations of queries and documents. In 2019, Reimers et al. [24] proposed a Sentence-BERT model that uses two BERT models with shared parameters to model queries and candidate sentences, but may suffer from inconsistent training and prediction. In 2021, Wan and Ye [29] proposed a twin CNN model based on TinyBERT to compute the similarity of judgments, but the representational-based network may improve retrieval efficiency at the expense of retrieval accuracy. In 2022, Costa and Pedrosa [3] proposed a model of legal document representation for legal document retrieval. The model uses weighted histograms of concepts generated from word distances and similar terms in a thesaurus. The process involves creating concepts by grouping word vectors from a neural network model. Then, the frequency of these concept clusters is used to represent document vectors. These approaches all attempt to capture the semantic relationships between queries and texts and compute the similarity between them through deep neural networks. However, representational-based networks sacrifice retrieval accuracy while improving retrieval efficiency.

In general, both network models have their advantages. The interaction-based network allows for better aggregation of relationships between queries and candidate documents, resulting in a richer document representation, but at the price of retrieval efficiency. In contrast, the representation-based network has a higher retrieval efficiency and its potential would be further unlocked if it could be further improved to obtain higher quality document embedding representations. In this regard, our model combines the advantages of both representational and interactive encoders by incorporating an attention mechanism on the one hand, which allows for better interaction between queries and candidate sentences compared to representational encoders. On the other hand, separate encoding of queries and candidate sentences allows for offline caching of the encoding vectors of candidate sentences, which provides a significant improvement in matching speed compared to interactive encoders. Moreover, unlike the one-way shallow interaction, our approach performs a deep interaction with two-way information fusion on the query side and the candidate document side. It helps to improve the model's ability to understand and model the semantic relationships between queries and documents, which in turn improves the retrieval performance.

4.3 Legal Case Retrieval

Early work on LCR mainly used word-to-word matching methods (e.g., TF-IDF [18], BM25 [25], and LMIR [35]), which estimate similarity through calculating the number of overlapping words between two cases. The work of Zhong *et al.* [39] in 2019 is a variant of word-to-word matching for LCR. It selects past cases with the maximum number of the exact vital words and the minimum number of different unimportant words as a query case. Their model's strength is that it can explain why closing a case rather than another. However, word-matching based LCR methods cannot identify case documents with high literal overlap but semantic dissimilarity.

Thus, some researchers tried to consider semantics of two cases in word-matching based LCS methods. For example, In 2019, Liu, Luo, and Yang [16] used a semantic and structure-based approach to calculate the similarity between the criminal factual texts of two legal cases, thus enabling the recommendation of similar legal cases. In 2020, Shao *et al.* [26] proposed a paragraph-level semantic interaction model that accomplishes document matching by aggregating the interaction information of paragraphs. In 2021, Xiao *et al.* [33] proposed a new model dedicated to the legal domain that captures the remote dependencies between words to accomplish semantic matching of long documents.

To address the issue that word-matching LCS methods cannot consider the semantics of cases, some researchers are on another way: using neural networks for LCR. For example, in 2019, Ranera, Solano, and Oco [22] used the document embedding technique (Doc2Vec) to calculate the semantic similarity between two cases. In 2021, Dhanani, Mehta, and Rana [6] proposed an LCR model based on graph clustering in order to find semantically similar cases in clustered clusters. However, these methods lack long-range semantic modelling capabilities. In contrast, our model uses the BERT model as the backbone to capture the semantic information of the cases and obtain the long-range modelling capability. In 2022, Zhu, Luo, and Wu [40] devised a BERT-based Chinese Legal Case Retrieval (LCR) system which optimises candidate selection, BERT fine-tuning, and case ranking through the BM25 ranking function, auxiliary learning with pointwise method, and pairwise method respectively, thereby surpassing other methods and securing a second-place finish in a relevant 2021 AI law competition. In 2023, Li and Wang [13] proposed an intelligent legal text analysis and information retrieval model. The model uses Bert and the Attention mechanism to extract semantic information and align element examples with document sentences.

These methods have achieved excellent performance in LCR tasks. However, their methods use an interaction-based model architecture, which suffers from slow inference and excessive resource consumption. Therefore, they are difficult to be applied in practical LCR scenarios. In contrast, our model uses a representation-based model architecture, which has a much faster inference speed than the interaction-based architecture. Besides, we use a late bidirectional interaction network to accomplish the semantic interaction between a query and its case candidates, which can help the lack of precision of the representation-based architecture.

5   Conclusions

This paper presented a late bidirectional interaction attention model based on pertained language model BERT for similar legal case retrieval. The critical point of our work is to add a bidirectional interaction network to the dual BERT model to fuse the feature information of a query and its case candidates. We have conducted extensive experiments on the LeCaRD dataset, showing that our model balances retrieval speed and text representation quality. It is faster than existing interaction-based legal case retrieval models and has better accuracy than representation-based retrieval models. In the future, it is worth studying how to use deeper information about the interaction between a query and its case candidates to generate higher-quality document embedding representations for legal case retrieval.

Acknowledgements. This work was partially supported by a Research Fund of Guangxi Key Lab of Multi-source Information Mining Security (22-A-01-02) and a Graduate Student Innovation Project of School of Computer Science, Engineering, Guangxi Normal University (JXXYYJSCXXM-2021-001) and the Middle-aged and Young Teachers' Basic Ability Promotion Project of Guangxi (No. 2021KY0067).

References

1. Arnold, S., van Aken, B., Grundmann, P., Gers, F.A., Löser, A.: Learning contextualized document representations for healthcare answer retrieval. In: WWW '20: Proceedings of the Web Conference 2020, pp. 1332–1343 (2020)
2. Beltagy, I., Peters, M.E., Cohan, A.: Longformer: the long-document transformer. arXiv preprint arXiv:2004.05150 (2020)
3. Costa, W.M., Pedrosa, G.V.: A textual representation based on bag-of-concepts and thesaurus for legal information retrieval. In: Anais do X Symposium on Knowledge Discovery, Mining and Learning, pp. 114–121. SBC (2022)
4. Cui, Y., Che, W., Liu, T., Qin, B., Wang, S., Hu, G.: Revisiting pre-trained models for Chinese natural language processing. In: Findings of the Association for Computational Linguistics: EMNLP 2020, pp. 657–668 (2020)
5. Devlin, J., Chang, M.W., Lee, K., Toutanova, K.: BERT: pre-training of deep bidirectional transformers for language understanding. In: Proceedings of the 17th Annual Conference of the North American Chapter of the Association for Computational Linguistics: Human Language Technologies, vol. 1, pp. 4171–4186 (2019)
6. Dhanani, J., Mehta, R., Rana, D.: Legal document recommendation system: a cluster based pairwise similarity computation. J. Intell. Fuzzy Syst. **41**(5), 5497–5509 (2021)
7. Ding, M., Zhou, C., Yang, H., Tang, J.: CogLTX: applying BERT to long texts. In: Proceedings of the 34th International Conference on Neural Information Processing Systems, vol. 33, pp. 12792–12804 (2020)
8. Huang, P.S., He, X., Gao, J., Deng, L., Acero, A., Heck, L.: Learning deep structured semantic models for web search using clickthrough data. In: Proceedings of the 22nd ACM International Conference on Information & Knowledge Management, pp. 2333–2338 (2013)

9. Huang, Q., Luo, X.: State-of-the-art and development trend of artificial intelligence combined with law. Comput. Sci. **45**(12), 1–11 (2018)

10. Humeau, S., Shuster, K., Lachaux, M.A., Weston, J.: Poly-encoders: architectures and pre-training strategies for fast and accurate multi-sentence scoring. In: International Conference on Learning Representations (2020)

11. Jiang, Z., El-Jaroudi, A., Hartmann, W., Karakos, D., Zhao, L.: Cross-lingual information retrieval with BERT. In: Proceedings of the Workshop on Cross-Language Search and Summarization of Text and Speech, pp. 26–31 (2020)

12. Khattab, O., Zaharia, M.: ColBERT: efficient and effective passage search via contextualized late interaction over BERT. In: Proceedings of the 43rd International ACM SIGIR Conference on Research and Development in Information Retrieval, pp. 39–48 (2020)

13. Li, B., Wang, M.: Design of intelligent legal text analysis and information retrieval system based on BERT model. https://doi.org/10.21203/rs.3.rs-2994403/v1 (2023)

14. Li, M., Gaussier, E.: KeyBLD: selecting key blocks with local pre-ranking for long document information retrieval. In: Proceedings of the 44th International ACM SIGIR Conference on Research and Development in Information Retrieval, pp. 2207–2211 (2021)

15. Liu, J., Wu, J., Luo, X.: Chinese judicial summarising based on short sentence extraction and GPT-2. In: Qiu, H., Zhang, C., Fei, Z., Qiu, M., Kung, S.-Y. (eds.) KSEM 2021. LNCS (LNAI), vol. 12816, pp. 376–393. Springer, Cham (2021). https://doi.org/10.1007/978-3-030-82147-0_31

16. Liu, Y., Luo, X., Yang, X.: Semantics and structure based recommendation of similar legal cases. In: 2019 IEEE 14th International Conference on Intelligent Systems and Knowledge Engineering (ISKE), pp. 388–395. IEEE (2019)

17. Ma, Y., Shao, Y., Wu, Y., Liu, Y., Zhang, R., Zhang, M., Ma, S.: LeCaRD: a legal case retrieval dataset for Chinese law system. In: Proceedings of the 44th International ACM SIGIR Conference on Research and Development in Information Retrieval, pp. 2342–2348 (2021)

18. Mao, W., Chu, W.W.: Free-text medical document retrieval via phrase-based vector space model. In: Proceedings of the AMIA Symposium, pp. 489–493 (2002)

19. Nguyen, H.T., Phi, M.K., Ngo, X.B., Tran, V., Nguyen, L.M., Tu, M.P.: Attentive deep neural networks for legal document retrieval. Artificial Intelligence and Law, pp. 1–30 (2022)

20. Palangi, H., et al.: Deep sentence embedding using long short-term memory networks: analysis and application to information retrieval. IEEE/ACM Trans. Audio, Speech, Lang. Process. **24**(4), 694–707 (2016)

21. Paszke, A., et al.: PyTorch: an imperative style, high-performance deep learning library. Adv. Neural. Inf. Process. Syst. **32**, 8026–8037 (2019)

22. Ranera, L.T.B., Solano, G.A., Oco, N.: Retrieval of semantically similar Philippine supreme court case decisions using doc2vec. In: 2019 International Symposium on Multimedia and Communication Technology, pp. 1–6 (2019)

23. Reimers, N., Gurevych, I.: Sentence-BERT: sentence embeddings using Siamese BERT-networks. In: Proceedings of the 2019 Conference on Empirical Methods in Natural Language Processing and the 9th International Joint Conference on Natural Language Processing, pp. 3982–3992 (2019)

24. Reimers, N., Gurevych, I.: Sentence-BERT: sentence embeddings using Siamese BERT-networks. In: Proceedings of the 2019 Conference on Empirical Methods in Natural Language Processing and the 9th International Joint Conference on Natural Language Processing, pp. 3982–3992 (2019)

25. Robertson, S.E., Walker, S.: Some simple effective approximations to the 2-poisson model for probabilistic weighted retrieval. In: Proceedings of the 17th Annual International ACM SIGIR Conference on Research and Development in Information Retrieval, pp. 232–241 (1994)

26. Shao, Y., Mao, J., Liu, Y., Ma, W., Satoh, K., Zhang, M., Ma, S.: BERT-PLI: modeling paragraph-level interactions for legal case retrieval. In: Proceedings of the 29th International Conference on International Joint Conferences on Artificial Intelligence, pp. 3501–3507 (2020)

27. Shen, Y., He, X., Gao, J., Deng, L., Mesnil, G.: Learning semantic representations using convolutional neural networks for web search. In: Proceedings of the 23rd International Conference on World Wide Web, pp. 373–374 (2014)

28. Sun, K., Luo, X., Luo, M.Y.: A survey of pretrained language models. In: Knowledge Science, Engineering and Management: KSEM 2022, Lecture Notes in Computer Science, vol. 13369, pp. 442–456 (2022)

29. Wan, Z., Ye, N.: Similarity calculation method of siamese-CNN judgment document based on TinyBERT. In: 2021 International Conference on Intelligent Computing, Automation and Applications, pp. 27–32 (2021)

30. Wu, J., Liu, J., Luo, X.: Few-shot legal knowledge question answering system for Covid-19 epidemic. In: 2020 3rd International Conference on Algorithms, Computing and Artificial Intelligence, pp. 1–6 (2020)

31. Wu, J., Luo, X.: Alignment-based graph network for judicial examination task. In: Qiu, H., Zhang, C., Fei, Z., Qiu, M., Kung, S.-Y. (eds.) KSEM 2021. LNCS (LNAI), vol. 12817, pp. 386–400. Springer, Cham (2021). https://doi.org/10.1007/978-3-030-82153-1_32

32. Wu, Z., Mao, J., Liu, Y., Zhan, J., Zheng, Y., Zhang, M., Ma, S.: Leveraging passage-level cumulative gain for document ranking. In: WWW'20: Proceedings of The Web Conference 2020, pp. 2421–2431 (2020)

33. Xiao, C., Hu, X., Liu, Z., Tu, C., Sun, M.: Lawformer: a pre-trained language model for Chinese legal long documents. AI Open **2**, 79–84 (2021)

34. Yu, P., Fei, H., Li, P.: Cross-lingual language model pretraining for retrieval. In: WWW'21: Proceedings of the Web Conference 2021, pp. 1029–1039 (2021)

35. Zhai, C., Lafferty, J.: A study of smoothing methods for language models applied to ad hoc information retrieval. In: ACM Special Interest Group on Information Retrieval, vol. 51, issue 2, pp. 268–276 (2017)

36. Zhang, T., et al.: Feature-level deeper self-attention network for sequential recommendation. In: Proceedings of the 28th International Conference on International Joint Conferences on Artificial Intelligence, pp. 4320–4326 (2019)

37. Zhang, X., Luo, X.: A machine-reading-comprehension method for named entity recognition in legal documents. In: Neural Information Processing: ICONIP 2022, Communications in Computer and Information Science, vol. 1793, pp. 224–236. Springer, Singapore (2022). https://doi.org/10.1007/978-981-99-1645-0_19

38. Zhong, H., Zhang, Z., Liu, Z., Sun, M.: Open Chinese language pre-trained model zoo. Tech. rep., Tsinghua University (2019). https://github.com/thunlp/openclap

39. Zhong, Q., Fan, X., Luo, X., Toni, F.: An explainable multi-attribute decision model based on argumentation. Expert Syst. Appl. **117**, 42–61 (2019)

40. Zhu, J., Luo, X., Wu, J.: A BERT-based two-stage ranking method for legal case retrieval. In: Knowledge Science, Engineering and Management: KSEM 2022, Lecture Notes in Computer Science, vol. 13369, pp. 534–546. Springer, Cham (2022). https://doi.org/10.1007/978-3-031-10986-7_43

41. Zhu, J., Wu, J., Luo, X., Liu, J.: Semantic matching based legal information retrieval system for COVID-19 pandemic. Artif. Intell. Law, 1–30 (2023). https://doi.org/10.1007/s10506-023-09354-x

42. Zhu, M., Ahuja, A., Wei, W., Reddy, C.K.: A hierarchical attention retrieval model for healthcare question answering. In: WWW'19: The World Wide Web Conference, pp. 2472–2482 (2019)

Learning Discriminative Semantic and Multi-view Context for Domain Adaptive Few-Shot Relation Extraction

Minghui Zhai[1,2], Feifei Dai[1(✉)], Xiaoyan Gu[1,2], Haihui Fan[1], Dong Liu[1], and Bo Li[1]

[1] Institute of Information Engineering, Chinese Academy of Sciences, Beijing, China
{zhaiminghui,daifeifei,guxiaoyan,fanhaihui,liudong,libo}@iie.ac.cn
[2] School of Cyberspace Security, University of Chinese Academy of Sciences, Beijing, China

Abstract. Few-shot relation extraction enables the model to extract new relations and achieve impressive success. However, when new relations come from new domains, semantic and syntactic differences cause a dramatic drop in model performance. Therefore, the domain adaptive few-shot relation extraction task becomes important. However, existing works identify relations more by entities than by context, which makes it difficult to effectively distinguish different relations with similar entity semantic backgrounds in professional domains. In this paper, we propose a method called multi-view context representation with discriminative semantic learning (MCDS). This method learns discriminative entity representations and enhances the use of relational information in context, thus effectively distinguishing different relations with similar entity semantics. Meanwhile, it filters partial entity information from the global information through an information filtering mechanism to obtain more comprehensive global information. We perform extensive experiments on the FewRel 2.0 dataset and the results show an average gain of 2.43% in the accuracy of our model on all strong baselines.

Keywords: Discriminative entity semantic · Multi-view context · Domain adaptive few-shot relation extraction

1 Introduction

Relation Extraction (RE) aims to identify the relation between two specified entities in a sentence, which has been widely used in question answering, knowledge graph construction, and many other applications [1,3,11]. Recently, many works have achieved significant success by training RE models on large amounts of labeled data [14,17,31]. However, it is difficult to obtain sufficient labeled data in real-world scenarios, which limits the performance of RE models. Inspired by few-shot learning, many studies have introduced the few-shot relation extraction (FSRE) task to extract new relations with sparsely labeled

© The Author(s), under exclusive license to Springer Nature Singapore Pte Ltd. 2024
B. Luo et al. (Eds.): ICONIP 2023, CCIS 1969, pp. 283–296, 2024.
https://doi.org/10.1007/978-981-99-8184-7_22

Relation	result_in	classified_as
Sentence	[E1] Human influenza [\E1] is closely related to [E2]viruses [\E2].	A study of [E1] nerve pain [\E1], especially [E2] toothache [\E2].
	The [E1] bacteria [\E1] may cause [E2] infection [\E2].	[E1] Cancer [\E1] is a common malignant [E2] tumor [\E2].

Fig. 1. An example of different relations in the biomedical domain with similar entity semantics, where the entities are mostly related to diseases or biomedicine. Blue and red represent the head and tail entities respectively. Green represents key information in the context that identifies the relation. [E1], [\E1], [E2] and [\E2] are virtual tokens used to highlight entities in existing works. (Color figure online)

instances [6,12,20,29]. This can alleviate the performance degradation caused by data scarcity. However, when new relations come from another domain with different professional knowledge, semantic and syntactic differences severely reduce the ability of the model to extract new relations. Therefore, it becomes crucial to make the model capable of adapting to new domains, and domain adaptive few-shot relation extraction (DAFSRE) becomes an important research direction.

The DAFERE task is mostly handled with the meta-learning framework [5,24,27]. It enables the model to learn meta knowledge from multiple RE tasks sampled in a large-scale dataset (source domain) and apply it to the target domain. Prototype network [25] is an effective algorithm based on meta-learning. It takes class centers of the same class samples as prototypes of relations and classifies unknown samples into their closest prototypes. For example, [7] uses the prototype network to handle DAFSRE tasks and achieves some success. The performance of the prototype network is closely related to the quality of relation prototypes. To improve the quality of relation prototypes, some works learn a large amount of prior knowledge by pre-training, which can help to understand the meaning of relations [10,22,23,26]. However, prior knowledge does not directly relate to relations, which can only provide limited assistance in relation extraction. Therefore, to learn knowledge directly related to relations, some works use the descriptive information of relations to learn high-quality relation prototypes [4,9,19,21,30]. Further, [18] introduces prototype discriminative information to effectively distinguish different relation prototypes.

Albeit much progress, their approaches to learning relational knowledge still limit the adaptation of the model to new domains. Most existing works train models to distinguish different relations using an entity-centric matching paradigm, which focuses on learning entities rather than the context. However, entities in professional domains often have similar semantic backgrounds. In this scenario, models that weaken relational knowledge in context have difficulty distinguishing different relations with similar entity semantics. As shown in Fig. 1, two types of relational entities such as "human influenza" and "nerve pain" are related to disease or biomedicine. The method of using virtual tokens to mark entities weakens the contextual "related to" and "especially" to help understand relations. This makes it more difficult for the matching paradigm to distinguish between the relations "result in" and "classified as".

To solve the above problems, we propose a new model: multi-view context representation with discriminative semantic learning (MCDS). The training process is shown in Fig. 2, we first propose a discriminative semantic learning module to reduce the interference caused by entities with similar semantics. We use prompt templates [13] to extract the semantics of entities from the pre-trained language model and amplify the differences in the semantics of different entities through contrastive learning. This allows the model to obtain discriminative entity representations and thus effectively distinguish different relations with similar entity semantics in professional domains. Second, to enhance the ability of the model to learn relational information in context, we propose a multi-view context representation module. It extracts local and global information from the context to construct multi-view context representation and use it as relation information. This enables the model to enhance the use of relational knowledge in context to distinguish different relations. In addition, it avoids over-focusing on entities by filtering partial entity information from the global information and thus extracts more comprehensive global information to effectively learn the relation in context.

The main contributions of our work are as follows:

- We design a discriminative semantic learning module, which makes the model capable of learning discriminative entity representations to distinguish different relations with similar entity semantics in professional domains.
- We design a multi-view context representation module, which extracts local and global information to enhance the use of relational information in context. Thus, the model can effectively use context to distinguish different relations.
- We propose an information filtering mechanism, which filters partial entity information from the global information to avoid over-focusing on entities and thus extracts more comprehensive global information.
- We construct a large number of experiments on the FewRel 2.0 dataset and demonstrate that our model has significant performance gains on the DAFSRE task.

2   Related Work

Traditional RE models cannot be effectively adapted to scenarios that lack labeled data. Especially in many professional domains, differences in knowledge further degrade the performance of the model. Therefore, DAFSRE has become an important research direction. The prototype network is a widely used algorithm in existing works. For example, Proto-BERT [7] uses the prototype network to classify query instances into their most similar relation prototypes. The performance of the prototype network depends heavily on the quality of relation prototypes. Therefore, to improve the quality of relation prototypes, some works learn the prior knowledge that can help to understand relations through pre-training methods. For example, CP [22] and MTB [26] use entities masked contrastive learning [8,28] to improve the quality of prototypes. ALDAP [23] learns prior knowledge by predicting masked words on the target domain data. LDUR [10] obtains unbiased prototypes by supervised contrastive pre-training and sentence entity-level prototype network.

Prior knowledge is not directly related to relations, thus many works introduce descriptive information of relations to assist in learning prototypes. HCRP [9] uses descriptive information of relations to learn fine-grained prototypes. FAEA [4] uses descriptive information to extract function words thus improving the prototype quality. PRM [19] proposes a gating mechanism that corrects the relation prototype distribution. Further, GTPN [18] introduces prototype discriminative information to distinguish different relation prototypes. Besides these, BERT-PAIR [7] proposes an instance matcher for computing the similarity score between query instance and relation prototype. GMGEN [16] uses topology and graph convolution methods to capture cross-task general knowledge. DaFeC [2] generates pseudo-labels using clustering on unlabeled target domain data to learn knowledge in the target domain directly.

However, existing works have difficulty distinguishing different relations with similar entity semantics. Therefore, we propose a multi-view context representation with discriminative semantic learning method. It learns discriminative entity representations and enhances the use of relational information in context to effectively distinguish different relations with similar entity semantics.

3   Task Definition

Given a source domain dataset D_S and a target domain dataset D_T. D_S contains rich base classes, each with a large number of labeled instances. D_T contains new relation classes, each with only a small number of instances. The new classes and base classes are separated from each other. The DAFSRE aims to train the model in source domain and apply it to target domain. We follow the typical few-shot task setup named N-way-K-shot, which consists of a support set S and a query set Q. Specifically, in each iteration of training we randomly select N classes each with K instances from the base class to form the support set $S = \{s_k^i; i = 1, ..., N, k = i, ..., K\}$. Then we randomly select G instances from the remaining instances of the N classes to form the query set $Q = \{q_j; j = 1, ..., G\}$. The model predicts relation classes of instances in Q so that it can extract different classes of relations. The instances in S and Q consist of (x, e, y), with x denoting the sentence, e consisting of (e_1, e_2) denoting the specified head and tail entities, and y denoting the label of the relation between entities.

4   The Proposed Method

In this section, we describe MCDS in detail. Figure 2 shows the overall framework of MCDS. The inputs are a support set with $N \times K$ sentences and a query set with G sentences. MCDS consists of two parts. The discriminative semantic

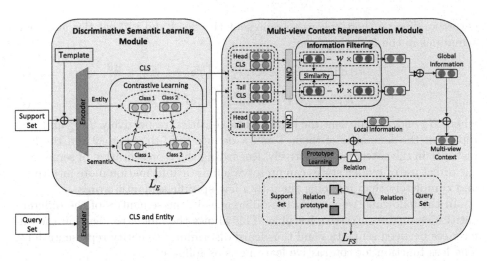

Fig. 2. The overall structure of MCDS. The total loss of the model consists of two parts (L_E and L_{FS}). The "$\oplus$" indicates concatenation, "Head" and "Tail" indicate the head and tail entities in a sentence, and "CLS" indicates the information of the whole sentence.

learning module learns discriminative entity representations by amplifying the differences in the semantics of different entities. The multi-view context representation module extracts local and global information from the context to represent relational information. Meanwhile, the information filtering mechanism filters part of the entity information to avoid over-focusing on entities and thus obtains more comprehensive global information. The above can effectively distinguish different relations with similar entity semantics.

4.1 Discriminative Semantic Learning Module

To distinguish different relations that have similar entity semantic backgrounds more effectively, we use contrastive learning to learn discriminative entity representations by amplifying differences in the semantics of different entities.

To extract the semantics of head and tail entities, we preprocess the sentences before inputting them into the pre-trained language model BERT [15]. First, inspired by the work in MTB [26], we use virtual tokens ([E1], [\E1], [E2], and [\E2]) to mark head and tail entities in the sentence. This can effectively capture entities from sentences. Then, to obtain semantic information of entities, we design the prompt template and splice it at the end of the sentence. The form of the template is "Entity means [M]". We give a complete sentence after preprocessing: "[E1] Newton [\E1] served as the president of [E2] the Royal Society [\E2]. Newton means [M1], the Royal Society means [M2]". At this point, we input sentences into the BERT encoder to get the embedding information.

We use the tokens [E1], [\E1], [E2], and [\E2] to represent entities and the tokens [M1] and [M2] to represent semantics. We combine the hidden states corresponding to these tokens to represent the head and tail entities and their semantics in the following form:

$$H_k^i = [E1] + [\backslash E1] \in \mathbb{R}^d, T_k^i = [E2] + [\backslash E2] \in \mathbb{R}^d, \tag{1}$$

$$HS_k^i = [M1] \in \mathbb{R}^d, TS_k^i = [M2] \in \mathbb{R}^d, \tag{2}$$

where H_k^i and T_k^i represent the head and tail entities, respectively, HS_k^i and TS_k^i represent the semantics of them, i and k denote the i-th category and the k-th instance in the category, respectively. Regarding the representation of semantics, the direct use of hidden states can preserve semantic information more intuitively and completely than mapping them to a few specific semantic words.

In the following, we amplify the differences in the semantics of the different entities through contrastive learning. This allows the model to focus on the differences between semantics and thus learn discriminative entity representations. The loss function for contrastive learning is as follows:

$$L_E = -\frac{1}{K} \times \sum_{k=1}^{K} \sum_{i=1}^{N} \log \left[\frac{\exp\left(d\left(H_k^i, HS_k^i\right)\right)}{\sum\limits_{j=1}^{N} \exp\left(d\left(H_k^i, HS_k^j\right)\right)} \times \frac{\exp\left(d\left(T_k^i, TS_k^i\right)\right)}{\sum\limits_{j=1}^{N} \exp\left(d\left(T_k^i, TS_k^j\right)\right)} \right], \tag{3}$$

where d is the distance function.

4.2   Multi-view Context Representation Module

To enhance the use of relational information in context for distinguishing different relations with similar entity semantic backgrounds, we design a multi-view context representation module. It extracts local and global information from the context and combines them into a multi-view context representation to serve as relational information. Local information indicates the essential semantic connection of head and tail entities, and global information indicates the information related to the entities in the whole sentence.

To extract the local information, we input head and tail entities together into a convolutional neural network (CNN) for mining the essential connections between entity semantics. First we obtain the representation of entities:

$$Head = [E1] \oplus [\backslash E1] \in \mathbb{R}^{2d}, Tail = [E2] \oplus [\backslash E2] \in \mathbb{R}^{2d}, \tag{4}$$

where "$\oplus$" indicates concatenation, "$Head$" and "$Tail$" indicate the head and tail entities respectively. Then the local information LI is extracted by CNN:

$$LI = CNN\left(Head \oplus Tail\right) \in \mathbb{R}^{2d}. \tag{5}$$

In the following, we use the built-in token "CLS" in BERT to represent the information of the whole sentence. Then, it is fed into CNN together with entities to extract the global information related to entities in the sentence:

$$\overline{HGI} = CNN\left(Head \oplus CLS\right) \in \mathbb{R}^d, \tag{6}$$

$$\overline{TGI} = CNN\left(Tail \oplus CLS\right) \in \mathbb{R}^d, \tag{7}$$

where $\overline{HGI}$ and $\overline{TGI}$ are the global information corresponding to the head entity and tail entity, respectively.

To avoid over-focusing on head and tail entities and thus extracting more comprehensive global information, we propose an information filtering mechanism. This mechanism filters out partial information of head and tail entities in the global information according to the similarity between $\overline{HGI}$ and $\overline{TGI}$, and the steps are as follows:

$$HGI = \overline{HGI} - w \times \overline{TGI} \in \mathbb{R}^d, \tag{8}$$

$$TGI = \overline{TGI} - w \times \overline{HGI} \in \mathbb{R}^d, \tag{9}$$

$$w = \frac{\overline{HGI}\left(\overline{TGI}\right)^T}{\left\|\overline{HGI}\right\|_2 \left\|\overline{TGI}\right\|_2}, \tag{10}$$

where w is the similarity, HGI and TGI are the information after filtering. When w is large, it indicates that both $\overline{HGR}$ and $\overline{TGR}$ focus excessively on head and tail entities, which obscures contextual information that can represent relations. At this point we need to filter out partial information about entities to highlight contextual information. After that, we obtain the global information:

$$GI = HGI \oplus TGI \in \mathbb{R}^{2d}. \tag{11}$$

Finally, we combine the local and global information to obtain the multi-view context representation:

$$MC = LI \oplus GI \in \mathbb{R}^{4d}, \tag{12}$$

and combine MC with the entity information to obtain the final relation of each sentence:

$$R = Head \oplus Tail \oplus MC \in \mathbb{R}^{8d}. \tag{13}$$

4.3 Training Objective

To make the model extract different classes of relations, we optimize the model by categorizing relations in the query set into their nearest relation prototypes.

First, we obtain the relation prototype P in the support set by prototype learning:

$$P = \{p_i, i = 1, ..., N\}, \tag{14}$$

$$p_i = \frac{1}{K} \sum_{k=1}^{K} s_k^i \in \mathbb{R}^{8d}, \tag{15}$$

where p_i denotes the prototype of the i-th relation category. Then the loss function of the training objective is defined as the following:

$$L_{FS} = -\sum_{j=1}^{G} \log \frac{\exp\left(d\left(q_j, p_q\right)/\tau\right)}{\sum_{i=1}^{N} \exp\left(d\left(q_j, p_i\right)/\tau\right)}, \tag{16}$$

where q_j is the j-th instance in the query set, p_q represents the relation prototype that the instance belongs to, τ is the temperature hyperparameter, and d is the distance function.

The loss function of our entire model is as follows:

$$L = \lambda_1 \times L_E + \lambda_2 \times L_{FS}, \tag{17}$$

where λ_1 and λ_2 are the weight hyperparameters.

5   Experiments

5.1   Data Set and Experiment Details

We conduct experiments on the widely used FewRel 2.0 dataset, which is the largest dataset available for the domain adaptive few-shot challenge. Its training set is from the general domain (Wiki data), and the validation and test sets are from the biomedical domain. We evaluate our model in four few-shot scenarios: 5-way-1-shot, 5-way-5-shot, 10-way-1-shot, and 10-way-5-shot. Since labels of the test set are not publicly available, we submit the predictions of the test set in the official to obtain the final results. The ablation study is conducted on the validation set.

We use the BERT-base-uncased model as the encoder. The CNN has 4 layers, and the number of convolutional kernels in each layer is 256, 128, 64, and 1. The convolutional kernel size is set to 5. The temperature parameter τ is set to 0.5, and the normalized cosine similarity is used as the distance function. The weight parameters λ_1 and λ_2 in the loss function L are set to 2 and 1, respectively. All hyperparameters are adjusted on the validation set.

5.2   Baselines

We compare MCDS with fifteen baseline methods. Based on the encoder, we divide baselines into CNN-based and BERT-based models. We further divide BERT-based models into two parts according to whether descriptive information of relations is used. Specifically, CNN-based models: Proto-CNN [7], Proto-CNN-ADV [7], DaFec-Proto-CNN [2]. BERT-based models that do not use description information: Proto-BERT [7], BERT-PAIR [7], ALDAP [23], DaFeC [2], MTB [26], CP [22], GTPN [18], LDUR [10]. BERT-based models that use relational description information: HCRP [9], PRM [19], FAEA [4], GMGEN [16].

5.3   Overall Results

Table 1 reports the results of our model compared to baselines on the test set. From the results we can observe that:

Table 1. Accuracy (%) on the FewRel 2.0 test set. Bolded indicates the highest accuracy, and underlined indicates the second highest accuracy.

Encoder	Model	5-way 1-shot	5-way 5-shot	10-way 1-shot	10-way 5-shot	AVG
CNN	Proto-CNN [7]	35.09	49.37	22.98	35.22	35.67
	Proto-CNN-ADV [7]	42.21	58.71	28.91	44.35	43.55
	DaFeC-Proto-CNN [2]	48.58	65.80	35.53	52.71	50.66
BERT	Proto-BERT [7]	40.12	51.50	26.45	36.93	38.75
	BERT-PAIR [7]	67.41	78.57	54.59	66.85	66.85
	ALDAP [23]	55.35	79.01	40.49	66.90	60.44
	DaFeC [2]	61.20	76.99	47.63	64.79	62.65
	MTB [26]	74.70	87.90	62.50	81.10	76.55
	CP [22]	79.70	84.90	68.10	79.80	78.13
	GTPN [18]	<u>80.00</u>	<u>92.60</u>	<u>69.25</u>	<u>86.90</u>	<u>82.20</u>
	LDUR [10]	79.33	91.59	67.48	85.70	81.03
	HCRP [9]	76.34	83.03	63.77	72.94	74.02
	PRM [19]	73.98	88.38	62.72	79.43	76.13
	FAEA [4]	73.58	90.10	62.98	80.51	76.79
	GMGEN [16]	76.67	91.28	64.19	84.84	79.25
	MCDS	**83.75**	**93.49**	**73.15**	**88.10**	**84.63**

- Our model outperforms all strong baselines. In four few-shot task scenarios, our model improves by 3.75, 0.89, 3.90, and 1.20 points, respectively. The average improvement is 2.43 points. This demonstrates the effectiveness of our innovation in learning discriminative entity representations and enhancing the use of context by extracting multi-view context representation. Notably, our model improves by an average of 3.83 points in the 1-shot scenario, which proves that our model is more suitable for tasks with fewer samples.
- MCDS is more effective than GTPN. This is because GTPN uses additional markers to distinguish different relations rather than exploring differences in relations at the semantic level. In contrast, MCDS achieves better results by amplifying differences between the semantics of different entities and thus obtaining discriminative entity representations.
- While MTB and CP use Wikipedia data to pre-train the model, MCDS works better. This is because MCDS extracts both local and global information and thus mines more contextual knowledge. This reflects the high utilization of data by MCDS, which means that more knowledge can be mined from the limited data.
- Models that introduce relational descriptions in the training process, such as HCRP and PRM, do not achieve significant performance breakthroughs. This is because introducing relational descriptions can complicate the model, which leads to overfitting of the source domain relations to some extent in cross-domain tasks.

Table 2. Ablation study of MCDS on the FewRel 2.0 validation set. We use accuracy (%) to evaluate the performance of the model in four task scenarios as well as the average performance.

Model	5-way-1-shot	5-way-5-shot	10-way-1-shot	10-way-5-shot	AVG
MCDS	83.93	93.55	75.35	88.23	85.27
-semantic	82.98	93.20	74.28	87.85	84.56(-0.71)
-local	82.70	93.11	73.93	88.07	84.45(-0.82)
-global	82.26	93.08	73.87	87.78	84.23(-1.04)
-multi	81.08	92.72	72.82	87.16	83.42(-1.85)
-filter	82.43	93.30	73.96	87.95	84.41(-0.86)

5.4   Ablation Study

To analyze the role of each component in MCDS, we perform the ablation study on the validation set of FewRel 2.0 and the results are reported in Table 2.

Analysis of Discriminative Semantic Learning. As shown in Table 2, after removing this module (-semantic), the average performance decreases by 0.71 points. This is because it can learn discriminative entity representations and thus effectively distinguish different relations with similar entity semantics.

To further demonstrate the effectiveness of learning discriminative entity representations, we split the validation set into two parts. Specifically, we put the samples with significant biomedical semantics for both head and tail entities into U_a and the remaining samples into U_b. Compared to U_b, the entities of different relations in U_a have more similar semantic backgrounds to each other. The number of samples in U_a and U_b accounts for 72% and 28% of the validation set, respectively. The datasets U_a and U_b are incomplete, which only supports the validation of the model in a 5-way-1-shot scenario.

In the following, we validate the effect of -semantic and MCDS on the validation set, U_a and U_b, respectively. The results are reported in Table 3. We can observe that this module improves the model by 1.10 points in U_a. This indicates that learning discriminative entity representations can effectively distinguish different relations with similar entity semantics. Moreover, we find that the effect of this module is less enhanced in U_b. This is because entities in U_b have some semantic differentiation, which is consistent with the purpose of this module. Therefore, this weakens the effectiveness of this module to some extent. Finally, we randomly select 5 classes of relations from U_a and visualize their distribution. The results are shown in Fig. 3, we can find that after removing this module, the similar entity semantics leads to unclear boundaries among some relations. When discriminative entity representations are learned through this module, there are obvious boundaries among relations.

Analysis of Multi-view Context Representation. As shown in Table 2, after removing this module (-multi), the average performance decreases by 1.85

Table 3. Accuracy (%) of -semantic and MCDS on the validation set as well as on U_a and U_b.

	Validation set	U_a	U_b
-semantic	82.98	81.98	84.61
MCDS	83.93 (+0.95)	83.08 (+1.10)	85.03 (+0.42)

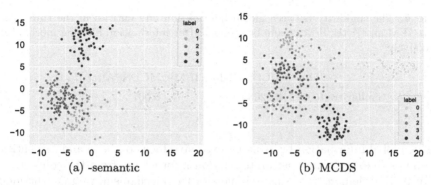

(a) -semantic   (b) MCDS

Fig. 3. The PCA visualization of the sample distribution in -semantic and MCDS models.

points. This is because this module enhances the ability to learn relational information in context thus effectively using the context to distinguish different relations. From Table 2, we can find that the performance decreases when only removing the local (-local) or the global information (-global). This indicates that both types of information are useful. Moreover, the performance of the model is further improved when the two are combined into a multi-view context representation. This suggests that local and global information are mutually reinforcing.

To demonstrate the effectiveness of the module for extracting relation information in context, we perform a visualization experiment. We take the sentence "These tumors are the most common non-epithelial neoplasms of gastric wall" in the target domain as an example. The head and tail entities in the sentence are "tumors" and "non-epithelial neoplasms", and the relation is "classified as ". In the following, we input the sentence into -multi and MCDS. The result is shown in Fig. 4. We can find that the multi-view context representation module makes the model focus on the words "are", "the", and "common". These are the most relevant information for "classified as" in the context. Moreover, the module does not focus on the remaining unimportant information such as "most" and "gastric".

Analysis of Information Filtering Mechanism. As shown in Table 2, after removing the information filtering mechanism (-filter), the average performance decreases by 0.86 points. This is because the mechanism obtains more compre-

Fig. 4. The attention scores of both -multi and MCDS models to relational information in context.

Table 4. The similarity between global information and entities in the three cases. Before IFM and After IFM denote before and after information filtering mechanisms, respectively.

	-filter	MCDS (Before IFM)	MCDS (After IFM)
Similarity	0.0572	0.0208	0.007

hensive global information by avoiding excessive focus on head and tail entities. This helps to extract more accurate relation information from the context.

To further demonstrate the usefulness of the mechanism, we take the model without the information filtering mechanism as a comparison. We calculate the similarity between global information and entities in both -filter and MCDS models, and the results are shown in Table 4. We can find that MCDS can effectively reduce the similarity between global information and entities by the information filtering mechanism. Moreover, the result of MCDS (Before IFM) is also significantly lower than the result of the -filter. This suggests that the process of filtering partial entity information can also constrain the model to focus less on entities once the global information is extracted.

6   Conclusion

In this paper, we propose MCDS, which can efficiently handle DAFSRE tasks. MCDS learns discriminative entity representations by amplifying the semantic differences between entities and enhances the use of relational information in context by extracting multi-view context representation. It also obtains more comprehensive global information through information filtering mechanisms. The above allows for effective distinction between different relations with similar entity semantics in professional domains. Finally, we achieve obvious effect improvement on the FewRel 2.0 dataset and demonstrate the effectiveness of MCDS along with its modules.

Acknowledgements. This work was supported by No. XDC02050200.

References

1. Bach, N., Badaskar, S.: A review of relation extraction. Lit. Rev. Lang. Stat. **II**(2), 1–15 (2007)

2. Cong, X., Yu, B., Liu, T., Cui, S., Tang, H., Wang, B.: Inductive unsupervised domain adaptation for few-shot classification via clustering. In: Machine Learning and Knowledge Discovery in Databases: European Conference, ECML PKDD 2020, Ghent, Belgium, 14–18 September 2020, Proceedings, Part II, pp. 624–639 (2021)

3. Distiawan, B., Weikum, G., Qi, J., Zhang, R.: Neural relation extraction for knowledge base enrichment. In: Proceedings of the 57th Annual Meeting of the Association for Computational Linguistics, pp. 229–240 (2019)

4. Dou, C., Wu, S., Zhang, X., Feng, Z., Wang, K.: Function-words adaptively enhanced attention networks for few-shot inverse relation classification. In: Proceedings of the Thirty-First International Joint Conference on Artificial Intelligence, pp. 2937–2943 (2022)

5. Finn, C., Abbeel, P., Levine, S.: Model-agnostic meta-learning for fast adaptation of deep networks. In: International Conference on Machine Learning, pp. 1126–1135 (2017)

6. Gao, T., Han, X., Liu, Z., Sun, M.: Hybrid attention-based prototypical networks for noisy few-shot relation classification. In: Proceedings of the AAAI Conference on Artificial Intelligence, vol. 33, pp. 6407–6414 (2019)

7. Gao, T., et al.: FewRel 2.0: towards more challenging few-shot relation classification. In: Proceedings of the 2019 Conference on Empirical Methods in Natural Language Processing and the 9th International Joint Conference on Natural Language Processing (EMNLP-IJCNLP), pp. 6250–6255 (2019)

8. Gao, T., Yao, X., Chen, D.: SimCSE: simple contrastive learning of sentence embeddings. In: Proceedings of the 2021 Conference on Empirical Methods in Natural Language Processing, pp. 6894–6910 (2021)

9. Han, J., Cheng, B., Lu, W.: Exploring task difficulty for few-shot relation extraction. In: Proceedings of the 2021 Conference on Empirical Methods in Natural Language Processing, pp. 2605–2616 (2021)

10. Han, J., Cheng, B., Nan, G.: Learning discriminative and unbiased representations for few-shot relation extraction. In: Proceedings of the 30th ACM International Conference on Information & Knowledge Management, pp. 638–648 (2021)

11. Han, J., Cheng, B., Wang, X.: Two-phase hypergraph based reasoning with dynamic relations for multi-hop KBQA. In: IJCAI, pp. 3615–3621 (2020)

12. Han, Y., et al.: Multi-view interaction learning for few-shot relation classification. In: Proceedings of the 30th ACM International Conference on Information & Knowledge Management, pp. 649–658 (2021)

13. Jiang, T., et al.: PromptBERT: improving BERT sentence embeddings with prompts. arXiv preprint arXiv:2201.04337 (2022)

14. Jin, L., et al.: Relation extraction exploiting full dependency forests. In: Proceedings of the AAAI Conference on Artificial Intelligence, vol. 34, pp. 8034–8041 (2020)

15. Kenton, J.D.M.W.C., Toutanova, L.K.: BERT: pre-training of deep bidirectional transformers for language understanding. In: Proceedings of NAACL-HLT, pp. 4171–4186 (2019)

16. Li, W., Qian, T.: Graph-based model generation for few-shot relation extraction. In: Proceedings of the 2022 Conference on Empirical Methods in Natural Language Processing, pp. 62–71 (2022)

17. Li, Y., Liu, Y., Gu, X., Yue, Y., Fan, H., Li, B.: Dual reasoning based pairwise representation network for document level relation extraction. In: 2022 IEEE International Conference on Multimedia and Expo (ICME), pp. 1–6. IEEE (2022)

18. Liu, F., et al.: From learning-to-match to learning-to-discriminate: global prototype learning for few-shot relation classification. In: Li, S., et al. (eds.) CCL 2021. LNCS (LNAI), vol. 12869, pp. 193–208. Springer, Cham (2021). https://doi.org/10.1007/978-3-030-84186-7_13

19. Liu, Y., Hu, J., Wan, X., Chang, T.H.: Learn from relation information: towards prototype representation rectification for few-shot relation extraction. In: Findings of the Association for Computational Linguistics: NAACL 2022, pp. 1822–1831 (2022)

20. Liu, Y., Hu, J., Wan, X., Chang, T.H.: A simple yet effective relation information guided approach for few-shot relation extraction. In: Findings of the Association for Computational Linguistics: ACL 2022, pp. 757–763 (2022)

21. Liu, Y., et al.: Powering fine-tuning: learning compatible and class-sensitive representations for domain adaption few-shot relation extraction. In: International Conference on Database Systems for Advanced Applications, pp. 121–131 (2023)

22. Peng, H., et al.: Learning from context or names? an empirical study on neural relation extraction. In: Proceedings of the 2020 Conference on Empirical Methods in Natural Language Processing (EMNLP), pp. 3661–3672 (2020)

23. Qian, W., Zhu, Y.: Adversarial learning with domain-adaptive pretraining for few-shot relation classification across domains. In: 2021 IEEE 6th International Conference on Computer and Communication Systems (ICCCS), pp. 134–139 (2021)

24. Ravi, S., Larochelle, H.: Optimization as a model for few-shot learning. In: International Conference on Learning Representations (2017)

25. Snell, J., Swersky, K., Zemel, R.: Prototypical networks for few-shot learning. In: Proceedings of the 31st International Conference on Neural Information Processing Systems, pp. 4080–4090 (2017)

26. Soares, L.B., Fitzgerald, N., Ling, J., Kwiatkowski, T.: Matching the blanks: distributional similarity for relation learning. In: Proceedings of the 57th Annual Meeting of the Association for Computational Linguistics, pp. 2895–2905 (2019)

27. Vinyals, O., Blundell, C., Lillicrap, T., Wierstra, D.: Matching networks for one shot learning. In: Advances in Neural Information Processing Systems 29 (2016)

28. Yan, Y., et al.: ConSERT: a contrastive framework for self-supervised sentence representation transfer. In: Proceedings of the 59th Annual Meeting of the Association for Computational Linguistics and the 11th International Joint Conference on Natural Language Processing (Volume 1: Long Papers), pp. 5065–5075 (2021)

29. Yang, K., et al.: Enhance prototypical network with text descriptions for few-shot relation classification. In: Proceedings of the 29th ACM International Conference on Information & Knowledge Management, pp. 2273–2276 (2020)

30. Zhang, P., Lu, W.: Better few-shot relation extraction with label prompt dropout. arXiv preprint arXiv:2210.13733 (2022)

31. Zhou, P., et al.: Attention-based bidirectional long short-term memory networks for relation classification. In: Proceedings of the 54th Annual Meeting of the Association for Computational Linguistics (volume 2: Short papers), pp. 207–212 (2016)

ML2FNet: A Simple but Effective Multi-level Feature Fusion Network for Document-Level Relation Extraction

Zhi Zhang[1,2], Junan Yang[1,2(✉)], and Hui Liu[1,2]

[1] College of Electronic Engineering, National University of Defense Technology, Hefei 230037, China
{zhangzhi,liuhui17c}@nudt.edu.cn, yangjunan@ustc.edu
[2] Anhui Province Key Laboratory of Electronic Restriction, Hefei 230037, China

Abstract. Document-level relation extraction presents new challenges compared to its sentence-level counterpart, which aims to extract relations from multiple sentences. Current graph-based and transformer-based models have achieved certain success. However, most approaches focus only on local information, independently on a certain entity, without considering the global interdependency among the relational triples. To solve this problem, this paper proposes a novel relation extraction model with a **Multi-Level Feature Fusion Network** (**ML2FNet**). Specifically, we first establish the interaction between entities by constructing an entity-level relation matrix. Then, we employ an enhanced U-shaped network to fuse the multi-level feature of entity pairs from local to global. Finally, the relation classification of entity pairs is performed by a bilinear classifier. We conduct experiments on three public document-level relation extraction datasets, and the results show that ML2FNet outperforms the other baselines. Our code is available at https://github.com/zzhinlp/ML2FNet.

Keywords: Document-level Relation extraction · Nature Language Process · Transformer · Feature Fusion

1 Introduction

Relation extraction (RE) aims to identify the relation between two entities in long texts and is an essential topic in information extraction and natural language processing. Existing research has focused on sentence-level RE [1,2], i.e., relations between entities in a single sentence. In the real world, however, just like relation facts from Wikipedia articles and biomedical literature, more than 40.7% of relations can only be identified by multiple sentences [3]. This problem, often called document-level RE, requires models that capture complex interactions between entities throughout a document. The document is a more realistic setting that is receiving increasing attention in the field of information extraction.

© The Author(s), under exclusive license to Springer Nature Singapore Pte Ltd. 2024
B. Luo et al. (Eds.): ICONIP 2023, CCIS 1969, pp. 297–312, 2024.
https://doi.org/10.1007/978-981-99-8184-7_23

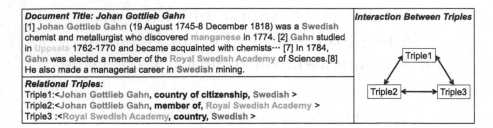

Fig. 1. An example from DocRED. Note that mentions of the same entity are marked with identical color.

In a sentence-level RE dataset, for example, WebNLG [4], the scenarios of task are limited to just one sentence, while the sentence is usually short. However, for document-level RE, the main challenges are as follows. (1) The complexity of the document-level RE task grows squarely with the number of entities. For a document contains N entities, $N(N-1)$ entity pairs must be classified for relations, while most entity pairs do not contain any relations. (2) A text contains multiple relational triples that are closely related to each other. For example, as shown in Fig. 1, intra-sentence triples are easily identified because their subjects and entities appear in one sentence, for example, triple 1 <*Johan Gottlieb Gahn, country of citizenship, Swedish*> in the first sentence and triple 2 <*Johan Gottlieb Gahn, member of, Royal Swedish Academy*> in the seventh sentence. However, the inter-sentence triples whose entities are mentioned to appear in different sentences are difficult to identify, such as triple 3 <*Royal Swedish Academy, country, Swedish*>, due to their long-distance dependencies.

To identify the relation between inter-sentence entity pairs, on the one hand, many approaches construct document graphs with entities and sentences as nodes and associations as edges based on graph neural networks. These approaches perform graph inference through heuristics, guided attention, or correlation dependency strategies to identify the inter-sentence relations. However, the graphs constructed by such approaches may ignore some important information in the text, and the complex inference operations on graphs prevent the model from capturing the text structure. On the other hand, thanks to the ability of pre-trained language models (PLMs) to implicitly establish long-distance dependencies of text, many methods directly use PLMs instead of graph inference structures, and these methods obtain entity representations through PLMs and then classify the relations on the entity pair representations. Although some success has been achieved, most of the methods only obtain entity representations that focus on the local information where the entities are located and ignore the global information of the whole document. And these methods cannot obtain the global interdependency between relational triples.

Capturing global interdependency between relational triples is effective because multiple triples in a document are usually related in some way. After identifying one or more triples, it is possible to exclude that some entity pairs must not have certain relations while facilitating the identification of other triples. For example, in Fig. 1, the inter-sentence relation between entity pairs <*Royal Swedish Academy, Swedish*> is usually difficult to identify, while in the case of the other two triples identified, it is easy to infer the triple <*Royal Swedish Academy, country, Swedish*>, which shows that the global interdependency between triples is extremely important for reasoning about inter-sentence relations in long texts.

To capture the local-to-global multi-level interdependency among the multiple relation triples, this paper proposes a novel RE model with a Multi-Level Feature Fusion Network. Specifically, First, we establish the interaction between entities by constructing an $N \times N$ entity-level relation matrix, in which each item represents an entity pair, and each item can be classified to obtain multiple relational triples. And in order to capture the local-to-global interdependency of the triples, we employ an enhanced U-shape network to capture different degrees of local-to-global interdependency among the triples. Finally we use a bilinear classifier to predict the relation between entity pairs. We conducted experiments on three datasets, DocRED [3], CDR [5], and GDA [6]. Experimental results show that our model consistently outperforms competitive baselines. The main contributions of this paper are as follows:

- We propose a novel document-level RE model ML2FNet that captures different degrees of local to global interdependency of relational triples, and such interdependency enhances the ability of model to infer inter-sentence relations.
- As far as we know, the proposed ML2FNet is the first model that combines enhanced U-shape module in computer vision with the relation extraction problem.
- We conducted experiments on three public RE datasets. Experimental results demonstrate that our model outperforms many state-of-the-art methods. At the same time, we conducted ablation experiments and case study to analyze the performance of our model and its reason.

2 Related Work

Relation extraction can be classified according to granularity into sentence-level [1,2] and document-level [3,7]. Early research focused on sentence-level RE, which mainly performs the prediction of relations in a one-sentence scenario. [8,9] proposed traditional rule-based and feature-based approaches. [10,11] proposed the use of neural networks to mine the features of entities and relations. With PLMs [12] achieving good performance in numerous downstream tasks of natural language processing, methods based on PLMs have become a popular method for RE, especially the paradigm of advanced pre-training followed by fine-tuning. Although they are effective, in practical situations, relations of real-world can only be reasoned through multiple sentences, and multiple mentions of entities are also distributed throughout the document, i.e., document-level relations. In recent years, researchers have gradually shifted from sentence-level to document-level RE [3]. In document-level RE, sophisticated inference techniques and inter-sentence information are required to predict relations of entity pairs. Depending on the line of research taken, document-level RE methods are divided into three main categories.

Sequence-Based Models. Sequence-based models encode entire document through sequences composed of words and sentences. These models use neural networks such as CNN [10] and BiLSTM [13] to encode and then obtain entity embeddings, where the embeddings of two entities relationships of entity pairs are predicted by bilinear functions.

Graph-Based Models. Graph-based models usually build a graph with mentions, entities, or sentences as nodes, and infers relations by reasoning about the graph, and the graph-based approach is widely used due to its effectiveness and advantages in relational reasoning. [14] first attempted to use tokens and dependency information as nodes and edges to build a document graph, after which a lot of work [7,15–17] emerged to extend this idea by applying different Graph neural networks (GNNs) [18] architectures. Specifically, [15] He created a graph with multiple types of nodes and edges. [16] proposed a latent structure refinement model that uses latent structure induction strategies to induct dependency trees in documents. [17] used a bipartite graph to process intra-sentence entity pairs and inter-sentence entity pairs separately. [19] suggested an novel classifier model HeterGSAN, which attempts to construct path dependencies from graph representations. The graph model can bridge sentences and entities of the full text, thus alleviating long distance dependencies and achieving promising performance. However, the graphs constructed by these models may ignore some important information in the text. The complex reasoning process of graphs can also prevent these methods from capturing the text structure.

Transformer-Based Models. Transformer-based models has the ability to model implicitly remote dependencies in text and have become a popular and promising approach in document-level RE. [20] proposed a two-step training methodology to determine whether two entities are related before predicting a specific relation. [21] proposes a hierarchical inference network, which effectively integrates reasoning information from three different granularities: entity, sentence, and document. [19] incorporated entity structure dependencies into attention network and encoding phase to generate attention biases to adaptively adjust its attention flow. [22] proposed adaptive thresholding and local context to solve problems multi-entity mentions and multi-relations classification respectively. However, these methods can only capture long-distance dependencies between tokens and cannot obtain global interdependency between entity pairs and relational triples.

Different from the previous approaches, in order to improve the ability of model to capture local and global interdependency, a multi-level feature fusion network is proposed to capture the local-to-global interdependency of relational triples, which improves the global interaction between triples and further improves the inference ability of model about inter-sentence relations.

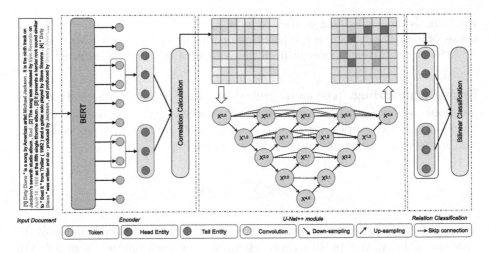

Fig. 2. Architecture of our proposed model.

3   Method

In this section, we first give a definition of document-level RE and then detail our proposed approach with the model structure shown in Fig. 2.

3.1   Problem Definition

Given a text document D that contain sentences $\mathcal{X}_D = \{x_i\}_{i=1}^{|\mathcal{X}_D|}$ and entities $\mathcal{E}_D = \{e_i\}_{i=1}^{|\mathcal{E}_D|}$, the task is to identify all possible relations between all entity pairs (e_s, e_o) from $\mathcal{R}$, where $\mathcal{R}$ is a pre-defined relation set, e_s, e_o are subject and object entities, respectively. For each entity $e \in \mathcal{E}_D$, it appears at least once in D, and may also be mentioned multiple times and scattered in D, with all its proper-noun mentions denoted as $\mathcal{M}_e = \{m_i\}_{i=1}^{|\mathcal{M}_e|}$. The entity pairs are marked as NA if they do not exhibit any relations. Each entity pairs (e_s, e_o) can hold multiple relations. Note entities and relations could occur simultaneously in different triples.

3.2   Encoder

Text Encoder. We leverage a Transform-based PLMs to obtain token embeddings and cross-token dependencies. Although the last layer is used in all other methods, the method in this paper uses the average of the last three layers. Specifically, before encoding, we insert special token "$*$" at the beginning and end of all entities mentions to indicate the location of entity. Then, tokens $\mathcal{T}_D = \{t_i\}_{i=1}^{|\mathcal{T}_D|}$ is fed into PLMs to obtain the d hidden dimensions tokens embedding $\boldsymbol{H} = [\mathbf{h}_1, ..., \mathbf{h}_{|\mathcal{T}_D|}]$ and the matrix of cross token attention $\boldsymbol{A}$:

$$\boldsymbol{H}, \boldsymbol{A} = \text{PLMs}(\mathcal{T}_D) \tag{1}$$

where $\mathbf{h}_i$ is the embedding of the token t_i, $\boldsymbol{H} \in \mathbb{R}^{d \times |\mathcal{T}_D|}$ averages over hidden states of each token from in the last three transformer layer of PLMs and $\boldsymbol{A} \in \mathbb{R}^{H \times |\mathcal{T}_D| \times |\mathcal{T}_D|}$ is attention weight from the average of last three transformer layer.

Entity Embedding. The entity embedding $\mathbf{h}_e \in \mathbb{R}^d$ for each entity e with mentions $\mathcal{M}_e = \{m_i\}_{i=1}^{|\mathcal{M}_e|}$ is computed by collecting information. Specifically, for a entity e, we leverage logsumexp pooling to obtain an entity embedding $\mathbf{h}_e$ that incorporates different entity mention embeddings:

$$\mathbf{h}_e = \log \sum_{i=1}^{|\mathcal{M}_e|} \exp(\boldsymbol{H}_{m_i}) \tag{2}$$

where $|\mathcal{M}_e|$ is the numbers of mentions for entity e, $\boldsymbol{H}_{m_i}$ is the embedding of the mention of entity. This pooling approach incorporates information from different references in the file and shows better performance than the average collection. Finally, we get the aggregated feature embedding $\mathbf{h}_e$ of entity e.

Localized Context Embedding. In previous works [19,22,23], it has been proven that textual information is essential for RE tasks, and we also adopt Localized Context Pooling [22]. Specifically, $\boldsymbol{A}_{ijk}$ in attention matrix $\boldsymbol{A}$ is the attention weight from token j to token k in the ith head (H represents the number of attention heads). For each entity e, We first obtain different mention-level attention $\boldsymbol{a}_{m_i} \in \mathbb{R}^{H \times |\mathcal{T}_D|}$ which represents the attention of mention for other tokens. Then average the attention of all mentions to obtain the entity-level attention $\boldsymbol{A}^e = \sum_{i=1}^{|\mathcal{M}_e|}(\boldsymbol{a}_{m_i})$. Then given an entity pair (e_s, e_o), we find the local context that is significant to e_s and e_o by multiplying their entity-level attention, yielding the localized context embedding $\boldsymbol{c}^{(s,o)}$. The specific operations are as follows:

$$\begin{aligned} \boldsymbol{q}^{(s,o)} &= \sum_{i=1}^{H}(\boldsymbol{A}_i^{e_s} \cdot \boldsymbol{A}_i^{e_o}) \\ \boldsymbol{a}^{(s,o)} &= \boldsymbol{q}^{(s,o)}/\mathbf{1}^{\mathsf{T}}\boldsymbol{q}^{(s,o)} \\ \boldsymbol{c}^{(s,o)} &= \boldsymbol{H}^{\mathsf{T}}\boldsymbol{a}^{(s,o)} \end{aligned} \tag{3}$$

where $\boldsymbol{A}^{e_s}, \boldsymbol{A}^{e_o} \in \mathbb{R}^{H \times |\mathcal{T}_D|}$ are the aggregated attention weight. $\boldsymbol{q}^{(s,o)}$ is the mean-pooled attention output for entity pair. $\boldsymbol{H} \in \mathbb{R}^{d \times |\mathcal{T}_D|}$ is the contextual representation. The context vector $\boldsymbol{c}^{(s,o)} \in \mathbb{R}^d$ is then combined with the entity embeddings.

3.3   Enhanced U-Shape Module

We use the obtained entity-to-entity correlations $\boldsymbol{c}^{(s,o)}$ to construct the entity-level relation matrix $\boldsymbol{M}$,Then we fuse the local-to-global interdependency of entity pairs and triples through a multi-level feature fusion module. Specifically, by counting all samples of the dataset and calculating the maximum number of entities N in each document, we construct an entity-level relation matrix $M \in \mathbb{R}^{N \times N \times D}$. After that, we utilize a enhanced U-shape module [24], which is an effective framework in semantic segmentation for computer vision task.

As shown in the Fig. 2, the module is an enhanced U-shaped network, where each node in the module represents a convolution block. They are connected by a series of nested dense convolution blocks, with downward arrows indicating down-sampling, upward arrows indicating up-sampling, and dotted arrows indicating skip connections. From the overall architecture, enhanced U-shaped module is a combination of U-Nets [25] of different depths into a unified architecture. The original skip connections are removed and every two adjacent nodes are connected with short skip connection, so that the deeper decoder sends supervisory signals to the shallower decoder. The skip connections are densely connected, making the features propagate densely along the skip connections and the decoder nodes have more flexible feature fusion. Finally we incorporate an encoder and the enhanced U-shape module to capture both local-to-global interdependency information Y as follows:

$$X^{i,j} = \begin{cases} \sigma\left(D(x^{i-1,j})\right), & j = 0 \\ \sigma\left(\left[[x^{i,k}]_{k=0}^{j-1}, U(x^{i+1,j-1})\right]\right), & j > 0 \end{cases} \tag{4}$$
$$Y = \text{UPlus}(W_u M)$$

where X denotes the convolution operation of the circle in the figure. D denotes down-sampling layer. U denotes up-sampling layer. UPlus denotes the entire module, and $Y \in \mathbb{R}^{N \times N \times D'}$ denotes the processed entity-level interaction matrix. W_u is the learnable weight matrix.

The overall structure is seen from the horizontal structure, which is able to combine local and global information at the same scale, and from the vertical structure, which is able to focus local to global information at different scales. This multi-level feature aggregation fuses the local information where entities are located as well as the global information to improve the entity-to-entity connection and thus the accuracy of relation classification.

3.4 Relation Classification Module

Given entity-level relation matrix M and the initial representation H of the document, we map the entities to hidden states z of entity e_s and e_o with a linear layer followed by non-linear activation, then calculate the probability of relation r by bilinear function and sigmoid activation. The specific operations are as follows:

$$z_s = \tanh(W_s h_{e_s} + Y_{(s,o)})$$
$$z_o = \tanh(W_o h_{e_o} + Y_{(s,o)}) \tag{5}$$
$$P(r|e_s, e_o) = \sigma(z_s^\mathsf{T} W_r z_o + b_r)$$

where $Y_{(s,o)}$ is the entity pair embeddings of (e_s, e_o) in Y. $W_s, W_o, W_r \in \mathbb{R}^{d \times d}, b_r \in \mathbb{R}$ are learned weight and bias. To minimize the number of bilinear classifier parameters, we use the group bilinear [26], which divides the embedding dimensions into k groups of equal size and applies bilinear within the groups:

$$z_s = [z_s^1; ...; z_s^k;]$$
$$z_o = [z_o^1; ...; z_o^k;]$$
$$P(r|e_s, e_o) = \sigma(\sum_{i=1}^{k} z_s^{i\top} W_r^i z_o^i + b_r) \tag{6}$$

where $P(r|e_s, e_o)$ is the probability that relation r is existed in the entity pair (e_s, e_o). By this way, the number of parameters of the classifier is reduced by a factor of k, from d^2 to d^2/k.

Since previous work [20] discovered an imbalanced relation distribution for RE, that is, some entity pairs have multiple relations while most of the entity pairs have no relations. In the training process, we employ a balanced softmax loss [27]. Specifically, introducing a special balanced label bl, we want the probability P_{pos} of the positive label to be higher than the probability P_{bl} of the balanced label and the probability P_{neg} of the negative labels to all be lower than P_{bl}. The specific operations are as follows:

$$\mathcal{L}_1 = \log(e^{P_{bl}} + \sum_{r \in \{pos\}} e^{P_r})$$
$$\mathcal{L}_2 = \log(e^{-P_{bl}} + \sum_{r \in \{neg\}} e^{P_r}) \tag{7}$$
$$\mathcal{L} = \mathcal{L}_1 + \mathcal{L}_2$$

where $\mathcal{L}_1, \mathcal{L}_2$ denote the loss of positive label and negative label prediction results, respectively. $\{pos\}, \{neg\}$ denote the set of positive and negative labels respectively.

$\mathcal{L}_1$ pushes the logit of all positive labels above the bl label. $\mathcal{L}_2$ pulls the logit of all negative labels below the bl label. The two parts of the total loss are simply summed.

4   Experiments

4.1   Datasets

Our model ML2FNet is examined using three public document-level RE datasets. Table 1 displays the datasets statistics.

- **DocRED** [3] is a benchmark in the task of document-level RE, built from Wikipedia data through large-scale crowdsourcing. DocRED contains 3053, 1000 and 1000 articles for training, validation and testing, respectively. It contains 97 types of relations. The DocRED dataset is complex in terms of entity types, relation types, inference types, and inter-sentence relations, which is quite a test for RE models.

Table 1. Statistics of the experimental datasets.

Statistics/Datasets	DocRED	CDR	GDA
# Train	3053	500	23353
# Dev	1000	500	5839
# Test	1000	500	1000
# Relation types	97	2	2
# Avg.#mentions per Ent	1.4	2.7	3.3
# Avg.#entities per Doc	19.5	7.6	5.4
# Avg.#senteces per Doc	8.0	9.7	10.2

- **CDR** [5] is a chemical-disease RE dataset in biomedicine, constructed using PubMed abstracts, containing a total of 1,500 human-annotated files, divided equally into train, valid and test sets. CDR is a binary classification task that is used in biomedical research to identify induced relationships between chemical entities and disease entities, which is important to the study of medical texts.
- **GDR** [6] is a large dataset in the biomedical field. It includes 29192 articles. Its task is to predict binary interactions between genes and disease concepts. We followed [15] to split the training set into an 80/20 training and development set.

4.2 Implementation Details

We implemented ML2FNet with the Pytorch version of Huggingface Transformers [28]. We use cased BERT [29] as the encoder on DocRED, and SciBERT [30] on CDR and GDA. The training process is based on Apex library[1]. Our model is optimized with AdamW [31] using learning rates 2e-5, with a linear warmup [32] for the first 6% steps followed by a linear decay to 0. We apply dropout [33] between layers with rate 0.1 and clip the gradients of model parameters to a max norm of 1.0. We perform early stopping based on the F1 score on the development set. We set the entity-level relation matrix size $N = 48$.

All training and testing is completed on a computer with an Inter Xeon(R) Gold 6230R CPU @ 2.10Ghz, 128G memory, an NVIDIA GeForce RTX 3090 GPU and Windows 10 professional edition. Training with a BERT-based encoder takes roughly 2 h and 45 min for the DocRED dataset. Training with SciBERT encoder takes 20 min for CDR datasets and 1 h and 10 min for GDA datasets.

4.3 Main Result

On the benchmark dataset DocRED, we compared ML2FNet with twelve baseline models. The experiment results are shown in Table 2. Sequence-based methods include CNN [3] and BiLSTM [3]. Graph-based models include AGGCN [34],

---

[1] https://github.com/NVIDIA/apex.

LSR [16], GLRE [35], GAIN [17], HeterGSAN [36]. Transformer-based models include Two-Step [20], HIN [21], Coref [37], SSAN [19], ATLOP [22]. We evaluated our model with indicators Ign F1 and F1 following [3]. The Ign F1 represents the F1 score without the relational information shared by the train, dev, and test sets.

Table 2. Experimental results (%) on dev and test set of the DocRED dataset. The best scores are in **Bold**.

Model	Dev		Test	
	Ign F1	F1	Ign F1	F1
Sequence-based Models				
CNN [3]	41.58	43.45	40.33	42.26
BiLSTM [3]	48.87	50.94	48.78	51.06
Graph-based Models				
AGGCN [34]	46.29	52.47	48.89	51.45
LSR-BERT$_{base}$ [16]	52.43	59.00	56.97	59.05
GLRE-BERT$_{base}$ [35]	52.43	59.00	56.97	59.05
GAIN-BERT$_{base}$ [17]	52.43	59.00	56.97	59.05
HeterGSAN-BERT$_{base}$ [36]	52.43	59.00	56.97	59.05
Transformer-based Models				
Two-Step-BERT$_{base}$ [20]	–	54.42	–	53.92
HIN-BERT$_{base}$ [21]	54.29	56.31	53.70	55.60
Coref-BERT$_{base}$ [37]	55.32	57.51	54.54	56.96
SSAN-BERT$_{base}$ [19]	57.03	59.19	55.84	58.16
ATLOP-BERT$_{base}$ [22]	59.22	61.09	59.31	61.30
Our Model				
ML2FNet-BERT$_{base}$	**59.58**	**61.58**	**59.72**	**61.64**

As shown in the Table 2, ML2FNet outperforms the current baseline approach for all metrics. Compared with these three types of models, ML2FNet has a significant improvement in F1 and Ign F1 are significantly improved. Among the many baselines, SSAN [19] and ATLOP [22] are the two most relevant to our model. Compared to SSAN-BERT$_{base}$, the performance of ML2FNet-BERT$_{base}$ improves roughly about 2.55% for Ign F1 and 2.39% for F1 in dev set. ML2FNet-BERT$_{base}$ also brings about 0.36% for Ign F1 and 0.49% for F1 enhancement compared to ATLOP-BERT$_{base}$, which demonstrates the efficacy of our suggested approach.

4.4 Results on Biomedical Datasets

Table 3 displays the experimental results of ML2FNet on the medical dataset, as well as the comparison with the eight baseline model. These baseline models include: CNN [38] and BRAN [7] are both sequence-based models employ CNN and attention network as encoder. EoG [15], LSR [16], DHG [39] and GLRE [35] are all graph-based models. SSAN [19] and ATLOP [22] are both transformer-based models. ML2FNet-SciBERT$_{base}$ produced new state-of-the-art results on these two datasets. Our ML2FNet-SciBERT$_{base}$ model is significantly better than the previous best method ATLOP-SciBERT$_{base}$. Further improving the F1 scores of CDR and GDA on the test set by 14.3% and 2.4% to 83.5% and 86.3% respectively,

Table 3. Test F1 score (%) on CDR and GDA dataset.

Model	CDR	GDA
Sequence-based Models		
BRAN [7]	62.1	–
CNN [38]	62.3	–
Graph-based Models		
EoG [15]	63.6	81.5
LSR-SciBERT$_{base}$ [16]	64.8	82.2
DHG-BERT$_{base}$ [39]	65.9	83.1
GLRE-SciBERT$_{base}$ [35]	68.5	–
Transformer-based Models		
SSAN-SciBERT$_{base}$ [19]	68.7	83.7
ATLOP-SciBERT$_{base}$ [22]	69.2	83.9
Ours Model		
ML2FNet-SciBERT$_{base}$	**83.5**	**86.3**

4.5   Ablation Study and Analysis

Table 4. Ablation study of ML2FNet on DocRED.

Our Model	Ign F1	F1
w/ Enhanced U-shape module	**59.58**	**61.58**
w/o Enhanced U-shape module	57.89	59.98

We conducted ablation experiments to verify the effectiveness of ML2FNet, and the results are shown in Table 4. After removing the multilevel feature fusion module, the model decreases by 1.69% and 1.60% on Ign F1 and F1, respectively, illustrating the superiority of ML2FNet.

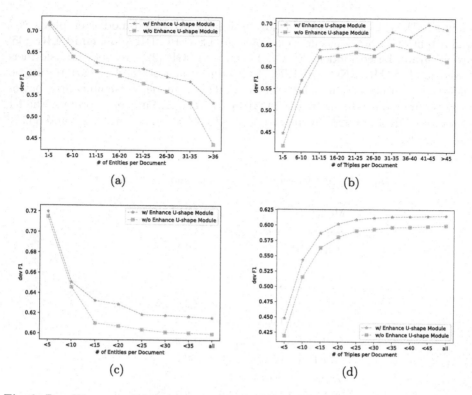

Fig. 3. Dev F1 score of documents with the different number of entities and triples on DocRED. (a) and (c) are dev F1 score of documents with the different number of entities on DocRED. (b) and (d) are dev F1 score of documents with the different number of triples on DocRED.

ML2FNet is proposed to fuse local and global information and enhance the global dependence of entity pairs and triples. To verify this idea, we perform ablation analysis in the perspective that each document contains different number of entities and each document contains different number of triples, respectively, and the results are shown in Fig. 3.

Specifically, in Sub-Fig. 3(a) and 3(c), with the increasing number of entities, the model w/ enhanced U-shaped module always outperforms the model w/o enhanced U-shaped module. The performance of the model w/o enhanced U-shaped module decreases significantly with the increasing number of entities, and the model w/ enhanced U-shaped module has an increasing advantage over the model w/o enhanced U-shaped module. Because with the increasing number of entities, the model w/ enhanced U-shaped module is able to learn the interdependency between multiple entities, thus improving the dependencies between entity representations.

In addition, Sub-Fig. 3(b) and 3(d), the model w/ enhanced U-shaped module is also always due to the model w/o enhanced U-shaped module. The performance of the model w/o enhanced U-shaped module starts to show a decreasing trend when the number of triples is greater than 30, while the model w/ enhanced U-shaped module still shows excellent performance. Intuitively, the more models the more difficult it is to reason. However, enhanced U-shaped module captures the global dependencies among relational triples, and the more the number of triples, the richer and tighter the interdependency are, which makes the model w/ enhanced U-shaped module perform well despite the large number of triples, which fully demonstrates that capturing the global dependencies among the triples plays a crucial role in reasoning about inter-sentence relations.

4.6 Case Study

Case One:		
[1] Johan Gottlieb Gahn (19 August 1745-8 December 1818) was a Swedish chemist and metallurgist who discovered manganese in 1774. [2] Gahn studied in Uppsala 1762-1770 and became acquainted with chemists··· [7] In 1784, Gahn was elected a member of the Royal Swedish Academy of Sciences.[8] He also made a managerial career in Swedish mining.		
Subject: **Johan Gottlieb Gahn** Object: Swedish Relation: **country of citizenship**	Subject: **Johan Gottlieb Gahn** Object: Royal Swedish Academy Relation: **member of**	Subject: Royal Swedish Academy Object: Swedish Relation: **country**
ATLOP: True ML2FNet: True	ATLOP: True ML2FNet: True	ATLOP: True ML2FNet: **None**

Case Two:
[0] IBM Research Brazil is one of twelve research laboratories comprising IBM Research , its first in South America. [1] It was established in June 2010 , with locations in So Paulo and Rio de Janeiro. [2] Research focuses on Industrial Technology and Science, systems of engagement and insight, social data...
Subject: Johan Gottlieb Gahn Object: Swedish Relation: **country of citizenship**
ATLOP: False   ML2FNet: False

Fig. 4. Two case study on the DocRED dev set. The results predicted by ATLOP and our ML2FNet. Entities are colored accordingly.

We compared ML2FNets with the ATLOP model in some cases from dev set and the results are shown in Fig. 4. In Case one, the first two regular triples are identified by ATLOP and ML2FNet, while the third inter-sentence triple needs to be reasoned out by establishing dependencies through the other two triples. Obviously, ML2FNet can capture the global dependencies among the triples and reason out other triples based on the identified triads, effectively solving the problem of reasoning about inter-sentence triples. In Case two, the reasoning of the triple <South America, continent, So Paulo> requires a combination of background information and common sense knowledge. Only through external commonsense knowledge can we know that South America is a continent and So Paulo is a city, which is a useful piece of information to help determine their relation. Reasoning cases like this are more common in real situations, and this kind of problem we think can be solved by adding some external knowledge, and we make it our future work.

5   Conclusion

In this paper, we introduce an enhanced U-shape module and propose a multi-level feature fusion network (ML2FNet) to solve the task of document-level relation extraction. Our approach first establishes the interactions between entities by constructing an entity-level relation matrix, and later improves the ability of the model to solve inter-sentence relation extraction by capturing the local-to-global dependencies of the relational triple through the enhanced U-shape module. Experiments on three public datasets demonstrate that our ML2FNet significantly outperforms existing baseline models and achieves the new state-of-the-art results.

Acknowledgements. This work was supported by the Anhui Provincial Science Foundation (NO.1908085MF202) and the Independent Scientific Research Program of National University of Defense Science and Technology (NO. ZK18-03-14).

References

1. Wang, L., Cao, Z., de Melo, G., Liu, Z.: Relation classification via multi-level attention CNNs. In: Proceedings of the 54th Annual Meeting of the Association for Computational Linguistics, pp. 1298–1307 (2016)
2. Wei, Z., Su, J., Wang, Y., Tian, Y., Chang, Y.: A novel cascade binary tagging framework for relational triple extraction. In: Proceedings of the 58th Annual Meeting of the Association for Computational Linguistics, pp. 1476–1488 (2020)
3. Yao, Y., et al.: DocRED: a large-scale document-level relation extraction dataset. In: Proceedings of the 57th Annual Meeting of the Association for Computational Linguistics, pp. 764–777 (2019)
4. Gardent, C., Shimorina, A., Narayan, S., Perez-Beltrachini, L.: Creating training corpora for NLG micro-planning. In: 55th Annual Meeting of the Association for Computational Linguistics, pp. 179–188 (2017)
5. Li, J., et al.: Biocreative v CDR task corpus: a resource for chemical disease relation extraction. Database: J. Biol. Datab. Curat. (2016)
6. Wu, Y., Luo, R., Leung, H.C.M., Ting, H.F., Lam, T.W.: Renet: a deep learning approach for extracting gene-disease associations from literature. In: Research in Computational Molecular Biology, pp. 272–284 (2019)
7. Verga, P., Strubell, E., McCallum, A.: Simultaneously self-attending to all mentions for full-abstract biological relation extraction. In: Proceedings of the 2018 Conference of the North American Chapter of the Association for Computational Linguistics, pp. 872–884 (2018)
8. Pantel, P., Pennacchiotti, M.: Espresso: leveraging generic patterns for automatically harvesting semantic relations. In: Proceedings of the 21st International Conference on Computational Linguistics and 44th Annual Meeting of the Association for Computational Linguistics, pp. 113–120 (2006)
9. Mintz, M.D., Bills, S., Snow, R., Jurafsky, D.: Distant supervision for relation extraction without labeled data. In: Annual Meeting of the Association for Computational Linguistics, pp. 1003–1011 (2009)
10. Zeng, D., Liu, K., Lai, S., Zhou, G., Zhao, J.: Relation classification via convolutional deep neural network. In: International Conference on Computational Linguistics, pp. 2335–2344 (2014)

11. Zhou, P., et al.: Attention-based bidirectional long short-term memory networks for relation classification. In: Proceedings of the 54th Annual Meeting of the Association for Computational Linguistics, pp. 207–212 (2016)

12. Vaswani, A., et al.: Attention is all you need. In: Proceedings of the 31st International Conference on Neural Information Processing Systems (NIPS 2017), pp. 6000–6010 (2017)

13. Schuster, M., Paliwal, K.K.: Bidirectional recurrent neural networks. IEEE Trans. Signal Process. **45**, 2673–2681 (1997)

14. Quirk, C., Poon, H.: Distant supervision for relation extraction beyond the sentence boundary. In: Conference of the European Chapter of the Association for Computational Linguistics, pp. 1003–1011 (2016)

15. Christopoulou, F., Miwa, M., Ananiadou, S.: Connecting the dots: document-level neural relation extraction with edge-oriented graphs. In: Conference on Empirical Methods in Natural Language Processing, pp. 4925–4936 (2019)

16. Nan, G., Guo, Z., Sekulic, I., Lu, W.: Reasoning with latent structure refinement for document-level relation extraction. In: Annual Meeting of the Association for Computational Linguistics, pp. 1546–1557 (2020)

17. Zeng, S., Xu, R., Chang, B., Li, L.: Double graph based reasoning for document-level relation extraction. In: Proceedings of the 2020 Conference on Empirical Methods in Natural Language Processing, pp. 1630–1640 (2020)

18. Scarselli, F., Gori, M., Tsoi, A.C., Hagenbuchner, M., Monfardini, G.: The graph neural network model. IEEE Trans. Neural Netw. **20**, 61–80 (2009)

19. Xu, B., Wang, Q., Lyu, Y., Zhu, Y., Mao, Z.: Entity structure within and throughout: modeling mention dependencies for document-level relation extraction. In: Proceedings of the AAAI Conference on Artificial Intelligence, vol. 35, pp. 14149–14157 (2021)

20. Wang, H., Focke, C., Sylvester, R., Mishra, N., Wang, W.: Fine-tune Bert for DocRED with Two-step Process. arXiv preprint arXiv:1909.11898 (2019)

21. Tang, H., et al.: HIN: hierarchical inference network for document-level relation extraction. Adv. Knowl. Discov. Data Mining **12084**, 197–209 (2020)

22. Zhou, W., Huang, K., Ma, T., Huang, J.: Document-level relation extraction with adaptive thresholding and localized context pooling. In: Proceedings of the AAAI Conference on Artificial Intelligence, vol. 35, pp. 14612–14620 (2021)

23. Peng, H., et al.: Learning from context or names? An empirical study on neural relation extraction. In: Proceedings of the 2020 Conference on Empirical Methods in Natural Language Processing, pp. 3661–3672 (2020)

24. Zhou, Z., Siddiquee, M.M.R., Tajbakhsh, N., Liang, J.: Unet++: redesigning skip connections to exploit multiscale features in image segmentation. IEEE Trans. Med. Imaging **39**(6), 1856–1867 (2020)

25. Ronneberger, O., Fischer, P., Brox, T.: U-net: convolutional networks for biomedical image segmentation. In: Medical Image Computing and Computer-Assisted Intervention, pp. 234–241 (2015)

26. Zheng, H., Fu, J., Zha, Z.J., Luo, J.: Learning deep bilinear transformation for fine-grained image representation. In: Advances in Neural Information Processing Systems, vol. 32, pp. 4279–4288 (2019)

27. Zhang, N., et al.: Document-level relation extraction as semantic segmentation. In: IJCAI, pp. 3999–4006 (2021)

28. Wolf, T., et al.: Transformers: state-of-the-art natural language processing. In: Proceedings of the 2020 Conference on Empirical Methods in Natural Language Processing: System Demonstrations, pp. 38–45 (2020)

29. Devlin, J., Chang, M.W., Lee, K., Toutanova, K.: BERT: pre-training of deep bidirectional transformers for language understanding. In: Proceedings of the 2019 Conference of the North American Chapter of the Association for Computational Linguistics, pp. 4171–4186 (2019)
30. Beltagy, I., Lo, K., Cohan, A.: Scibert: pretrained language model for scientific text. In: EMNLP, pp. 3615–3620 (2019)
31. Loshchilov, I., Hutter, F.: Decoupled weight decay regularization. In: ICLR (2017)
32. Goyal, P., et al.: Accurate, large minibatch SGD: training imagenet in 1 hour. arXiv preprint arXiv:1706.02677 (2017)
33. Srivastava, N., Hinton, G.E., Krizhevsky, A., Sutskever, I., Salakhutdinov, R.: Dropout: a simple way to prevent neural networks from overfitting. J. Mach. Learn. Res. **15**, 1929–1958 (2014)
34. Guo, Z., Zhang, Y., Lu, W.: Attention guided graph convolutional networks for relation extraction. In: ACL, pp. 241–251 (2019)
35. Wang, D., Hu, W., Cao, E., Sun, W.: Global-to-local neural networks for document-level relation extraction. In: EMNLP, pp. 3711–3721 (2020)
36. Xu, W., Chen, K., Zhao, T.: Document-level relation extraction with reconstruction. In: The Thirty-Fifth AAAI Conference on Artificial Intelligence (2021)
37. Ye, D., Lin, Y., Du, J., Liu, Z., Sun, M., Liu, Z.: Coreferential reasoning learning for language representation. In: EMNLP, pp. 7170–7186 (2020)
38. Nguyen, D.Q., Verspoor, K.M.: Convolutional neural networks for chemical-disease relation extraction are improved with character-based word embeddings. In: Proceedings of the BioNLP 2018 Workshop, pp. 129–136 (2018)
39. Zhang, Z., et al.: Document-level relation extraction with dual-tier heterogeneous graph. In: Proceedings of the 28th International Conference on Computational Linguistics, pp. 1630–1641 (2020)

Implicit Clothed Human Reconstruction Based on Self-attention and SDF

Li Yao⑩, Ao Gao⑩, and Yan Wan$^{(\boxtimes)}$⑩

School of Computer Science and Technology, Donghua University,
Shanghai 201620, China
winniewan@dhu.edu.cn

Abstract. Recently, implicit function-based approaches have advanced 3D human reconstruction from a single-view image. However, previous methods suffer from problems such as artifacts, broken limbs, and loss of surface details when dealing with challenging poses. To address these issues, this paper proposes a novel neural network model based on PaMIR. Firstly, the Signed Distance Function (SDF) is introduced to define an implicit function, which improves the generalization ability of the model. Secondly, a feature volume encoding network with self-attention mechanism is designed to extract voxel-aligned features and provide richer geometric information, further improving the accuracy of shape topology structure. Through validation on the CAPE dataset, our method exceeds the PaMIR by 50.9% and 30.6% reduction in Chamfer and Point-to-Surface Distances respectively, and 18.2% reduction in normal estimation errors.

Keywords: Single-view human reconstruction · Implicit Function · Self-attention · SDF

1 Introduction

Recovering the 3D surface of a human body wearing clothing from a single 2D image is a critical task in computer vision. This task has various applications, including 3D games, VR/AR, autonomous driving, motion capture, and virtual fitting. Due to significant progress in deep learning techniques, recent methods [1,2] focus on learning-based approaches for reconstructing a 3D human body model from a single-view image. Among these methods, implicit function-based approaches have attracted considerable attention.

In 2019, Staio et al. [1] proposed PIFu, in which they define an implicit function using pixel-aligned features and depth-Z features to predict the continuous occupancy field of query points and extract 3D meshes through Marching Cubes [3]. However, due to the inherent depth ambiguity in monocular reconstruction, PIFu reconstructs mesh surfaces that exhibit a significant amount of shape artifacts in natural in-the-wild images. The most recent notable implicit model is PaMIR, proposed by Zheng et al. [4], which introduces the parameterized model SMPL [5] as a shape proxy and improves the depth ambiguity issue

© The Author(s), under exclusive license to Springer Nature Singapore Pte Ltd. 2024
B. Luo et al. (Eds.): ICONIP 2023, CCIS 1969, pp. 313–324, 2024.
https://doi.org/10.1007/978-981-99-8184-7_24

with voxel-aligned features. The latest implicit function-based method, SuRS [6], attempts to establish a two-level multi-layer perceptron (MLP) to predict high-resolution meshes from low-resolution outdoor images in a refinement manner. However, these methods still suffer from shape artifacts and broken limbs under challenging poses.

To address these issues, we propose an improved deep-learning implicit model for 3D clothed human reconstruction based on PaMIR. Firstly, we address the problem of depth ambiguity by introducing the Signed Distance Function (SDF) and combining it with pixel-aligned features and voxel-aligned features to jointly encode the query points. The SDF describes the signed distance between points in 3D space, defining the geometric shape of objects and enhancing the robustness of the reconstruction results. Secondly, we introduce a self-attention mechanism and design a feature volume encoding network to extract rich geometric information from SMPL. These two improvements address the problem of depth ambiguity. Additionally, recent research works [2,7] have demonstrated that normal maps can capture local details and generate more realistic clothing folds on the front and back of the human body. Therefore, we also incorporate normal maps as additional inputs to improve clothing fold details. In summary, this paper contributes as follows:

(1) We introduce SDF for query points encoding and define an implicit function based on it, which effectively addresses the issue of depth ambiguity in monocular views.
(2) Based on the self-attention mechanism, we design a feature volume encoding network to extract voxel-aligned features, which further reducing shape topology structure errors.

2   Related Work

2.1   Single-View Clothed Human Reconstrcution

Many methods have adopted different representation approaches to reconstruct clothed human mesh models. Voxel is an effective means of describing geometric objects in 3D space. DeepHuman [8] predicts human voxels from images using a multi-scale image and volume estimation network. Although it can simulate human poses and clothing, voxel-based methods result in lower-resolution model due to the high memory requirements, which cannot effectively express the detailed surface of the 3D human body. Meanwhile, implicit functions have received significant attention due to their small memory footprint and high expressive power. The implicit function-based method PIFu [1] encodes query points using pixel-aligned features and depth z-values, performing well only in standing postures. PIFuHD [2] improves the input image resolution and introduces normal maps to enhance details. However, these methods exhibit noisy artifacts due to the lack of geometric regularization.

To overcome these issues, subsequent methods started to incorporate geometric information. Geo-PIFu [9] estimates 3D voxels from 2D images and then uses voxel-aligned features and pixel-aligned features to estimate query point occupancy. Unlike the Geo-PIFu method, PaMIR introduces the parameterized model SMPL as a geometric prior, using the SMPL model to provide more robust voxel-aligned features. These methods provide geometric information to implicit functions through additional voxel-aligned features. However, due to limitations in existing hardware devices, the voxel resolution used in these methods is low, resulting in incomplete utilization of geometric information. There are also methods that avoid using low-resolution voxels. ARCH [10] and ARCH++ [11] transform query points from standard space to pose space using SMPL and generate clothing details using skinning weights. Some methods enhance the expressive power of the implicit function by introducing additional image features. IntegratedPIFu [7] introduces normal maps, depth maps and human parsing maps to the implicit function. StereoPIFu [12] extends the recovery of three-dimensional human body surfaces from multiple views. However, multiple views require camera registration, making them less portable in outdoor environments.

In summary, these methods either lack geometric information in the features used or have insufficient ability to capture features, leading to poor generalization of the model in outdoor conditions. To address these issues, this paper introduces SDF to provide sufficient geometric information to the implicit function. Additionally, the volume feature encoding module captures voxel-aligned features from the parameterized model SMPL using self-attention mechanism, thereby improving the generalization ability in outdoor conditions.

2.2 Self-attention for Reconstruction

Recently, attention mechanisms have gained increasing attention in the field of computer vision due to their ability to capture global relationships of the target. Bhattacharyya et al. [13] proposed Sa-det3d for 3D object detection by combining self-attention features with convolutional features. Li et al. [14] introduced Transformer and used self-attention mechanism to design a feature extractor for point cloud data classification. Li et al. [15] proposed 3D-VRVT, which utilizes self-attention mechanism to reconstruct 3D voxels from a single image. Liu et al. [16] developed a model that combines convolutional neural networks with self-attention mechanism to reconstruct 3D faces. In a recent study [17], the Transformer model was employed for reconstructing human SMPL model. These works reveal the important role of self-attention mechanisms in 3D vision. However, the application of self-attention mechanisms in the implicit reconstruction of clothed human has not been explored yet.

3 Methodology

3.1 Overview

The network architecture of our method is shown in Fig. 1. Although PaMIR attempts to provide voxelized SMPL as a geometric prior for the implicit function,

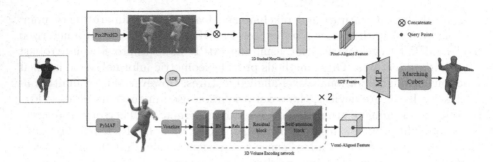

Fig. 1. The overall network architecture of this article. Given an image, we first predict the front and back normal maps and the SMPL model respectively. Then, we concatenate the front and back normal maps with the initial image and input them into the 2D Stacked HourGlass [18] to obtain the 2D feature maps. Additionally, we voxelize the SMPL model and input it into the 3D volume encoding network to obtain the 3D feature volume. Subsequently, the query points are projected onto the 2D feature maps and 3D feature volume, to obtain pixel-aligned features and voxel-aligned features respectively. Furthermore, the query points also calculate the SDF from the SMPL model. Finally, the three features are concatenated along the channel dimension and fed into the MLP to predict the occupancy values.

the voxel feature resolution is low ($32 \times 32 \times 32$) due to computer hardware limitations. We believe that the geometric prior provided by PaMIR is crude, which leads to poor reconstruction ability in outdoor conditions, while aggressively increasing voxel resolution would cause memory explosion. Additionally, we observe that the self-attention mechanism can capture correlations between features and extract richer information compared to ordinary convolutional layers.

Therefore, we make four improvement points based on PaMIR: (1) We use the PyMAF method [19] to estimate the parameterized SMPL model from the image. The PyMAF method provides more accurate estimation under outdoor conditions compared to the GCMR [20] used by PaMIR; (2) We use the pix2pixHD network [21] to predict the front and back normal maps of the human body as additional inputs; (3) We introduce the SDF to define the implicit function by combining pixel-aligned features and voxel-aligned features; (4) We introduce the self-attention mechanism to design an 3D volume encoding network, which obtains richer geometric information. This network consists of one convolutional layer, one BN layer, one ReLU layer, and a Residual Block, as well as a 3D Self-attention block. We repeat this structure twice to output 3D feature volume.

3.2   Implicit Function

Implicit functions take the features of query points as input and output occupancy, which indicates whether the point is inside or outside the surface, thus implicitly modeling a 3D object. In this paper, for each query point $P \in \mathbb{R}^3$, its features consist of three parts: pixel-aligned features, voxel-aligned features and SDF. The implicit function takes these three features as input and outputs occupancy value in the range of $[0, 1]$. The implicit function in this paper can be formulated as follows:

$$f(\mathcal{F}_I, \mathcal{F}_V, \mathcal{F}_{SDF}) \mapsto [0, 1] \tag{1}$$

where f represents the implicit function represented by a multi-layer perceptron (MLP), $\mathcal{F}_I$ represents pixel-aligned feature, $\mathcal{F}_V$ represents voxel-aligned feature and $\mathcal{F}_{SDF}$ represents the signed distance from a query point P to the nearest point of the parametric model SMPL, which is donated as $\mathcal{M}(\beta, \theta)$.

Pixel-Aligned Feature $\mathcal{F}_I$. The front and back normal maps can provide the model with rich front and back folding [2]. Thus, we predict front and back normal maps and feed these to the network as additional input to provide the pixel-aligned feature. We describe the pixel-aligned feature with normal maps as:

$$\mathcal{F}_I = \mathcal{B}(\pi(P), g_I(I, N_f, N_b)) \tag{2}$$

where I, N_f, N_b are inputs with a resolution of 512×512, representing the RGB image, the front-side normal map and back-side normal map, respectively. Pixel-aligned features $\mathcal{F}_I$ are obtained from the 2D feature map $g_I(I, N_f, N_b)$ by bilinear interpolation $\mathcal{B}(\cdot, \cdot)$ at the query point P. π using the orthogonal projection.

Voxel-Aigned Feature $\mathcal{F}_V$. To combine the robustness advantages of parametric models, we obtain voxel-aligned features from the parametric model SMPL $\mathcal{M}$. The voxel-aligned features can be formulated as:

$$\mathcal{F}_V = \mathcal{T}(P, g_V(V), \tag{3}$$

where V donates the feature volume obtained from $\mathcal{M}$. Voxel-aligned features $\mathcal{F}_V$ are obtained from the 3D feature map $g_V(V)$ by trilinear interpolation $\mathcal{T}(\cdot, \cdot)$ at the query point P.

SDF $\mathcal{F}_{SDF}$. We describe SDF as the signed distance between a query point and the closest point of the parametric model SMPL $\mathcal{M}$, and the SDF can be described as $\mathcal{F}_{SDF}$.

$$\mathcal{F}_{SDF} = sign(\|P - Q\|_2) \tag{4}$$

where $\|\cdot\|_2$ denotes the distance from the query point P to the closest body point $Q \in \mathcal{M}$. $sign(\cdot)$ indicates that $\mathcal{F}_{SDF}$ is positive if query point P is inside the SMPL $\mathcal{M}$, and negative otherwise.

3.3 Self-attention Block

As mentioned in Sect. 3.1, although PaMR incorporates SMPL as a shape proxy in the implicit function, leading to improved robustness compared to PIFu, the resolution of the utilized SMPL feature volume is low. This limitation results in inadequate fine geometric information and the inability to reconstruct reasonable models in challenging poses. To reconstruct realistic human meshes, not only accurate shape proxy are needed, but also richer geometric features are required. Therefore, we propose a SMPL volume feature encoding network with a self-attention mechanism to extract voxel-aligned features. By leveraging the ability of the attention mechanism to capture relationships between features, we enhance the geometric representation of voxel-aligned features, enabling the implicit function to accurately predict the occupancy value of query points.

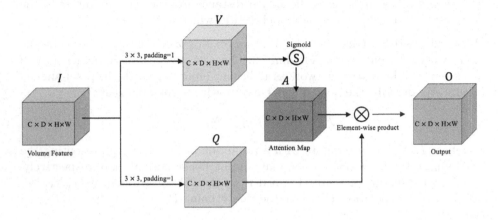

Fig. 2. The self-attention mechanism module structure.

The structure of the self-attention mechanism module is shown in Fig. 2. Given the volume feature I, we first send it to two 3D convolutional layers (with a convolutional kernel size of 3×3 and padding=1), obtaining voxel features V and Q, respectively. It is worth noting that I, V and Q have the same resolution and channel dimensions. Then, we use the sigmoid function to generate the attention map A based on the voxel features V. We denote the attention map A as follows:

$$A = \frac{1}{(1 + e^{-V})} \tag{5}$$

after the attention map A is given, we multiply A by Q to obtain the voxel feature output O, which has the same dimensions as I. We denote voxel feature O as:

$$O = A \times Q \tag{6}$$

3.4   Loss Fuction

Our method can be divided into two parts: normal maps estimation and implicit function reconstruction. We set different loss functions to supervise the training of these two parts, which is the basis of the effectiveness of our method.

Normal Maps Estimation Loss $\mathcal{L}_\mathcal{N}$. We use the loss function $\mathcal{L}_\mathcal{N}$ following PIFuHD [2]. The architecture of the front and back normal maps prediction networks utilizes the extended L1 loss function:

$$\mathcal{L}_\mathcal{N} = \mathcal{L}_{l1} + \lambda \mathcal{L}_{VGG} \tag{7}$$

where $\mathcal{L}_{l1}$ is an L1 loss function established between predicted normals and ground-truth normals. The $\mathcal{L}_{VGG}$ loss is a perceptual loss proposed by Johnson et al. [22]. λ is used to balance these two losses, which is set to 5.0 in our experiments.

Implicit Function Reconstruction Loss $\mathcal{L}_R$. We define the prediction of an implicit function for a 3D query point as binary occupancy value. Therefore, we use the extended BCE loss to supervise the reconstruction of the implicit function, which can be described as:

$$\mathcal{L}_R = \sum_{P \in S} f(\hat{P}) log f(P) + (1 - f(\hat{P}))(1 - log f(P)) \tag{8}$$

where S represents the set of sampling points, $f(\cdot) \in [0, 1]$ represents the ground-truth occupancy value of the query point P.

4   Experiments

4.1   Datasets and Evaluation Metrics

Datasets. Our method is mainly trained on the THUman2.0 dataset [8] and tested on the CAPE dataset [23]. The THUman2.0 dataset consists of 526 high-quality 3D clothed human scans in various poses, captured by a dense DSLR rig. We select the first 500 models as the training set and the last 26 models as the supplementary test set. The CAPE dataset is a public dataset generated based on the SMPL body model using vertex offset, and it has broad applications in 3D clothing capture and human body reconstruction. We use 150 models as the primary test dataset.

Metrics. We report Chamfer Distance(CD), Point-to-Surface(P2S) and Normal estimation errors (Normal) metrics to evaluate the effectiveness of our algorithm. P2S measures the distance between the ground-truth scan points and the closest reconstructed surface points. Both CD and P2S are used to assess the accuracy of geometric reconstruction, while Normal is employed to evaluate the high-frequency geometric details.

4.2   Implements Details

When extracting pixel-aligned features, the normal maps and initial images are maintained at a resolution of 512×512. The 2D image feature output has a total

Table 1. Ablation experiments on the four improvements points of this paper. (P=PyMAF, N=normal maps, S=SDF, A=Self-attention)

Model	CD↓	P2S↓	Normal↓
(1) PaMIR	2.122	1.495	0.088
(2) PaMIR+P	1.440	1.253	0.099
(3) PaMIR+P+N	1.581	1.228	0.104
(4) PaMIR+P+N+S	1.146	1.129	0.081
(5) PaMIR+P+N+S+A	**1.041**	**1.037**	**0.072**

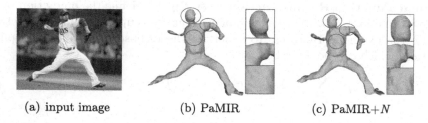

(a) input image   (b) PaMIR   (c) PaMIR+N

Fig. 3. Ablation experiment of introducing normal maps. After the introduction of the normal maps, the facial expression, clothing folds have been significantly improved.

of two layers with a resolution of [6, 128, 128] and [6, 128, 128], respectively. The volume encoding network takes the SMPL feature volume as input, with a resolution of $128 \times 128 \times 128$. The voxel feature output also has two layers, with dimensions of [8, 32, 32, 32] and [8, 32, 32, 32]. During training, losses are calculated for both the stacked hourglass network and the volume encoding network outputs. However, during inference, only the output of the final layer is used. Based on the above description, we obtain pixel-aligned features with channel numbers of 6 and 8, voxel-aligned features with channel number of 1 for the SDF. Therefore, we set the number of neurons in the MLP as (14, 1024, 512, 256, 128, 1).

In terms of hyperparameter settings, we set the batch size to 4 and the learning rate to 1.0×10^{-4}. Additionally, we employ the RMSProp weight optimizer with a weight decay of 0.0 and a momentum of 0.0 for pre-training. It is worth noting that we adopt the same 3D spatial sampling strategy as PIFu and PaMIR, using a combination of non-uniform sampling and importance sampling.

4.3   Ablation Study

We conducted ablation experiments on four improvement points based on the PaMIR model. As shown in the second row of Table 1, after the introduction of PyMAF, the geometrical error indicators CD and P2S further decreased. CD and P2S decreased by 32.1% and 16.2% respectively, while Normal did not show significant changes. As shown in Fig. 3, the introduction of normal maps has

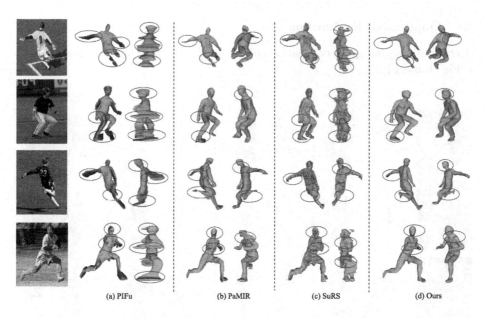

(a) PIFu (b) PaMIR (c) SuRS (d) Ours

Fig. 4. Compared with state-of-the-art methods, including (a) PIFu [1], (b) PaMIR [4] and (c) SuRS [6]. As shown in the comparison regions highlighted in the red boxes, our method accurately reconstructs challenging poses in the wild, like sports, without creating artifacts or broken limbs in both frontal and side views.

Table 2. Comparison with state-of-the-art methods.

Methods	THuman 2.0			CAPE		
	CD↓	P2S↓	Normal↓	CD↓	P2S↓	Normal↓
PIFu [1]	1.518	1.647	0.122	3.627	3.729	0.116
PIFuHD [2]	1.032	1.046	0.112	3.237	3.123	0.112
PaMIR [4]	1.713	1.818	0.134	2.122	1.495	0.088
SuRS [6]	**0.931**	**1.151**	0.167	–	–	–
Ours	1.054	1.152	**0.091**	**1.041**	**1.037**	**0.072**

improved the facial expressions and clothing folds of the model. As shown in the fourth row of Table 1, after introducing SDF, compared to the original PaMIR, all indicators showed significant improvement. CD, P2S, and Normal decreased by 46.0%, 24.5%, and 8.0% respectively. Moreover, it has improved the CD deterioration caused by introducing normal maps. Finally, with the introduction of the self-attention mechanism (the fifth row of Table 1), the model achieves optimal performance. Compared to Model(4), CD, P2S, and Normal each show additional reduction of 9.1%, 8.1%, and 11.1% respectively. The results show that the four improvement points in this paper are effective.

4.4 Qualitative Evaluation

We conduct a qualitative comparison of our method with other existing methods in Fig. 4 and Fig. 5.

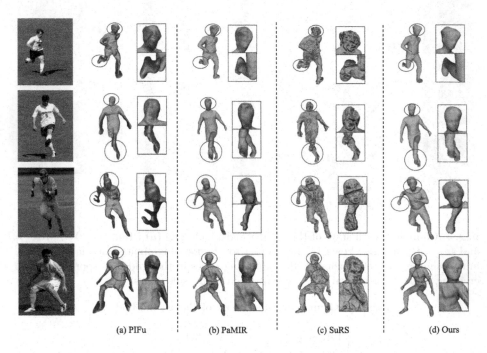

(a) PIFu (b) PaMIR (c) SuRS (d) Ours

Fig. 5. Compared with state-of-the-art methods, including (a) PIFu [1], (b) PaMIR [4] and (c) SuRS [6]. As shown in the local part's enlargements labeled with red boxes, our method can accurately reconstruct 3D clothed human bodies with more realistic surface details in various poses in the wild. (Color figure online)

In Fig. 4, we observe that the reconstructed meshes from the other three methods exhibit artifacts and broken limbs. While PaMIR reconstructs more reasonable models compared to PIFu and SuRS, noticeable incomplete body parts are still present. In contrast, our method demonstrates better robustness, preserving complete body parts such as arms, head and legs, without any artifacts. In Fig. 5, we can observe noisy artifacts in PIFu and PaMIR, while SuRS presents unrealistic facial features and clothing wrinkles. In contrast, our method exhibits clear and distinguishable facial expressions with realistic surface details.

4.5 Quantitative Evaluation

We report the experimental results of our approach in Table 2 and quantitatively evaluate it against the state-of-the-art methods.

As shown in Table 2, our approach outperforms PIFu, PIFuHD, and PaMIR in all metrics, with PaMIR serving as our baseline method. Although the SuRS method exhibits slight superiority in terms of CD and P2S values, our method achieves the best results in the Normal metric. Overall, the performance of our method is comparable to these state-of-the-art techniques.

5 Conclusion

We propose an improved implicit function-based neural network model for 3D clothed human reconstruction. Firstly, we introduce SDF to define the implicit function and enhance robustness. Secondly, we introduce self-attention mechanism on the SMPL volume feature encoding network to extract voxel-aligned features with richer geometric information. Additionally, we enhance local details through the introduction of normal maps. Through qualitative and quantitative testing, our approach improves robustness in outdoor conditions. The reconstructed mesh exhibits complete limbs and realistic surface details.

References

1. Saito, S., Huang, Z., Natsume, R., Morishima, S., Kanazawa, A., Li, H.: Pifu: pixel-aligned implicit function for high-resolution clothed human digitization. In: Proceedings of the IEEE/CVF International Conference on Computer Vision, pp. 2304–2314 (2019)
2. Saito, S., Simon, T., Saragih, J., Joo, H.: Pifuhd: multi-level pixel-aligned implicit function for high-resolution 3d human digitization. In: Proceedings of the IEEE/CVF Conference on Computer Vision and Pattern Recognition, pp. 84–93 (2020)
3. Lorensen, W.E., Cline, H.E.: Marching cubes: a high resolution 3d surface construction algorithm. In: Seminal Graphics: Pioneering Efforts that Shaped the Field, pp. 347–353 (1998)
4. Zheng, Z., Yu, T., Liu, Y., Dai, Q.: Pamir: parametric model-conditioned implicit representation for image-based human reconstruction. IEEE Trans. Pattern Anal. Mach. Intell. **44**(6), 3170–3184 (2021)
5. Loper, M., Mahmood, N., Romero, J., Pons-Moll, G., Black, M.J.: Smpl: a skinned multi-person linear model. In: Seminal Graphics Papers: Pushing the Boundaries, vol. 2, pp. 851–866 (2023)
6. Pesavento, M., Volino, M., Hilton, A.: Super-resolution 3D human shape from a single low-resolution image. In: Avidan, S., Brostow, G., Cissé, M., Farinella, G.M., Hassner, T. (eds.) Computer Vision – ECCV 2022: 17th European Conference, Tel Aviv, 23–27 October 2022, Proceedings, Part II, pp. 447–464. Springer, Cham (2022). https://doi.org/10.1007/978-3-031-20086-1_26
7. Chan, K.Y., Lin, G., Zhao, H., Lin, W.: IntegratedPIFu: integrated pixel aligned implicit function for single-view human reconstruction. In: Avidan, S., Brostow, G., Cissé, M., Farinella, G.M., Hassner, T. (eds.) Computer Vision – ECCV 2022: 17th European Conference, Tel Aviv, Israel, 23–27 October 2022, Proceedings, Part II, pp. 328–344. Springer, Cham (2022). https://doi.org/10.1007/978-3-031-20086-1_19

8. Zheng, Z., Yu, T., Wei, Y., Dai, Q., Liu, Y.: Deephuman: 3d human reconstruction from a single image. In: Proceedings of the IEEE/CVF International Conference on Computer Vision, pp. 7739–7749 (2019)
9. He, T., Collomosse, J., Jin, H., Soatto, S.: Geo-pifu: geometry and pixel aligned implicit functions for single-view human reconstruction. Adv. Neural. Inf. Process. Syst. **33**, 9276–9287 (2020)
10. Huang, Z., Xu, Y., Lassner, C., Li, H., Tung, T.: Arch: animatable reconstruction of clothed humans. In: Proceedings of the IEEE/CVF Conference on Computer Vision and Pattern Recognition, pp. 3093–3102 (2020)
11. He, T., Xu, Y., Saito, S., Soatto, S., Tung, T.: Arch++: animation-ready clothed human reconstruction revisited. In: Proceedings of the IEEE/CVF International Conference on Computer Vision, pp. 11046–11056 (2021)
12. Hong, Y., Zhang, J., Jiang, B., Guo, Y., Liu, L., Bao, H.: Stereopifu: depth aware clothed human digitization via stereo vision. In: Proceedings of the IEEE/CVF Conference on Computer Vision and Pattern Recognition, pp. 535–545 (2021)
13. Bhattacharyya, P., Huang, C., Czarnecki, K.: Sa-det3d: self-attention based context-aware 3d object detection. In: Proceedings of the IEEE/CVF International Conference on Computer Vision, pp. 3022–3031 (2021)
14. Li, Y., Cai, J.: Point cloud classification network based on self-attention mechanism. Comput. Electr. Eng. **104**, 108451 (2022)
15. Li, X., Kuang, P.: 3d-vrvt: 3d voxel reconstruction from a single image with vision transformer. In: 2021 International Conference on Culture-Oriented Science & Technology (ICCST), pp. 343–348. IEEE (2021)
16. Liu, Q.M., Jia, R.S., Zhao, C.Y., Liu, X.Y., Sun, H.M., Zhang, X.L.: Face super-resolution reconstruction based on self-attention residual network. IEEE Access **8**, 4110–4121 (2019)
17. Lin, K., Wang, L., Liu, Z.: End-to-end human pose and mesh reconstruction with transformers. In: Proceedings of the IEEE/CVF Conference on Computer Vision and Pattern Recognition, pp. 1954–1963 (2021)
18. Newell, A., Yang, K., Deng, J.: Stacked hourglass networks for human pose estimation. In: Leibe, B., Matas, J., Sebe, N., Welling, M. (eds.) ECCV 2016. LNCS, vol. 9912, pp. 483–499. Springer, Cham (2016). https://doi.org/10.1007/978-3-319-46484-8_29
19. Zhang, H., et al.: Pymaf: 3d human pose and shape regression with pyramidal mesh alignment feedback loop. In: Proceedings of the IEEE/CVF International Conference on Computer Vision, pp. 11446–11456 (2021)
20. Kolotouros, N., Pavlakos, G., Daniilidis, K.: Convolutional mesh regression for single-image human shape reconstruction. In: Proceedings of the IEEE/CVF Conference on Computer Vision and Pattern Recognition, pp. 4501–4510 (2019)
21. Wang, T.C., Liu, M.Y., Zhu, J.Y., Tao, A., Kautz, J., Catanzaro, B.: High-resolution image synthesis and semantic manipulation with conditional gans. In: Proceedings of the IEEE Conference on Computer Vision and Pattern Recognition, pp. 8798–8807 (2018)
22. Johnson, J., Alahi, A., Fei-Fei, L.: Perceptual losses for real-time style transfer and super-resolution. In: Leibe, B., Matas, J., Sebe, N., Welling, M. (eds.) ECCV 2016. LNCS, vol. 9906, pp. 694–711. Springer, Cham (2016). https://doi.org/10.1007/978-3-319-46475-6_43
23. Ma, Q., et al.: Learning to dress 3d people in generative clothing. In: Proceedings of the IEEE/CVF Conference on Computer Vision and Pattern Recognition, pp. 6469–6478 (2020)

Privacy-Preserving Federated Compressed Learning Against Data Reconstruction Attacks Based on Secure Data

Di Xiao[✉][iD], Jinkun Li, and Min Li

College of Computer Science, Chongqing University, Chongqing 400044, China
{dixiao,jkli,minli}@cqu.edu.cn

Abstract. Federated learning is a new distributed learning framework with data privacy preserving in which multiple users collaboratively train models without sharing data. However, recent studies highlight potential privacy leakage through shared gradient information. Several defense strategies, including gradient information encryption and perturbation, have been suggested. But these strategies either involve high complexity or are susceptible to attacks. To counter these challenges, we propose to train on secure compressive measurements by compressed learning, thereby achieving local data privacy protection with slight performance degradation. A feasible method to boost performance in compressed learning is the joint optimization of the sampling matrix and the inference network during the training phase, but this may suffer from data reconstruction attacks again. Thus, we further incorporate a traditional lightweight encryption scheme to protect data privacy. Experiments conducted on MNIST and FMNIST datasets substantiate that our schemes achieve a satisfactory balance between privacy protection and model performance.

Keywords: Federated learning · Data reconstruction attack · Privacy-preserving · Compressed learning · Lightweight encryption

1 Introduction

Artificial Intelligence (AI) is a transformative technology that enables machines to mimic human-like thinking, learning, and reasoning capabilities. The performance of AI systems heavily relies on the availability of a substantial amount of high-quality data. However, due to growing public concern over privacy issues and increasingly stringent data protection regulations, collecting training data that may contain private information has become a significant challenge. To tackle these challenges, federated learning (FL) [9] has emerged as a promising solution. FL offers a powerful approach to addressing the issues of "data islands" and privacy, making it a prominent area of research in the field of AI technology. Unlike traditional centralized machine learning training, FL aggregates model parameter updates from local devices to a central server, so as to realize model training and updating without sharing raw data.

B. Luo et al. (Eds.): ICONIP 2023, CCIS 1969, pp. 325–339, 2024.
https://doi.org/10.1007/978-981-99-8184-7_25

Unfortunately, this approach falls short in guaranteeing data privacy sufficiently. Numerous studies have demonstrated the ability to infer sensitive information from the shared gradient information. Membership inference attack [12] is the pioneering research that reveals privacy leakage in FL, by analyzing whether a specific sample was present in the training data. Building upon this, subsequent research has shown that user-level information can be reconstructed from the gradient [16]. Later, the data reconstruction attack that recovers the original data through the gradient is produced, fundamentally challenging the notion that FL adequately protects local data privacy. In this paper, we focus on data reconstruction attacks, which represent a type of inference attack.

Recently, there have been several data reconstruction attack methods [4,6, 18–20] proposed, which have demonstrated the ability of an attacker to reconstruct the local training data by exploiting the shared gradient. These methods primarily achieve data reconstruction by minimizing the distance between the gradient generated by the virtual data and label, and the gradient generated by the real data and label. To tackle the problem of data reconstruction attacks, various defense strategies have been proposed. These strategies can be broadly classified into two categories. The first category includes cryptographic-based methods, such as multiparty secure computation. These methods employ secure aggregation protocols, such as homomorphic encryption [1,8,13], to safeguard the original gradient information from exposure. The second category involves gradient perturbation methods, with differential privacy [15,17] being a prominent example. Furthermore, [20] proposes that gradient compression can effectively mitigate gradient privacy concerns. Another defensive approach suggested by Sun et al. [14], called Sotera, involves perturbing the data before gradient calculation. Unfortunately, the method proposed in [6] can reconstruct similar data even when gradient perturbation strategies such as gradient clipping, additive noise to gradient, gradient sparsification, and Sotera are employed.

Li et al. [6] propose to utilize the latent space of a generative adversarial network (GAN) trained on a public image dataset to compensate for information loss caused by gradient degradation. However, training on a dataset that is not relevant to the public image dataset can prevent GAN from learning relevant prior information. Hence, we are inspired to address the data reconstruction attack problem in FL by using secure data. Compressed sensing (CS) [3] is a mathematical framework for efficient signal acquisition and robust recovery. In addition to compression, CS measurements are encrypted. Calderbank et al. [2] introduce the concept of compressed learning (CL), where the inference system is directly built on CS measurements. Subsequent researches [7,10,11,21] have further improved CL, with [10] demonstrating that CL can achieve performance almost comparable to the original image domain. To the best of our knowledge, CL has not yet been explored in the context of FL to address data reconstruction attacks, making it a promising area for investigation.

In this paper, we design a privacy-preserving framework for FL based on CL, which exploits the secrecy property of CS measurements to achieve the defensive effect against data reconstruction attacks. Previous studies [10,21] have shown that jointly optimizing the sampling matrix and the inference network can improve model performance. However, as the sampling matrix is part of the model during the training process, it remains vulnerable to data reconstruction attacks. To address this issue, we introduce traditional cryptography to encrypt the training data, which protects the data from the threat of data reconstruction attacks. Different from the first two categories of defense methods based on gradient encryption and gradient perturbation, our approach focuses on training with secure data and is orthogonal to the previous methods. Experimental results further validate the feasibility of our proposed method.

Our main contributions can be summarized as follows:

- We propose to apply CL to address the privacy leakage problem caused by data reconstruction attacks in FL.
- To further enhance performance, we jointly optimize the sampling matrix and the inference network. Additionally, we introduce lightweight encryption novel methods in both the spatial and frequency domains to protect the training data.
- Experimental results demonstrate the effectiveness of our proposed scheme in achieving satisfactory privacy and utility.

2   Related Work

2.1   Data Reconstruction Attacks

Recently, Zhu et al. [20] propose to reconstruct the original data and label by making the ℓ_2 norm of the gradient of a pair of virtual data and label close to that of the gradient generated by the real data and label. A follow-up work [19] proposes to use the gradient of the last fully connected network layer to recover the true label. However, these approaches encounter difficulties when applied to large-scale and discontinuous network models. Then Geiping et al. [4] introduce a recovery method that is independent of the gradient magnitude. They utilize cosine similarity to measure the similarity between gradients, demonstrating that more complex data can be recovered even in deeper models. Later, Yin et al. [18] introduce a group consistency regularization term, enabling the reconstruction of a large batch of images after gradient averaging.

2.2   Defenses Against Data Reconstruction Attacks

The existing defense schemes for FL can be classified into two categories, one is based on gradient encryption, and the other is based on gradient perturbation.

Homomorphic encryption serves as an effective secure aggregation scheme for gradient protection. In [1], an asynchronous stochastic gradient descent scheme leveraging additive homomorphic encryption is proposed. To address the high computational complexity of homomorphic encryption aggregation on the server side, [13] introduces a calculation provider third party. Additionally, [8] proposes homomorphic encryption with multiple keys to resist collusion attacks. However, these methods often impose considerable computational complexity, rendering them unsuitable for resource-constrained users.

Another line of research explores perturbing the gradient to degrade it, thereby preventing privacy leaks. Differential privacy, which introduces random noise, can quantify and limit the disclosure of user privacy. The convergence of FL with the introduction of differential privacy is proven in [17]. Moreover, [15] incorporates local differential privacy into FL, and customizes improvements suitable for FL. While these gradient perturbation methods offer lower computational complexity, they still result in a decrease in model accuracy.

2.3   Compressed Sensing

Compressed sensing (CS) [3] is a mathematical paradigm that exploits signal redundancy to accurately reconstruct signals from a significantly reduced number of measurements compared to the Shannon sampling rate.

Specifically, given a signal $x \in \mathbb{R}^N$, the CS measurement $y \in \mathbb{R}^N$ is obtained by the sampling process $y = \Phi x$, where $\Phi \in \mathbb{R}^{M \times N}$ is the sampling matrix, $M \ll N$, and the sampling rate is defined as $\gamma = M/N$. Since the number of unknowns is much larger than the number of equations, it is often impossible to reconstruct x from the observations y. But if x is sparse in some basis Ψ, then x can be reconstructed from y, which is CS theory.

There are many studies devoted to solving the above reconstruction problem, and traditional optimization algorithms usually reconstruct the original signal x by solving the following optimization problem:

$$\min_{x} ||\Psi x||_1 \quad s.t. \ \ y = \Phi x, \tag{1}$$

where Ψ denotes some sparse basis on which x is sparse.

2.4   Compressed Learning

Compressed learning (CL) aims to perform inference tasks directly in the measurement domain without signal recovery. The concept is initially proposed by Calderbank et al. [2], who provide the first theoretical foundation for CL and demonstrate the feasibility of performing inference tasks in the compressed domain. Subsequently, Lohit et al. [7] utilize Convolutional Neural Network (CNN) for classification in the compressed domain. Building upon this, Adler et al. [21] devise an end-to-end deep learning solution for classification in the compressed domain, jointly optimizing the sampling matrix and inference operator. In [11], the authors demonstrate the robustness of the CL scheme by performing inference using partially observed image data. Recently, [10] employs an elaborate transformer network to conduct inference tasks in the measurement domain.

3   Methodology

In our approach, we utilize a CL scheme to project the measurement back into the image space, generating a noise-like proxy image in Sect. 3.1. This proxy image is directly used as the training data for the FL client to resist data reconstruction attacks. To further enhance model performance, we jointly optimize the sampling matrix and the subsequent inference network. However, during training, the original training data is re-input into the model, which will suffer from data reconstruction attacks. To address this, we propose encrypting the training data before training. In Sect. 3.2, we introduce a spatial domain encryption scheme. Additionally, to enhance the security of the encryption scheme, we propose a frequency domain encryption scheme in Sect. 3.3.

3.1 Training with Proxy Image

Motivation. In the scenario where we possess CS measurements, obtained through an imaging device like a single-pixel camera, these measurements often manifest as noise distributions. We can adopt the framework of CL and use the secure measurements as the training data of the client in FL, so that an honest but curious server is unable to extract any information about the original image, even if it obtains the measurements through data reconstruction attack methods. Furthermore, let us assume that the server has acquired our sampling matrix through some means, enabling them to potentially reconstruct the original image using CS reconstruction algorithms. However, it is essential to note that the distribution of CS measurements is closely linked to the image and is not identical, even when the same sampling matrix is applied. Therefore, before training the network model, we normalize all measurements, as this step is a crucial part of the neural network training process. Consequently, even if the normalized measurements are inferred through a data reconstruction attack, the inability to restore them to their original distribution prevents the recovery of the corresponding original images using CS reconstruction algorithms.

We propose a CL scheme to resist the data reconstruction attack and apply Fig. 1 to illustrate the whole process of the client during one communication round. For an image, we flatten it into a one-dimensional vector and obtain the measurement by CS sampling. The measurement obtained is a one-dimensional vector, which is usually projected back to the original image dimension and then used by CNN to extract features for image classification tasks. To restore the measurement back to the original image dimension, we can employ a matrix related to the sampling matrix (such as its transpose) or utilize a fully connected network layer. Consider that the data reconstruction attack method we used in the subsequent experiments is more effective for recovering image data, because of the addition of the total variation regularization term in its loss function. Therefore, we project the measurement back into the image space using the transpose of the sampling matrix and reshape it to obtain the proxy image. Subsequently, the proxy image undergoes normalization to ensure consistency in the gray value interval [0, 255]. This step is necessary due to significant changes in the pixel distribution of the proxy image. The normalized proxy image is then employed as training data for the network model. The server can only perform data reconstruction attacks using shared gradients to recover the proxy image, but can not obtain information about the original image.

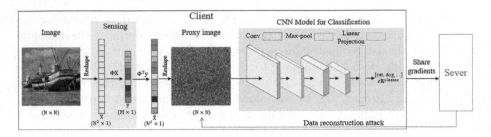

Fig. 1. The whole process of client local training when taking one picture as an example. Red arrows represents the data reconstruction attack. (Color figure online)

- Step 1: The client performs CS on the local dataset D. For one $N \times N$ image from D, it is first flattened into a one-dimensional vector. Then the flattened image is sampled by $y = \Phi X$ to obtain the measurement y, where $\Phi \in \mathbb{R}^{M \times N}$ and the sampling rate is M/N.
- Step 2: Next, the one-dimensional vector projected back into the image space is obtained by $\widehat{X} = \Phi^T y$, and the proxy image is obtained by reshaping it to $N \times N$.
- Step 3: The proxy image is normalized before being fed into the CNN model for image classification task training.
- Step 4: After the local training, the selected clients of this communication round upload the model updates to the server.
- Step 5: The server aggregates all the received model updates and transmits them back to all clients.

3.2   Training with Encrypted Data in the Spatial Domain

The scheme in Sect. 3.1 uses a fixed sampling matrix for all the training data. However, optimizing the sampling matrix with the subsequent inference network concurrently will further improve the network's inference performance. To achieve this, the original training data needs to be fed into the network during the training process. However, an honest but curious server could potentially utilize a data reconstruction attack method to access the original training data.

Therefore, in this section, we propose a novel approach to counter the data reconstruction attack by encrypting the image data prior to training in the neural network. However, it is crucial to strike a balance between usability and privacy, as excessive encryption of the images can render them untrainable. To address this concern, we explore a two-dimensional random permutation scheme for encrypting the image data.

For a two-dimensional image $X \in \mathbb{R}^{N \times N}$, multiplying the left and right sides of it by a row random permutation matrix P and a column random permutation matrix Q, respectively, and the encryption process can be expressed as

$$E(X) = PXQ. \tag{2}$$

The matrices P and Q are $N \times N$ square matrices with only one '1' in each row and column and zeros in the rest. We can use a chaotic system such as Logistic map to generate these two square matrices. By controlling the keys k1 and k2 as the initial values of the chaotic system, Algorithm 1 describes the process of generating P and Q using Logistic chaotic map. We apply Fig. 2 to illustrate the whole process of the client during one communication round.

Algorithm 1: Logistic chaotic system generates two random permutation matrices.

Input: Initial value x_0,y_0;Number of iterations N;Control parameter
$\mu = 3.56995$.

1 Initial chaotic sequence $S1 = \{\}; S2 = \{\}$;
2 **for** $i = 1$ *to* $i = N$ **do**
3 $\quad x_i = \mu \times x_{i-1} \times (1 - x_{i-1})$;
4 $\quad y_i = \mu \times y_{i-1} \times (1 - y_{i-1})$;
5 $\quad S1 = S1 \bigcup x_i$;
6 $\quad S2 = S1 \bigcup y_i$;
7 Sort $S1$ and $S2$ separately;
8 Denote their corresponding position indexes in the new ordered sequences as $indexS1$ and $indexS2$, which are in the range of $[1, N]$;
9 $P_{i,j} = \begin{cases} 1, & i = indexS1(j), \\ 0, & others \end{cases}, Q_{i,j} = \begin{cases} 1, & j = indexS2(i), \\ 0, & others \end{cases}$;
Output: Row random permutation matrix P; Column random permutation matrix Q.

- Step 1: The client encrypts each $N \times N$ image from the local dataset D, using the two-dimensional random permutation.
- Step 2: The encrypted images are normalized and then input into the sensing module.
- Step 3: The sensing module samples the flattened one-dimensional vector of each image with a learnable sampling matrix to obtain the measurements.
- Step 4: The measurements are first activated by ReLU and then projected back to the one-dimensional vector of the original image dimension using a fully connected network layer. The vector is reshaped to be input into the subsequent CNN module.
- Step 5: After the local training, the selected clients of this communication round upload their model updates to the server.
- Step 6: The server aggregates all the received model updates and transmits them back to all clients.

3.3   Training with Encrypted Data in the Frequency Domain

To enhance the security and resistance against statistical analysis and other attacks, it is beneficial to convert the image into a transform domain, such

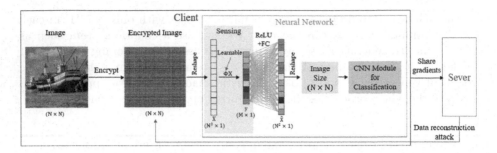

Fig. 2. The whole process of client local training when taking one picture as an example. Red arrows represents the data reconstruction attack. (Color figure online)

as the frequency domain, before encrypting it. Therefore, in this section, we propose a two-dimensional random permutation encryption scheme for the two-dimensional discrete cosine transform coefficients of the image.

For a two-dimensional image $X \in \mathbb{R}^{N \times N}$, we first apply the two-dimensional discrete cosine transform (2D-DCT) on it and obtain a transformed coefficient matrix $D \in R^{N \times N}$. This process can be expressed as

$$D = \Psi X \Psi^T, \tag{3}$$

where Ψ represents the sparse basis matrix of the discrete cosine transform. Next, the coefficient matrix D is encrypted using the two-dimensional random permutation scheme described in Sect. 3.2. This encryption process results in the encrypted coefficient matrix E. Finally, the encrypted coefficient matrix E is transformed back to the image domain using the two-dimensional inverse discrete cosine transform. The process can be expressed as

$$E = PDQ, \tag{4}$$

$$X' = \Psi^T E \Psi, \tag{5}$$

where P and Q represent the row random permutation matrix and column random permutation matrix, respectively, X' denotes the encrypted image after transforming back to the image domain. Due to the scrambling of the 2D-DCT coefficient matrix, the distribution range of pixel values undergoes significant changes after the shuffled coefficient matrix is transformed back to the image domain. Therefore, after encrypting, all encrypted images are normalized to the same gray value interval [0, 255], which further improves the security of the frequency domain encryption scheme. The overall client process represented by this scheme can also refer to Fig. 2.

4   Experiment

In this section, we begin by introducing our experimental settings in Sect. 4.1. The data reconstruction attack results on different datasets are presented in

Sect. 4.2. We conduct a security performance evaluation in Sect. 4.3, followed by the presentation of model performance results in Sect. 4.4. For ease of reference, we assign the names Scheme I, Scheme II, and Scheme III to the schemes discussed in Sects. 3.1, 3.2, and 3.3, respectively.

4.1 Experimental Settings

Datasets and Evaluation Metrics. We use the MNIST and FMNIST datasets for our experiments. These two datasets are commonly used image classification datasets and are widely used for deep learning algorithms. Our evaluation focuses on defense and performance. To assess the effectiveness of defense techniques, we employ the peak signal-to-noise ratio (PSNR) as a metric. A lower PSNR value indicates a greater visual difference. In terms of performance, we evaluate the test accuracy of model across different datasets.

Training Details. In our simulation experiments, the FL system consists of a central server and one hundred clients. The total number of training rounds is 100, and the local epoch in each round is 5. In each communication round, 50 clients are randomly selected to participate in FL training. For the MNIST dataset, we use the LeNet5 model [5], and the local batch size is set to 32. For the FMnist dataset, we use a small ConvNet model which contains three convolutional layers and a fully connected layer. The number of channels in the convolutional layer is 16,32,64, respectively, and the size of the convolution kernel is 5. The ReLU activation function and Max pooling are used after each convolution, and the local batch size is set to 64. We train our models on PyTorch using a 1080Ti card, and all models are optimized using SGD optimizer with momentum set to 0.9. The learning rate is initially set to 0.01, and the learning rate is reduced to 0.001 after 30 rounds of communication.

4.2 Defense Effect Against Data Reconstruction Attacks

We apply the Inverting Gradients (IG) method proposed in [4] to recover a single input image. The recovery process consists of 24,000 iterations, with a total variation term weight of 0.0001. To ensure the independence of recovery results from the initial seed, we conduct 10 repetitions of the experiment for each attack and report the result with the lowest loss.

Figure 3 displays the results of IG attack against the MNIST and FMNIST datasets. In the original FL scenario without privacy protection, the original data is fully exposed. However, in Scheme I, the recovered image is only a proxy image, which resembles the noise distribution, thereby making its recovery more challenging. In Scheme II, it can be observed that the recovered image is relatively clear. However, since the image itself is encrypted, the obtained image remains an unrecognizable encrypted form. Scheme III demonstrates that the recovered image has no correlation with the original image, indicating that the encryption effect is superior to that of Scheme II.

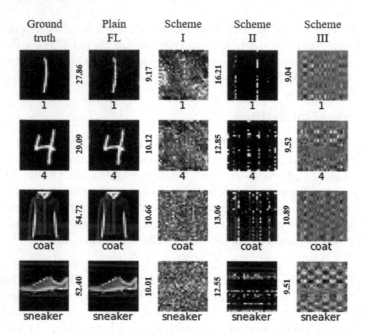

Fig. 3. Recovery results of IG attack on the MNIST and FMNIST datasets for different methods. For our schemes, the sampling rate is set to 0.1. The PSNR between the image and its ground truth is displayed on the left side of the image.

We also conduct IG attack experiments on the gradient compression method proposed in [20]. The defense effect under different sparsity ratios P is illustrated in Fig. 4. It is observed that the image remains recognizable until P = 0.01. It is worth noting that [20] claims resistance against data reconstruction attacks when the sparsity ratio approaches 0.8. However, while [20] utilizes a reconstruction attack based on minimizing the ℓ_2 norm between gradients, our paper employs IG attack based on minimizing the cosine similarity between gradients.

4.3   Secure Performance Evaluation

In Scheme I, we employ a linear transformation operation on the image. It is important to consider that if the CS sampling matrix is stolen, an attacker may utilize the least square method to reconstruct the original image based on the proxy image recovered through the data reconstruction attack, this process can be described as

$$\hat{X} = (\Phi^T \Phi)^{-1} X, \tag{6}$$

where $X \in \mathbb{R}^{N \times N}$ represents the proxy image recovered through the data reconstruction attack, and $\hat{X}$ denotes the final original image reconstructed using the least square method. After projecting the CS measurement back into the original image space, the proxy image is then normalized to the gray value interval of [0, 255]. So, if the pixel distribution of the proxy image is not restored, the attacker

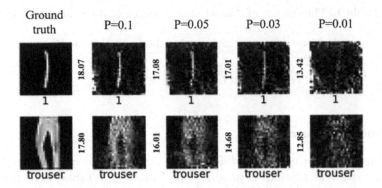

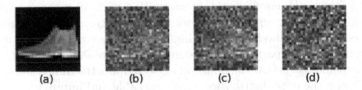

Fig. 4. Recovery results of IG attack at different sparsity ratios. The PSNR between the image and its ground truth is displayed on the left side of the image.

Fig. 5. (a) original image, (b) proxy image with a sampling rate of 0.25, (c) image inferred from the IG attack, (d) image reconstruct from the inferred image.

cannot obtain a relatively clear image using the least square method. This claim is further validated by the example depicted in Fig. 5.

In Scheme II and Scheme III, we employ the two-dimensional random permutation encryption method. In Scheme II, the encryption is applied to the original image, while in Scheme III, it is applied to the corresponding 2D-DCT coefficients of the original image. The key space for an $N \times N$ dimensional image is $(N!)^2$, which provides sufficient resistance against brute-force search attacks. It is worth noting that even if the encryption key is compromised, Scheme III remains secure due to the normalization of all encrypted images to the same gray value interval $[0, 255]$.

We employ information entropy to evaluate the level of chaos in the datasets associated with the three schemes. Given the strong correlation between adjacent pixels, image data tends to exhibit low entropy, indicating higher predictability. As depicted in Table 1, the corresponding datasets of Scheme I and Scheme III demonstrate relatively higher entropy, rendering the data less predictable. In Scheme II, only the pixel positions in the image are shuffled, while the distribution of pixel values remains unchanged, resulting in the same entropy as the original dataset. Consequently, Scheme III significantly enhances the security performance compared to Scheme II.

Table 1. The entropy of dataset corresponding to different schemes.

Dataset	Original	Scheme I	Scheme II	Scheme III
MNIST	1.60	7.13	1.60	6.89
FMNIST	4.12	7.19	4.12	7.06

4.4 Model Performance

The accuracy of the model on different test datasets for various compression sampling rates and schemes is presented in Table 2 and Table 3. For Scheme I, the model demonstrates improved performance with higher compression sampling rates. At a sampling rate of 0.25, the model performs reasonably well, slightly below the federated average (FedAvg) algorithm without protection. Turning to Scheme II, employing only encryption without a sensing module at a sampling rate of 1, an interesting observation emerges. When a sensing module is added, the final model achieves higher accuracy compared to the scheme with encryption alone, particularly at sampling rates of 0.25, 0.1, and 0.05. One potential explanation is that the addition of CS reduces the dimensionality of the data, rendering the data distribution more traceable and improving the model's performance. Scheme II outperforms Scheme I across all sampling rates. Introducing Scheme III to further protect the training data, we observe that, similar to Scheme II, the model's performance with encryption followed by CS surpasses that of the model with encryption alone (sampling rate of 1) at sampling rates of 0.25 and 0.1. Both Scheme II and Scheme III exhibit relatively minor performance degradation compared to the unprotected FedAvg algorithm.

In other privacy-preserving approaches in FL, the adoption of gradient encryption methods, such as homomorphic encryption, significantly increases computational complexity and communication overhead. On the other hand, gradient perturbation methods such as differential privacy approaches, often lack specific demonstrations of their protective effects and primarily focus on different privacy budget scenarios. For instance, in the study conducted by [15], the optimal test accuracy on the FMNIST dataset is reported as 86.93%, whereas the FedAvg algorithm achieves approximately 90% accuracy. Notably, some of our proposed schemes achieve comparable results at an appropriate sampling rate.

We evaluate the performance of the gradient compression (GC) method proposed in [20] with a prune ratio of 0.99. The accuracy achieved by GC on the MNIST and FMNIST test datasets is 96.78% and 85.74%, respectively. These results indicate that our schemes all outperform GC at some sampling rates.

Table 2. Model accuracy of different schemes on the MNIST test dataset.

Sampling Rate	Scheme I	Scheme II	Scheme III	FedAvg
1	–	96.30	96.23	
0.25	97.53	97.57	96.84	
0.1	95.32	97.34	96.53	98.98
0.05	91.55	96.94	95.96	
0.01	56.68	93.45	93.05	

Table 3. Model accuracy of different schemes on the FMNIST test dataset.

Sampling Rate	Scheme I	Scheme II	Scheme III	FedAvg
1	–	87.26	87.54	
0.25	86.69	87.98	87.98	
0.1	85.32	87.88	87.72	89.89
0.05	83.68	87.55	87.43	
0.01	58.69	85.27	84.57	

5   Conclusion

In this paper, we propose a novel privacy-preserving FL framework based on CL. Our approach introduces CL as a mechanism to address gradient leakage privacy concerns in FL, and we demonstrate its feasibility. Additionally, we propose the utilization of lightweight encrypted data as a protective scheme against data reconstruction attacks. Through simulation results, we validate the effectiveness of our schemes in resisting attacks, with slight impact on accuracy under suitable compression rates. In future research, we plan to explore additional protection schemes derived from CL for integration into FL. Furthermore, we aim to develop protection solutions tailored for resource-constrained device scenarios, ensuring their suitability and practicality.

Acknowledgements. The work was supported by the National Key R&D Program of China under Grant 2020YFB1805400, the National Natural Science Foundation of China under Grant 62072063 and the Project Supported by Graduate Student Research and Innovation Foundation of Chongqing, China under Grant CYB22063.

References

1. Aono, Y., Hayashi, T., Wang, L., Moriai, S., et al.: Privacy-preserving deep learning via additively homomorphic encryption. IEEE Trans. Inf. Forensics Secur. **13**(5), 1333–1345 (2017)
2. Calderbank, R., Jafarpour, S., Schapire, R.: Compressed learning: Universal sparse dimensionality reduction and learning in the measurement domain. preprint (2009)

3. Donoho, D.L.: Compressed sensing. IEEE Trans. Inf. Theory **52**(4), 1289–1306 (2006)
4. Geiping, J., Bauermeister, H., Dröge, H., Moeller, M.: Inverting gradients-how easy is it to break privacy in federated learning? Adv. Neural. Inf. Process. Syst. **33**, 16937–16947 (2020)
5. LeCun, Y., Bottou, L., Bengio, Y., Haffner, P.: Gradient-based learning applied to document recognition. Proc. IEEE **86**(11), 2278–2324 (1998)
6. Li, Z., Zhang, J., Liu, L., Liu, J.: Auditing privacy defenses in federated learning via generative gradient leakage. In: Proceedings of the IEEE/CVF Conference on Computer Vision and Pattern Recognition, pp. 10132–10142 (2022)
7. Lohit, S., Kulkarni, K., Turaga, P.: Direct inference on compressive measurements using convolutional neural networks. In: 2016 IEEE International Conference on Image Processing (ICIP), pp. 1913–1917. IEEE (2016)
8. Ma, J., Naas, S.A., Sigg, S., Lyu, X.: Privacy-preserving federated learning based on multi-key homomorphic encryption. Int. J. Intell. Syst. **37**(9), 5880–5901 (2022)
9. McMahan, B., Moore, E., Ramage, D., Hampson, S., y Arcas, B.A.: Communication-efficient learning of deep networks from decentralized data. In: Artificial Intelligence and Statistics, pp. 1273–1282. PMLR (2017)
10. Mou, C., Zhang, J.: TransCL: transformer makes strong and flexible compressive learning. IEEE Trans. Pattern Anal. Mach. Intell. **45**(4), 5236–5251 (2023)
11. Nair, A., Liu, L., Rangamani, A., Chin, P., Bell, M.A.L., Tran, T.D.: Reconstruction-free deep convolutional neural networks for partially observed images. In: 2018 IEEE Global Conference on Signal and Information Processing (GlobalSIP), pp. 400–404. IEEE (2018)
12. Nasr, M., Shokri, R., Houmansadr, A.: Comprehensive privacy analysis of deep learning: passive and active white-box inference attacks against centralized and federated learning. In: 2019 IEEE Symposium on Security and Privacy (SP), pp. 739–753. IEEE (2019)
13. Park, J., Lim, H.: Privacy-preserving federated learning using homomorphic encryption. Appl. Sci. **12**(2), 734 (2022)
14. Sun, J., Li, A., Wang, B., Yang, H., Li, H., Chen, Y.: Soteria: provable defense against privacy leakage in federated learning from representation perspective. In: Proceedings of the IEEE/CVF Conference on Computer Vision and Pattern Recognition, pp. 9311–9319 (2021)
15. Truex, S., Liu, L., Chow, K.H., Gursoy, M.E., Wei, W.: LDP-fed: federated learning with local differential privacy. In: Proceedings of the Third ACM International Workshop on Edge Systems, Analytics and Networking, pp. 61–66 (2020)
16. Wang, Z., Song, M., Zhang, Z., Song, Y., Wang, Q., Qi, H.: Beyond inferring class representatives: user-level privacy leakage from federated learning. In: IEEE INFOCOM 2019-IEEE Conference on Computer Communications, pp. 2512–2520. IEEE (2019)
17. Wei, K., et al.: Federated learning with differential privacy: algorithms and performance analysis. IEEE Trans. Inf. Forensics Secur. **15**, 3454–3469 (2020)
18. Yin, H., Mallya, A., Vahdat, A., Alvarez, J.M., Kautz, J., Molchanov, P.: See through gradients: image batch recovery via gradinversion. In: Proceedings of the IEEE/CVF Conference on Computer Vision and Pattern Recognition, pp. 16337–16346 (2021)
19. Zhao, B., Mopuri, K.R., Bilen, H.: iDLG: improved deep leakage from gradients. arXiv preprint arXiv:2001.02610 (2020)

20. Zhu, L., Liu, Z., Han, S.: Deep leakage from gradients. In: Advances in Neural Information Processing Systems 32: Annual Conference on Neural Information Processing Systems 2019, NeurIPS 2019, Vancouver, BC, Canada, 8–14 December 2019, pp. 14747–14756 (2019)
21. Zisselman, E., Adler, A., Elad, M.: Compressed learning for image classification: a deep neural network approach. In: Handbook of Numerical Analysis, vol. 19, pp. 3–17. Elsevier (2018)

An Attack Entity Deducing Model
for Attack Forensics

Tao Jiang, Junjiang He$^{(\boxtimes)}$, Tao Li, Wenbo Fang, Wenshan Li, and Cong Tang

Sichuan University, Chengdu 610065, Sichuan, China
hejunjiang@scu.edu.cn

Abstract. The forensics of Advanced Persistent Threat (APT) attacks, known for their prolonged duration and utilization of multiple attack methods, require extensive log analysis to discern their attack steps. Facing the massive amount of data, researchers have increasingly turned to extended machine learning methods to enhance attack forensics. However, the limited number of attack samples used for training and the inability of the data to accurately represent real-world scenarios pose significant challenges. To address these issues, we propose ASAI, an attack deduction model that leverages auxiliary strategies and dynamic word embeddings. Firstly, ASAI tackles the problem of data imbalance through a sequence sampling method enhanced by a custom auxiliary strategy. Subsequently, the sequences are transformed into dynamic vectors using dynamic word embedding. The model is trained to capture the spatio-temporal characteristics of entities under diverse contextual conditions by employing these dynamic vectors. In this paper, ASAI is evaluated using ten real-world APT attacks executed within an actual virtual environment. The results demonstrate ASAI's ability to successfully recover the key steps of the attacks and construct attack stories, achieving an impressive F1 score of up to 99.70%-a significant 16.98% improvement over the baseline which uses one-hot embedding after resample.

Keywords: Causal graphs · Attack forensics · Natural language processing · Entity sequences

1 Introduction

APTs are notorious for their stealth, precision, and targeted nature. They utilize sophisticated, persistent, and highly effective threats, often leveraging zero-day exploits. As a result, conducting forensics on the symptoms of these attacks becomes a vital and indispensable tool in preventing similar future assaults. In a broader context, attack forensics methods involve the collection of diverse audit logs from multiple hosts, applications, and network interfaces. These vast volumes of logs are typically analyzed offline or monitored in real-time to diagnose system failures, as well as identify and address intricate threats and vulnerabilities.

B. Luo et al. (Eds.): ICONIP 2023, CCIS 1969, pp. 340–354, 2024.
https://doi.org/10.1007/978-981-99-8184-7_26

In recent years, the accurate and efficient identification of attack entities within logs has garnered significant attention among scholars. Extensive research has been conducted on constructing causal dependency graphs from audit logs [10,12], enabling the use of query systems to locate key attack phases such as compromised processes or malicious payloads [5,20]. Causal graphs provide a depiction of the data flow relationships between system entities [11] enriched with contextual information. By leveraging these causal dependency graphs, some research systems employ advanced machine learning (ML) techniques to extract features or sequences from graphs, facilitating automated intrusion and fault detection [18,26]. Other techniques focus on identifying associations between different log events through event correlation [28], while additional studies concentrate on extracting dependency subgraphs [24,27] to represent and detect attack subgraphs, thus aiding in locating the attacking entities.

However, the above approaches primarily face two challenges struggling to accurately pinpoint the key attack steps. Research [13] reviews that attacking entities only account for 0.29% of non-attacking entities, and this proportion continues to decrease over time. So model training is severely affected by the unbalanced dataset problem [19]. Secondly, there is a issue regarding the representational validity of sequences, specifically the impact of sequence length and representation on deducing attacking entities. Therefore, we need to design reasonable sequence sampling and representation techniques, this can better describe attacker's behavioral patterns and attack strategies within sequence modeling, ultimately improving the model's ability to deduce attack entities.

To address the aforementioned challenges, we propose ASAI, an Attack Deduction model based on Auxiliary Strategies and Dynamic Word Embeddings. The approach involves several key steps. Firstly, sequences are extracted from the provenance graph, and the data imbalance issue is tackled through the auxiliary strategy sequence sampling method. Next, to ensure effective sequence characterization, the problem of variable sequence lengths is addressed using the adaptive performance of Doc2Vec. Subsequently, a bidirectional word embedding model is employed to capture the spatiotemporal characteristics of the sequences, generating dynamic vectors for characterization. These dynamic vectors serve as inputs to the model, thereby optimizing the deduction effect of attack entities for downstream tasks. The main contributions of this paper are summarized as follows:

- We propose an auxiliary strategy sampling method that enriches attack details, increases the distinction between positive and negative samples, and improves the model's generalization ability.
- We introduce a novel entity sequence representation method based on dynamic word embeddings, which enhances the vector space with more contextual semantics.
- By combining the aforementioned methods, we propose ASAI, an attack entity deduction model. Evaluation results of the event-based model demonstrate an impressive F1 score of 99.68%.

2   Related Work

ASAI's work on attack forensics is related to three main directions, which are, causality analysis, anomaly-based analysis, and application of machine-learning techniques.

Causality Analysis. Numerous studies have made significant contributions to the field of causality analysis [7,17,22]. These studies have focused on streamlining provenance graphs using techniques like software reverse engineering or runtime dynamic tracing. However, it is important to acknowledge that the flow of units can vary across different applications, and reversing a program can impose a heavy workload. Additionally, this process typically requires kernel-level daemon development, making it less portable and adaptable to different environments. In order to address challenges associated with causality analysis, researchers have proposed various approaches [5,8,25]. These strategies aim to mitigate issues such as coarse-grained analysis and dependency explosion. Heuristics such as log fusion, threat scoring methods, label decay, and other rules have been employed. While these techniques provide some relief, it is crucial to note that the rules themselves need to be regularly updated and maintained. As time passes and scenarios evolve, managing and updating these rules becomes prohibitively expensive.

Anomaly-Based Analysis. Anomaly-based approaches have proven effective in detecting unknown attacks by learning the normal behavior of a system to identify deviations. However, they are susceptible to higher false positive rates, because system instability and bias in the training dataset caused by prolonged operations. Priotracker [16] proposes a forward-backward causal tracker for anomaly detection, analyzing the scarcity of events to detect anomalies. Nodoze [6] computes anomaly scores based on event frequency and reduces false alarms by aggregating these scores. Provdetector [23] identifies stealth malware by learning the sequence of normal execution paths from the provenance graph. Log2vec [15] customizes the representation of system logs to identify invisible anomaly sequences in a clustered manner to detect stealth malware. Unicorn [3] constructs graph sketches by learning models from normal source graphs to identify anomalies. In contrast to anomaly-based approaches that primarily focus on learning user behavior, ASAI goes beyond by learning both attack and non-attack sequences. It utilizes their temporal and causal relationships to effectively reduce both false positives and false negatives. By considering the context of attacks and leveraging the interplay between different sequences, ASAI achieves a more comprehensive understanding of attack entities, resulting in improved detection accuracy and lower false alarm rates.

Learning-Based Analysis. Learning-based attack forensics [1,4,14,24] utilize machine learning techniques to train detection models using logs. Sigl [4] introduces an LSTM-based self-encoder that captures node context during malware installation. However, the node-level approach is susceptible to the dependency explosion problem, limiting its effectiveness. Atlas [1] combines causal analysis, natural language processing, and machine learning techniques to build sequence-based models. Nevertheless, it struggles to accurately represent actual attack scenarios due to data sampling limitations. Prographer [24] extracts temporal snapshots from logs to detect attacks, while logkernel [14] employs graph kernels and clustering methods to distinguish attacks from benign activities. Additionally, Shadewatcher [27] identifies network threats by predicting the preferences of system entities using graph neural networks. However, these graph-level approaches [2,9,24,27] often yield coarse-grained results that do not fully capture the entire attack process. In contrast, ASAI only requires the extraction and training of sequences. By utilizing the obtained model, ASAI can pinpoint attack entities at the node level, providing more fine-grained insights into the attack process.

3 Proposed Method

Initially, we construct the provenance graph from the log data, followed by sampling the sequences derived from the graph. Subsequently, we proceed to perform the characterization of the processed sequences, culminating in model training. The ASAI (Auxiliary Strategy based Attack Investigation) architecture in Fig. 1 is described in detail.

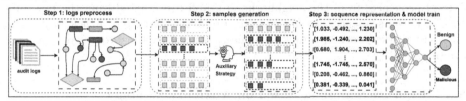

Fig. 1. Overview of ASAI architecture.

3.1 Definitions

Neighborhood Graph. A neighborhood graph refers to a subset of the causal graph that encompasses all entities and the relationships (edges) between them. This type of graph is commonly employed to depict the localized processes associated with the activities of an entity.

Event. An event is extracted from the neighborhood graph to delineate a particular activity taking place within the system's operation. A fundamental event comprises four essential elements $(E_{src}, Operation, E_{dest}, T)$, where E_{src} represents the source entity initiating the event, while E_{dest} denotes the target entity receiving the operation. The *operation* signifies the specific action being performed between the entities. Lastly, T denotes the timestamp indicating the occurrence of the event.

Sequence. We extract a sequence related to a specific entity from the causal graph. The sequence comprises all the events present in the neighborhood graph of the entity, arranged in chronological order.

3.2 Sequence Extraction

In the process of extracting training data for the model, we depend on attack entities to isolate attack and non-attack sequences. During the simulation of an attack, entities involved in the malicious operation, including files, processes, etc., are labeled as malicious. On the other hand, entities not directly associated with the attack are marked as benign.

Attack Sequence Extraction. The attack sequence extraction steps are as follows. We 1) obtains the set N of all attacking entities from the provenance graph, 2) traverse the subset of attacking entities in the set $N_1\{n_1, n_1\}$, ... , $N_i\{n_j, ... , n_k\}$. 3) For each attack entity subset N_i, ASAI extracts its neighborhood graph, followed by a sequence S from the established neighborhood graph, and the sequence S consists of attack events E arranged in chronological order. 4) The source or target nodes in the event belong to N, then the event E is defined as an attack event. If all the events in the sequence are attack events, then the sequence is defined as an attack sequence.

Non-attack Sequence Extraction. The goal of sequence model learning is to accurately learn and identify the boundary between malicious and non-malicious activities. Thus for each non-attack entity, the non-attack sequences are extracted by going through the following steps. We 1) add a non-attacking entity to each subset of attacking entities to form a new subset of entities, 2) extract sequences from the neighborhood graph constructed from the new subset. 3) If the sequences do not match any of the extracted attack sequences, they are marked as non-attack sequences, otherwise, the processed sequences are discarded.

3.3   Sequence Sampling Based on Auxiliary Strategy

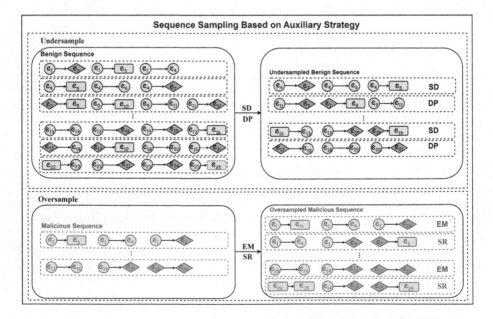

Fig. 2. Sequence Sampling Based on Auxiliary Strategy. SD, DP, EM, and SR stand for Similarity Deduplication, Distance Preservation, Entity Mutation, Sequence Recombination, respectively.

We introduce an auxiliary strategy to address the issue of data imbalance. This strategy consists of two components: the attack sequence oversampling strategy and the non-attack sequence undersampling strategy. Illustrated in Fig. 2, the original attack sequence comprises a limited number of attack stages. To tackle this limitation, we use the available attack events to create additional attack sequences that represent different attack phases. This approach helps generate more diverse samples of attack sequences with unique meanings. Regarding non-attack sequences, it is crucial not only to reduce their overall count but also to maintain differentiation from the attack sequences. To achieve this, we propose two techniques. Firstly, we use a non-attack sequence deduplication method based on the edit distance algorithm, which can remove redundant non-attack sequences. Secondly, we introduce a non-attack sequence optimization approach based on the sequence comparison algorithm, this method refines the non-attack sequences and ensures their distinctiveness from the attack sequences. By combining these strategies, we aim to mitigate the data imbalance problem, enhance the representation of both attack and non-attack sequences, and foster a more effective training process for our model.

3.3.1 Attack Sequence Oversampling Strategy Entity Mutation Oversampling Method.

Sequence contains four major types of process, file, network, and operation after sequence lemmatization [21], as shown in Table 1. For the extracted attack sequences, one word is randomly changed to another word of the same type based on lemmatization. This replacement does not change the mutation sequence, but it adds some similar sequences.

Table 1. Vocabulary set for lemmatization.

Vocabulary Types	Vocabulary List
Process	system_process, lib_process, user_process, etc.
File	system_file, lib_file, user_file, combined_files, etc.
Network	ip_address, domain, url, connection, session, etc.
Operation	read, write, delete, execute, invoke, fork, refer, bind, etc.

Sequence Recombination Oversampling Method. The sequence recombination oversampling method is founded on the observation, which shows an attacker's attack process can be divided into four crucial phases: defensive breakthrough, channel establishment, lateral penetration, and information stealing. Each of these phases involves distinct attack strategies that result in different attack sequences. By combining attack sequences from various phases, new attack scenarios can be formed, leading to the generation of recombinant sequences. For example, $\{S_1, S_2, ..., S_n\}$ represents the channel establishment phase, and by Sequence recombination we can get $\{S_i, S_j, ..., S_m\}$, which is a sequence that has not appeared in other attack phases with the same sequence.

The recombinant sequences effectively contribute to the enrichment of the attack sample, significantly enhancing its diversity and representativeness.

3.3.2 Non-attack Sequence Undersampling Strategy Similarity Filtering Undersampling Method.

The non-attack sequence filtering method, based on the edit distance algorithm (LD), aims to assess the similarity between non-attack sequences after word reduction. When the similarity between two non-attack sequences exceeds a defined threshold, these sequences are filtered out. The LD algorithm calculates the minimum number of transformations required to convert one sequence into another, considering possible operations such as character insertion, deletion, or replacement. To calculate the similarity between two sequences using LD, we determine the shortest path in the edit distance matrix from the top left corner to the bottom right corner. Each step in the path incurs a cost of 1, and the length of the path corresponds to the value of the LD. By utilizing LD, we can calculate the shortest distance by inserting, deleting, and replacing words in the vocabulary. If $|S_1|$ and $|S_2|$ are used to denote the lengths of S_1, S_2 non-attack sequences respectively, then their LD are $lev_{S_1,S_2}(|S_1|,|S_2|)$, which is calculated as

$$lev_{S_1,S_2}(i,j) = \begin{cases} max(i,j) & if min(i,j) = 0 \\ min \begin{cases} lev_{S_1,S_2}(i-1,j)+1 \\ lev_{S_1,S_2}(i,j-1)+1 \\ lev_{S_1,S_2}(i-1,j-1)+1_{(a_i \neq b_j)} \end{cases} & otherwise \end{cases} \tag{1}$$

where $lev_{S_1,S_2}(i,j)$ denotes the LD between the first i characters of S_1 and the first j characters of S_2.

Distance-Preserving Undersampling Method. The non-attack sequence optimization method, based on the sequence comparison algorithm (SW), filters out non-attack sequences that exhibit excessive local similarity to the attack sequence. This is achieved by calculating the local sequence correlation between the attack and non-attack sequences. To illustrate this method, let's consider the attack sequence $A = (a_1, a_2, ..., a_n)$ and the non-attack sequence $B = (b_1, b_2, ..., b_m)$ with lengths n and m respectively. After establishing the permutation matrix and null penalty score, a score matrix H with dimensions $(n+1)$ rows and $(m+1)$ columns is created. The initial elements of the first row and column in the score matrix are set to 0, while the remaining positions are scored iteratively from left to right and top to bottom using Eq. 2. Subsequently, backtracking is performed from the element with the highest score towards the element with a score of 0, moving from right to left. During this process, the locally similar segments encountered form $Similarclips(A, B)$. Each non-attack sequence is compared with all attack sequences to calculate $Similarclips(others, B)$. The resulting local sequences are stored in the list $list\_Similarclips$. The local sequence with the longest length in the list represents the implied attack feature value of the non-attack sequence. This calculation is performed for each non-attack sequence, allowing us to obtain the implied attack feature value for each of them. Subsequently, the non-attack sequences are optimized in descending order based on their feature values. Finally, the non-attack sequences are optimized in accordance with the number of attack sequences, prioritizing those ranked at the top.

$$H_{i,j} = \begin{cases} H_{i-1,j-1} + S(a_i, b_j) \\ H_{i-k,j} - W_1 \\ H_{i,j-l} - W_1 \\ 0 \end{cases} \tag{2}$$

where W_1 is 2 (can be set as needed), $S(a_i, b_j) = \begin{cases} +3, a_i = b_j \\ -3, a_i \neq b_j \end{cases}$.

3.4 Sequential Representation Based on Dynamic Word Embedding

We introduce the dynamic word embedding method to characterize the sequence, which includes two parts: static word embedding through Doc2Vec and dynamic word embedding through Bi-GRU, and the dynamic word embedding method is shown in Fig. 3.

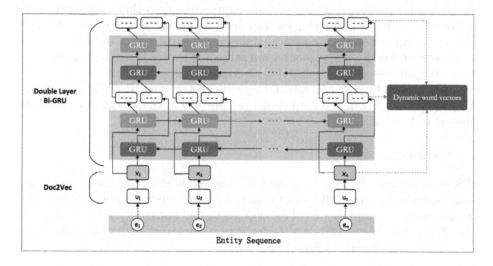

Fig. 3. Dynamic word embedding model.

To begin, both attack and non-attack sequences are transformed into static word vector representations using the word embedding feature of Doc2Vec. This static word embedding process is adaptable to sequences of varying lengths. These converted static vectors are then inputted into a two-layer Bidirectional Gated Recurrent Unit (Bi-GRU) network. The purpose of this network is to capture the contextual information and spatio-temporal features of the sequences. By employing a dynamic word embedding process, the Bi-GRU network can provide a more comprehensive characterization of the sequences. Finally, the sequence features are utilized for the downstream task of entity deduction.

3.5   Model Training and Application

Model Training Phase. First, the attacking and non-attacking entities (labeled by manual tagging) are extracted from the audit logs by preprocessing, the sets of attacking and non-attacking entities are extracted from the causal graph, next, the corresponding attacking and non-attacking events are extracted from the neighborhood graph, which is constructed using the obtained sets of attacking and non-attacking entities. These events are then converted into sequences based on the timestamped order of their occurrence. Then, the attack sequences and non-attack sequences are made relatively balanced by the auxiliary sampling strategy. Finally, dynamic word embedding encode the contextual information and temporal relationships to vectors, which serve as inputs to the classification model, enabling it to learn and make predictions.

Attack Entity Deduction Phase. Attack entities are automatically discovered by querying a trained model that relies on a known attack entity. The attack entity derivation process is as follows: first, and based on the defined attack entities, the entities in the causal graph are divided into attack entities and unknown entities, and a subset containing only one unknown entity is constructed. Then, the attacking entities are added to each entity subset, so that each entity subset contains all known attacking entities and one unknown entity, and all entity subsets are converted into sequence, and several sequences of different lengths are generated. Finally, the sequences are subjected to word form reduction, word embedding process and then passed to the trained model to predict whether each sequence is an attack or non-attack by prediction scoring. If the sequences are classified as attack sequences, the unknown entities in the subset of entities are inferred to be attack entities.

4   Experiments

This section primarily consists of the following validation experiments: (i) evaluation experiments on the efficacy of sequence sampling using auxiliary strategies, (ii) assessment experiments on the effectiveness and optimality of dynamic word embedding, and (iii) validation experiments on the attack entity inference capability of the attack entity inference model.

4.1   Datasets

In this paper, we conducted experiments using ten attacks based on real-world attack reports and generated audit logs in a controlled test environment [1]. To simulate diverse normal user activities, we manually generated benign user activities that encompassed browsing different websites, executing various applications (such as reading emails and downloading attachments), and connecting to other hosts. We aimed to replicate the typical activities of a workday, and these operations were randomly performed within an 8-h time window during the day. Table 2 provides an overview of the specific attacks for which evaluation data were generated. Attack IDs S-1 to S-4 indicate attacks executed on a single host, while IDs M-1 to M-6 represent attacks executed across multiple hosts.

4.2   Environment

The experimental runtime environment in this section includes local code client and remote server, local host and remote server hardware configurations are amd ryzen 5 2400G and intel Core i9-12900kf-32GB, debugging code in pycharm in the local host, remote connection to the server side application container engine docker through ssh, server side running environment: python version is 3.7.7, tensorflow version is 2.3.0, keras version is 2.4.3, matplotlib version is 2.2.5, numpy version is 1.19.5, fuzzywuzzy version is 0.18.0, networkx version is 2.2, word embedding maximum input the number of features is 31, the embedding dimension is 128, the longest input length of the sequence is 400, the batch of the double-layer gru algorithm is 100, the number of iterations epochs is 10, and the output dimension is 256.

Table 2. Evaluation data generated by implementing attacks.

Attack ID	Entity			Event			Sequence		
	Attack	Benign	Total	Attack	Benign	Total	Attack	Benign	Total
S-1	22	7,445	7,468	4,598	90,464	95,062	42	13,698	13,740
S-2	12	34,008	34,021	15,073	382,879	397,952	42	13,366	13,408
S-3	26	8,972	8,998	5,165	123,152	128,317	21	8,077	8,098
S-4	21	13,016	13,037	18,062	107,551	125,613	32	11,693	11,725
M-1	28	17,565	17,599	8,168	243,507	251,675	95	30,626	30,721
M-2	36	24,450	24,496	34,965	249,365	284,330	98	31,143	31,241
M-3	36	24,424	24,481	34,979	299,157	334,136	98	31,542	31,640
M-4	28	15,378	15,409	8,236	250,512	258,748	89	29,787	29,876
M-5	30	35,671	35,709	34,175	667,337	701,512	90	28,523	28,613
M-6	42	19,580	19,666	9,994	344,034	354,028	82	25,730	25,812

4.3   Evaluation Method

To assess the effectiveness of the attack entity deduction method, the model training and attack evaluation will be conducted separately for each of the ten different attacks. For instance, when identifying an attack on S-1, the log data from S-2, S-3, and S-4 will be utilized as training data for the attack deduction model. Similarly, when identifying an attack on M-1, the log data from the other multiple hosts will be used as the training data for the model.

The main two perspectives in the model evaluation phase include entity-based evaluation and event-based evaluation.

Entity-Based Evaluation. Firstly, a number of sequences are generated based on the combination of the input attack entities and unknown entities, and then, the generated sequences are checked to see if they are predicted by the model as attack sequences or non-attack sequences, and then Analyze the unknown entities are related to attack events to determine whether the unknown entities are attack entities.

Event-Based Evaluation. Most of the real-world attack forensics are presented in the form of attack events. To further help security personnel in attack forensic analysis, it is necessary to identify and locate attack events, traverse all events in the audit logs, and mark the event as an attack event if the subject or object of the event matches the attack entity, which is identified through the entity deduction model.

Finally, entities and events classified as attack and non-attack are compared with their baseline labels, and the model deduction effectiveness is evaluated using Accuracy, Precision, Recall, and F1-score evaluation metrics.

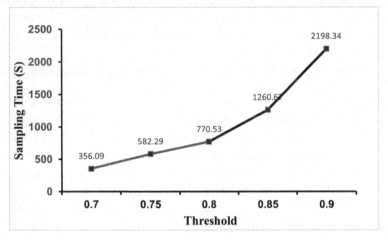

Fig. 4. Effect of Threshold Selection on Efficiency.

4.4   Efficiency Experiment of Sequence Sampling

The attack sequence performs sequence oversampling with the strategy of random combination of sequences and mutation of individual entities, which is a computationally lightweight sampling method, on the contrary, the non-attack sequence performs sequence undersampling with the strategy of calculating the similarity of the individual sequences, which is a computationally intensive sampling method, and the experiments show that the non-attack sequence undersampling time consumes more than 98% of the whole sampling time. In addition, the undersampling time is closely related to setting the sampling threshold. The results of the average sampling time and threshold selection experiments are shown in Fig. 4. The sampling time is positively correlated with the threshold value, and the time overhead grows faster when the threshold value is greater than 0.8, therefore, the final choice of this experiment is to set the threshold value to 0.8.

4.5   Comparison Experiment of Sequence Sampling

The experiments in this subsection focus on evaluating the optimality of the validation experiments. By keeping other parameters and configurations unchanged, the effects of using three different sampling methods are examined: undersampling only, oversampling only, and the auxiliary strategy sequence sampling method. The results are compared in Table 3. The results show that the model achieves the best detection performance when the auxiliary sampling strategy is employed. This approach consistently achieves scores of over 99% across all evaluation metrics.

Table 3. Experimental results of different sampling methods.

Sampling methods	Average accuracy	Average Recall	F1-score
Only undersampling method	96.72%	82.61%	89.10%
Only oversampling method	98.33%	85.54%	91.49%
Auxiliary strategy sequence sampling	**99.66%**	**99.70%**	**99.68%**

4.6   Ablation Experiment of Word Embedding

We apply different representation methods to the same model training, and finally use the obtained models for attack entity deduction on datasets, and the obtained data are the highest of the experimental results as shown in Fig. 5.

The one-hot embedding method for word embedding exhibits poor detection performance, in particular, the recall rate is only 82.72%. This performance gap can be attributed to the limitations of one-hot embedding, which is a discrete data embedding method that does not consider the semantic relationships between different words. On the other hand, static word embedding methods alleviate this problem by mapping the words in a sequence to a fixed vector space. Specifically, the static word embedding method achieves 97.42% accuracy, 96.33% recall, and 96.87% F1 score, respectively. However, it is still inferior to dynamic word embedding. In contrast, the dynamic word embedding method

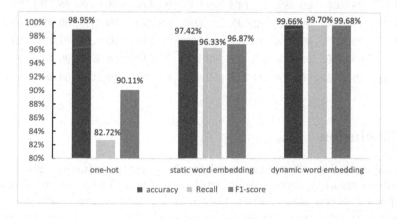

Fig. 5. Experimental results of different word embedding methods.

enhances the feature description and identification of the attack sequences. The experimental results demonstrate that our model achieves the best detection performance among the tested methods. All its scores exceeded 99%.

4.7 Model Evaluation Experiment

Experiments in this subsection is to validate the model's effectiveness in extrapolating unknown entities from audit logs, given the attack entities. Additionally, it aims to evaluate the model's ability to perform attack event identification based on the extrapolated attack entities. The results of the evaluation, conducted using both entity-based and event-based metrics, are presented in Table 4.

Based on the results, it is evident that the average scores for all metrics are above 94%. The event-based model evaluation achieves an impressive score of 99%, indicating that the majority of attack entity samples can be successfully detected. Furthermore, the method exhibits a lower number of false positives while successfully recovering key entities. The reason event-based works better than entity-based is because event-based evaluations introduce more semantics at the event level, allowing the model to capture more information.

Table 4. Model evaluation results.

Model	Accuracy		Precision		Recall		F1-score	
	Entity	Event	Entity	Event	Entity	Event	Entity	Event
SM-1	99.96%	99.92%	95.24%	99.11%	90.91%	99.26%	93.02%	99.19%
SM-2	100.00%	99.98%	92.31%	99.56%	100.00%	99.99%	96.00%	99.77%
SM-3	99.97%	99.99%	92.59%	99.92%	96.15%	99.73%	94.34%	99.83%
SM-4	99.98%	99.94%	95.00%	99.70%	90.48%	99.88%	92.68%	99.79%
MM-1	99.99%	99.98%	96.43%	99.84%	96.43%	99.60%	96.43%	99.72%
MM-2	99.99%	99.99%	94.59%	99.96%	97.22%	99.96%	95.89%	99.96%
MM-3	99.99%	99.98%	97.14%	99.96%	94.44%	99.87%	95.77%	99.92%
MM-4	99.97%	99.97%	92.59%	99.69%	89.29%	99.21%	90.91%	99.45%
MM-5	99.99%	99.98%	96.55%	99.75%	93.33%	99.93%	94.92%	99.84%
MM-6	99.98%	99.96%	95.35%	99.10%	97.62%	99.59%	96.47%	99.35%
Average	99.97%	99.98%	94.78%	99.66%	94.59%	99.70%	94.68%	99.68%

5   Conclusion

In this paper, we introduce a framework for threat entity deduction that leverages an auxiliary strategy and dynamic word embedding. The proposed framework encompasses several stages. First, the audit logs are preprocessed, and events are converted into sequences to facilitate model learning. Then, an auxiliary sampling strategy is employed to enhance the diversity of attack sequences while reducing noise interference from non-attack sequences. Additionally, a dynamic word embedding method is applied to capture the spatio-temporal characteristics embedded within the sequences. Through model training, the framework enables the recognition of attack patterns formed by both known and unknown attack entities. Experimental results demonstrate that the deduction accuracy of our approach, referred to as ASAI, reaches an impressive 99.66%.

While ASAI demonstrates promising results using Windows Event, DNS resolution logs, and Firefox browsing logs, it is important to note that experiments have not been conducted on other datasets. As a result, the deduction ability of ASAI for different application logs remains unverified. This presents an opportunity for future research and the next steps in our work.

Acknowledgements. This work was supported in part by the National Key Research and Development Program of China (No. 2020YFB1805400); in part by the National Natural Science Foundation of China (No. U19A2068, No. 62032002, and No. 62101358); in part by Youth Foundation of Sichuan (Grant No. 2023NSFSC1395); in part by the China Postdoctoral Science Foundation (No. 2020M683345); Fundamental Research Funds for the Central Universities (Grant No. 2023SCU12127); in part by Joint Innovation Fund of Sichuan University and Nuclear Power Institute of China (Grant No. HG2022143).

References

1. Alsaheel, A., et al.: Atlas: a sequence-based learning approach for attack investigation. In: USENIX Security Symposium, pp. 3005–3022 (2021)
2. Gilmer, J., Schoenholz, S.S., Riley, P.F., Vinyals, O., Dahl, G.E.: Neural message passing for quantum chemistry. In: International Conference on Machine Learning, pp. 1263–1272. PMLR (2017)
3. Han, X., Pasquier, T., Bates, A., Mickens, J., Seltzer, M.: Unicorn: runtime provenance-based detector for advanced persistent threats. arXiv preprint arXiv:2001.01525 (2020)
4. Han, X., et al.: SIGL: securing software installations through deep graph learning. In: USENIX Security Symposium, pp. 2345–2362 (2021)
5. Hassan, W.U., Bates, A., Marino, D.: Tactical provenance analysis for endpoint detection and response systems. In: 2020 IEEE Symposium on Security and Privacy (SP), pp. 1172–1189. IEEE (2020)
6. Hassan, W.U., et al.: NODOZE: combatting threat alert fatigue with automated provenance triage. In: Network and Distributed Systems Security Symposium (2019)
7. Hassan, W.U., Noureddine, M.A., Datta, P., Bates, A.: OmegaLog: high-fidelity attack investigation via transparent multi-layer log analysis. In: Network and Distributed System Security Symposium (2020)
8. Hossain, M.N., Sheikhi, S., Sekar, R.: Combating dependence explosion in forensic analysis using alternative tag propagation semantics. In: 2020 IEEE Symposium on Security and Privacy (SP), pp. 1139–1155. IEEE (2020)
9. Kapoor, M., Melton, J., Ridenhour, M., Krishnan, S., Moyer, T.: PROV-GEM: automated provenance analysis framework using graph embeddings. In: 2021 20th IEEE International Conference on Machine Learning and Applications (ICMLA), pp. 1720–1727. IEEE (2021)

10. Khoury, J., Upthegrove, T., Caro, A., Benyo, B., Kong, D.: An event-based data model for granular information flow tracking. In: Proceedings of the 12th USENIX Conference on Theory and Practice of Provenance, p. 1 (2020)
11. Kwon, Y., et al.: MCI: modeling-based causality inference in audit logging for attack investigation. In: NDSS, vol. 2, p. 4 (2018)
12. Lagraa, S., Amrouche, K., Seba, H., et al.: A simple graph embedding for anomaly detection in a stream of heterogeneous labeled graphs. Pattern Recogn. **112**, 107746 (2021)
13. Landauer, M., Skopik, F., Wurzenberger, M., Hotwagner, W., Rauber, A.: Have it your way: generating customized log datasets with a model-driven simulation testbed. IEEE Trans. Reliab. **70**(1), 402–415 (2020)
14. Li, J., Zhang, R., Liu, J., Liu, G., et al.: LogKernel: a threat hunting approach based on behaviour provenance graph and graph kernel clustering. In: Security and Communication Networks 2022 (2022)
15. Liu, F., Wen, Y., Zhang, D., Jiang, X., Xing, X., Meng, D.: Log2vec: a heterogeneous graph embedding based approach for detecting cyber threats within enterprise. In: Proceedings of the 2019 ACM SIGSAC Conference on Computer and Communications Security, pp. 1777–1794 (2019)
16. Liu, Y., et al.: Towards a timely causality analysis for enterprise security. In: NDSS (2018)
17. Ma, S., Zhai, J., Wang, F., Lee, K.H., Zhang, X., Xu, D.: MPI: multiple perspective attack investigation with semantic aware execution partitioning. In: USENIX Security Symposium, pp. 1111–1128 (2017)
18. Michael, N., Mink, J., Liu, J., Gaur, S., Hassan, W.U., Bates, A.: On the forensic validity of approximated audit logs. In: Annual Computer Security Applications Conference, pp. 189–202 (2020)
19. Milajerdi, S.M., Gjomemo, R., Eshete, B., Sekar, R., Venkatakrishnan, V.: Holmes: real-time apt detection through correlation of suspicious information flows. In: 2019 IEEE Symposium on Security and Privacy (SP), pp. 1137–1152. IEEE (2019)
20. Nieto, A.: Becoming JUDAS: correlating users and devices during a digital investigation. IEEE Trans. Inf. Forensics Secur. **15**, 3325–3334 (2020)
21. Plisson, J., Lavrac, N., Mladenic, D., et al.: A rule based approach to word lemmatization. In: Proceedings of IS, vol. 3, pp. 83–86 (2004)
22. Tabiban, A., Zhao, H., Jarraya, Y., Pourzandi, M., Zhang, M., Wang, L.: ProvTalk: towards interpretable multi-level provenance analysis in networking functions virtualization (NFV). In: The Network and Distributed System Security Symposium 2022 (NDSS 2022) (2022)
23. Wang, Q., et al.: You are what you do: hunting stealthy malware via data provenance analysis. In: NDSS (2020)
24. Yang, F., Xu, J., Xiong, C., Li, Z., Zhang, K.: PROGRAPHER: an anomaly detection system based on provenance graph embedding (2023)
25. Yu, L., et al.: ALchemist: fusing application and audit logs for precise attack provenance without instrumentation. In: NDSS (2021)
26. Zeng, J., Chua, Z.L., Chen, Y., Ji, K., Liang, Z., Mao, J.: WATSON: abstracting behaviors from audit logs via aggregation of contextual semantics. In: NDSS (2021)
27. Zengy, J., et al.: SHADEWATCHER: recommendation-guided cyber threat analysis using system audit records. In: 2022 IEEE Symposium on Security and Privacy (SP), pp. 489–506. IEEE (2022)
28. Zhu, T., et al.: General, efficient, and real-time data compaction strategy for APT forensic analysis. IEEE Trans. Inf. Forensics Secur. **16**, 3312–3325 (2021)

Semi-supervised Classification on Data Streams with Recurring Concept Drift Based on Conformal Prediction

ShiLun Ma[1], Wei Kang[1], Yun Xue[2], and YiMin Wen[1(✉)]

[1] Guangxi Key Laboratory of Image and Graphic Intelligent Processing,
Guilin University of Electronic Technology, Guilin 541004, China
ymwen@guet.edu.cn
[2] School of Municipal and Surveying Engineering, Hunan City University,
Yiyang 413000, China

Abstract. In this article, we consider the problem of semi-supervised data stream classification. The main difficulties of data stream semi-supervised classification include how to jointly utilize labeled and unlabeled samples to adress concept drift detection and how to use unlabeled to update trained classifier. Existing algorithms like the CPSSDS method constantly retrain a new classifier when concept drift is detected, it is very consuming and wasteful. In this paper, the algorithm of data stream semi-supervised classification with recurring concept drift named as CPSSDS-R is proposed. First, the labeled samples in the first data block are used to initialize a classifier, which is added into a pool and actived for classification. While a new data block arrives, concept drift is detected by computing conformal prediction results. If no concept drift is detected, the pseudo-labeled samples in the previous data block are added with the labeled samples in the current data block to incrementally train the active classifier. If a new concept is detected, a new classifier is trained on the labeled samples of the current data block and added into the pool and actived for classification, else if a recurring concept is detected, the pseudo-labeled samples and labeled samples in the current data block are used to incrementally update the classifier corresponding to the recurring concept in the pool and actived for classification. The proposed algorithm is tested on multiple synthetic and real datasets, and its cumulative accuracy and block accuracy at different labeling ratios demonstrate the effectiveness of the proposed algorithm. The code for the proposed algorithm is available on https://gitee.com/ymw12345/cpssds-r.

Keywords: Semi-supervised classification · Ensemble learning · Model reuse · Concept drift

---

Supported by the National Natural Science Foundation of China (62366011), the Key R&D Program of Guangxi under Grant (AB21220023), and Guangxi Key Laboratory of Image and Graphic Intelligent Processing (GII P2306).

B. Luo et al. (Eds.): ICONIP 2023, CCIS 1969, pp. 355–366, 2024.
https://doi.org/10.1007/978-981-99-8184-7_27

1   Introduction

With the development of technology, a large amount of data are generated in form of data stream from real computing devices [1]. For example, applications like social networks, network intrusion detection, and weather forecasts [2]. More seriously, the number of labeled data is limited in many cases, the use of machine learning methods to effectively handle semi-supervised data stream classification is very challenging.

In CPSSDS [3], conformance prediction is used to calculate the confidence of unlabeled samples in adjacent data blocks and used for concept drift detection. If concept drift occurs, the existing model is eliminated, and a new model is trained on the current data block. One of the main advantages of this paper is that it uses conformal prediction to both concept drift detection and self-labeled data selection for updating the trained modal. However, when a concept reappears, a new model needs to be retrained, and no historical model cannot be reused [4], which affects the classification accuracy of CPSSDS.

In response to the above problem, this paper proposes a semi-supervised classification algorithm called CPSSDS-R based on CPSSDS [3]. The proposed algorithm maintains a classifier pool, which stores classifiers trained using different concept data. It detects whether a concept drift has occurred between the current data block and the previous data block. If so, the conformance prediction output of the current data block is compared with the conformance prediction outputs of the component classifiers in the classifier pool to detect reoccurring concept. If detected, the component classifiers corresponding to the reoccurring concept is updated and actived for classification, otherwise, initialize a new classifier on the labeled data of the current data block for classification and add it into the pool. The main innovations of this paper are as follows:

1) Combining the idea of ensemble learning, maintaining a classifier pool that retains classifiers trained with different concept data, thus solving the problem of CPSSDS's inability to improve classification accuracy using historical model.
2) Proposing a method for detecting reoccurring concept based on conformance prediction. By comparing the conformal Prediction prediction outputs of the component classifiers in the classifier pool with the conformal Prediction prediction output of the current data block, recurring concept drift can be effectively detected.

2   Related Work

In this section, we first introduce algorithms related to concept drift detection and then discuss related work on semi-supervised data stream classification.

Concept drift detection algorithms identify change points or change time intervals by quantifying statistical measures of change [5]. Based on the test statistics they use, they can be roughly divided into two categories: methods based on classifier accuracy and methods based on data distribution. In accuracy-based drift detection methods [6], changes in performance are detected through statistical tests. The second category of detection algorithms quantifies the dissimilarity between previous and current data distributions [7]. The disadvantage of accuracy-based drift detection methods is that concept drift can only be detected when the classifier's performance declines. This article proposes a distribution-based concept drift detection method that utilizes conformed prediction output of unlabeled data. Recurring concepts can be detected.

Wen et al. first proposed a review on semi-supervised data stream classification algorithms [8]. Next, we will introduce some of the more classical semi-supervised data stream classification algorithms. In semi-supervised classification algorithms for data streams, SmSCluster [9] and ReaSC [10] adopt a semi-supervised approach to construct a collection of classifiers for classification, discarding the worst-performing models from both the old and newly trained models. Going further, SPASC [11] uses the EM algorithm to update classifiers based on labeled data in data blocks and performs sequential classification on instances in the block through a dynamic weighting adjustment mechanism. Addressing the limitations of SPASC, Liu et al. [12] proposed SCBELS, which uses BIRCH ensembles and local structure mapping. [13] maintains a set of clusters as learning models to capture the distribution of data streams and uses KNN for classification of data streams. Each cluster is assigned a weight that dynamically changes based on real-time classification accuracy. Another algorithm similar to SSE-PBS [14] is a high-confidence-based unlabeled sample selection strategy, which can effectively mine and utilize information from unlabeled samples and implicitly adapt to concept drift. Zheng et al. [15] proposed the ESCR algorithm, which calculates the Jensen-Shanon divergence between two distributions, detects significant changes in classifier confidence, and detects repeated concept drift. Khezri et al. [16] proposed the STDS algorithm based on a self-training algorithm, which uses the Kullback-Leibler divergence to measure the distribution difference between consecutive blocks and detect concept drift. Tanha et al. [3] proposed CPSSDS, which uses incremental classifiers as base learners and a self-training framework to augment labeled samples, thereby addressing the scarcity of labeled samples.

3   Proposed Algorithm

This section introduces the algorithm framework of CPSSDS-R, and then provides a more detailed description of the proposed method.

3.1   CPSSDS-R

Like CPSSDS, the CPSSDS-R algorithm is also a new semi-supervised self-training [17] data stream classification framework. In a semi-supervised data stream environment, only a small portion of samples in the data chunk may be labeled, and a data stream is represented as $D_1, D_2, D_3, ..., D_\infty$. Each data

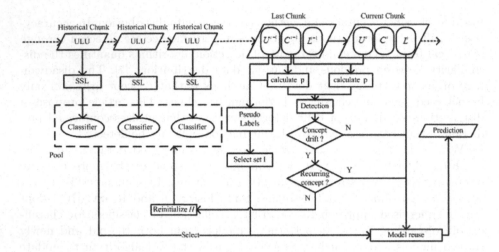

Fig. 1. The framework of CPSSDS-R algorithm

block consists of labeled samples and unlabeled samples, and the data block is divided into a training set (80%) and a test set (20%), where the labeled samples are further divided into a labeled sample set L (70%) and a validation set C (30%). The CPSSDS-R algorithm framework is shown in Fig. 1, and its technical details will be provided in the following sections. Table 1 provides commonly used symbols.

The pseudocodes for the CPSSDS-R algorithm are shown in Algorithm 1. The algorithm uses an incremental Hoeffding tree as base classifier. The steps of the algorithm are as follows: lines 1–2 initialize the model and classifier pool. Lines 6–7 perform inductive conformance prediction on the unlabeled samples of adjacent data blocks, with the method shown in the Algorithm 2. Line 8 indicates that concept drift detection is performed as shown in Algorithm 3. Line 9–10 indicates that if concept drift has occurred, Then Algorithm 4 is used for recurring concept drift detection. Lines 12–16 indicate that if a reoccurring concept is detected, the component classifier with highest similarity to the current data block concept is selected and incrementally updated. Lines 18–22 indicate that if no recurring concept is detected, a new classifier is created using the labeled samples from the current data block and added into the classifier pool. Lines 24–25 indicate that if the test value is greater than or equal to 0.05, the Algorithm 5 is used to select a pseudo-labeled sample set with high information gain and add it to the labeled sample set of the current data block. Line 26 indicates that the model is being updated incrementally.

3.2 Inductive Conformal Prediction

Inductive Conformal Prediction (ICP) proposed in CPSSDS [3] is often used as a framework to determine the set of prediction labels for the relevant confidence

Table 1. Symbols Table

Variables	Descriptions
D_t	The tth data chunk
L^t	Labeled set of the tth chunk
C^t	Calibration set of the tth chunk
U^t	Unlabeled set of the tth chunk
A	Non-conformity measure
H^t	Trained model on the tth chunk
M	Classifier pool
Y	class set
ϵ	Significance level
p_t	p-values of U^t
p_i	the p-value of the ith classifier in the M
p_t^y	The p-value of the sample in the tth data block with predicted category y
p_{t,x_i}^y	p-values of pair (x_i, y) in the tth chunk
S_u	Confidence level for predicted label of sample u
$\alpha_{x_i}^y$	Non-conformity value of pair (x_i, y)
I	Informative samples set

values of test samples. The core components of ICP are the inconsistency function and p-values. The inconsistency function measures the degree of consistency
of each (x, y) pair relative to other samples. The latter value measures the proportion of training samples that are different from the class y for a new sample.
Next, we will introduce the relevant steps of ICP. The Algorithm 2 presents its
pseudocode.

1) Using the labeled sample set L in the data chunk to train an incremental
   Hoeffding Tree classifier;
2) Calculate the inconsistency value $\alpha_{x_c}^y$ of the calibration set C. The calculation
   is shown in the Eq. (1),

$$\alpha_{x_i}^y = 1 - \frac{1 + g(y)}{K + \sum_{j=1}^K g(y_j)} \tag{1}$$

Where $g(y)$ is the number of samples belonging to the y class, which are in
the same cotyledon as x_i, and K is the number of samples for class y.;
3) Calculate the inconsistency value $\alpha_{x_i}^y$ of each sample in the unlabeled sample
   set one by one, and then compare it with the inconsistency value of the
   calibration set. Calculate p-values using the Eq. (2),

$$p_{x_i}^y = \frac{|\{c = 1, \ldots, n \mid \alpha_{x_c}^y \geq \alpha_{x_i}^y\}|}{n} \tag{2}$$

Where n is the number of in the calibration set, and the superscript y represents the p-value of sample x_i as class y.
4) Select the unlabeled samples with p-value greater than or equal to the significance level parameter ϵ and label them with pseudo labels.

Algorithm 1: CPSSDS-R

Input: $D = \{D_1, D_2, \cdots, D_t, \cdots\}$, significance level ϵ, Y

1 *Initialize the classifier H^1 based in L^1, $M = \emptyset$*
2 $M \leftarrow M \cup H^1$
3 $t \leftarrow 2$
4 **while** *data chunk D_t is available* **do**
5       $p_{t-1} \leftarrow ICP(H^{t-1}, C^{t-1}, U^{t-1})$
6       $p_t \leftarrow ICP(H^{t-1}, C^{t-1}, U^t)$
7       $drift \longleftarrow Drift\_Detection(p_{t-1}, p_t, Y)$
8   **if** *drift* **then**
9           $recurring\_concept = Recurring\_Detection(M, U^t, Y)$
10   **if** *recurring_concept* **then**
11              $I = select\_information\_sample(U^t)$
12              $L^t \leftarrow L^t \cup I$
13              $H' \leftarrow \text{argmax}_{H^i \in M}(p_i, p_t)$
14              $H' \leftarrow incremental\_train(H', L^t)$
15   H' actived for classification
16   **else**
17   Initialize H^t based on L^t
18   **if** $len(M) == max\_pool\_size$ **then**
19   remove the earliest classifier in M
20                  $M \leftarrow M \cup H^t$
21   H^t actived for classification
22   **else**
23          $I = select\_information\_sample(U^{t-1})$
24          $L^t \leftarrow L^t \cup I$
25          $H^t \leftarrow incremental\_train(H^{t-1}, L^t)$
26      $t \leftarrow t + 1$

3.3 Concept Drift Detection

Concept drift detection consists of two steps. The first step is to detect whether there is a concept drift between the current data block and the previous concept drift. The second step is to detect whether the current data block is a recurring concept.

The Algorithm 3 presents the process of concept drift detection. Line 4 represents using the KS (Kolmogorov Smirnov) test for each class y. Lines 5–6 represents calculating the p-value distribution of the adjacent unlabeled sample sets of two data blocks, then accumulating them. Lines 7–8 indicate that concept drift is detected if $R/|Y|$ is less than 0.05.

The pseudocode for recurring concept drift detection is presented in the Algorithm 4. Line 3 computes the p-value of each unlabeled sample in U^i to each class label for the classifier H^i and the calibration set C^i, while Line 4 computes the p-value of each unlabeled sample in U^t to each class label for the classifier

---

Algorithm 2: ICP

Input: H, C, U

Output: p-values

1   $p\_values = \emptyset$

2   $C = \{Z_1, Z_2 \cdots Z_n\} = \{(x_1, y_1), (x_2, y_2), \cdots (x_n, y_n)\}$

3   **foreach** $x_i \in U$ **do**

4   **foreach** *class* $y \in Y$ **do**

5         $N_i = (x_i, y)$

6         $\alpha_{x_i}^y = 1 - \dfrac{1+g(y)}{K+\sum_{j=1}^{K} g(y_j)}$

7   **foreach** $x_c \in C$ **do**

8            $\alpha_{x_c}^y = 1 - \dfrac{1+g(y)}{K+\sum_{j-1}^{K} g(y_j)}$

9      $p_{x_i}^y = \dfrac{\left| \left\{ c=1,\ldots,n \mid \alpha_{x_c}^y \geq \alpha_{x_i}^y \right\} \right|}{n}$

10 **return** p_values

Algorithm 3: $Drift\_Detection$

Input: p_{t-1}, p_t, Y

Output: ks

1   *boolean* $drift = false$

2   $R = 0$

3   **foreach** $y \in Y$ **do**

4      $pVal \leftarrow KStest(p_{t-1}^y, p_t^y)$

5      $R \leftarrow R + pVal$

6   **if** $(R/|Y| < 0.05)$ **then**

7      $drift = true$

8 **return** $drift$

H^i and the calibration set C^i. Lines 5–7 represent that after all component classifiers are computed, the *index* and *mas_ks* value of the component classifier that is most similar to the concept of the current data block are obtained. Lines 8–9 represent that if the *mas_ks* value is greater than 0.1, the *index* value is returned, indicating the detection of a recurring concept.

3.4   Information Sample Selection Strategy

The key step in the self-training framework is the information sample selection step, where the most reliable unlabeled samples are chosen to improve the generalization ability of the classifier. The pseudocode for selecting samples is shown in Algorithm 5. For each sample in the unlabeled sample set, line 2 represents calculating the largest p-value for all class labels. Line 3 represents calculating the confidence of the sample belonging to class y. Lines 4–5 represent that if the confidence is greater than the significance level, the sample is added to the

---

Algorithm 4: *Recurring_Detection*

Input: M, U^t, Y
Output: classifier index
1 $index = 0, max\_ks = 0$
2 **foreach** $H^i \in M$ **do**
3 $\quad$ $p_i \leftarrow ICP(H^i, C^i, U^i)$
4 $\quad$ $p_t \leftarrow ICP(H^i, C^i, U^t)$
5 $\quad$ $ks \leftarrow ks\_test(p_i, p_t, Y)$
6 $\quad$ **if** $max\_ks < ks$ **then**
7 $\quad\quad$ $index = i$
8 $\quad\quad$ $max\_ks = ks$

9 **if** $mas\_ks > 0.1$ **then**
10 $\quad$ **return** $index$

Algorithm 5: *select_information_sample*

Input: U
Output: I
1 **foreach** $x_u \in U$ **do**
2 $\quad$ $y_u = argmax_y \{p^y_{t,x_u}\}$
3 $\quad$ $S_u = p^{y_u}_{t,x_u}$
4 $\quad$ **if** $S_u \geq \epsilon$ **then**
5 $\quad\quad$ $I = I \cup x_u$

6 **return** I

---

informative sample set. Finally, after all unlabeled samples are calculated, the informative sample set is return.

4   Experiments

4.1   Datasets

To verify the performance of the algorithm, we used 3 real and 6 synthetic datasets [18]. All synthetic datasets are class balanced. The detailed information of these 9 datasets is shown in Table 2.

4.2   Experimental Setup

We use a block-based model to evaluate the proposed framework, where a fixed portion of instances is selected as labeled at the beginning of each block's training set, set to 10% in the experiments. The experiments are conducted under the following parameter settings: for CPSSDS-R, the significance level for adaptive predictions is tested and adjusted for each dataset. The parameters for comparing algorithms are fixed and use the default values from the original paper. All experiments in the paper are the average results of 10 runs.

Table 2. Properties of the Datasets

datasets	Attributes	Instances	Classes	Chunk Size
Agrawal	9	130000	2	5000
RandomRBF	10	300000	4	5000
Stagger	3	250000	2	5000
FG2C2D	2	200000	2	5000
GEAR2C2D	2	200000	2	5000
MG2C2D	2	200000	2	5000
Electricity	8	45312	2	5000
SLDD	48	58509	11	5000
Shuttle	9	845228	10	4000

4.3   Algorithm Accumulative Accuracy and Result Analysis

In this section, the accumulative accuracy of CPSSDS-R algorithm compared to CPSSDS algorithm on 9 datasets with different labeled ratio was analyzed. The average results of ten runs are shown in Table 3, and paired t-tests with a 95% confidence level are performed on the results. In order to observe the performance of the proposed algorithm more intuitively, the Run chart of the accumulative accuracy of the algorithm on each dataset versus the number of data chunks is shown in Fig. 2.

Table 3. Accuracy comparison for CPSSDS and CPSSDS_R on the experimental datasets with a 10% labeling ratio.

datasets	CPSSDS	CPSSDS-R
Agrawal	53.01	**55.24**
RandomRBF	57.79	**58.98**
Stagger	91.75	**97.28**
FG2C2D	88.94	**91.14**
GEAR2C2D	96.07	**96.79**
MG2C2D	91.37	**92.17**
Electricity	65.16	**65.41**
SLDD	46.45	**49.51**
Shuttle	90.51	**92.49**

The cumulative accuracy comparison of datasets with a labeling rate of 10% is shown in Table 3. It can be seen that, on all datasets, the cumulative accuracy of the CPSSDS-R algorithm is significantly better than that of the CPSSDS algorithm. Figure 2 shows the cumulative accuracy comparison of algorithms for

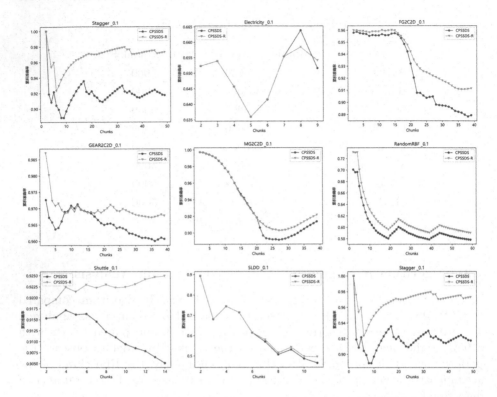

Fig. 2. Accuracy comparison on the experimental datasets with 10% labeled each

each dataset with a labeling rate of 10%. From the figure, it can be observed that, on the synthetic datasets, the cumulative accuracy of the proposed algorithm is consistently higher than that of the CPSSDS algorithm. On the real-world datasets, however, due to the unknown location and type of concept drift, the proposed algorithm has slightly lower cumulative accuracy than the CPSSDS algorithm in the early stages. However, as the number of data blocks increases, the cumulative accuracy of CPSSDS-R gradually surpasses that of the CPSSDS algorithm, which is more pronounced in the Shuttle dataset.

In Tables 4, the CPSSDS-R algorithm outperforms other algorithms in classification performance on most datasets, indicating that detecting recurring concept drift through conformal prediction can more effectively address the problem of continuous model initialization, reduce the cold start problem when recurring concepts, and achieve better classification performance on different datasets. But on the GEAR2C2D dataset, the accuracy of the SCBELS algorithm is higher because the component classifier of the SCBELS algorithm is an integrated classifier, which performs better on this dataset. In contrast, the CPSSDS-R algorithm has a slightly lower accuracy than the SCBELS algorithm due to its ability to detect recurrence concepts in a timely manner.

Table 4. Accuracy comparison with other semi-supervised classification algorithms on the experimental datasets with a 10% labeled each

datasets	SPASC	SCBELS	CPSSDS-R
Agrawal	49.89	51.81	**55.24**
RandomRBF	55.11	55.39	**58.98**
Stagger	80.89	78.38	**97.28**
FG2C2D	79.98	88.11	**91.14**
GEAR2C2D	94.45	**97.35**	96.79
MG2C2D	55.27	84.86	**92.17**
Electricity	53.56	58.96	**65.41**
SLDD	18.16	32.95	**49.51**
Shuttle	91.16	88.56	**92.49**

5   Conclusions

In this paper, we propose the semi-supervised data stream classification algorithm CPSSDS-R. Firstly, the Hoeffding tree model is trained using labeled sample sets in data chunk, and on the basis of the self training framework, conformal prediction is used to label unlabeled samples. Samples with high information content are selected to expand the labeled sample set. Secondly, by maintaining a classifier pool and calculating the KS statistical values of the current data chunk and component classifiers in the classifier pool, the recurrent concept can be detected, and the occurrence of recurrent drift detected before the classifier can be updated. The appropriate component classifier can be selected in a timely manner to adapt to the recurrence concept drift. Finally, the algorithm is validated on multiple synthetic and real datasets for its effectiveness.

Future work include: on the one hand, the classifier trained by incremental Hoeffding tree is always updated incrementally, which may lead to overfitting. It should consider designing a reasonable pruning strategy based on the characteristics of semi-supervised data stream to reuse trained model more effectively. On the other hand, based on conformal prediction, the algorithm selects pseudo labeled samples with high confidence to expand the labeled sample set. Generally, samples far away from the Decision boundary are used to update the classifier. Failure to take into account the global distribution may lead to a decline in the performance of classifier.

References

1. Kuo, Y.-H., Kusiak, A.: From data to big data in production research: the past and future trends. Int. J. Prod. Res. **57**(15–16), 4828–4853 (2019)
2. Tan, C.H., Lee, V.C., Salehi, M.: Information resources estimation for accurate distribution-based concept drift detection. Inf. Process. Manage. **59**(3), 102911 (2022)

3. Tanha, J., Samadi, N., Abdi, Y., Razzaghi-Asl, N.: CPSSDS: conformal prediction for semi-supervised classification on data streams. Inf. Sci. **584**, 212–234 (2022)
4. Zhao, P., Zhou, Z.: Learning from distribution-changing data streams via decision tree model reuse. Scientia Sinica Informationis **51**(1), 1–12 (2021)
5. Lu, J., Liu, A., Dong, F., Gu, F., Gama, J., Zhang, G.: Learning under concept drift: a review. IEEE Trans. Knowl. Data Eng. **31**(12), 2346–2363 (2018)
6. Lu, N., Lu, J., Zhang, G., De Mantaras, R.L.: A concept drift-tolerant case-base editing technique. Artif. Intell. **230**, 108–133 (2016)
7. Gu, F., Zhang, G., Lu, J., Lin, C.-T.: Concept drift detection based on equal density estimation. In: 2016 International Joint Conference on Neural Networks (IJCNN), pp. 24–30. IEEE (2016)
8. Wen, Y., Liu, S., Miao, Y., Yi, X., Liu, C.: Survey on semi-supervised classification of data streams with concepts. J. Softw. **33**(4), 1287–1314 (2022)
9. Masud, M.M., Gao, J., Khan, L., Han, J., Thuraisingham, B.: A practical approach to classify evolving data streams: training with limited amount of labeled data. In: 2008 Eighth IEEE International Conference on Data Mining, pp. 929–934. IEEE (2008)
10. Masud, M.M., et al.: Facing the reality of data stream classification: coping with scarcity of labeled data. Knowl. Inf. Syst. **33**, 213–244 (2012)
11. Hosseini, M.J., Gholipour, A., Beigy, H.: An ensemble of cluster-based classifiers for semi-supervised classification of non-stationary data streams. Knowl. Inf. Syst. **46**, 567–597 (2016)
12. Wen, Y.-M., Liu, S.: Semi-supervised classification of data streams by birch ensemble and local structure mapping. J. Comput. Sci. Technol. **35**, 295–304 (2020)
13. Din, S.U., Shao, J., Kumar, J., Ali, W., Liu, J., Ye, Y.: Online reliable semi-supervised learning on evolving data streams. Inf. Sci. **525**, 153–171 (2020)
14. Khezri, S., Tanha, J., Ahmadi, A., Sharifi, A.: A novel semi-supervised ensemble algorithm using a performance-based selection metric to non-stationary data streams. Neurocomputing **442**, 125–145 (2021)
15. Zheng, X., Li, P., Hu, X., Yu, K.: Semi-supervised classification on data streams with recurring concept drift and concept evolution. Knowl.-Based Syst. **215**, 106749 (2021)
16. Khezri, S., Tanha, J., Ahmadi, A., Sharifi, A.: STDS: self-training data streams for mining limited labeled data in non-stationary environment. Appl. Intell. **50**, 1448–1467 (2020)
17. Zhu, X., Goldberg, A.B., Brachman, R., Dietterich, T.: Synthesis lectures on artificial intelligence and machine learning. In: Introduction to Semi-Supervised Learning, vol. 3, no. 1, pp. 1–130. Morgan & Claypool (2009)
18. Montiel, J., Read, J., Bifet, A., Abdessalem, T.: Scikit-multiflow: a multi-output streaming framework. J. Mach. Learn. Res. **19**(1), 2914–2915 (2018)

Zero-Shot Relation Triplet Extraction via Retrieval-Augmented Synthetic Data Generation

Qing Zhang[1,2], Yuechen Yang[1,2], Hayilang Zhang[1,2], Zhengxin Gao[1,2], Hao Wang[1,2], Jianyong Duan[1,2(✉)], Li He[1,2], and Jie Liu[1,3]

[1] School of Information Science and Technology, North China University of Technology, Beijing 100144, China
duanjy@ncut.edu.cn
[2] CNONIX National Standard Application and Promotion Lab, Beijing 100144, China
[3] China Language Intelligence Research Center, Capital Normal University, Beijing, People's Republic of China

Abstract. In response to the challenge of existing relation triplet extraction models struggling to adapt to new relation categories in zero-shot scenarios, we propose a method that combines generated synthetic training data with the retrieval of relevant documents through a rank-based filtering approach for data augmentation. This approach alleviates the problem of low-quality synthetic training data and reduces noise that may affect the accuracy of triplet extraction in certain relation categories. Experimental results on two public datasets demonstrate that our model exhibits stable and impressive performance compared to the baseline models in terms of precision, recall, and F1 score, resulting in improved effectiveness for zero-shot relation triplet extraction.

Keywords: Relation triplet extraction · Retrieval-augmented · Zero-shot

1 Introduction

Relation triplet extraction is a fundamental aspect of knowledge graph construction, aiming to extract entities and their relationships from various data sources. Existing relation triplet extraction models mostly rely on pipeline-based extraction methods [1], which divide the task into two subtasks: entity extraction and relation extraction. As a result, many models face the risk of error propagation. Some joint extraction models have proposed parameter sharing methods, which increase the correlation between the two tasks through multi-task learning and

This work is supported by the National Natural Science Foundation of China (61972003), R&D Program of Beijing Municipal Education Commission (KM202210009002), the Beijing Urban Governance Research Base of North China University of Technology (2023CSZL16), and the North China University of Technology Startup Fund.

joint optimization. However, they do not truly solve the problem of error propagation. Since 2018, generation-based relation triplet extraction models have been gradually proposed, which are based on RNN and utilize traditional sequence-to-sequence deep learning generation frameworks. However, these generative models have not shown significant improvement compared to extraction-based models. It was not until recent years that generation-based pre-training models (such as GPT, T5, and BART were widely used, and effective generation-based triplet extraction models emerged. Generation-based relation triplet extraction models have stronger transferability and scalability compared to extraction-based models, making them more effective for extraction tasks in low-resource scenarios.

Relation triplet extraction, as a challenging structured prediction task, has seen limited research in the context of few-shot or zero-shot scenarios [3]. Chen and Li [2] introduced attribute representation learning, projecting sentences and relation descriptions into an embedding space and minimizing the distance in the embedding space to achieve zero-shot relation extraction. This work represents the first attempt at relation extraction in the domain of unseen relations. Zero-shot relation triplet extraction tasks require models to extract triplet information of relation categories that were not seen during training, without providing explicit head and tail entities (Fig. 1). Subsequently, Chia et al. [3] were inspired by the use of large-scale language models as knowledge bases and redefined the zero-shot relation triplet extraction task. They achieved training by generating a large amount of synthetic data for unseen relations. However, ensuring the quality of synthetic data is challenging, especially when the knowledge of desired relations does not exist in the pre-training language model or when the model cannot accurately understand the semantics of relation labels. In such cases, the generated synthetic data can introduce significant noise to the model.

Building upon previous research that utilized pre-trained language models as knowledge bases [6], we consider the method of data augmentation for zero-shot tasks using large-scale language models as innovative and effective. However, when augmenting data for each relation, it is crucial to consider whether the generation model has the capability to understand the semantics of the relations and generate sufficiently high-quality synthetic data. To address this issue, we not only use language models to generate synthetic data but also employ retrieval models to search for text passages relevant to relation labels. We ensure the quality of data augmentation by combining these two data sources using a zero-shot reordering approach. In our work, we focus on the task of relation triplet extraction and propose a method that combines retrieval-based augmentation and reordering to further enhance the knowledge correctness, consistency, and diversity of synthetic data. Specifically, we first train synthetic data generator and relation extractor using manually labeled data. During the synthetic data generation, we refer to previous retrieval-based works, where we use the relation labels from the validation and test sets as search sources and employ Dense Passage Retriever (DPR) [5] as a neural retriever to search for relevant documents in non-parametric memory (dense vector index of Wikipedia). The retrieved documents are combined with relation labels to create structured data. Additionally, using relation labels as prompts, we generate a certain amount of synthetic data using a well-trained auto-regressive data generation model. Since the quality of synthetic data varies for each relation, we use a language model to re-rank and select the retrieved passages and synthetic data based on their relevance to the relation labels. This approach provides great cross-attention between relation labels and passages, effectively complementing the synthetic data with external knowledge and removing low-quality data noise, thus improving the quality of data augmentation.

Finally, the final relation extractor is retrained on the synthetic data. Such an approach endows the model with the capability to transfer from the seen relation domain to the unseen relations, while also leveraging synthetic data for supervised training, thus enabling more accurate and efficient extraction of relation triplets. The main contributions of our work can be summarized as follows:

- We introduce a retrieval-based data augmentation method for zero-shot relation triplet extraction tasks, utilizing both the internal knowledge of the generative language model and external retrieval passages to expand training data for zero-shot tasks.
- We use a zero-shot relation generation method to calculate the relation entailment probability conditioned on the retrieved and generated passages, and reorder both parts of synthetic data to improve the quality of augmented training data for downstream extraction tasks.
- We conduct experiments on two large-scale datasets, the results demonstrate that our model performs well in zero-shot relation triplet extraction tasks and exhibits excellent knowledge transfer ability.

Fig. 1. Samples in train dataset and test dataset. The relation labels "league", "screenwriter" and "residence" decide which relation class the sample belongs to. Each sample consists of a sentence and a relation triplet contained within the sentence, but the samples in the two datasets belong to different relation classes.

2   Related Work

Zero-Shot/Few-Shot Relation Triplet Extraction
To handle different types of relation extraction flexibly, Cabot and Navigli [9] proposed a generative model that treats relation triplet extraction as a seq2seq task, reducing the number of epochs required for fine-tuning on downstream tasks. For zero-shot learning, RelationPrompt [3] also uses a seq2seq model that generates structured templates for triplets based on the given source context. However, the quality and consistency of synthetic data are not guaranteed since it cannot be ensured that the data generation model contains sufficient relevant knowledge. Kim et al. [10] abandoned the pipeline and data augmentation approaches and redefined the task as a template filling problem. However, it faces difficulties in distinguishing approximate relations in the context of unseen relations and can lead to confusion.

Retrieval Augmentation. In recent years, retrieval-augmented methods have shown significant advantages in text generation tasks. RAG [4] uses a dual-encoder retrieval module to retrieve relevant documents based on the input text. The retrieved documents are then concatenated with the input text and used as input to the generation module to generate corresponding results. We recognize the importance of a strong correlation between the retrieval documents and the original text. If the quality of these passages is poor, it can introduce noise and affect the overall performance of the model when converting passages into synthetic training samples.

Controllable Text Generation. Controllable text generation refers to the generation of text with specific attributes given the model source sequence. Keskar and McCann [7] introduced control codes and prompts as control signals during model pre-training to influence the domain, topic, entities, and style of the generated text. However, it requires adjusting all model parameters, which can be computationally expensive. To make full use of common-sense knowledge to improve the novelty and coherence of generated passages, Yang and Li [8] integrated external knowledge from knowledge bases into the generator using a dynamic memory mechanism to assist the model in generating text. For us, we included a reordering module to selectively generate sentences that better align with the provided control information. This dynamic combination of internal and external knowledge allowed for a more intuitive approach to generate the text that we want.

3   Methodology

The proposed retrieval-augmented zero-shot relation triplet extraction model is illustrated in Fig. 2, which consists of four components: synthetic data generator, document retrieval module, passage ranking module, and triplet extractor.

The extraction of relation triplets is ultimately achieved through the triplet extractor. However, after training the model with training data, we also use the relation labels in the validation and test sets as text prompts, generating synthetic training data containing domain-specific knowledge corresponding to the relations and retrieving relevant passages from an external knowledge corpus. Once these two parts of data are processed into similar structured texts, they are fed into the passage ranking module for reordering and filtering, resulting in a certain amount of high-quality data, which is then used to retrain the triplet extractor, enabling it to learn relation-specific knowledge that it had not encountered during the training phase and extract triplets of these relations.

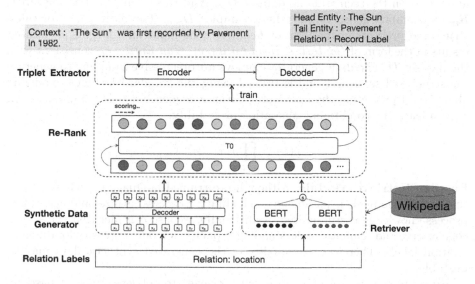

Fig. 2. Our model structure. The bottom half of the image illustrates the data augmentation process in our approach, while the topmost part showcases the input and output content of the relation triplet extraction task.

Task Definition. In the task of relation triplet extraction, given an input sentence $S = x_1, x_2, x_3, x_4, ..., x_n$, we aim to extract relation triplets $Tri = (h, t, r)$ from the sentence S, where n represents the length of the string sequence, and Tri, h and t in the triplet represent the head entity and tail entity of an entity pair, while r represents the relation between them. The objective of zero-shot relation triplet extraction task is to learn from the seen training dataset D_{tr} and generalize to the unseen test dataset D_{te}. D_{tr} and D_{te} are split from the complete dataset $D = (S, Tri, R)$, where S represents input sentences, Tri represents relation triplets, and R represents the set of relation labels present in D. The label sets in D_{tr} and D_{te} are pre-defined and denoted as $R_{tr} = R_{tr}^1, R_{tr}^2, ..., R_{tr}^n$ and $R_{te} = R_{te}^1, R_{te}^2, ..., R_{te}^m$, respectively, where n and m represent the sizes of the label sets in D_{tr} and D_{te}. Since these two datasets are disjoint, namely $R_{tr} \cap R_{te} = \emptyset$, each sample contains a sentence $s \in S$, which corresponds to one or more relation triplets Tri.

Synthetic Data Generator. It has been shown in many studies that language models can be used as knowledge bases [6], utilizing the knowledge stored in them to generate data that is beneficial for downstream tasks. Moreover, pre-trained language models can implicitly perform zero-shot generalization based on pre-training. Additionally, designing control signals or prompts to use autoregressive language models for controllable text generation has been proven to be feasible and effective [7]. Therefore, we use the relation labels of unseen relation set as prompts to input into the language model for generating synthetic data samples.

We use the causal language model GPT-2 as the data generator. Firstly, it is fine-tuned on the seen training dataset D_{tr}. And then using the relation labels R_{tr} as prompts, generating synthetic samples D_{gen}. The data generator takes structured prompts in the form of *Relation: r* as input and outputs structured results in the form of *Context: s, Head Entity: e_h, Tail Entity: e_t*, consistent with the dataset D. The data generator employs autoregressive sampling to generate the structured sequences and is trained using the standard language modeling objective of predicting the next word. Given a sequence x $= [x_1, x_2, x_3, x_4, ..., x_n]$, to learn the conditional generation probability:

$$p(x) = \prod_{i=1}^{n} p\left(x_i \mid x_{<i}\right) \tag{1}$$

Passage Retrieve Module. Although pre-trained language models store factual knowledge in their parameters, the ability to access, utilize, and accurately process knowledge is still limited. Additionally, we also consider that the knowledge reserve and semantic understanding capacity of model may vary for different relation labels. Therefore, the quality of the generated synthetic data may be unstable.

To address this, we employ a text retriever to retrieve relevant passages from an external knowledge corpus to supplement the missing knowledge in the synthetic data. We use the dense vector index of Wikipedia, denoted as z, as the external knowledge source for document retrieval, and a pre-trained neural retriever to access this knowledge. For the label set R_{te} of unseen relation set, we first construct the question template *What is the relation r?*, where r belongs to R_{te}, and we use the question as the source text x for retrieval.

Specifically, we leverage two separate BERT dual-tower models to encode the text x and the documents z, which are divided into segments of 100 words each, where the segmented documents amount to 21 million in total. We employ a document encoder to calculate the embeddings of each document and use Faiss [11] to build a single MIPS (Maximum Inner Product Search) index with HNSW (Hierarchical Navigable Small World) approximation [12] for efficient retrieval. As shown in Fig. 2, the retriever retrieves relevant documents using the relation label as the source text x. We initialize our retriever and build the document index using a pre-trained dual encoder from DPR (Dense Passage Retriever) [5]:

$$p_\eta(z \mid x) \propto \exp\left(\mathbf{d}(z)^\top \mathbf{q}(x)\right) \tag{2}$$

$$\mathbf{d}(z) = \text{BERT}_d(z), \quad \mathbf{q}(x) = \text{BERT}_q(x) \tag{3}$$

In the given context, $d(z)$ refers to the dense representation of documents generated by the $BERT_{Base}$ document encoder, while $q(x)$ represents the query representation of the source text generated by the source text encoder, which is also based on $BERT_{Base}$. By combining these two representations, we employ the Maximum Inner Product Search (MIPS) method to compute a list of K documents z with the highest prior probability $P_\eta(z|x)$, which can be approximated in sub-linear time [11]. Currently, we retrieve 150 relevant documents for each relation label, which are used in conjunction with the synthesized training data to improve the quality of data augmentation. Effect of retrieved document size is analyzed in experiment section.

Data Ranking Module. Although pre-trained language models contain a wealth of knowledge, we cannot guarantee that the language model will generate high-quality data for every relation label during the synthesis of data. Training a triplet extractor using such data can introduce a significant amount of noise and diminish the extraction performance. From another perspective, the triplet extraction task is a structured prediction task, and our synthesized data must fully adhere to the structure of the original training data. Otherwise, it would be challenging to evaluate the extraction results accurately.

In Sects. 3.2 and 3.3, we generated structured synthetic data for each one in the unseen relation set R_{te} based on the relation label r (where r belongs to R_{te}) via a data generator and retrieved the relevant documents using a neural retriever. Firstly, we processed the retrieved documents to match the structure of the synthetic data, which includes sentences and relation triplets. Next, we combined these two parts of data in different proportions to achieve a complementary effect, ensuring the quality of the data used to train the triplet extractor. When combining the synthetic data and retrieved documents, we employed an unsupervised passage ranking module to mix and rank these two parts, ultimately using the top-k ranked passages as augmented training data for that relation class.

Firstly, for each relation, we combine the two parts of previously generated data, totaling n instances, into a set of passages $P = p_1, p_2, ..., p_n$. The objective of the passage ranking module is to reorder the passages in set P so that the passages containing the correct knowledge corresponding to the relation class are ranked as high as possible. For each passage $pi \in P$, the ranking is calculated based on the relevance score $p(p_i|r)$. Our passage reordering approach is unsupervised and does not require any task-specific training. As shown in Fig. 3, we use a pre-trained language model T0 to score the probability of the potential entailment relation r given the passage text p. The entailment generation model is zero-shot, allowing independent reordering regardless of the dataset, and it incorporates cross-attention between the relation and passage tokens by forcing the model to interpret each token in the input relation label. Therefore, our passage ranking module is more interpretable than using a dense retriever alone, even though both methods are built upon very similar pre-trained models.

Specifically, we estimate the relevance of each passage to its corresponding relation label, denoted as $P(p_i|r)$, by calculating the probability of the relation

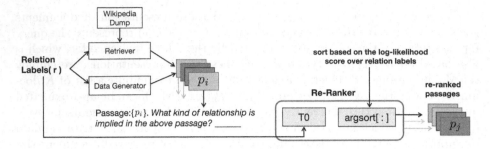

Fig. 3. An illustration of the different components in data ranking module. For more details, please refer to text.

being implied given the passage text, namely $P(r|p_i)$. We appropriately apply Bayesian rule to compute $P(p_i|r)$:

$$\log p(p_i|r) = \log p(r|p_i) + \log p(p_i) + c \qquad (4)$$

Where $p(p_i)$ represents the prior probability of the retrieved passage and c is a common constant for all pi. To simplify this assumption, we consider the prior probability $\log p(p_i)$ of the passages to be consistent and can be disregarded during reordering. Therefore, the above expression can be simplified as:

$$\log p(p_i|r) \propto \log p(r|p_i), \forall p_i \in P \qquad (5)$$

We utilize a pre-trained language model to compute the average log-likelihood of relation tokens conditioned on passages, thereby estimating $\log p(r|p_i)$:

$$\log p\left(\boldsymbol{r} \mid \boldsymbol{p}_i\right) = \frac{1}{|\boldsymbol{r}|} \sum_t \log p\left(r_t \mid \boldsymbol{r}_{<t}, \boldsymbol{p}_i; \Theta\right) \qquad (6)$$

Where θ represents the parameters of the pre-trained language model, $|r|$ denotes the number of question tokens. As shown in Fig. 3, we employ the language model in a zero-shot manner by simply concatenating the natural language prompt *What kind of relation is implied in the above passage?* with the passage tokens to generate potential implied relations, without any fine-tuning process.

Finally, based on the $\log p(r|p_i)$, we rank both the generated and retrieved passages for each relation class separately, and extract a certain amount of data as training samples for the relation triplet extractor. This reordering approach reasonably determines the proportion of combining the two parts of enhanced data, providing knowledge that covers the target relation domain as much as possible and reducing noise from low-quality data. The experimental section below demonstrates how our method offers effective data augmentation strategies for zero-shot tasks.

Triplet Extractor. We can train a zero-shot relation triplet extractor with the synthetic data samples generated from the unseen relation set in Sect. 3.4. The relation extractor is first fine-tuned on the seen dataset D_{tr} and then trained on the synthetic data D_{gen}. Finally, it predicts and extracts relation triplets Tri_{te} from the sentence set S_{te} in the unseen dataset D_{te}. We employ a seq2seq learning approach, which has proven to be effective for structured prediction tasks.

As shown in Fig. 1, the relation triplet extractor takes a sentence s in the form of *Context: s* as input, which then generates a structured output sequence containing a pair of entity mentions that satisfy the relation y, in the form of *Head Entity: h, Tail Entity: t, Relation: r*. We use standard sequence-to-sequence objectives for training and employ greedy decoding for generation.

4 Experiment

Datasets. We conducted experiments using two datasets, FewRel [13] and Wiki-ZSL [2]. FewRel is a dataset used for few-shot relation classification. We employed the transformed samples from this dataset to extract zero-shot relation triplets [3]. Wiki-ZSL is a subset of Wiki-KB, constructed through distant supervision using Wikipedia articles and the Wikidata knowledge base for zero-shot relation extraction. We split both datasets into disjoint sets of relation labels to make them suitable for zero-shot settings. The statistics of the datasets are shown in Table 1. We adopted the zero-shot settings proposed by Chia et al. [3]: 1) all sets of relations for training, validation, and testing are disjoint; 2) we conducted experiments with three different settings to investigate the performance of methods under different settings of the unseen label set size, where m is set to 5, 10, and 15, respectively; 3) we generated different data splits based on five different random seeds to mitigate the influence of experimental noise, resulting in five different data splits for each label set size setting.

Table 1. Dataset statistics. "Average Sentence Length" refers to the average number of words in each sentence.

Datasets	Sample Size	entity count	relation count	Average Sentence Length (in words)
Wiki-ZSL	94,383	77,623	113	24.85
FewRel	56,000	72,854	80	24.95

Experimental Settings. For the data generator and triplet extractor, we fine-tuned pre-trained models with 124M parameters of GPT-2 and 140M parameters of BART, respectively. During the fine-tuning process of these two models, we performed up to five epochs of fine-tuning on the training set, with early stopping based on validation loss. In the experiments, the batch size was set to 128, the learning rate to 3×10^{-5}, and the AdamW optimizer was utilized to optimize all models, with a linear warm-up for the first 20% of training steps. For the retrieval module, we used the DPR model based on BERT, and employed the T0 model with a parameter size of 3B to score the passages based on relation labels for the re-ranking module.

Experimental Results

Main Experiment. We compared the performance of our model with three triplet extraction methods:

1) Our triplet extractor can perform zero-shot relation triplet extraction without fine-tuning on synthetic samples. Since it has been fine-tuned on the training set of visible relations, we believe it has some knowledge transfer capabilities. We denote this model as "w/o Gen" indicating that it does not use generated synthetic samples for training.

2) TableSequence [14] is a joint extraction model where two separate encoders perform relation extraction and named entity recognition simultaneously. Since TableSequence is designed for supervised learning, we trained Table-Sequence using the generated synthetic data from our model and reported the results.

3) RelationPrompt [3] solely uses the generative model to generate data for unseen relations and does not incorporate retrieval from the knowledge base or evaluate the quality of the generated synthetic data.

We validated the superiority of our method compared to other baseline methods on the FewRel dataset in Table 2. In terms of single triplet evaluation, our method consistently outperforms other methods in accuracy. For multi-triplets, our method achieves a significantly higher F1 score compared to baseline methods, especially when the number of unseen labels is 5 and 15. Our method outperforms the previous state-of-the-art (SOTA) by 4.73 and 2.01% points, respectively. Additionally, our method shows good performance in precision and recall. Although we did not discover a consistent advantage across all metrics, the baseline methods that perform well in either precision or recall for multi-triplet extraction fail to achieve a balanced precision-recall trade-off, resulting in poor F1 results. Table 3 presents the comparison results of our method against baselines on the Wiki-ZSL dataset, which show similar trends as FewRel. While our method does not exhibit overwhelming advantages in each metric under different numbers of unseen relations, overall, our method achieves comparable results to the previous SOTA in relation triplet extraction, which effectively demonstrates that the combination of synthetic data and retrieval of document segments helps eliminate some noise and reduces the possibility of model overfitting. The difference in results between "w/o Gen" and our method also indicates the crucial importance of using generated and retrieved synthetic samples since, in most cases, our method achieves an F1 score that is higher than double that of "w/o Gen".

Table 2. Results for Zero-Shot Relation Triplet Extraction on FewRel

Unseen Labels	m = 5				m = 10				m = 15			
Models	Single	Multi			Single	Multi			Single	Multi		
	Acc.	P	R	F1	Acc.	P	R	F1	Acc.	P	R	F1
w/o Gen	11.49	9.45	**36.74**	14.57	12.4	6.4	**41.7**	11.02	10.93	4.61	**36.39**	8.15
TableSequence	11.82	15.23	1.91	3.4	12.54	**28.93**	3.6	6.37	11.65	**19.03**	1.99	3.48
RelationPrompt	22.27	20.8	24.32	22.34	23.18	21.59	28.68	**24.61**	18.97	17.73	23.2	20.08
Ours	**24.14**	**25.61**	29.04	**27.07**	**23.62**	23.5	25.48	24.35	**19.93**	18.1	24.55	**22.09**

Table 3. Results for Zero-Shot Relation Triplet Extraction on Wiki-ZSL

Unseen Labels	m – 5				m = 10				m = 15			
Models	Single	Multi			Single	Multi			Single	Multi		
	Acc.	P	R	F1	Acc.	P	R	F1	Acc.	P	R	F1
w/o Gen	9.05	15.58	**43.23**	22.26	7.1	9.63	**45.01**	15.7	6.61	7.25	**44.68**	12.34
TableSequence	14.47	**43.68**	3.51	6.29	9.61	**45.31**	3.57	6.4	9.2	**44.43**	3.53	6.39
RelationPrompt	**16.64**	29.11	31	30.01	16.48	30.2	32.31	31.19	**16.16**	26.19	32.12	**28.85**
ours	16.59	35.43	30.93	**32.75**	**17.16**	31.66	34.32	**32.91**	12.226	21.852	23.09	22.43

Ablation Study. In order to validate the effectiveness of each module in our proposed model for zero-shot triplet extraction tasks, we conducted ablation experiments using the FewRel dataset with 5 unseen labels, as shown in Table 4. In this experiment, we considered four conditions: full method, removing the data ranking module, using only model-generated data without retrieved passages, and just leveraging document retrieval as data augmentation. The performance drop in the latter three conditions compared to the full method indicates that both parts contributing to synthetic data are indispensable. When not retrieving documents for relation labels, the noise in the generated data affects the model's prediction performance for certain relation classes. When using only the retrieved document for structured processing as data augmentation, there is a more significant performance drop because it hampers the model's ability to make structured predictions since we cannot provide triplet information during synthetic data generation based on the content of the retrieved documents alone. Furthermore, the data ranking module also makes a significant contribution to the final performance. It effectively improves the ranking of high-quality data, serving as a data filtering mechanism.

Table 4. Ablation Study. We conducted ablation experiments by individually removing the data ranking module, passage retrieve module, and synthetic data generator from the entire model.

Model	Single	Multi
	Acc.	F1
Full Method	24.14	27.07
–Data Ranking Module	23.02	24.54
–Passage Retrieve Module	22.19	25.12
–Synthetic Data Generator	18.66	21.9

Analysis. We further investigated how the quantity of retrieved documents affects the performance of zero-shot relation triplet extraction. Based on the research by Chia et al. [3], we generated 250 synthetic data instances for each relation. Additionally, we conducted re-ranking experiments using a varying number of retrieval documents, ranging from 100 to 500, combined with the synthetic data. The results indicated that while retrieved documents were beneficial for the triplet extraction task, excessive amounts could lead to overfitting due to noise. Moreover, the synthetic data transformed from retrieved documents does not possess strict triplet-structured information, which could potentially hamper the model's ability to make structured predictions at that level. Ultimately, we chose to retrieve 150 documents per relation.

5   Conclusion

In this work, we proposed a zero-shot relation triplet extraction method via retrieval-augmented synthetic data generation. By combining the results of generative language model and retrieval model, we generate synthetic data for the seq2seq-based triplet extraction model. This approach enables our model to extract triplets with unseen relation labels in the zero-shot setting. Experimental results on FewRel and Wiki-ZSL demonstrate that our model effectively generates high-quality synthetic data for this structured prediction task and achieves data augmentation improvements.

References

1. Nayak, T., Majumder, N., Goyal, P., et al.: Deep neural approaches to relation triplets extraction: a comprehensive survey. Cogn. Comput. **13**(5), 1215–1232 (2021)
2. Chen, C.-Y., Li, C.-T.: ZS-BERT: towards zero-shot relation extraction with attribute representation learning. In: Proceedings of the 2021 Conference of the North American Chapter of the Association for Computational Linguistics: Human Language Technologies, pp. 3470–3479 (2021)
3. Chia, Y.K., Bing, L., Poria, S., Si, L.: RelationPrompt: leveraging prompts to generate synthetic data for zero-shot relation triplet extraction. In: Findings of the Association for Computational Linguistics: ACL 2022, Dublin, Ireland, pp. 45–57 (2022)
4. Lewis, P., Perez, E., Piktus, A., et al.: Retrieval-augmented generation for knowledge-intensive NLP tasks. In: Advances in Neural Information Processing Systems (NeurIPS 2020), vol. 33, pp. 9459–9474 (2020)
5. Karpukhin, V., Oguz, B., Min, S., et al.: Dense passage retrieval for open-domain question answering. In: Proceedings of the 2020 Conference on Empirical Methods in Natural Language Processing (EMNLP), pp. 6769–6781 (2020)
6. Petroni, F., Rocktäschel, T., Lewis, P., et al.: Language models as knowledge bases? In: Proceedings of EMNLP-IJCNLP (2019)
7. Keskar, N.S., McCann, B., Varshney, L.R., et al.: CTRL: a conditional transformer language model for controllable generation, arXiv preprint arXiv:1909.05858 (2019)

8. Yang, P., Li, L., Luo, F., et al.: Enhancing topic-to-essay generation with external commonsense knowledge. In: Proceedings of the 57th Annual Meeting of the Association for Computational Linguistics, pp. 2002–2012 (2019)
9. Cabot, P.-L.H., Navigli, R.: REBEL: relation extraction by end-to-end language generation, In: Findings of the Association for Computational Linguistics: EMNLP 2021, Punta Cana, Dominican Republic, pp. 2370–2381 (2021)
10. Kim, B., Iso, H., Bhutani, N., Hruschka, E., Nakashole, N.: Zero-shot triplet extraction by template infilling. arXiv preprint arXiv:2212.10708 (2022)
11. Johnson, J., Douze, M., Jégou, H.: Billion-scale similarity search with GPUS. IEEE Trans. Big Data **7**(3), 535–547 (2019)
12. Malkov, Y.A., Yashunin, D.A.: Efficient and robust approximate nearest neighbor search using hierarchical navigable small world graphs. IEEE Trans. Pattern Anal. Mach. Intell. **42**(4), 824–836 (2016)
13. Han, X., et al.: FewRel: a large-scale supervised few-shot relation classification dataset with state-of-the-art evaluation. In: Proceedings of the 2018 Conference on Empirical Methods in Natural Language Processing, Brussels, Belgium, pp. 4803–4809 (2018)
14. Wang, J., Lu, W.: Two are better than one: joint entity and relation extraction with table-sequence encoders. In: Proceedings of the 2020 Conference on Empirical Methods in Natural Language Processing, pp. 1706–1721. Association for Computational Linguistics (2020)

Parallelizable Simple Recurrent Units
with Hierarchical Memory

Yu Qiao[1,2], Hengyi Zhang[1,2], Pengfei Sun[3], Yuan Tian[4], Yong Guan[1,2],
Zhenzhou Shao[1,2(✉)], and Zhiping Shi[1]

[1] College of Information Engineering, Capital Normal University,
Beijing 100048, China
{2211002024,2211002081,guanyong,zshao,shizp}@cnu.edu.cn
[2] Beijing Key Laboratory of Light Industrial Robot and Safety Verification,
Capital Normal University, Beijing 100048, China
[3] Beijing Smartchip Microelectronics Technology Company Limited, Beijing, China
sunpengfei1@sgchip.sgcc.com.cn
[4] Industrial and Commercial Bank of China Limited Beijing Branch,
Beijing 100032, China

Abstract. Recurrent neural networks and its many variants have been
widely used in language modeling, text generation, machine translation,
speech recognition and so forth, due to the excellent ability to process
sequence data. However, the above-mentioned networks are constructed
in a multi-layer stacking way, which makes the memory-dependent information in the distant past continuously decay. To this end, this paper
proposes a parallelizable simple recurrent unit with hierarchical memory (PSRU-HM) to preserve more long-term historical information for
inference. It is achieved by the nested SRU structure, and realizes the
information interaction between inner and outer memory cell through
the connection between inner and outer layers. The depth of network
can be dynamically adjusted according to the task complexity. Meanwhile, a jump connection that combines high-level and low-level features
is added to the outermost layer. It maximizes the utilization of effective input information. In order to accelerate the training and inference
of the network, the weights of PSRU-HM are reorganized to enable the
parallelization deployment in the CUDA framework. Extensive experiments are conducted to verify the proposed method using several public datasets, including text classification, language modeling and question answering. Experimental results show that PSRU-HM outperforms
the traditional methods and achieves 2× speed-up compared to cuDNN-
optimized LSTM.

Keywords: Recurrent Neural Network · Simple Recurrent Unit ·
Hierarchical Memory · Parallelization

Supported by the Natural Science Foundation of China (62272322, 62002246,
62272323), the Project of Beijing Municipal Education Commission (KM202010028010)
and the Applied Basic Research Project of Liaoning Province (2022JH2/101300279).

B. Luo et al. (Eds.): ICONIP 2023, CCIS 1969, pp. 380–392, 2024.
https://doi.org/10.1007/978-981-99-8184-7_29

1   Introduction

Due to the excellent ability to deal with the sequence data, recurrent neural networks (RNNs) have been widely applied in several applications, such as text classification [2], machine translation [3], language modeling [4], etc. In recent years, many RNN variants are presented in literature, e.g., Long Short-Term Memory (LSTM) [7] and Gated Recurrent Unit (GRU) [8]. In particular, it is still a challenging task to satisfy the requirements on both time and memory costs on the premise of performance guarantee.

Lei *et al.* proposes the Simple Recurrent Unit (SRU) [9]. It demonstrates the faster speed, compared to cuDNN-optimized LSTM on text classification and question answering datasets, and outperforms both LSTM and convolutional neural network models. Therefore, many researchers have made improvements based on SRU [10, 19–21]. Z *et al.* fused SRU with CNN for signal classification. Neishi *et al.* utilized SRU as part of the relative positional encoder and combined it with Transformer. Cui *et al.* proposed a simplified SRU network structure to reduce training time. Ullah *et al.* introduce a method based on stacked SRU, complex spectral mapping and complex scale masking, which improves the accuracy and robustness of the network. However, these networks are usually organized in a stacked manner, as shown in Fig. 1(b). Their approach consists of a simple stacking of multiple layers of neural network layers in sequence. Each layer receives the output of the previous layer as input. Such a structure is difficult to capture the input information of long-term memory when faced with long input. Since the memory in the distant past is prone to be forgotten. Therefore, it is necessary to find a recursive network that can catch more long-term memory information.

To solve above problem, we propose a Parallelizable Simple Recurrent Unit with Hierarchical Memory (PSRU-HM), A simple diagram of his structure is as shown in Fig. 1(c). Different from the simple stack structure of Z et al. Neishi et al. Cui et al. ignoring the long-term memory. Our method utilizes a depth-

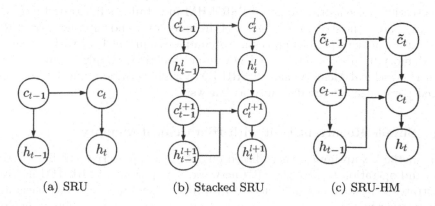

(a) SRU   (b) Stacked SRU   (c) SRU-HM

Fig. 1. Comparison diagram of memory information propagation between SRU, Stacked SRU, and SRU-HM.

hierarchical integrated network, which can dynamically adjusts the depth based on task complexity. The network consists of inner and outer layers, allowing selectively access to memory information and processed data between layers. Using the proposed hierarchical memory, the network extends the memory preserving capacity of cell units. Additionally, the skip connections designed in the outermost layer retain original features.

To improve the computational speed of the network, we deploy a hierarchical recurrent neural network on the GPU using CUDA programming for parallel computation. This enables batch processing of the matrix multiplication operations within the recurrent units. Additionally, we optimize the matrix multiplication by performing element-wise operations, significantly enhancing the training speed of the model. The training time of Our method is reduced by half compared to cuDNN-optimized LSTM. The main contributions of this paper are three-fold:

- We propose a Simple Recurrent Unit with Hierarchical Memory (SRU-HM) that addresses the issues of long training time and poor long-term memory performance in traditional recurrent neural networks, when handling long sequence inputs.
- While ensuring accuracy, we have deployed all weights and multiplication operations to CUDA, enabling parallel processing and achieving faster training speed than traditional RNNs and its Variants.
- The proposed m1ethod is evaluated on various natural language processing datasets, including text classification, character-level language modeling, and sentence question answering. Experimental results demonstrates that the proposed method outperforms the traditional RNNs and its Variants.

2   PSRU-HM: Parallelizable Simple Recurrent Unit with Hierarchical Memory

This section presents the proposed PSRU-HM in detail. As illustrated in Fig. 2, the gray part is the structural diagram of the inner layer and outer layer of PSRU-HM and we achieve parallel processing of Simple Recurrent Unit with Hierarchical Memory by implementing a CUDA kernel function. Simple Recurrent Unit with Hierarchical Memory, named SRU-HM for abbreviation, and CUDA optimized parallelization are described as follows.

2.1   Simple Recurrent Unit with Hierarchical Memory

This method is an improvement based on SRU, as it preserves the basic structure and operation logic of the SRU network. The portion of SRU-HM in Fig. 2 illustrates a network design structure with hierarchical memory. The basic unit of network structure consists of two parts: outer memory unit and inner memory unit. An innovation of this method is the hierarchical storage of memory units in the inner and outer layers through nesting.

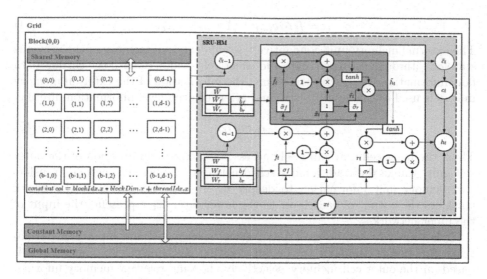

Fig. 2. Overall structure of PSRU-HM.

The logical structure design of the inner and outer layers is as follows: outer–inner–outer. In this structure, the network embedded in the inner layer performs extensive selection based on the information in the outer memory unit, employing gated random reads and writes. The outer layer serves as the fundamental gated selection unit. The forward propagation of the outer layer is expressed as Eqs. (1–5).

$$X_t = Wx_t, \tag{1}$$

$$f_t = \sigma(W_f x_t + b_f), \tag{2}$$

$$r_t = \sigma(W_r x_t + b_r), \tag{3}$$

where x_t represents the input, and X_t in Eq. (1) is a linear transformation of x_t, the purpose of which is to extract effective features of the input data; Eq. (2) and Eq. (3) respectively represent the calculation process of forget gate and reset gate. They are designed to filter through their information and select what is needed. The input here contains only the data x_t to be entered at the current time t. Compared to traditional SRU, the forget gate and the reset gate remove the $v \odot c_t$, but keep the highway connection to maintain optimal performance of the deep network. In the context, f_t represents the forget gate, while r_t represents the reset gate. W, W_f, W_t are the parameter matrix correspond respectively to the weights of self-connections, forget gate, and reset gate of the outer layer network and b_f, b_t are the parameters of forget gate and update gate respectively which both needed to be learned in the external structure. σ represents the activation function sigmoid.

$$c_t = f_t \odot c_{t-1} + (1 - f_t) \odot X_t. \tag{4}$$

The external memory unit c_t adaptively selects external memory of the previous state c_{t-1} through the forget gate and effectively combines with the current data flow information X_t. At the same time, the information of the external memory unit c_t is used as the input of the inner layer for deep level information filtering.

$$\tilde{x}_t = c_t. \tag{5}$$

Thus, the inner cell unit can freely access the memory information of the outer cell unit. Unlike the traditional stacked network, the nesting method transfers the memory unit information to deeper layers, whereas the stacked network transfers the output of each layer to the lower layer. $\tilde{x}_t$ represents the input of the inner network.

It is primarily composed of forget gates, reset gates, and input transformations, functioning similarly to the outer gating mechanism. The inner input is based on the outer cell memory, selectively choosing relevant memory information from the outer layer. Through the inner gating mechanism, it further filters and refines the current relevant memory information. The inner cell unit also filters out irrelevant historical inner layer memory through forget gates, adds new useful outputs, and updates the inner layer's memory information. After activation and selection by the reset and activation gates, the inner layer's output is fed back to the outer layer network, updating the content of the outer cell memory. The forward propagation of the inner layer can be represented as Eqs. (6–11).

$$\tilde{X}_t = \widetilde{W}_t \tilde{x}_t, \tag{6}$$

$$\tilde{f}_t = \tilde{\sigma}(\widetilde{W}_f \tilde{x}_t + \tilde{b}_f), \tag{7}$$

$$\tilde{r}_t = \tilde{\sigma}(\widetilde{W}_r \tilde{x}_t + \tilde{b}_r), \tag{8}$$

$$\tilde{c}_t = \tilde{f}_t \odot \tilde{c}_{t-1} + (1 - \tilde{f}_t) \odot \tilde{X}_t. \tag{9}$$

Information of inner memory unit $\tilde{c}_t$ depends on how much the inner forget gate filters the state $\tilde{c}_{t-1}$ of the inner memory unit at the previous time and the Information flow $\tilde{X}_t$ at the current moment. By storing memory information in hierarchical memory units, the long-term sequence dependent information can be retained to the maximum extent, in order to ensure that memory can be used in future situations, thereby improving the network's predictive ability.

$$\tilde{h}_t = \tilde{r}_t \odot \tilde{g}(\tilde{c}_t), \tag{10}$$

$$c_t = \tilde{h}_t. \tag{11}$$

According to Eq. (10) and Eq. (11), the output $\tilde{h}_t$ obtained by inner memory unit $\tilde{c}_t$ is filtered through the reset gate $\tilde{r}$ of inner layer, and $\tilde{h}_t$ updates the memory unit c_t of outer layer, $\tilde{g}$ is the activation function tanh. In other words, the outer memory unit has the ability to access the relevant information needed by the inner memory unit. Through information filtering, all long-term memory information related to the current situation in the hierarchical memory unit will be finally filtered by the outer reset gate r to maintain the validity of long-term memory. Hierarchical memory involves dividing a network into inner and outer layers. The outer layer functions as a fundamental memory unit that interacts with external information and input data through selective reading and writing. The inner layer then refines this information by selectively reading and filtering it based on the outer layer's output. This hierarchical approach enables the network to capture and store information at various levels of abstraction, enhancing its ability to handle long-term dependencies in lengthy sequential data.

$$h_t = r_t \odot g(c_t) + (1 - r) \odot X_t. \tag{12}$$

Meanwhile, the highway network component [5] improved training of deep network based on gradient. In Eq. (12), $(1 - r) \odot X_t$, s a skip connection, so that the information of X_t is directly combined with the memory information screened by the outer reset gate, which improves the network scalability.

2.2   CUDA Optimized Network Parallelization

Although the state composition of SRU-HM is dependent on time, each dimension of state is independent, considering the design of the forget gate and reset gate in both inner and external structures. It facilitates the parallel computation for each dimension in the network. By executing the operations within SRU-HM simultaneously on different threads for each dimension, it efficiently harnesses the advantages of GPU parallel processing. In this paper, the CUDA programming technique is utilized to optimize the computation of network.

In the both inner and outer structures, the forget gate and reset gate involve the multiplication of weight matrix by the same input. To enable the effective parallel deployment, Eqs. (1–3) and (6–8) are combined as augmented matrices, respectively. In the outer layer, given the input sequence $[x_1, x_2, \ldots, x_L]$, Eqs. (1–3) can be rewritten as

$$\mathbf{U} = \sigma \left\{ \begin{pmatrix} W \\ W_f \\ W_r \end{pmatrix} [x_1, x_2, \ldots, x_L] \right\}, \tag{13}$$

where L represents the length of the input sequence, and d is the dimension of the input sequence. The matrices W, W_f and W_r correspond to the weights of the self-connection, forget gate, and reset gate of the outer layer network, respectively. According to Eq. (13), three weight matrices are combined into one matrix and undergo batch matrix multiplication with the input sequence, resulting in the tensor U. Given the batch size of the input is B, the dimension of

---

Algorithm 1. Parallel feed forward connection of SRU-HM network

---

1: **Indices:** Sequence length is L, where $l = 1, 2, \ldots L$; Batch size is B, where $i = 1, 2, \ldots B$; The dimension of h of network hidden state is d, where $j = 1, 2, \ldots d, j' = 1, 2, \ldots 3d, j'' = 1, 2, \ldots 3d$.

2: **Input:** $x[l, i, j]$, the tensor of the outer mixing matrix $U[l, i, j']$,the tensor of the outer mixing matrix $\tilde{U}[l, i, j'']$,the initial outer layer and inner layer memory are both $[i, j]$ and the bias values are $b_f, b_r, \tilde{b}_f, \tilde{b}_r$.

3: **Output:** Hidden state h and outer memory unit information c.

4: **Initializes:** Tensor h and c with dimensions $L \times B \times d$.

1) **for all** $i[1, B]$, $j[1, d]$ **do**

2)   **for** $l[1, L]$ **do**

3) $f = \sigma(U[l, i, j + d] + b_f)$

4) $r = \sigma(U[l, i, j + d \times 2] + b_r)$

5) $\tilde{x} \leftarrow c = c \times f + (1 - f) \times U[l, i, j']$ //Outer memory acts as the inner

    input.

6) $\tilde{f} = \sigma(\tilde{U}[l, i, j + d] + \tilde{b}_f)$

7) $\tilde{c} = \tilde{c} \times \tilde{f} + (1 - \tilde{f}) \times \tilde{U}[l, i, j'']$

8) $\tilde{r} = \sigma(\tilde{U}[l, i, j + d \times 2] + \tilde{b}_r)$

9) $c \leftarrow \tilde{h} = \tilde{r} \times \tilde{c}$ //The inner hidden outputs are used to update the

    external memory.

10) $h = r \times c + (1 - r) \times x[l, i, j]$ //highway connection

11) $c[l, i, j] = c$ and $h[l, i, j] = h$

12)   **end for**

13) **end for**

14) **return** $h[\cdot, \cdot, \cdot]$ and $c[\cdot, \cdot, \cdot]$

---

tensor U is $(L, B, 3d)$. After undergoing the gating mechanism of the outer layer, the memory of outer layer c is obtained, following the calculation of $\tilde{x} \in \mathbb{R}^{L \times B \times d}$ as the input of inner layer. Similarly, Eqs. (6–8) is reformulated by Eq. (14) for parallel computation.

$$\tilde{\mathbf{U}} = \tilde{\sigma} \left\{ \begin{pmatrix} \widetilde{W} \\ \widetilde{W}_f \\ \widetilde{W}_r \end{pmatrix} [\tilde{x}_1, \tilde{x}_2, \ldots, \tilde{x}_L] \right\}. \tag{14}$$

This method compiles all point-wise operations into a fused CUDA kernel, and parallelizes computation on each dimension of hidden state according to Algorithm 1. The time complexity of each step of the sequence is $O(L \times B \times d)$,which is an order of magnitude faster than the time complexity $O(L \times B \times d \times d)$ of LSTM without parallel computation.

3   Experimental Results and Discussion

In this section, three sets of experiments are conducted to evaluate the performance of proposed method in typical applications, including text classification, language model and question answering. In each set of experiments, PSRU-HM is compared with the state-of-the-art methods, i.e., CNN, GRU, LSTM and SRU with respect to the accuracy and effectiveness. Our experimental hardware configuration is as shown in the Table 1.

Table 1. Hardware Configuration.

Category	Specification
Host	DGX-1
Operating system	Linux
CPU	Intel (R) Xeon (R) CPU E5-2698 v4 @ 2.20 GHz
GPU	Tesla P100-SXM2
Graphics memory	16 GB

3.1 Performance in Text Classification

Benchmark Datasets and Evaluation. 6 sets of natural language text classification experiments are conducted to verifiy the superiority of PSRU-HM. It is compared with CNN, LSTM, GRU and SRU in the network classification accuracy and training speed. We use six text classification benchmarks as our dataset: Customer reviews polarity (CR) [15], movie review sentiment (MR) [16], opinion polarity (MPQA) [17], question type (TREC) [14], sentence subjectivity (SUBJ) [18], and the Stanford sentiment treebank (SST) [1].

Implementation Details. We stack two and three PSRU-HM layers and use the last output state to predict the class label for a given sentence. In the experiments, We train for 100 epochs and record time cost of each epoch and total epochs. We perform 10-fold cross validation for datasets that do not have a standard train-evaluation split. We use Adam with the default learning rate 0.001, a weight decay 0 and a hidden dimension of 128.

Results. Table 2 and Table 3 compare the accuracy and training time of 5 three-layer models and 5 two-layer models across 6 tasks. From Table 2, it can be observed that PSRU-HM performs the best in most tasks, while the performance gain of three-layer models over two-layer models is not significant. However, compared to the other four models, three-layer PSRU-HM performs exceptionally well. Table 3 represents the training time of the models. It is visually evident that the two-layer PSRU-HM is almost as fast as CNN. Therefore, two-layer networks are sufficient to demonstrate the advantages of SRU-HM.

3.2 Performance in Language Modeling

Benchmark Datasets and Evaluation. In language modeling experiment, we use Wikipedia (Enwik8) and Pennsylvania Tree Bank (PTB) [13]. The dataset

Table 2. The accuracy of two-layer and three-layer CNN, GRU, LSTM, SRU and PSRU-HM in text classification tasks. A subscript ∗ in the upper right corner indicates three-layers, while the absence of a subscript indicates two-layers.

Accuracy	cr	mr	mpqa	trec	subj	sst
CNN	81.75%	78.44%	90.20%	94.20%	**95.00%**	85.39%
LSTM	79.63%	77.69%	89.63%	94.00%	94.10%	86.99%
GRU	82.54%	79.76%	90.29%	93.20%	94.00%	86.66%
SRU	81.48%	83.30%	89.73%	93.80%	94.00%	89.07%
PSRU-HM	**84.88%**	**83.51 %**	**90.95%**	**94.40%**	94.40%	**89.35%**
CNN*	83.07%	78.54%	89.35%	92.00%	**94.50%**	85.61%
LSTM*	82.54%	79.85%	89.63%	93.40%	**94.50%**	86.88%
GRU*	82.54%	80.51%	90.01%	93.60%	93.90%	87.53%
SRU*	84.66%	83.02%	90.39%	94.60%	93.30%	88.85%
PSRU − HM*	**85.68%**	**85.00%**	**90.95%**	**95.00%**	94.10%	**89.07%**

Table 3. The training time of two-layer and three-layer CNN, GRU, LSTM, SRU and PSRU-HM in text classification tasks. A subscript ∗ in the upper right corner indicates three-layers, while the absence of a subscript indicates two-layers.

Time(s)	cr	mr	mpqa	trec	subj	sst
CNN	34.61	91.76	82.87	52.13	85.82	746.10
LSTM	76.52	181.44	108.03	68.77	199.52	1269.14
GRU	82.59	194.84	99.10	78.36	197.59	1379.04
SRU	38.59	98.30	101.31	56.50	94.44	882.21
PSRU-HM	37.76	102.14	96.48	54.80	87.44	833.25
CNN*	39.33	102.58	94.24	46.44	79.62	747.26
LSTM*	124.29	281.83	124.57	85.05	315.68	1971.19
GRU*	138.05	317.07	127.01	100.79	344.59	2217.63
SRU*	42.30	119.95	109.53	63.84	124.75	915.29
PSRU − HM*	50.11	113.90	111.13	71.82	113.24	963.58

uses the model's perplexity (ppl) and tests per bits (bpc) to verify the efficiency of the PSRU-SM network training speed and the long-term memory capacity of the hierarchical network.

Implementation Details. We use the first 90M letters in Eniki8 for training and the remaining 10M for validation and test. We use a batch size of 128 and 200 letters to expand so that each training instance has a longer context. The model is trained for 100 steps. Finally, test the bpc of different recurrent network models and the training time corresponding to each epoch on the Enwik8 dataset to verify the performance of the PSRU-HM network. The ptb corpus has a batch size of 32 and a character extension of 35. The remaining parameters are similar to the Enwik8 dataset. The experimental training steps are 300 steps. The ptb corpususes ppl and bpc to evaluate the performance of language models.

Results. We collect the performance results of the Enwik8 language model based on three methods: LSTM, SRU, and PSRU-HM. As shown in Table 4, bpc and ppl of the three-layer PSRU-HM network are smaller than the three-layer LSTM and SRU network. Therefore, the language model based on the PSRU-HM network is more accurate. The Enwik8 language model of the 3-layer PSRU-HM has smaller bpc and ppl values than the Enwik8 language model of the 2-layer PSRU-HM. This shows that the advantages of deep networks in memory information processing. In the time aspect, the three-layer PSRU-HM network has advantages over LSTM and SRU networks in terms of training speed. In particular, PSRU-HM trains twice as fast as LSTM. The ptb language model further validates the advantages of the PSRU-HM network. As shown in Table 5 shows that the hierarchical network has a strong advantage for long-term memory ability and can better predict the language model. From the perspective of training time, the training speed of PSRU-HM is three times of LSTM. Based on the impact of network depth on the language model, the ppl and bpc of the 3-layer PSRU-HM are reduced. It shows that deep networks have stronger learning capabilities.

Table 4. Establish Enwik8 language model by using different methods.

Method	Layer	dev_ppl	dev_bpc	Time/Epoch
LSTM	3	0.8858	1.28	62.2 min
SRU	3	0.8811	1.27	37.5 min
PSRU-HM	2	1.0990	1.58	18.7 min
PSRU-HM	3	0.8409	1.21	34.3 min

Table 5. Establish ptb language model by using different methods.

Method	Layer	train_ppl	dev_ppl	dev_bpc	Time/Epoch
LSTM	3	35.05	69.19	6.11	73.2 s
SRU	3	34.97	66.75	6.06	26.4 s
PSRU-HM	2	51.88	75.27	6.23	17.4 s
PSRU-HM	3	34.44	65.16	6.02	25.8 s

3.3   Performance in Question Answering

Benchmark Datasets and Evaluation. We use the Stanford Question Answering Dataset in question answering experiment. SQuAD is a large machine comprehension dataset that includes over 100K question-answer pairs extracted from Wikipedia articles [12]. We use two evaluation indicators: Exact Match (EM) and F_1 scores to evaluate the performance of the established question answering model.

Implementation Details. In this experiment, we compared PSRU-HM with SRU, GRU, and LSTM networks to verify performance.We use a three-layer PSRU-HM network to build a question answering model. The batch size is 32, the hidden dimension is 128, the learning rate is 0.001, the optimizer is Adamax, and the number of training steps is 40.

Table 6. Different methods to establish performance of question answering models.

Method	Layer	Epoch	EM	F_1	Time/Epoch
LSTM	3	40	67.68 ± 0.34	77.50 ± 0.26	778 s
GRU	3	40	67.52 ± 0.38	77.12 ± 0.32	558 s
SRU	3	40	68.45 ± 0.37	77.85 ± 0.32	468 s
PSRU-HM	3	40	69.33 ± 0.35	78.50 ± 0.33	482 s

Results. The larger values of EM and F1 represent better performance of question answer performance. As shown in Table 6 Both of EM and F1 scores obtained on the test set based on the PSRU-HM question answering model exceed the results of other three models. We record training time of each step. The training speed of PSRU-HM model is close to the SRU network which is 38.05% and 13.80% shorter than the training time of LSTM and GRU. The experimental results show that the structurally parallel design of the network improves the training speed of the network, while the hierarchical memory storage can enhance the ability to process long-term information.

4   Conclusion

This paper primarily focused on addressing the limitations of traditional recurrent neural networks in handling long-term sequences. We propose a parallel simple recurrent unit with hierarchical memory (PSRU-HM), We validate our method on various public datasets, including text classification, language modeling and question answering. Experimental proof our method enhances the capability of handling long sequence information, improves accuracy, and reduces training time compared to existing methods. By modifying the internal structure of SRU and enabling parallel computation, PSRU-HM achieves superior performance.

However, it is worth noting that our research still falls short, and there is untapped potential to explore further. We did not conduct comparisons with deeper model architectures in our experiments. We can carry out in-depth comparisons of accuracy and training/inference times, thus providing a comprehensive evaluation and analysis while also offering valuable directions for future research. In summary, although we have made positive strides in the realm of long sequence processing, we acknowledge that there is more potential to be unearthed, and we look forward to expanding and enhancing our work in the future.

References

1. Socher, R., et al.: Recursive deep models for semantic compositionality over a sentiment treebank. In: Proceedings of the 2013 Conference on Empirical Methods in Natural Language Processing, pp. 1631–1642 (2013)
2. Liu, P., Qiu, X., Huang, X.: Recurrent neural network for text classification with multi-task learning. arXiv preprint arXiv:1605.05101 (2016)
3. Jean, S., Cho, K., Memisevic, R., Bengio, Y.: On using very large target vocabulary for neural machine translation. arXiv preprint arXiv:1412.2007 (2014)
4. Zaremba, W., Sutskever, I., Vinyals, O.: Recurrent neural network regularization. arXiv preprint arXiv:1409.2329 (2014)
5. Srivastava, R.K., Greff, K., Schmidhuber, J.: Training very deep networks. Adv. Neural Inf. Process. Syst. **28** (2015)
6. Lei, T.: When attention meets fast recurrence: training language models with reduced compute. arXiv preprint arXiv:2102.12459 (2021)
7. Hochreiter, S., Schmidhuber, J.: Long short-term memory. Neural Comput. **9**(8), 1735–1780 (1997)
8. Cho, K., et al.: Learning phrase representations using RNN encoder-decoder for statistical machine translation. arXiv preprint arXiv:1406.1078 (2014)
9. Lei, T., Zhang, Y., Wang, S.I., Dai, H., Artzi, Y.: Simple recurrent units for highly parallelizable recurrence. arXiv preprint arXiv:1709.02755 (2017)
10. Ullah, R., et al.: End-to-end deep convolutional recurrent models for noise robust waveform speech enhancement. Sensors **22**(20), 7782 (2022)
11. Pan, J., Lei, T., Kim, K., Han, K.J., Watanabe, S.: SRU++: pioneering fast recurrence with attention for speech recognition. In: ICASSP 2022–2022 IEEE International Conference on Acoustics, Speech and Signal Processing (ICASSP), pp. 7872–7876. IEEE (2022)
12. Rajpurkar, P., Zhang, J., Lopyrev, K., Liang,P.: Squad: 100,000+ questions for machine comprehension of text. arXiv preprint arXiv:1606.05250 (2016)
13. Miltsakaki, E., Prasad, R., Joshi, A.K., Webber, B.L.: The penn discourse treebank. In: LREC (2004)
14. Li, X., Roth, D.: Learning question classifiers. In: COLING 2002: The 19th International Conference on Computational Linguistics (2002)
15. Hu, M., Liu, B.: Mining and summarizing customer reviews. In: Proceedings of the Tenth ACM SIGKDD International Conference on Knowledge Discovery and Data Mining, pp. 168–177 (2004)

16. Pang, B., Lee, L.: Seeing stars: exploiting class relationships for sentiment categorization with respect to rating scales. arXiv preprint arxiv:cs/0506075 (2005)
17. Wiebe, J., Wilson, T., Cardie, C.: Annotating expressions of opinions and emotions in language. Lang. Res. Eval. **39**, 165–210 (2005)
18. Pang, B. Lee, L.: A sentimental education: sentiment analysis using subjectivity summarization based on minimum cuts. arXiv preprint arxiv:cs/0409058 (2004)
19. Lyu, Z., Wang, Y., Li, W., Guo, L., Yang, J., Sun, J., Liu, M., Gui, G.: Robust automatic modulation classification based on convolutional and recurrent fusion network. Phys. Commun. **43**, 101213 (2020)
20. Neishi, M., Yoshinaga, N.: On the relation between position information and sentence length in neural machine translation. In: Proceedings of the 23rd Conference on Computational Natural Language Learning (CoNLL), pp. 328–338 (2019)
21. Cui, X., Chen, Z., Yin, F.: Speech enhancement based on simple recurrent unit network. Appl. Acoust. **157**, 107019 (2020)

Enhancing Legal Judgment Prediction with Attentional Networks Utilizing Legal Event Types

Yaoyao Yu[ID] and Yihui Qiu[✉][ID]

School of Economics and Management, Xiamen University of Technology, Xiamen, China
yaoyaoyu@s.xmut.edu.cn, qiuyihui@xmut.edu.cn

Abstract. Legal Judgment Prediction (LJP) is a critical task that aims to predict charges, articles, and terms of penalty based on the fact descriptions provided in the cases. However, current LJP methods often fail to fully utilize the important aspect of legal event information, leading to suboptimal predictions. To address this issue, our proposed model introduces a legal event type attention mechanism, which effectively identifies key event information within the fact descriptions. By combining event-aware and event-free representations, our framework enables a comprehensive understanding of the fact descriptions, leading to better performance on LJP. Importantly, our approach outperforms the state-of-the-art models, achieving an average improvement of 3.86% in the prediction of articles, 1.82% in the prediction of charges, and 5.24% in the prediction of terms of penalty.

Keywords: Legal Judgment Prediction · Legal event information · Attention mechanisms

1 Introduction

Legal Judgment Prediction (LJP) is a critical task that involves predicting charges, legal articles, and the terms of penalty based on fact descriptions provided in the cases [13,19]. This application of natural language processing (NLP) in the legal domain serves to assist legal professionals and laypeople in comprehending and examining legal cases, while also providing valuable insights for judicial decision-making. Figure 1 visually represents the process of LJP within the judicial system.

Accurate extraction of legal event information from fact descriptions is crucial for making precise judgments in LJP [4,17]. Legal events involve actions or events that impact legal relationships, such as murder and robbery in criminal activities. These events typically consist of a trigger, indicating the occurrence of the event, and arguments, which include participants and attributes such as time and location. Different legal events result in varying legal consequences.

B. Luo et al. (Eds.): ICONIP 2023, CCIS 1969, pp. 393–404, 2024.
https://doi.org/10.1007/978-981-99-8184-7_30

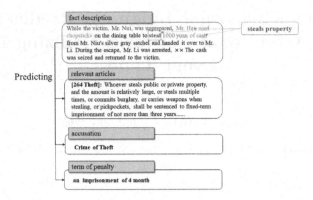

Predicting

Fig. 1. Illustration depicting LJP, with highlighted words representing key event information associated with the steal property legal event. The word in red serves as the event trigger, while other colored words indicate event arguments. (Color figure online)

For example, a person switching price labels in a supermarket could be charged with fraud under Article 266 and face a penalty of three to ten years imprisonment. On the other hand, if the person takes an item without paying, the charge would be theft under Article 264, with a maximum penalty of ten years imprisonment. These examples highlight the complexity and specificity of legal events, emphasizing the significance of legal event information in LJP.

Existing research in LJP often overlooks the importance of locating crucial event information in fact descriptions or fails to effectively utilize this information. Some previous works [1,11] primarily rely on bag of words models to interpret fact descriptions, which neglect word order and potentially overlook key event information. Additionally, these methods treat all words equally, disregarding the varying importance of different information in fact descriptions. Other studies [2,5,6,8] have attempted to leverage external legal knowledge for improved prediction accuracy and transparency. However, these methods face challenges in adapting to different scenarios due to their heavy reliance on manual annotation of external legal knowledge. Recent studies [4,17] have incorporated legal event information into LJP models, but they have not fully exploited the potential of this information to enhance performance.

In contrast to existing methods, we propose a novel attentional framework, specifically designed to incorporate legal event types, aiming to enhance the performance of LJP. Unlike previous studies that neglect or underutilize legal event information, our model effectively leverages both event-aware and event-free information within the fact descriptions for LJP. Our approach designs a legal event types attention mechanism, which precisely identifies the relevant parts of the fact descriptions associated with specific legal event types. By combining event-aware and event-free representations, our model achieves a comprehensive interpretation of the fact descriptions, leading to more accurate judgment predictions. Our framework consists of four main modules: (1) a legal text encoder module that encodes legal event types and fact descriptions, (2) a legal event types attention mechanism that extracts crucial legal event information from the fact descriptions, (3) a legal representation fusion module that integrates event-aware and event-free representations, and (4) a multi-task layer for predicting judgment results.

To assess the effectiveness of our framework, we evaluate it on a standard LJP dataset. Our evaluation includes quantitative analysis as well as case studies. The results clearly demonstrate the efficacy of our approach in leveraging legal event types to identify key event information and improve the performance of LJP.

The main contributions of this paper are as follows:

1. We propose a novel multi-task learning framework to enhance the performance of LJP. To achieve this, we design a legal event types attention mechanism to learn event-aware fact representations.
2. By concatenating event-aware and event-free representations, our approach provides a comprehensive fact description representation for judgment prediction.
3. We conduct efficient experiments on a real-world dataset, and our model significantly outperforms other baselines.

2   Related Work

LJP is a task that involves predicting the outcome of a legal case based on its fact description, with three subtasks including predicting law articles, charges, and terms of penalty in criminal cases. Over the past few decades, LJP has been the subject of extensive research.

One direction of research focuses on utilizing novel embedding-based models to encode similar fact descriptions [14–16]. For example, Xu et al. [14] proposed a method for distinguishing similar law articles using a graph neural network and a novel attention mechanism. Zhong et al. [20] pretrained BERT with millions of criminal and civil legal documents to acquire contextual legal embeddings for distinguishing similar fact descriptions.

Another direction explores effective ways to leverage external legal knowledge or exploit the topological dependencies of the subtasks to differentiate confusing and similar charges and laws. Zhong et al. [18] introduced a topological multi-task learning framework that focuses on the dependencies among subtasks. Hu et al. [5] incorporated discriminative attributes of charges as additional information for few-shot charges, providing effective signals for distinguishing confusing charges. Lyu et al. [9] proposed a reinforcement learning-based framework called the Criminal Element Extraction Network (CEEN), which addresses confusing fact descriptions and law articles in LJP simultaneously.

More recently, there has been a growing interest in integrating event information into LJP. Yao et al. [17] introduced LEVEN, a large-scale Chinese legal event detection dataset, and demonstrated the positive impact of event information on downstream legal tasks such as LJP. Feng et al. [4] proposed the Event-based Prediction Model (EPM), which uses event extraction to leverage key event details from fact descriptions and employs consistency constraints across subtasks to enhance LJP.

While these studies highlight the importance of incorporating event information into LJP models, they do not fully capitalize on the potential of utilizing this information to improve both performance and interpretability. In contrast, our model designs a legal event attention mechanism that effectively captures event-related details from fact descriptions, generating event-aware information within the context of the facts. By integrating this event-aware information with the event-free information, our approach achieves superior performance, distinguishing it from existing methodologies in the field of LJP.

3   Method

In this section, we propose a novel attentional neural network that utilizes legal event types to assist LJP within a multi-task framework. Similar to previous works [4, 17], we recognize the significance of locating and extracting legal event information from fact descriptions to ensure accurate judgments.

As shown in Fig. 2, our model takes the fact description f as input and generates two important outputs: the fact description representation H and $\hbar$, which we refer to as the event-free fact representation. Additionally, we employ the same encoder to generate a representation u_i for each legal event type $\hat{e}_i$. Then the legal event type representations and the fact description representation are then passed through an attention layer to compute the event-aware fact representation g_{event}. Finally, we concatenate g_{event} with $\hbar$ to make predictions regarding the appropriate charges, articles, and terms of penalty for the given case.

3.1   Legal Text Encoder

To obtain contextual embeddings for factual descriptions and legal event types, we employ the pre-trained bert-base-chinese model[1] as a legal text encoder. This allows us to capture rich contextual information within the legal event types and fact descriptions, and we refer to this model as BERT in the subsequent discussion.

The fact description f is initially provided as input and processed by it to generate the fact description representation H_f and the event-free fact representation $\hbar$. The fact description f consists of a sequence of tokens denoted as $\{x_1, x_2, \ldots, x_{l_f}\}$. l_f is the length of the fact description. The encoding process is calculated as follows:

$$H_f = \{h_1, h_2, \ldots, h_{l_f}\} = \text{BERT}(x_1, x_2, \ldots, x_{l_f}) \tag{1}$$

Here, $h_i \in \mathbb{R}^k$ represents the context vector of token x_i in the fact description f, and k is the dimension of h_i.

---

[1] https://huggingface.co/bert-base-chinese.

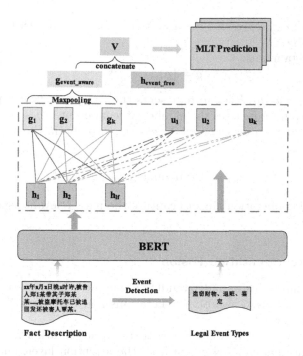

Fig. 2. Overview of the proposed model

To obtain the event-free fact representation $\hbar$, we utilize the pooler output from BERT, which captures the overall semantic information of the fact description.

Simultaneously, the predicted event types $\{\hat{e}_1, \hat{e}_2, \ldots, \hat{e}_n\}$ are also encoded as follows:

$$\{u_1, u_2, \ldots, u_n\} = \text{BERT}(\hat{e}_1, \hat{e}_2, \ldots, \hat{e}_n) \tag{2}$$

Here, $u_i \in \mathbb{R}^k$ corresponds to the contextual representation of the i-th event type, and k is the dimension of u_i.

3.2 Legal Event Types Attention

To extract key event information from the fact descriptions, we propose a legal event types attention mechanism, inspired by the work of Hu et al. [5] and Bao et al. [2]. This attention mechanism allows the model to focus on crucial event details within the fact descriptions, generating the event-aware fact representation.

As depicted in Fig. 2, the attention layer takes H_f and $\{u_1, u_2, \ldots, u_n\}$ as input. It calculates attention weights $a = \{a_1, \ldots, a_n\}$ for each u_i, where $a_i = [a_{i,1}, \ldots, a_{i,l_f}]$, $i \in [1, n]$. The attention weight $a_{i,j}$ is computed as follows:

$$a_{i,j} = \frac{\exp\left((h_j^T u_i)/\dim(u_i)\right)}{\sum_t^{l_f} \exp\left((h_t^T u_i)/\dim(u_i)\right)} \tag{3}$$

Here, $j \in [1, l_f]$, $a_{i,j}$ represents the attention weight of the i-th legal event type and the h_j in H_f.

Next, we obtain g as:

$$g = \{g_1, \ldots, g_n\} \tag{4}$$

where $g_i = \sum_{t=1}^{l_f} a_{i,t} h_t$. Each g_i represents the i-th legal event type aware fact representation, and n denotes the number of legal event types.

Finally, we apply a max-pooling layer over g to obtain the event-aware representation g_{event}:

$$g_{\text{event}} = \text{Maxpooling}\{g_1, g_2, \ldots, g_n\} \tag{5}$$

The max-pooling operation aims to capture the most relevant legal event information from g.

3.3   Legal Representation Fusion

To obtain a comprehensive understanding of fact descriptions, we combine both the event-free and event-aware representations and feed them into the output layer for prediction.

Specifically, we concatenate the event-free representation $\hbar$ and the event-aware representation g_{event} obtained from the attention layer, and use them to make the final prediction:

$$\mathbf{v} = \hbar \oplus g_{\text{event}} \tag{6}$$

Here, the symbol $\oplus$ signifies the concatenation of $\hbar$ and g_{event} to form the final fact representation $\mathbf{v}$.

3.4   The Output Layer

In the output layer, we have three subtasks: t_1, charge prediction; t_2, law article prediction; and t_3, term prediction. Given the final fact representation $\mathbf{v}$, we pass it through three separate linear layers for each task's prediction, the predicted distribution over all labels in subtask t_i is calculated as follows:

$$\hat{y}_i = w^i \mathbf{v} + b^i \tag{7}$$

Here, w^i and b^i represent the task-specific parameters for t_i.

3.5   Training

During training, we formulate t_1, t_2, and t_3 as multi-class classification tasks. We employ the cross-entropy loss function for each subtask. The loss $Loss_t$ of the t-th subtask is calculated as follows:

$$Loss_t = -\frac{1}{N} \sum_{i=1}^{N} \sum_{j=1}^{C} y_{ij} \log(\hat{y}_{ij}) \tag{8}$$

Here, N denotes the total number of samples, C represents the total number of classes, y_{ij} is the ground-truth label for whether sample i belongs to class j and $\hat{y}_{ij}$ is the predicted distribution that sample i belongs to class j. The total loss L is computed as the sum of the losses for all three subtasks:

$$L = \sum_{t=1}^{3} Loss_t \qquad (9)$$

4   Experiments

4.1   Datasets

We evaluate the performance of our model using the CAIL-small dataset, which is derived from the Chinese AI and Law challenge (CAIL2018) dataset [13]. Following the approach of Xu et al. [14] and Lyu et al. [9], we filter out samples with fewer than 10 meaningful words, as well as samples with multiple applicable law articles and charges. Additionally, in line with Zhong et al. [18], we retain only those law articles and charges that occur at least 100 times and categorize the terms of penalty into non-overlapping intervals. To obtain the legal event types, we employ the BERT+CRF model trained by Yao et al. [17] to detect the legal event types from fact descriptions in the datasets. The detailed dataset statistics are presented in Table 1.

Table 1. Distribution of CAIL-small

Dataset	CAIL-small
#Training Set Cases	101619
#Testing Set Cases	26749
#Law Articles	103
#Charges	119
#Term of Penalty	11
#Legal Event types	108

4.2   Baselines

We compare our model with several baselines and state-of-the-art models, including:

HARNN: An RNN-based document classification method with a hierarchical attention mechanism [16].

FLA: A rule-based charge prediction method that models correlations between fact descriptions and related law articles [8].

Few-Shot: An attribute-attentive prediction model that infers ten additional discriminative attributes and charges simultaneously [5].

TopJudge: A topological multi-task learning framework for LJP introduced by Zhong et al. [18].

MPBFN: A multi-task framework that utilizes result dependencies among the subtasks with forward predictions and backward verifications [15].

LADAN: A multi-task framework captures subtle differences within perplexing law articles,utilizing a graph neural network-based approach [14].

CEEN: A multi-task model proposed by Lyu et al. [9], which addresses confusing fact descriptions and law articles in LJP simultaneously using reinforcement learning.

Bert+event: A multi-task model proposed by Yao et al. [17], which leverages event information in the fact description by adding an event type embedding in the BERT encoder.

Bert: A transformer-based pre-trained language model pretrained by Devlin et al. [3].

4.3   Experiment Settings and Evaluation Metrics

We use THULAC [12] for word segmentation. And we use the word embeddings pretrained by Xu et al. [14] using the Skip-Gram model [10], where the model's embedding size and frequency threshold are set to 200 and 25 respectively. The maximum document length is set to 512 words for CNN-based models and 15 sentences for LSTM-based models, with a maximum sentence length of 100 words. For the Transformer-based model, we set the hidden size to 768 and the maximum document length to 512 words. We employ the Adam optimizer [7] for all models. During training, we set the learning rate to 1e-5, the weight decay to 1e-5, and the batch size to 18. After training each model for 16 epochs, we select the best model based on the validation set for testing. We use macro-precision (MP), macro-recall (MR), and macro-F1 (F1) metrics to evaluate the performance.

4.4   Experimental Results

Table 2. Experimental results of judgment prediction on CAIL-small

Model	Law articles			Charges			Term of penalty		
	MP	MR	F1	MP	MR	F1	MP	MR	F1
HARNN+MLT	75.26	76.79	74.90	82.44	82.78	82.12	34.66	31.26	31.40
FLA+MLT	75.32	74.36	72.93	79.25	77.61	76.94	30.94	28.40	28.00
Few-Shot+MLT	77.80	77.59	76.09	80.84	82.01	81.55	35.07	26.88	27.14
TopJudge	79.77	73.67	73.60	83.60	78.42	79.05	34.73	32.73	29.43
MPBFN	76.30	76.02	74.78	82.28	80.72	80.72	31.94	28.60	29.85
LADAN	78.24	77.38	76.47	83.42	82.52	82.74	36.16	32.49	32.65
CEEN	81.38	81.82	80.41	86.61	86.86	86.22	37.11	35.00	35.07
Bert+event	82.20	81.30	80.20	**86.60**	85.40	85.20	37.60	37.00	36.40
Bert	80.80	80.40	79.50	85.30	84.80	84.50	39.10	35.70	36.60
Ours	**82.50**	**84.10**	**82.30**	85.90	**87.30**	**86.30**	**39.50**	**39.10**	**39.10**

Table 2 presents the results of our method compared to other models. Compared to the state-of-the-art model CEEN, our model shows significant improvements in F1 scores for articles, charges, and terms of penalty prediction, with gains of 1.89%, 0.08%, and 4.03% respectively. Particularly notable is the remarkable enhancement in the terms of penalty prediction, highlighting the crucial role of legal event information in achieving an accurate terms of penalty prediction. In contrast, Bert+event, despite utilizing legal event information, falls short in all subtasks in terms of F1 scores when compared to our model. This indicates that Bert+event fails to fully leverage the potential of event information in fact descriptions for LJP. However, it's worth noting that our model slightly underperforms Bert+event in terms of macro-precision (MP) across charges prediction, which can be attributed to the challenges posed by the complex interactions in the legal event types attention layer. Additionally, when evaluating the performance under the same multi-task framework (FLA, Few-Shot, TopJudge, and MPBFN), our method demonstrates a distinct advantage, further reinforcing its superiority. These results emphasize the effectiveness of our model in leveraging event information from fact descriptions to enhance judgment prediction.

4.5   Ablation Test

In our approach, we propose a novel legal event types attentional framework for LJP. The legal event types attention mechanism and the fusion of event-aware and event-free information are essential to enhancing the performance of the framework. To evaluate the importance of these elements, we conduct four ablation studies. Firstly, we assess the significance of the legal event types attention mechanism by replacing the event-aware information with either the event-pooler or the y-context. The event-pooler utilizes the pooler output of the legal event types representation obtained from BERT, while the y-context employs the self-attention of the fact description. Secondly, we investigate the importance of combining event-aware and event-free information by conducting experiments using either the event-aware or event-free approaches.

Table 3. Results of ablation test

Model	Law articles			Charges			Term of penalty		
	MP	MR	F1	MP	MR	F1	MP	MR	F1
ours (event-free cat event-aware)	82.50	**84.10**	**82.30**	**85.90**	**87.30**	**86.30**	**39.50**	**39.10**	**39.10**
event-free cat event-pooler	**82.60**	82.80	81.40	84.40	85.40	84.20	39.00	36.30	36.50
event-free cat y-context	82.50	83.00	81.70	85.50	85.80	85.00	37.60	35.40	34.90
event-aware	81.50	82.10	80.70	85.50	85.70	85.00	36.50	36.50	36.00
event-free	80.80	80.40	79.50	85.30	84.80	84.50	39.10	35.70	36.60

Table 3 showcases the consistent superiority of our model over other configurations across all tasks and metrics. When the event-aware component is replaced with either the event-pooler or the y-context, the model's F1 score experiences a decrease of at least 0.9%, 2.1%, and 4.2% respectively. These findings highlight the significant contribution of the legal event types attention mechanism to the model's performance. However, it's important to note that when the event-aware representation is substituted with the event-pooler, our model slightly underperforms in terms of macro-precision (MP) across law articles prediction. This can be attributed to the challenges posed by the complex interactions in the legal event types attention layer during the generation of event-aware representation. Conversely, when only the event-aware or event-free information is included, the model's performance is inferior to the full model across all tasks. The F1 score drops by at least 2.8%, 1.8%, and 3.1% respectively. In summary, the results presented above demonstrate that each component positively impacts the model's performance, with the event-aware component potentially being the most effective.

5   Case Study

In this part, we select a representative case to show how legal event types attention works in locating the key event information. In this case, Fig. 3 is a part of the overall heatmap of the attention matrix. We enhance the clarity of the heatmap by removing all values less than 10^{-3}.

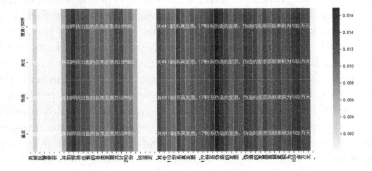

Fig. 3. Partial heatmap of the attention matrix. The vertical axis is the legal event types implied in the fact description and the horizontal axis is a fragment of the fact description.

Analysis of the heatmap reveals two areas of focus in the fact description. Firstly, the words 'seizure', 'waiting to be sold', and '290 various invoices' receive high attention weights, corresponding to the seizure and sale legal events types. This suggests these legal events information are closely examined in the fact description. Secondly, the forgery event shows high attention weights for '179 of them', 'forged invoices', and 'over 1.5 million yuan', indicating a close examination of these legal event details in the fact description. Besides, the heatmap with attention matrix provides valuable insights for legal professionals by highlighting key event information for judicial prediction. This tool can assist further analysis and decision-making in legal scenarios, proving beneficial for LJP.

6   Conclusion

In this paper, we propose a novel legal event types attentional network to enhance the performance of LJP. Our approach introduces a legal event types attention mechanism that captures event-aware fact representations. We effectively predict the case outcome by combining event-free and event-aware fact representations. Our experiments on a real-world dataset demonstrate the effectiveness of our model in focusing on essential legal event information in fact representation. For future research, we aim to explore the joint modeling of legal event detection and LJP tasks to further enhance the performance and transparency of LJP. By integrating these two tasks, we anticipate improved accuracy and a deeper understanding of the relationship between legal events and case outcomes.

Acknowledgements. We thank the anonymous reviewers for their thoughtful comments. This work is supported by the National Youth Science Foundation of China(71804157), the Provincial Natural Science Foundation of Fujian(2022J011261), and the Fujian Province Key Laboratory of Green Intelligent Cleaning Technology and Equipment.

References

1. Aletras, N., Tsarapatsanis, D., Preotiuc-Pietro, D., Lampos, V.: Predicting judicial decisions of the European court of human rights: a natural language processing perspective. PeerJ Comput. Sci. **2**, e93 (2016)
2. Bao, Q., Zan, H., Gong, P., Chen, J., Xiao, Y.: Charge prediction with legal attention. In: Natural Language Processing and Chinese Computing: 8th CCF International Conference. NLPCC 2019, pp. 447–458. Dunhuang, China (2019)
3. Devlin, J., Chang, M., Lee, K., Toutanova, K.: BERT: pre-training of deep bidirectional transformers for language understanding. In: Proceedings of the 2019 Conference of the North American Chapter of the Association for Computational Linguistics: Human Language Technologies, NAACL-HLT 2019, pp. 4171–4186. Minneapolis, MN, USA (2019)
4. Feng, Y., Li, C., Ng, V.: Legal judgment prediction via event extraction with constraints. In: Proceedings of the 60th Annual Meeting of the Association for Computational Linguistics (Volume 1: Long Papers), ACL 2022, pp. 648–664. Dublin, Ireland (2022)
5. Hu, Z., Li, X., Tu, C., Liu, Z., Sun, M.: Few-shot charge prediction with discriminative legal attributes. In: Proceedings of the 27th International Conference on Computational Linguistics, COLING 2018, pp. 487–498 (2018)
6. Jiang, X., Ye, H., Luo, Z., Chao, W., Ma, W.: Interpretable rationale augmented charge prediction system. In: Zhao, D. (ed.) COLING 2018, The 27th International Conference on Computational Linguistics: System Demonstrations, pp. 146–151. Santa Fe, New Mexico (2018)
7. Kingma, D.P., Ba, J.: Adam: a method for stochastic optimization. In: 3rd International Conference on Learning Representations, ICLR 2015. San Diego, CA, USA (2015)
8. Luo, B., Feng, Y., Xu, J., Zhang, X., Zhao, D.: Learning to predict charges for criminal cases with legal basis. In: Proceedings of the 2017 Conference on Empirical

Methods in Natural Language Processing, pp. 2727–2736. Copenhagen, Denmark (2017)

9. Lyu, Y., et al.: Improving legal judgment prediction through reinforced criminal element extraction. Inf. Process. Manag. **59**(1), 102780 (2022)

10. Mikolov, T., Sutskever, I., Chen, K., Corrado, G.S., Dean, J.: Distributed representations of words and phrases and their compositionality. In: Advances in Neural Information Processing Systems 26: 27th Annual Conference on Neural Information Processing Systems 2013, pp. 3111–3119. Lake Tahoe, Nevada, United States (2013)

11. Sulea, O., Zampieri, M., Malmasi, S., Vela, M., Dinu, L.P., van Genabith, J.: Exploring the use of text classification in the legal domain. In: Proceedings of the Second Workshop on Automated Semantic Analysis of Information in Legal Texts co-located with the 16th International Conference on Artificial Intelligence and Law (ICAIL 2017), vol. 2143, London, UK (2017)

12. Sun, M., Chen, X., Zhang, K., Guo, Z., Liu, Z.: THULAC: an efficient lexical analyzer for Chinese. Technical report (2016). https://github.com/thunlp/THULAC

13. Xiao, C., et al.: CAIL2018: a large-scale legal dataset for judgment prediction. CoRR abs/1807.02478 (2018)

14. Xu, N., Wang, P., Chen, L., Pan, L., Wang, X., Zhao, J.: Distinguish confusing law articles for legal judgment prediction. In: Proceedings of the 58th Annual Meeting of the Association for Computational Linguistics, pp. 3086–3095. Online (2020)

15. Yang, W., Jia, W., Zhou, X., Luo, Y.: Legal judgment prediction via multiperspective bi-feedback network. In: Proceedings of the Twenty-Eighth International Joint Conference on Artificial Intelligence, IJCAI 2019, pp. 4085–4091 (2019)

16. Yang, Z., Yang, D., Dyer, C., He, X., Smola, A., Hovy, E.: Hierarchical attention networks for document classification. In: Proceedings of the 2016 Conference of the North American Chapter of the Association for Computational Linguistics: Human Language Technologies, pp. 1480–1489. San Diego, California (2016)

17. Yao, F., et al.: LEVEN: A large-scale Chinese legal event detection dataset. In: Findings of the Association for Computational Linguistics: ACL 2022, pp. 183–201. Dublin, Ireland (2022)

18. Zhong, H., Guo, Z., Tu, C., Xiao, C., Liu, Z., Sun, M.: Legal judgment prediction via topological learning. In: Proceedings of the 2018 Conference on Empirical Methods in Natural Language Processing, pp. 3540–3549. Brussels, Belgium (2018)

19. Zhong, H., Xiao, C., Tu, C., Zhang, T., Liu, Z., Sun, M.: How does NLP benefit legal system: a summary of legal artificial intelligence. In: Proceedings of the 58th Annual Meeting of the Association for Computational Linguistics, pp. 5218–5230. Online (2020)

20. Zhong, H., Zhang, Z., Liu, Z., Sun, M.: Open Chinese language pre-trained model zoo. Technical report (2019). https://github.com/thunlp/openclap

MOOCs Dropout Prediction
via Classmates Augmented Time-Flow
Hybrid Network

Guanbao Liang, Zhaojie Qian, Shuang Wang, and Pengyi Hao[✉]

Zhejiang University of Technology, Hangzhou, China
haopy@zjut.edu.cn

Abstract. Massive Open Online Courses (MOOCs) provide learners with a platform for free learning. However, MOOCs have been criticized for high dropout rates in recent years. For the purpose of predicting users' potential dropout risk in advance, a novel framework named Classmates Augmented Time-Flow Hybrid Network (CA-TFHN) is proposed in this paper. TFHN, which takes advantage of LSTM and Self-Attention mechanism, is designed to generate user activity features by using user learning records. At the same time, an effective correlation calculation is defined based on user potential interests on courses with link prediction, bringing in relationships of classmates. Influences among classmates, modeled by a reconstructed user graph, are employed to augment the activity features of the user, resulting in an accurate prediction of dropout. Experiments on the XuetangX dataset demonstrate the effectiveness of CA-TFHN in predicting MOOCs dropout. The CA-TFHN codes are available from https://github.com/codeds27/CA-TFHN.

Keywords: Dropout prediction · MOOCs · LSTM · Self-Attention · Classmates Relationships

1 Introduction

Massive Open Online Courses (MOOCs) provide learners with a platform for free learning. One serious problem that prevents MOOCs from further development is high dropout rates [1], which also brings a big waste of resources and might let part of teachers lose their enthusiasm for providing excellent courses. The main factors of high dropout rates may be that users lack intrinsic motivation [2] and are often influenced by peers [3]. Therefore, it is important to predict users potential dropout risk in advance.

From the view of learning records, generally the learning statuses shown by users in different periods affect each other. It is sensible to predict that users will not give up studying in the following periods while they participated actively in the early periods [4]. Many scholars view this as a time series problem and utilize LSTM [5] to early identify dropout [6]. Many LSTM-based methods have been proposed in recent years [7,8]. However, the structure of the LSTM can

© The Author(s), under exclusive license to Springer Nature Singapore Pte Ltd. 2024
B. Luo et al. (Eds.): ICONIP 2023, CCIS 1969, pp. 405–416, 2024.
https://doi.org/10.1007/978-981-99-8184-7_31

not avoid inherent limitations. The hidden states which are primary information carriers in the LSTM have the same and fixed shape, in each recurrent operation the information is squeezed [9]. Therefore, the capacity of the LSTM to directly capture the global cross-period influences between learning statuses is restricted, especially with long time intervals. But in reality, as enthusiasm fades and with increasing difficulties in studying the course, whether the user will complete the course during the final periods largely depends on the initial ambitious and willingness. Compared to LSTM, the structure of the Self-Attention mechanism can easily discover these relationships between periods. Because, Self-Attention mechanism structure can calculate the similarities between different periods directly and express each period's output as a weighted sum of all periods [10]. Alternatively, all periods' data can be viewed as a single entity and thus the Self-Attention mechanism is applied to directly mine global relationships [11,12]. Some methods using the Self-Attention mechanism [13–15] perform well in MOOCs dropout prediction. But, the Self-Attention mechanism can not capture local continuous influences well in time series.

From the view of course selections, usually if two users have enrolled the same course, they are considered as classmates. Some researchers believe that classmates can affect each other, which should be considered in the dropout prediction [16]. Taking into account the influence of classmates is becoming increasingly common in models [17] designed to aid in predicting MOOCs dropout. For example, ref. [3] and ref. [11] both constructed a user-course bipartite graph based on course selections, then node features were extracted on the graph directly using DeepWalk and Markov chain Monte Carlo Negative Sampling respectively. The features of classmates were considered as supplementary information in dropout prediction. However, in real life, two users may come from different places, may take the same course out of interest. Obviously, they are less likely to affect each other in online studies. Therefore, the relationships among classmates derived from the course selections may not fully reflect the interrelationships among classmates. Real classmates usually exhibit greater similarity and share a wider range of interests reflected in their course selection patterns [17].

Building upon the aforementioned analysis, a novel framework of dropout prediction named Classmates Augmented Time-Flow Hybrid Network (CA-TFHN) is proposed in this paper. CA-TFHN addresses the issue of predicting dropout in two aspects. On the one hand, to generate user activity features, a time-flow hybrid network (TFHN) is designed by combining the advantages of the LSTM and Self-Attention mechanism. It is composed of blocks, and each block includes a LSTM layer and a Self-Attention layer, capable of capturing both local and global dependencies in time series problems. On the other hand, for employing the influences among classmates, a new calculation is designed by using the link prediction based on user's course preferences. Correlations between users can be measured, and their classmates relationships are reconstructed according to the strong correlations. The classmates graphs are then constructed using these relationships, and the influence between classmates is characterized as the aggregation of background information features on these graphs. Experiments on a public dataset not only show the excellent performance

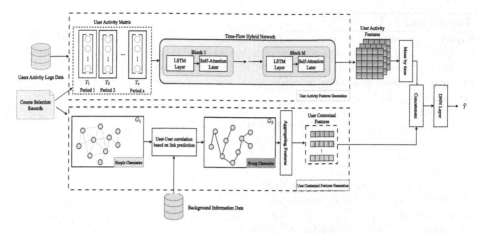

Fig. 1. The framework of CA-TFHN for dropout prediction.

of CA-TFHN, but also reveal that even a short period of learning records can also make predictions of dropout from courses. To sum up, the contributions of this paper are as follows, (1) a universal network structure called TFHN, is proposed for general time series problems. TFHN has been validated to show effective performance on MOOCs learning records that exhibit strong time series characteristics; (2) with the goal of reconstructing classmates relationships, a new correlation calculation is proposed, whose effectiveness is verified in user-user correlation of MOOCs.

2   Proposed Method

2.1   Problem Description

Given l course selection records generated by n users on m courses, combined with user activity log data over s weeks and background information data containing gender, education level, course category, etc., our task is to obtain dropout prediction $\hat{Y} = [\hat{y}_1, \dots, \hat{y}_r, \dots, \hat{y}_l]$ for all records. Let the r-th $(1 \leq r \leq l)$ record, which represents a user u enrolls the course c, be denoted as user-course pair (u, c) and calculate a probability $\hat{y}_r \in [0, 1]$ of the u dropping out of c. By mapping and standardization, the raw background information of users and courses are transferred to be initial contextual features I_u and I_c respectively. The structure of the proposed framework is shown in Fig. 1 including user activity feature generation based on the time-flow hybrid network given in Fig. 2 and user contextual feature generation based on user-user correlation calculation. The Algorithm 1 shows the details of CA-TFHN.

2.2   Time-Flow Hybrid Network

Suppose there are s matrices $T_1, T_2, \cdots, T_s$ arranged chronologically. For the s-th matrix $T_s \in \mathbb{R}^{l \times d_t}$, each row represents the d_t-dimension features of a record

---

Algorithm 1. CA-TFHN

Input: user activity matrices $T_1, T_2, \cdots, T_s$; initial features I_u, I_c; the number of blocks in TFHN M; the number of layers in fully connected networks $\mathcal{O}$; weights $W^1, W^2, \cdots, W^{\mathcal{O}}$ and bias $b^1, b^2, \cdots, b^{\mathcal{O}}$ of fully connected networks.

Output: dropout prediction $\widehat{Y} = [\widehat{y_1}, \widehat{y_2}, \cdots, \widehat{y_l}]$ for l course selections;

  1: initialize p to be 1;

  2: **repeat**

  3:     calculate $H_1^p, H_2^p, \cdots, H_s^p$ in LSTM layer of the p-th block;

  4:     calculate $Z_1^p, Z_2^p, \cdots, Z_s^p$ in Self-Attention layer of the p-th block;

  5:     increment p by 1;

  6: **until** $(p > M)$

  7: obtain user activity features $Z_1^M, Z_2^M, \cdots, Z_s^M$;

  8: obtain user contextual features R_u^2;

  9: concatenate $Z_1^M, Z_2^M, \cdots, Z_s^M$ with R_u^2 to obtain F;

10: initialize λ to be 1, initialize $\mathcal{H}^0$ to be F;

11: **repeat**

12:     calculate $\mathcal{H}^\lambda$ using formula $\mathcal{H}^\lambda = \sigma(W^\lambda \cdot \mathcal{H}^{\lambda-1} + b^\lambda)$;

13:     increment λ by 1;

14: **until** $(\lambda > \mathcal{O})$

15: sigmoid operation on $\mathcal{H}^{\mathcal{O}}$ to obtain dropout prediction $\widehat{Y} = [\widehat{y_1}, \widehat{y_2}, \cdots, \widehat{y_l}]$;

---

in the s-th period. The time-flow hybrid network (TFHN) is proposed to extract local continuous and global cross-period dependencies in these s matrices. As the Fig. 1 shows, TFHN is composed of M blocks, and the output generated by the preceding block serves as the input to the succeeding block. Specifically, the data flow sequentially passes through the LSTM layer and the Self-Attention layer in each block. To provide a clear description of TFHN, the processing of the first block is used to explain the process.

In the LSTM [5] layer of the first block, there is cell state C^1 and hidden state H^1. C^1 runs like a conveyor belt carrying and transferring information at different time steps. Additionally, forget gate, input gate and output gate are designed for keeping or discarding information to capture local continuous dependencies in s matrices. According to Fig. 2, at time s, T_s with two states H_{s-1}^1 and C_{s-1}^1 from the previous step $s - 1$ are inputted to obtain new values H_s^1 and C_s^1 for this time. Besides, in the Fig. 2, $*$ means the point-wise multiplication, σ and $tanh$ mean the sigmoid and hyperbolic tangent activation function respectively. The hidden states in different time steps $H_1^1, H_2^1, \ldots, H_s^1$ are collected as the input of the Self-Attention layer, each $H_s^1 \in \mathbb{R}^{l \times d_h}$, where d_h means the features dimension.

For the Self-Attention layer, sinusoidal position encoding [10] is employed. A set of s matrices supplemented with extra position information $H_1^1 + PE_1^1, H_2^1 + PE_2^1, \ldots, H_s^1 + PE_s^1$ is loaded into the the Self-Attention layer. PE_s^1 represents the extra position information on the s-th time step of hidden state H_s^1. Let $query_i^1 = key_i^1 = value_i^1 = H_i^1 + PE_i^1$ as the i-th $(1 \leq i \leq s)$ period's input, and they are mapped to a lower dimension d_a by fully connected networks to acquire Q_i^1, K_i^1, V_i^1.

$$\begin{cases} Q_i^1 = \sigma(W_Q^1 \cdot query_i^1 + b_Q^1) \\ K_i^1 = \sigma(W_K^1 \cdot key_i^1 + b_K^1) \\ V_i^1 = \sigma(W_V^1 \cdot value_i^1 + b_V^1) \end{cases} \tag{1}$$

where W_Q^1, W_K^1, W_V^1 are the weights, and b_Q^1, b_K^1, b_V^1 are the biases. Similarly, Q_j^1, K_j^1, V_j^1 are acquired as the j-th ($1 \le j \le s$) period's input. To better illustrate the Self-Attention mechanism, taking r-th ($1 \le r \le l$) record as an example. The r-th rows of Q_i^1, K_i^1, V_i^1 are denoted as q_i^1, k_i^1, v_i^1 respectively. q_j^1, k_j^1, v_j^1 can be obtained in the same way. The feature $\mathcal{L}_i^1(r) \in \mathbb{R}^{1 \times d_a}$ of the r-th record in the i-th period is calculated as

$$\mathcal{L}_i^1(r) = \sum_{j=1}^{s} \alpha_{i,j}^1 \cdot v_j^1 \tag{2}$$

where the attention weight $\alpha_{i,j}^1$ between the i-th period's input and the j-th period's input is calculated as

$$\alpha_{i,j}^1 = \frac{q_i^1 \cdot (k_j^1)'}{\sqrt{d_a}} \tag{3}$$

where $()'$ means the transpose operation, $\alpha_{i,j}^1 \in [0,1]$. Thus, the attention weights $\alpha_{i,1}^1, \cdots, \alpha_{i,s}^1$ between the i-th period's input and all periods' input can be computed in the same way, and then they are normalized by softmax. Besides, in the Fig. 2, $A_{i,1}^1$ denotes the attention weights between the i-th period's input and the first period's input for l records, $A_{i,2}^1, ..., A_{i,s}^1$ have the similar meanings. The features of l records in the i-th period in the first block can be calculated, which is denoted as $Z_i^1 = [\mathcal{L}_i^1(1), \mathcal{L}_i^1(2), ..., \mathcal{L}_i^1(r), ..., \mathcal{L}_i^1(l)] \in \mathbb{R}^{l \times d_a}$. In this way, $Z_1^1, Z_2^1, \cdots, Z_s^1$ are formed and fed into the next block.

2.3 User Activity Matrix Generation

The user activity logs for the r-th course selection during the i-th period can be utilized as a user activity matrix $X_i(r)$. The element in $X_i(r)$ indicates the frequencies of a specific activity in a particular day. Furthermore, we add some additional values on $X_i(r)$, as shown in the following,

$$X_i(r) = \begin{bmatrix} \chi_{1,1} & \cdots & \chi_{1,f} & \chi_{1,f+1} \\ \vdots & \cdots & \cdots & \vdots \\ \chi_{e,1} & \cdots & \chi_{e,f} & \chi_{e,f+1} \\ \chi_{e+1,1} & \cdots & \chi_{e+1,f} & \chi_{e+1,f+1} \end{bmatrix}_{(e+1) \times (f+1)}$$

where, the $f + 1$-th column represents the total number of all activities in a particular day, the $e + 1$-th row represents the total frequencies of a specific activity during the i-th period, and $\chi_{e,f}$ means the frequencies of the activity type f on the specific day e. This matrix is flattened to get $X_i(r) \in \mathbb{R}^{1 \times d_t}$,

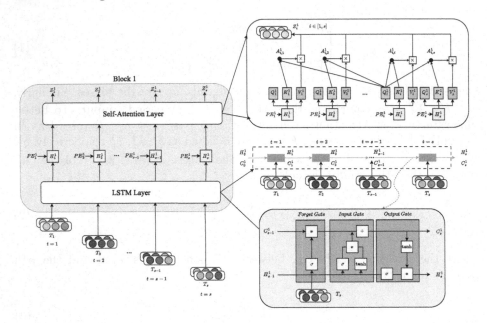

Fig. 2. The illustration of Time-Flow Hybrid Network.

where $d_t = (e+1) \times (f+1)$. The user activity logs of all of the l course selections in the i-th period are applied with the same operations to form a matrix $T_i = [X_i(1), X_i(2), ..., X_i(r), ..., X_i(l)] \in \mathbb{R}^{l \times d_t}$. In the same way, $T_1, T_2, \cdots, T_s$ are generated, which are fed into the TFHN. The outputs of the final block in TFHN, called as user activity features, are denoted as Z_1^M, Z_2^M, $\cdots$, Z_s^M, each $Z_s^M \in \mathbb{R}^{l \times d_a}$.

2.4   User Contextual Features Generation

Given course selections, classmates can be defined if two users have enrolled the same course. Hence, a simple graph of classmates can be constructed as $G_1 = (N, E)$. N is the set of all of the users. $E \in \mathbb{R}^{n \times n}$ is the user-user adjacency matrix that consists of 0 and 1, where 1 indicates the corresponding users are simple classmates and 0 stands on the contrary. Due to the presence of numerous weak and dense connections in G_1, a user-user correlation calculation based on link prediction is designed to find strong classmates. For two users u_x and u_z, their correlation $\rho(u_x, u_z)$ is computed as

$$\rho(u_x, u_z) = \frac{P_x \cdot (P_z)'}{|P_x| \times |P_z|} \tag{4}$$

where $\rho(u_x, u_z) \in [0, 1]$, $P \in \mathbb{R}^{n \times m}$ is the user's course preference matrix, P_x and P_z represent the course preferences for u_x and u_z, derived from the x-th and z-th rows of P respectively. P is acquired using the link prediction method inspired by

[18]. The binarization operations are applied on the P through setting a threshold η_1. In other words, the raw element in P, which indicates the probability that a user prefers a course based on interests, will be changed to 1 if its value is greater than η_1. Furthermore, the users u_x and u_z can be considered as strong classmates when their correlation $\rho(u_x, u_z)$ is greater than a correlation threshold η_2. Based on this, the graph G_1 is reconstructed as a graph $G_2 = (N, E')$, where E' is the new adjacency matrix that contains all these satisfied user pairs (u_x, u_z).

Let $\mathbb{C}_u$ denotes the set of courses enrolled by a user u. The initial feature B_u of the node u in G_2 can be formulated as $B_u = I_u \odot (\mu_{c \in \mathbb{C}_u} I_c)$, where $\odot$ means the concatenate operation and μ means the operation of calculating average values. The influences by classmates are characterized as the aggregation of node features in G_2. Here, the message passing mechanism with two graphsage convolution layers [19] are adopted. The formulas of the layers are listed as

$$R_u^1 = W_1^1 \cdot B_u + W_2^1 \cdot \mu_{u_\tau \in \mathcal{N}(u,1)} B_{u_\tau} \tag{5}$$

$$R_u^2 = W_1^2 \cdot R_u^1 + W_2^2 \cdot \mu_{u_\tau \in \mathcal{N}(u,1)} R_{u_\tau}^1 \tag{6}$$

where $\mathcal{N}(u, 1)$ means the first-order neighbors of the target node u, W_1^1, W_1^2, W_2^1 and W_2^2 are trainable parameters. When all user nodes finish message passing processes, the user contextual features $R_u^2 \in \mathbb{R}^{n \times d_g}$ are generated.

2.5   Prediction

Now the user contextual features R_u^2 is used to augment the user activity features $Z_1^M, Z_2^M, \cdots, Z_s^M$. Let $\mathcal{D}(r)$ denotes the fusion features for the r-th record, and it is formulated as

$$\mathcal{D}(r) = \mathcal{Z}^M(r) \odot R_{u_r}^2 \tag{7}$$

where $\mathcal{Z}^M(r) \in \mathbb{R}^{1 \times d_a}$ means the average values of the r-th rows of user activity features, $R_{u_r}^2 \in \mathbb{R}^{1 \times d_g}$ means the contextual features of the corresponding user u_r based on course selections for the r-th record. The fusion features of all of the l records can be denoted as $F = [\mathcal{D}(1), ..., \mathcal{D}(r), ..., \mathcal{D}(l)] \in \mathbb{R}^{l \times (d_a + d_g)}$.

In the prediction layer, we design $\mathcal{O}$-layers fully connected networks. For the λ-th ($1 \leq \lambda \leq \mathcal{O}$) layer, $\mathcal{H}^\lambda = \sigma(W^\lambda \cdot \mathcal{H}^{\lambda-1} + b^\lambda)$ where W^λ, b^λ are trainable parameters in the λ-th layer and $\mathcal{H}^0 = F$. The prediction $\widehat{Y} = [\widehat{y_1}, \widehat{y_2}, \cdots, \widehat{y_l}]$ is acquired by applying the sigmoid function to the $\mathcal{H}^\mathcal{O}$, which provides the probabilities of dropout for l course selection records. The true labels are denoted as $Y = [y_1, y_2, \cdots, y_l]$. Thus, the binary cross entropy can be utilized as the loss function to backward and update the parameters in training. This loss function can be formulated as

$$Loss(\theta) = \sum_{r=1}^{l} [\, y_r \cdot log(\widehat{y_r}) + (1 - y_r) \cdot log(1 - \widehat{y_r}) \,] \tag{8}$$

where θ denotes the trainable parameters for CA-TFHN.

3 Experiments

3.1 Dataset

To evaluate the performance of CA-TFHN, we employ XuetangX dataset [3] in the experiments. It contains $225,642$ course selections formed by $77,083$ users enrolling in 247 courses from June 2015 to June 2017. Besides, the XuetangX dataset has $42,110,397$ user activity logs, where there are four main types including *Click*, *Video*, *Assignment*, and *Forum*. The respective proportions of these four types of user activity are 59.2%, 25.8%, 14.7%, 0.3%. Specifically, these four main activity types consist of 22 specific activities like seeking video or clicking information. These logs are recorded within 36 days from the start of each course. Each course selection is labeled dropout or not according to whether the user gives up studying on this course in the next 10 days. The dataset also gives user background information and course background information. The user background information includes user ID, age, gender, education level, and the number of register courses. The course background information includes course ID, course category, and the number of registers. In the experiments, the whole dataset is divided into the training set and test set randomly. The size of the training set is $157,943$ while test set occupies the rest.

3.2 Implementation and Evaluations Metrics

The proposed CA-TFHN is implemented with PyTorch and PyTorch Geometric which is a graph neural network framework. In the experiments, a week ($e = 7$) is considered as one period, 35 days (five weeks, $s = 5$) and 22 types ($f = 22$) of user activity logs are selected, important dimensions are set as $d_t = 184$, $d_h = 64$, $d_a = 32$ and $d_g = 32$. Two thresholds η_1 and η_2 take the values 0.6 and 0.95 respectively. Moreover, the Adam optimizer with an initial learning rate 1×10^{-3}, and L_2 regularization with weight decay 1×10^{-5} are adopted in the training of CA-TFHN. The batch sizes for training set and test set are 256, 128 respectively, while epoch $= 15$ both for two sets.

Not only accuracy, precision, and recall are used to validate the method's effectiveness, but also the comprehensive ones like F1 score (F1) which takes into account the performance of precision and recall, and the area under the ROC curve (AUC) that is better suited for the situation of the sample imbalance.

3.3 Parameter Analysis

In the time-flow hybrid network, the number of blocks M is worth discussing. According to the Table 1, TFHN reaches its peak performance on the XuetangX dataset when M is set to 2, with an AUC of 0.8706 and an F1 score of 0.9077. Additionally, it shows an increase of 1.23% and 0.33% in accuracy, as well as a 0.78% and 1.13% increase in recall, compared to M=1 and M=3 respectively. However, in term of precision, it shows a decrease of 0.43% compared to the case of three blocks. This can be elucidated as when TFHN has more than 2 blocks, it has a higher tendency to predict the result as positive in the XuetangX dataset. At the same time, the TFHN with two or three blocks outperform the TFHN with only one block, which proves that the design of multi-block architecture is meaningful and truly increases the performance. In summary, we finally choose 2 blocks for our method in the later experiments.

Table 1. The performance of applying different amounts of blocks in TFHN.

	Accuracy	Precision	Recall	AUC	F1
$M = 1$	0.8422	0.8664	0.9361	0.8508	0.8999
$M = 2$	**0.8545**	0.8742	**0.9439**	**0.8706**	**0.9077**
$M = 3$	0.8512	**0.8785**	0.9326	0.8667	0.9047

To assess the impact of input data volume on the CA-TFHN, we discuss the value of s in various permutations of 5 periods with Table 2, for example, $\{1, 2, 3\}$ represents that the previous 3 periods of data are adopted. From this table, three noteworthy findings are revealed. (1) With more periods of data, the TFHN achieves better performance. In term of adopting five periods of data, it achieves highest accuracy, recall, AUC and F1, precision, compared to other four combinations of five periods. When adopting five periods of data, the accuracy is 1.86% to 5.35% higher, the precision is 1.83% to 5.08% greater, and the recall is 0.06% to 0.52% better than other four combinations. The AUC is improved ranging from 3.08% to 11.91%, and the F1 is improved ranging from 1.07% to 3.04% by using the five periods of data. However, the problems of collecting more data and costing much more computational resources can not be ignored at the same time. (2) In term of accuracy, the performance of adopting one or two periods of data are equivalent to 93.74%, 95.23% of adopting all periods of data respectively. This indicates that, at the commencement of the semester, or for a new user, when the data is limited, utilizing fewer data for dropout prediction by CA-TFHN can still yield satisfactory result that holds significant implications for MOOCs administrators. (3) In term of AUC, there is a 11.91% gap between the cases of adopting one period of data and adopting five periods of data. The reason for this phenomenon might be associated with the particularly unbalanced dataset and AUC's special calculation.

Table 2. The impact of different permutations of five periods.

	Accuracy	Precision	Recall	AUC	F1
$\{1\}$	0.8010	0.8234	0.9387	0.7515	0.8773
$\{1,2\}$	0.8137	0.8309	0.9431	0.7817	0.8847
$\{1,2,3\}$	0.8235	0.8427	0.9433	0.8110	0.8902
$\{1,2,3,4\}$	0.8359	0.8559	0.9421	0.8398	0.8970
$\{1,2,3,4,5\}$	**0.8545**	**0.8742**	**0.9439**	**0.8706**	**0.9077**

3.4 Comparison of Correlation Calculations

The proposed link prediction based user-user correlation calculation is compared to two alternative calculations, Contextual feature based correlation and Course selection based correlation. The contextual feature based correlation refers to the cosine similarity calculation between users' contextual features I_u. If the similarity score between two users exceeds 0.9, they can be considered as strong classmates due to their alike background information. The course selection based correlation refers to cosine similarity calculation between users' original course selections. For each user, a course choice vector that consists of binary values, where 0 represents the user did not select the corresponding course and 1 represents the opposite. If similarity score between two users exceeds 0.5, they can be considered as strong classmates. The proposed link prediction based correlation calculation given in the Sect. 2.4, that utilizes link prediction to reveal users' potential interests in courses. The three calculations generate their adjacency matrix based on the same condition that only user pairs meeting respective criteria are included in the matrix. Thus, the main difference among them is the different aspects that assess users' correlation. The dropout prediction results of applying these three correlation calculations are shown in Table 3.

It can be concluded that link prediction based correlation stands out the most. The number of user pairs generated by link prediction based correlation is only 20% and 50% of the pairs generated by the other two calculations, respectively. At the same time, link prediction based correlation brings an increase of 3.19% and 2.52% in AUC, as well as an increase of 4.54% and 3.15% in F1, compared to other two calculations. Hence, it not only reduces the computational resources required to store edges in the graph but also identifies strong classmates in predicting MOOCs dropout.

Table 3. Comparison with other correlation calculations.

	User Pairs	AUC	F1
Contextual feature based	49,617,866	0.8387	0.8623
Course selection based	19,341,858	0.8454	0.8762
Link prediction based	**10,211,795**	**0.8706**	**0.9077**

3.5 Comparison with Other Dropout Prediction Methods

The proposed CA-TFHN is compared with the current state-of-the-art methods both in machine learning [20] and deep learning [3,11,12]. We follow the parameters set in their papers and conduct experiments independently. SVM based prediction uses Gaussian kernel, where L_2 regularization penalty is set to be 1×10^{-1} and kernel function parameter is set to be 1×10^{-3}. In AdaBoost (AB) based prediction, a combination of 150 weak classifiers and random forest as the base estimator are utilized. Besides, the sum of user activity matrices is fed as input for these methods. On the other hand, Adam optimizer and an initial learning rate 1×10^{-4} are adopted for Cross-TabNet [12], CFIN [3], CEDN [11]. They are fed with user activity matrices and contextual features related to users and courses. The AUC and F1 are employed to measure these methods' performance. Table 4 summarizes the results.

We can see that the SVM based prediction has the lowest AUC and F1 among these methods. The dropout prediction based on AB also performs not well. This demonstrates that such methods can not grasp the effective information from the complex, sparse and large amount data. On the other hand, CA-TFHN achieves an F1 score of 0.9077 and an AUC of 0.8706, both outperforming other deep learning methods by 0.47% to 0.89% and 0.55% to 1.75%, respectively. The CA-TFHN incorporates the benefits of both LSTM and Self-Attention mechanism to extract user activity features and enhances impact between strong classmates, however these compared deep learning methods tend to treat all data as a single entity, disregarding the correlation between different periods of data, and failing to uncover more meaningful and robust relationships between classmates.

Table 4. Comparison with other dropout prediction methods.

Method	AUC	F1
SVM [20]	0.7925	0.8217
AB [20]	0.8236	0.8495
CFIN [3]	0.8531	0.9030
CEDN [11]	0.8548	0.8988
Cross-TabNet [12]	0.8651	0.9021
CA-TFHN	**0.8706**	**0.9077**

4 Conclusions

In this paper, a novel dropout prediction approach of MOOCs has been proposed, where Time-Flow Hybrid Network is designed to generate users' activity feature using their learning records in MOOCs. Besides, a correlation calculation is defined based on users' potential interests on courses that is obtained using link prediction, which gives strong correlations among users. Therefore, classmates graph is reconstructed using such strong correlations, bringing the mutual influences among classmates into dropout prediction. The parameters of the proposed CA-TFHN have been analyzed in the experiments on XuetangX dataset. The whole performance of the proposed dropout prediction has been evaluated, and the results have demonstrated the effectiveness of CA-TFHN.

Acknowledgements. This work is supported by Natural Science Foundation of Zhejiang Province of China under grants No. LR21F020002, and the First class undergraduate course construction project in Zhejiang Province of China.

References

1. Miladi F, Lemire D, Psyché V.: Learning engagement and peer learning in MOOC: a selective systematic review. In: ITS, pp. 324–332 (2023)
2. Nawrot I, Doucet a.: building engagement for MOOC students: introducing support for time management on online learning platforms. In: WWW, pp. 1077–1082 (2014)
3. Feng W, Tang J, Liu T X.: Understanding dropouts in MOOCs. In: AAAI (2019)
4. Fei, M., Yeung, D.Y.: Temporal models for predicting student dropout in massive open online courses. In: ICDMW, pp. 256–263 (2015)
5. Hochreiter, S., Schmidhuber, J.: Long short-term memory. Neural Comput. 9(8), 1735–1780 (1997)
6. Tang, C., Ouyang, Y., Rong, W., et al.: Time series model for predicting dropout in massive open online courses. In: AIED, pp. 353–357 (2018)
7. Mrhar, K., Benhiba, L., Bourekkache, S., et al.: A Bayesian CNN-LSTM model for sentiment analysis in massive open online courses MOOCs. Int. J. Emerg. Technol. Learn. 16(23), 216–232 (2021)
8. Qiu, L., Liu, Y., Hu, Q., et al.: Student dropout prediction in massive open online courses by convolutional neural networks. Soft. Comput. 23, 10287–10301 (2019)
9. Yin, S., Lei, L., Wang, H., et al.: Power of attention in MOOC dropout prediction. IEEE Access 8, 202993–203002 (2020)
10. Vaswani, A., Shazeer, N., Parmar, N., et al.: Attention is all you need[. In: Advances in Neural Information Processing Systems, vol. 30 (2017)
11. Wu, D., Hao, P., Zheng, Y., et al.: Classmates enhanced diversity-self-attention network for dropout prediction in MOOCs. In: ICONIP, pp. 609–620 (2021)
12. Pan, T., Feng, G., Liu, X., et al.: Using feature interaction for mining learners' hidden information in MOOC dropout prediction. In: ITS, pp. 507–517 (2023)
13. Fu, Q., Gao, Z., Zhou, J., et al.: CLSA: a novel deep learning model for MOOC dropout prediction. Comput. Electr. Eng. 94, 107315 (2021)
14. Zhang, J., Gao, M., Zhang, J.: The learning behaviours of dropouts in MOOCs: a collective attention network perspective. Comput. Educ. 167, 104189 (2021)
15. Zheng, Y., Shao, Z., Deng, M., et al.: MOOC dropout prediction using a fusion deep model based on behaviour features. Comput. Electr. Eng. 104, 108409 (2022)
16. Clow, D.: MOOCs and the funnel of participation. In: The International Conference on Learning Analytics and Knowledge, pp. 185–189 (2013)
17. Goel, Y., Goyal, R.: On the effectiveness of self-training in MOOC dropout prediction. Open Comput. Sci. 10(1), 246–258 (2020)
18. Zhang, C., Song, D., Huang, C., et al.: Heterogeneous graph neural network. In: ACM SIGKDD, pp. 793–803 (2019)
19. Hamilton, W., Ying, Z., Leskovec, J.: Inductive representation learning on large graphs. In: Advances in Neural Information Processing Systems, vol. 30 (2017)
20. Basnet, R.B., Johnson, C., Doleck, T.: Dropout prediction in MOOcs using deep learning and machine learning. Educ. Inf. Technol. 27(8), 11499–11513 (2022)

Multiclass Classification and Defect Detection of Steel Tube Using Modified YOLO

Deepti Raj Gurrammagari[1] , Prabadevi Boopathy[1]([✉]) ,
Thippa Reddy Gadekallu[2,3] , Surbhi Bhatia Khan[4], and Mohammed Saraee[5]

[1] School of Computer Science Engineering and Information Systems, VIT University,
Vellore, India
deeptiraj.g2020@vitstudent.ac.in, prabadevi.b@vit.ac.in
[2] Department of Electrical and Computer Engineering, Lebanese American
University, Byblos, Lebanon
thippareddy@ieee.org
[3] College of Information Science and Engineering, Jiaxing University, Jiaxing 314001,
China
[4] Department of Data Science, School of Science Engineering and Environement,
University of Salford, Manchester, UK
S.Khan138@salford.ac.uk
[5] School of science engineering and environment, University of Salford, Salford, UK
m.saraee@salford.ac.uk

Abstract. Steel tubes are widely used in hazardous high pressure environments such as petroleum, chemicals, natural gas and shale gas. Defects in steel tubes have serious negative consequences. Using deep learning object recognition to identify and detect defects can greatly improve inspection efficiency and drive industrial automation. In this work, we use a well-known YOLOv7(You Only Look Once version7) deep learning model and propose to improve it to achieve accurate defects detection of steel tube images. First, the classification of the dataset is checked using a sequential model and AlexNet. A Coordinate Attention (CA) mechanism is then integrated into the YOLOv7 backbone network to improve the expressive power of the feature graph. Additionally, the SIoU (SCYLLA-Intersection over Union) loss function is used to speed up convergence due to class imbalance in the dataset. Experimental results show that the evaluation index of the optimized and modified YOLOv7 algorithm outperforms other models. This study demonstrates the effectiveness of using this method in improving the model's detection performance and providing a more effective solution to steel tube defects.

Keywords: Classification · Defect Detection · YOLO · YOLOv7 · Coordinate Attention · SIoU loss

---

The original version of this chapter has been revised. The names of the first two authors of the paper have been corrected. A correction to this chapter can be found at https://doi.org/10.1007/978-981-99-8184-7_39

B. Luo et al. (Eds.): ICONIP 2023, CCIS 1969, pp. 417–428, 2024.
https://doi.org/10.1007/978-981-99-8184-7_32

1   Introduction

1.1   Need for Defect Classification and Detection

Classification and identification of defects is an essential process in many industries such as manufacturing, software development and quality control [1]. They
support the identification and classification of system, process or product errors
or defects. By detecting and classifying errors, organizations can analyze and fix
underlying problems to improve service. They are able to ensure customer satisfaction, satisfy regulatory requirements, and improve company reputation. Early
detection and classification of production or development process problems can
significantly reduce costs. By classifying and identifying defects, it becomes possible to understand the root cause of defects. It is also important to implement
timely and effective defect management strategies [2].

In industries such as medical, aerospace, and automotive, material defects
can lead to safety risks, product failures, recalls, legal liability, financial losses,
quality control issues, loss of customer confidence, regulatory compliance issues
and environmental impact. Manufacturers should prioritize quality control to
prevent such consequences. Industries are facing many challenges which forces
the manufacturing units move to upcoming transformations like predictive maintenance and defect detection [3].

1.2   Why YOLO for Object Detection

In the field of steel pipe defect detection, deep learning techniques offer many
advantages over traditional computer vision techniques. Convolutional neural
networks can achieve end-to-end input recognition and classified output without
the need to manually extract image features. YOLO is a popular deep learning
algorithm used in computer vision for object recognition. It is widely used in
various applications such as error and object detection. YOLO is known for
its incredible speed and efficiency, capable of processing images and videos in
real time [4]. Unlike some common object detection algorithms that use a two-
step technique (region recommendation and classification), YOLO uses a unified
approach, performing object detection and classification in a single step. This
reduces complexity and increases efficiency. When making predictions, YOLO
considers the whole picture rather than just focusing on a specific area of interest
[5]. This allows YOLO to gather more detailed information and context for
accurate object detection. YOLO uses anchor boxes to handle objects of different
sizes and aspect ratios.
The input image is divided into grids and the task is to predict the object
that fits in each grid cell assigned to each grid cell. This grid-based strategy
enables YOLO to properly detect multiple instances of the same object class in
images and effectively manage overlapping objects. The latest versions of YOLO
are YOLOv7 and YOLOv8. YOLOv7 was proposed in July 2022. YOLOv7 is
efficient in terms of error detection [6]. YOLOv7 is very popular in the industry
due to its excellent balance of speed and accuracy. YOLOv7 outperforms other
currently known target detectors in terms of accuracy and speed ranging from 5
FPS to 160 FPS. While YOLOv8 is being regarded as the new state-of-the-art
in object detection, an official paper has not been provided.

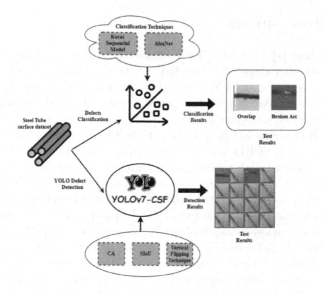

Fig. 1. Flow of the paper

1.3   Contribution of the Paper

Surface defects in steel tubes can be classified based on many properties, but it is difficult to obtain and exploit all these properties using conventional methods. This study, at first analyses classification techniques on the dataset. Then to balance accuracy and detection time, the YOLOv7 baseline model was improved. A CA module and SIoU loss function are added to the YOLOv7 model to provide a lightweight and optimized YOLOv7. Experimental results confirm that the proposed method outperforms original YOLOv7 on steel tube dataset.

The contributions of this paper are enumerated as follows:

1. Two classification techniques such as: Sequential multiclass classification model and Alexnet are used to check the classification of the classes in the dataset.
2. To improve detection accuracy, CA is integrated to YOLOv7. SIoU is used instead of traditional CIoU (Complete IoU) to speed up network convergence and address dataset imbalance issues.
3. The model is trained using the data enhanced with vertical flipping technique.

The rest of the paper is structured as follows. The second section presents literature survey. The third section describes classification techniques, background of YOLOv7 detection framework, working of CA mechanism, SIoU loss function and Modified-YOLOv7 model is presented with vertical flip data augmentation. The fourth section presents experimental datasets, evaluation indicators and metrics. Test results are described in the fifth section. Finally, conclusions are drawn and recommendations for further research are made.

2   Literature Survey

The authors from [7] used YOLOv5 to detect defects in steel pipes, the results were compared with those of Faster R-CNN (region based CNN), showing the effectiveness of single shot object detection. A method that combines advanced faster R-CNN and advanced ResNet50 was presented by [8] for automatic steel pipe defect detection to reduce average run time and increase accuracy. Based on the latest YOLO series YoloX, [9] suggests SCED-Net (Steel Coil End Surface Detection Network). SCED-Net achieved significantly better results by improving the data augmentation method, combining the proposed multi-information fusion module, and adopting focus loss and soft nms. [10] trains a CNN (convolutional Neural Networks) to extract features from an augmented image set using transfer learning methods. Then hierarchical model ensemble is used to detect errors based on location. CondenseNetV2 was proposed by [11] as a lightweight CNN-based model that excels at inspecting small defects and can run on low-frequency edge devices. By reusing the current set of Sparse Feature Reactivation modules, the authors achieve sufficient feature extraction with minimal computational effort. The model was thoroughly tested running on an edge device (NVIDIA Jetson Xavier Nx SOM) using two real-world data sets.

3   Methodology

3.1   Classification Techniques

By enabling automatic identification and classification of objects, patterns, and anomalies in image and video data, classification algorithms play an important role in computer vision and defect detection. These methods use machine learning algorithms to learn from labeled training data and then classify new unbiased data based on learned patterns and features. Computer vision tasks such as object detection, object recognition, image segmentation, and scene understanding all use classification techniques [12]. These techniques can be customized according to the problem. The feature selection phase considers interactions between variables and handles missing data. It helps speed up the work of classification algorithms, even on high-dimensional datasets [13]. This study uses Sequential Model and AlexNet for checking the classification of different classes in the dataset.

Sequential Model. Keras is a popular deep learning library that provides an easy-to-use interface for building neural network models. Sequential multiclass models in Keras make it easy to build and train neural networks for multiclass classification problems, allowing you to quickly classify input data into multiple categories [14]. This process involves defining the layers of the model using the Sequential class and specifying the number and type of layers to include. Dense (fully connected) and activation layers are commonly used for classification of multiple classes. Sequential multiclass models require labeled data for training, with each sample assigned a class label. The goal of internal parameter tuning of a model during training is to reduce the difference between predicted and actual class labels.

A softmax activation function is often used at the top level of a model. To estimate the probability of each class for a given input, Softmax converts the previous level's outputs to class probabilities. To classify objects, the class with the highest probability is considered as the predicted class. Once the model is trained, it can be used to predict new unknown data by giving inputs to the model and retrieving the predicted class labels.

AlexNet. A deep CNN architecture known as AlexNet was created in 2012 and is widely used for image classification [15]. Image classification, object recognition, transfer learning, feature extraction and visualization are some of its main uses. AlexNet's architecture and learning capabilities were critical to achieving excellence in image classification, and have influenced the creation of subsequent deep learning models in computer vision. It consists of 8 layers, including 3 fully connected layers and 5 convolutional layers. To improve generalization, AlexNet uses dropout regularization, pooling layers for downsampling, and ReLU (Rectified Linear Unit) activation. It was trained on the ImageNet dataset using backpropagation and stochastic gradient descent. The success of AlexNet demonstrated the use of deep learning to classify images and prompted improvements in CNN designs.

3.2 Background of YOLOv7

YOLOv7 introduced many architectural changes, such as additive scaling, EELAN (Extended Efficient Layer Aggregation Network), a bag of freebies with planned and redesigned convolution, coarseness for auxiliary losses, and fine lead losses [6] [16]. This set of features made the models better support mobile GPUs and GPUs. Therefore, we choose the backbone of YOLOv7 as the model feature extraction network.

Backbone. The original backbone is first stacked with four CBS(Convolutional layer, batch normalization and SiLU (Sigmoid Linear Unit) layer)s, and then the underlying features are found from the input image using four convolution operations. MP (Max Pooling block) and E-ELAN modules then extract fine-grained features.

Feature Fusion Zone. The purpose of the network feature fusion layer is to improve the network's ability to learn features taken from the underlying network. To learn as many image features as possible, features with different resolutions are learned independently and then centrally combined.

Detection Head. The YOLOv7 algorithm builds on the strengths of previous algorithms and includes three detection heads that are used to find and print the expected class probabilities, confidence and prediction frame coordinates of the target object. The three feature scales produced by the detection heads are 20×20, 40×40, and 80×80 [17]. Each of the three scales detects a different target scale, such as a large target, a medium target, and a small target.

3.3   Modified-YOLOv7

Transfer learning is used effectively in several fields such as computer vision, natural language processing, and speech processing [18]. In this work, the concept of transfer learning is used to get the weights from the original YOLOv7 model trained on the COCO dataset. Modified-YOLOv7 is formed by taking the YOLOv7 backbone, adding the CA attention mechanism, and the SIoU loss function. The model is then optimized using a vertical flipping data augmentation technique. It is found that compared with the original backbone network, the updated backbone network of YOLOv7 can extract the nonlinear features of the steel tubes in the image more completely and clearly, which shows the effectiveness of the modified approach.

Attention Mechanism. The attention mechanism used in neural networks, especially in computer vision applications, is the CA mechanism. It was first proposed by [19] with the title "An intriguing failing of convolutional neural networks and the coordconv solution". Traditional CNNs are widely used and very successful in computer vision tasks. However, they are not very good at managing spatial data or capturing remote dependencies. To overcome these limitations, the mechanism of coordinated attention was developed. Each convolutional layer of a CNN operates on small areas of the input image, and as the network deepens, the area of the receptive field expands. CNNs, on the other hand, do not explicitly model spatial coordinates, so they treat all spatial locations the same, regardless of their geometric relationships [20]. The CA technique adds coordinate channels to the feature maps that the CNN receives as input. The x- and y-coordinates of each pixel, or separation from the center of the image, are encoded as spatial data in these coordinate channels. The network obtains precise information about the spatial location of each pixel by adding coordinate information. Concatenation or element-based fusion is used to connect the coordinate channels to the original input channels, followed by network processing at higher layers. As a result, the network can be trained to focus on specific spatial locations depending on their coordinates. The network's performance is improved by CA, which helps it learn more accurate spatial representations. This can be particularly useful for applications such as object detection, image segmentation, or pose estimation that require accurate spatial reasoning.

SIoU Loss Function. By including the SIoU loss function in the training process, the object detection model can handle different object sizes better and improve the accuracy of bounding box prediction. The SIoU loss function is used to evaluate the quality of predicted bounding boxes in object detection tasks. According to [21], the addition of SIoU greatly aids the training process, as it causes prediction boxes to drift to the nearest axis rather quickly, requiring only regression of one coordinate, either X or Y. The bounding box regression loss function in the YOLOv7 source code is CIoU. CIoU takes into account the overlap area, centroid distance, and aspect ratio of the actual and predicted frames. SIoU was chosen instead of CIoU because SIoU can better reflect variations in width, height and confidence level. SIoU loss further improves by introducing a scale-sensitive adjustment factor. This factor is intended to account for size differences between objects and reduce the difficulty in optimizing the loss function for objects with large differences in size [22]. The SCYLLA IoU loss is calculated as:

$$SCYLLA\_IoU\_loss = 1 - IoU + \alpha \star (1 - IoU) + \beta \star (1 - C) \qquad (1)$$

where α and β are constants that control the relative importance of the SIoU term and the scale-sensitive balancing term, respectively. Typically these are set to 0.1 and 0.5. The scale-sensitive balancing term C is defined as

$$C = \log\left((S + 1) / (R + 1)\right) \qquad (2)$$

where S represents the maximum side length of the predicted bounding box and R represents the maximum side length of the ground truth bounding box. By taking the logarithm of the ratio between the predicted size and the ground truth size, the balancing term penalizes the scale between two boxes.

Optimization. Optimizing deep learning models is critical for improving performance, faster convergence, efficient use of resources, scalability, deployment to edge devices, interpretability, and transfer learning. Not only does it improve accuracy and reduce errors, but it also enables models to handle increasingly complex datasets. Various methods such as random scaling, rotation, flipping, and cropping are applied to the training data to optimize YOLO models through data augmentation. This extends the data range, supports the model's ability to generalize to new cases, and improves the efficiency of object detection. Additionally, data augmentation can be used to reduce class imbalance and overfitting, resulting in a more reliable and accurate YOLO model. In this work, we generalized the YOLOv7 model using the vertical flip data augmentation technique.

4   Dataset Description, Evaluation Indicators and Metrics

The dataset used in this study is the steel tube dataset. This dataset consists of a total of 3344 images classified into 8 defect classes: air hole, air hole hollow, bite edge, broken arc, crack, overlap, slag inclusion, and unfused defects. The processing platform is a desktop computer running Windows 10 operating system. CPU: Intel(R) Core(TM) i5-1035G Memory: 12 GB. A Google Colaboratory Notebook with an NVIDIA T4 Tensor Core GPU, Python version 3.7.13,

Table 1. Evaluation Indices for Classification

Parameter	Value
Input Image Size	180*180
Epochs	100
Batch	16
Optimizer	Adam
Initial, Final learning rate	1*10-3, 1*10-2

Table 2. Evaluation Indices for Detection

Parameter	Value
Input Image Size	640*640
Epochs	100
Batch	16
Weight Decay	5*10-4
Momentum	937*10-3
Weights	yolov7.pt
Optimizer	SGD
Initial, Final learning rate	1*10-2, 1*10-1

Torch Framework version 1.11.0 cu113 was used as the implementation platform for this work.

The metrics used in this study to demonstrate the improved performance of the modified YOLOv7 model include mAP, latency and F1-score. Latency is the time required for inference, and F1-score refers to weighted average of precision and recall. Tables 1 and 2 refers to the different evaluation indices used for classification and detection respectively.

5   Results

5.1   Classification

AlexNet and Sequential models are the two classification models used in this study. The classification recognition accuracy for the steel tube dataset by sequential model is 97.54% and by AlexNet is 97%. Both models provide virtually the same classification accuracy, but the sequential model classifies the dataset 0.54% more accurately. This indicates that the dataset has been classified correctly and can be used to search for defects.

5.2   Detection

Table 3 gives the comparison of r-cnn, faster r-cnn and YOLOv7 algorithms on steel tube dataset. Table 4 compares mAP, recall, F1 score and detection

Table 3. Comparison of Object Detection Algorithms

Model Name	mAP@0.5	Precision	Recall
R-CNN	84	81	80.8
Faster R-CNN	86.7	85	85.7
YOLOv7	**88.5**	**86.8**	**85.7**

inference between models. YOLOv7-CA integrates the CA mechanism into the YOLOv7 network structure. In YOLOv7-SIoU, SIoU loss function is used instead of traditional CIoU, and Modified-YOLOv7 is formed by taking CA, SIoU, hyper-parameter vertical flipping. As can be seen when compared with the original YOLOv7 model, the revised model Modified-YOLOv7 improved the average accuracy by 6.21%, improved the precision by 6.56%, and recall by 5.95%. Modified-YOLOv7 has the highest F1 score of 91.6% compared to other models. YOLOv7-SIoU has a higher recall of 75.1%, while other measures of Modified-YOLOv7 were much better. When computed at a confidence threshold 0.1, the detection inference of the original YOLOv7 model is shown to be 5.7 ms which is less than all other models. Even though the inference has been slightly raised, Modified-YOLOv7 can still detect steel strip defects in real-time engineering. The visual detection results on steel tube dataset with Modified-YOLOv7 are shown in Fig. 2.

Table 4. Metrics Comparison for different YOLO models

Model Name	mAP@0.5(%)	Precision(%)	Recall(%)	F1-Score(%)	Detection Inference(ms)
YOLOv7	88.5	86.8	85.7	86.2	**5.7**
YOLOv7-CA	91.4	87.7	86.3	86.9	7.8
YOLOv7-SIoU	93	90.6	**91**	90.7	11.3
YOLOv7-CA+SIoU	91.2	93.2	85.6	89.2	8.5
Modified-YOLOv7	94	**92.5**	90.8	**91.6**	8.2

Figure 3 show the results values graph for the Modified-YOLOv7 on the steel tube dataset.

According to the comparison and analysis of the above series of experiments, the improved YOLOv7 algorithm proposed in this work Modified-YOLOv7, shows obvious advantages in recognition accuracy. Even with a slight speed reduction, it is still capable of identifying steel tube defects in real time.

(a)   (b)   (c)

Fig. 2. Visual detection results on Steel tube Dataset. In sequence, the pictures are: (a) 7(unfused), (b) 4,0 (crack,air hole), (c) 5(overlap)

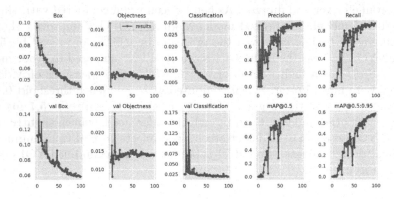

Fig. 3. Results Graph for Modified-YOLOv7

6   Conclusion

In this work, we propose an extended version of the YOLOv7 basic model, Modified-YOLOv7, to identify defects in steel tube datasets with complex backgrounds. We also validated the classification of the dataset using two classification methods, the sequential model and AlexNet, and found that both models had similar classification accuracies, but the sequential model performed better than AlexNet by 0.54% in classifying this steel tube dataset. We combined the CA mechanism in the header of YOLOv7 to improve the expressiveness of the feature graph, and used the SIoU loss function as opposed to the original CIoU to determine the gap between the actual and predicted boxes and to converge the network. to speed up. Additionally, a vertical flip data augmentation technique is added to fine-tune the model. Experimental results show that by incorporating the above tactics, the updated Modified-YOLOv7 improves precision, mAP@0.5 and recall by 6.56%, 6.21% and 5.95%, respectively, compared to the original YOLOv7. To achieve the industry standards for defect detection, this model strikes the right balance between detection accuracy and speed. This model can be used in many future scenarios to identify small targets in industry and agriculture. We plan to extend our research in the future to better understand the network model and pay more attention to the backbone. This study uses grayscale images of the dataset. Therefore, we plan to extend our investigation to the use of hyperspectral images and evaluate the model's performance.

References

1. Park, M., Jeong, J.: Design and implementation of machine vision-based quality inspection system in mask manufacturing process. Sustainability **14**(10), 6009 (2022)
2. Yenduri, G., et al.: GPT (generative pre-trained transformer)-a comprehensive review on enabling technologies, potential applications, emerging challenges, and future directions
3. Maddikunta, P.K.R., et al.: Industry 5.0: a survey on enabling technologies and potential applications. J. Ind. Inf. Integr. **26**, 100257 (2022)
4. Du, J.: Understanding of object detection based on CNN family and yolo. J. Phys. Conf. Ser. **1004**, 012029 (2018)
5. Diwan, T., Anirudh, G., Tembhurne, J.V.: Object detection using yolo: challenges, architectural successors, datasets and applications. Multimedia Tools Appl. **82**(6), 9243–9275 (2023)
6. Wang, C.Y., Bochkovskiy, A. and Liao, H.Y.M.: YOLOv7: trainable bag-of-freebies sets new state-of-the-art for real-time object detectors. In: Proceedings of the IEEE/CVF Conference on Computer Vision and Pattern Recognition, pp. 7464–7475 (2023)
7. Yang, D., Cui, Y., Zeyu, Yu., Yuan, H.: Deep learning based steel pipe weld defect detection. Appl. Artif. Intell. **35**(15), 1237–1249 (2021)
8. Wang, S., Xia, X., Ye, L., Yang, B.: Automatic detection and classification of steel surface defect using deep convolutional neural networks. Metals **11**(3), 388 (2021)
9. Li, Y., Lin, S., Liu, C., Kong, Q.: The defects detection in steel coil end face based on SCED-net. In: 2022 International Joint Conference on Neural Networks (IJCNN), pp. 1–6. IEEE (2022)
10. Zeng, W., You, Z., Huang, M., Kong, Z., Yu, Y., Le, X.: Steel sheet defect detection based on deep learning method. In: 2019 Tenth International Conference on Intelligent Control and Information Processing (ICICIP), pp. 152–157. IEEE (2019)
11. Rani, D.S., Burra, L.R., Kalyani, G., Rao, B., et al.: Edge intelligence with light weight CNN model for surface defect detection in manufacturing industry. J. Sci. Ind. Res. **82**(02), 178–184 (2023)
12. Brownlee, J.: A gentle introduction to object recognition with deep learning. Machine Learning Mastery, May 2019
13. Deepa, N., et al.: An AI-based intelligent system for healthcare analysis using ridge-adaline stochastic gradient descent classifier. J. Supercomput. **77**, 1998–2017 (2021)
14. Bedeir, R.H., Mahmoud, R.O., Zayed, H.H.: Automated multi-class skin cancer classification through concatenated deep learning models. IAES Int. J. Artif. Intell. **11**(2), 764 (2022)
15. Sharma, H., Zerbe, N., Klempert, I., Hellwich, O., Hufnagl, P.: Deep convolutional neural networks for automatic classification of gastric carcinoma using whole slide images in digital histopathology. Comput. Med. Imaging Graph. **61**, 2–13 (2017)
16. Gallo, I., Rehman, A.U., Dehkordi, R.H., Landro, N., La Grassa, R., Boschetti, M.: Deep object detection of crop weeds: performance of yolov7 on a real case dataset from UAV images. Remote Sens. **15**(2), 539 (2023)

17. Zhang, Y., Sun, Y., Wang, Z., Jiang, Y.: YOLOv7-RAR for urban vehicle detection. Sensors **23**(4), 1801 (2023)
18. Weiss, K., Khoshgoftaar, T.M., Wang, D.: A survey of transfer learning. J. Big Data **3**(1), 1–40 (2016)
19. Liu, R., et al.: An intriguing failing of convolutional neural networks and the coord-conv solution. In: Advances in Neural Information Processing Systems, vol. 31 (2018)
20. Chen, X., Girshick, R., He, K., Dollár, P.: TensorMask: a foundation for dense object segmentation. In: Proceedings of the IEEE/CVF International Conference on Computer Vision, pp. 2061–2069 (2019)
21. Gevorgyan, Z.: SIoU loss: more powerful learning for bounding box regression. arXiv preprint arXiv:2205.12740 (2022)
22. Carion, N., Massa, F., Synnaeve, G., Usunier, N., Kirillov, A., Zagoruyko, S.: End-to-end object detection with transformers. In: Vedaldi, A., Bischof, H., Brox, T., Frahm, J.-M. (eds.) ECCV 2020. LNCS, vol. 12346, pp. 213–229. Springer, Cham (2020). https://doi.org/10.1007/978-3-030-58452-8_13

GACE: Learning Graph-Based Cross-Page Ads Embedding for Click-Through Rate Prediction

Haowen Wang[✉], Yuliang Du, Congyun Jin, Yujiao Li, Yingbo Wang,
Tao Sun, Piqi Qin, and Cong Fan

Intelligence Department of Merchants Operation, AntGroup, Shanghai, China
{wanghaowen.whw,duyuliang.dyl,jincongyun.jcy,liyujiao.lyj,wangyingbo.wyb,
suntao.sun,piqi.qpq,fancong.fan}@antgroup.com

Abstract. Predicting click-through rate (CTR) is the core task of many ads online recommendation systems, which helps improve user experience and increase platform revenue. In this type of recommendation system, we often encounter two main problems: the joint usage of multi-page historical advertising data and the cold start of new ads. In this paper, we proposed GACE, a graph-based cross-page ads embedding generation method. It can warm up and generate the representation embedding of cold-start and existing ads across various pages. Specifically, we carefully build linkages and a weighted undirected graph model considering semantic and page-type attributes to guide the direction of feature fusion and generation. We designed a variational auto-encoding task as pre-training module and generated embedding representations for new and old ads based on this task. The results evaluated in the public dataset AliEC from RecBole and the real-world industry dataset from Alipay show that our GACE method is significantly superior to the SOTA method. In the online A/B test, the click-through rate on three real-world pages from Alipay has increased by 3.6%, 2.13%, and 3.02%, respectively. Especially in the cold-start task, the CTR increased by 9.96%, 7.51%, and 8.97%, respectively.

Keywords: embedding learning · cross-page · click-through rate prediction · graph neural network

1 Introduction

With the increase of APP pages and advertising frequency, there are some challenges in improving the efficiency of essential advertising CTR tasks of e-commerce platform [2,16] applications: For e-commercial platforms such as Taobao, Alipay, etc., we have recently seen an increase in recommendation channels for APPs on different pages. This means that the recommendation behavior and user interaction history information on multiple pages can be considered jointly. The recommendation performance on multiple pages can be improved by using the interaction information of multiple pages, including improving cold start advertising distribution efficiency.

© The Author(s), under exclusive license to Springer Nature Singapore Pte Ltd. 2024
B. Luo et al. (Eds.): ICONIP 2023, CCIS 1969, pp. 429–443, 2024.
https://doi.org/10.1007/978-981-99-8184-7_33

Over the years, deep-learning-based models [4,11,24,27] have been proven to improve the efficiency of ads distribution due to their powerful ability of feature intersection and fusion. Despite the remarkable success of these models themselves, their performance still depends primarily on the input of embedded vectors in practical applications. A high-quality ad embedding vector has been proven to improve the accuracy of CTR's prediction effectively [17,18,30].

In this paper, We believe differences between pages should be described and considered in a cross-page universal recommendation system. This concept should be applied to the item embedding generation process, which existing item embedding generation methods have yet to be paid attention to. We propose a graph-based cross-page ad embedding learning framework (GACE) from different perspectives, which is an improved variational graph auto-encoder [15]. One ad's features are mainly composed of Semantic Knowledge (advertising text), Page Knowledge, and User Interaction Knowledge. We first designed a weighted undirected graph network [20] that obtains links based on the semantic similarity of advertising texts and the similarity of page representations. Based on the graph attention network [26], we precisely designed an auto-encoding task [14] of an undirected weighted graph as the pre-training module, in which the variational graph attention encoder can adaptively extract information. Through such pre-training tasks, old and new ads can obtain embeddings considering cross-page neighbor information. The main contributions of this work can be summarized as follows:

- We build linkages based on the text information of the advertising content and page features to generate a weighted undirected graph to guide the process of ads information transfer.
- We designed the pre-training task based on the improved variational graph auto-encoder for the weighted undirected graph of the ad, which can generate the embedding considering the neighbor information for either the new or old ad.
- We have conducted extensive experiments on public large-scale real-world datasets AliEC and offline experimental datasets from Alipay and evaluated through online A/B testing of the Alipay platform, which proves that GACE has achieved significant CTR improvement on each page, especially in the advertising cold start scenario.

2 Related Works

Our work is committed to improving the click-through rate of the recommendation system by generating ad item embedding under the consideration of cross-page information. It mainly involves several research fields: General CTR prediction of recommendation systems and learning of advertisement embedding.

2.1 General CTR Recommendation Systems

The CTR prediction task in online advertising recommendation systems is to predict the click probability of a given user for a given advertisement, which plays an increasingly important role in various personalized online advertising recommendation services. In recent years, deep learning methods based on deep neural networks can interact and represent features in complex ways, and a series of deep neural network models have been proposed. The Wide&Deep [4] model attempts to capture high-level feature interactions through a simple multilayer perceptron (MLP) [9] network. DeepFM [11] and DCN [27] are proposed to handle complex feature interactions based on product operations. AutoInt [22] uses the Multi-head Self-Attention mechanism to produce high-level composite features. In addition, in the scenario with sequence characteristics, a series of neural networks based on sequence characteristics are proposed: Deep Interest Network (DIN) [30], Behavior Sequence Transformer (BST) [3], Deep Session Interest Network (DSIN) [8] and Bert4Rec [24] etc.

However, the performance of these models depends mainly on the quality of the input item and user embedding and cannot effectively solve the cold start problem of new advertisements. They often fail to perform satisfactorily in the sparse page scenario or when distributing ads not in the training set.

2.2   Item Embedding Generation

For item embedding, the first concept is to establish a one-hot vector for item id representation. The embedding learning methods based on collaborative filtering and matrix decomposition were subsequently proposed, but these methods have weak generalization ability and poor performance for items with sparse historical behavior. Researchers try to extend the concept of word embedding in the field of natural language process (NLP) [5] to the recommendation field, and propose Item2Vec [1] based on the concept of Word2Vec [6]. With the development of graph neural networks, embedding learning methods based on graph networks have also been proposed, such as DeepWalk [19], GraphSage [12], etc. However, new advertising embedding without historical interaction cannot be learned via these methods. In particular, for the embedded learning of cold-start new items, our standard solution is to put forward a series of neighbor embedding methods based on manifold learning. Some researchers have put forward some generative methods from the perspective of meta-learning, such as the MetaEmbedding model [21], autodis [10], etc. However, these methods only play a role in learning the embedded representation of new advertisements, and it is difficult to improve the recommendation efficiency of existing old items.

3   Preliminaries

The information stored in the item knowledge base mainly includes three parts as shown in Fig 1: semantic knowledge, user interaction knowledge, and page knowledge.

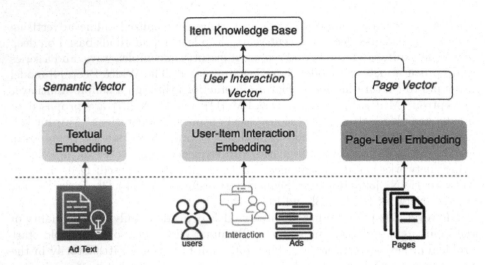

Fig. 1. Item Knowledge Base: semantic knowledge, user interaction knowledge and page knowledge

3.1   Item Knowledge Base

Semantic Knowledge Semantic knowledge is the content information of ads. Advertising recommendations usually include advertising text and corresponding attributes (such as font size, style, color, language, etc.)

User Interaction Knowledge. User interaction knowledge is a vital part of the dynamic attributes of advertising. Its indicators (UV, PV, UVCTR, PVCTR) can reflect the transformation effect of advertising history and user satisfaction.

    UV: unique visitors, the number of distinct individuals visiting an ad within the specified period (usually one day).

    PV: page views, the number of times a specific ad is accessed in a specific period (usually one day).

    UVCTR: click-through rate in UV, the percentage of unique people who see your ads and click it. The formula of UVCTR is clicks of unique visitors divided by impressions of unique visitors.

    PVCTR: click-through rate in PV, the percentage of page views who see your ad and click it. The formula of PVCTR is clicks of ads divided by impressions of ads views.

Page Knowledge. Page knowledge includes the identification of the page where this ad is distributed. We count the number of ads distributed on specific page channels and aggregate the average value of user interaction knowledge (UV, PV, UVCTR, PVCTR) of historical ads distributed on a specific page channel to serve as the embedded representation of the page.

3.2   Problem Formulation

The task of an online advertising system is to establish a prediction model to estimate the click probability of a specific user for a specific ad. Each instance includes multiple-field information: user information ('User ID', 'City', 'Age', etc.), item knowledge base, combined with the label from historical user interaction feedback.

4   Methodology

4.1   Overview

We proposed a graph-based cross-page ads embedding learning method (GACE) by fully exploiting useful information in the item knowledge base. Our model contains two steps: graph creation and pre-training embedding learning. In the graph creation step, three entities' embedding vectors of each item are extracted or encoded from the item knowledge base, representing latent knowledge in each domain. Then we established a weighted undirected graph structure based on the semantic knowledge and page knowledge and set the splicing of entities embeddings in the item knowledge base as the initial embedding for the graph node. In the pre-training embedding learning step, we proposed a pre-training task based on an improved graph auto-encoder for the weighted undirected graph established in the first step to learn the potential embedding representation of each advertising node in the graph. Accordingly, we obtain the optimization parameters of the ad embedding encoder and the potential vector representation of ad items. The graph creation and pre-training embedding learning steps will be further discussed in Sects. 4.2 and 4.3.

4.2   Graph Creation

Typically, for a graph $\mathcal{G} = (\mathcal{V}, \mathcal{E})$, where $\mathcal{V} = \{v_1, \ldots, v_n\}$ is a set of ad nodes and $\mathcal{E}$ is a set of edges with weightings, we combined and set the splicing of three entities' embeddings in the item knowledge base as the ad node's initial vector v. We proposed an adjacent weighting matrix $\mathbf{A} \in \mathbb{R}^{n \times n}$, where $\mathbf{A}_{i,j} \geq 0$ to represent the graph structure. If $\mathbf{A}_{i,j} > 0$, then there is an edge between ads item i and ads item j, where $\mathbf{A}_{i,j}$ represents the edge weighting. If $\mathbf{A}_{i,j} = 0$, then there is no connection between ads item i and ads item j. The Adjacent weighting matrix is calculated based on semantic knowledge and page knowledge. Since the semantic knowledge of advertising is mainly composed of sentences, here we use the output of the sentence transformer [7] as the semantic knowledge vectors $\mathbf{s} \in \mathbb{R}^k$. As mentioned in Sect. 2.1.3, we count the total number of ads placed on specific page channels and aggregate the average value of user interaction knowledge of historical ads placed on a specific page channel as Page knowledge vectors $\mathbf{s} \in \mathbb{R}^d$. The graph creation process is shown in Fig. 2.

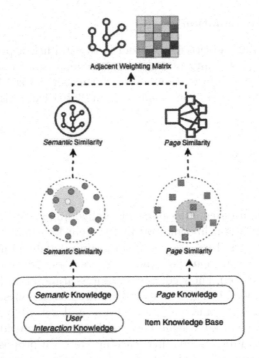

Fig. 2. The graph creation part for GACE using item knowledge base

We use dot product as the semantic knowledge similarity α and page similarity β, and calculate the advertising weight matrix $\mathbf{A}$ based on α and β as follows:

$$\alpha_{i,j} = s_i \cdot s_j \tag{1}$$

$$\beta_{i,j} = p_i \cdot p_j \tag{2}$$

$$ReLU(x) = max(0, x) \tag{3}$$

$$\mathbf{A}_{i,j} = ReLU(\alpha_{i,j}) \cdot ReLU(\beta_{i,j}) \tag{4}$$

where $\alpha_{(}i,j)$ refers semantic knowledge similarity between ads item i and ads item j, s_i and s_j are semantic knowledge vectors for ads item i and ads item j. Similarly, $\beta_{(}i,j)$ is page similarity between ads item i and ads item j, p_i and p_j are Page knowledge vectors for ads item i and ads item j. $ReLU$ is a non-linear activation function.

4.3   Pre-training Embedding Learning

In this section, to integrate the knowledge of different entities of ads nodes and their neighbors, we designed a variational graph auto-encoder for an elaborately designed self-encode task as a pre-training module to recover the undirected weighting graph structure, as shown in Fig. 3. After the pre-training task, the encoder can generate embedding representations Z for new and old ads.

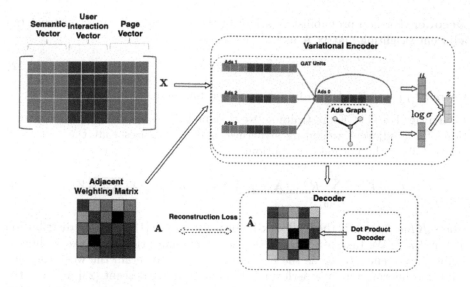

Fig. 3. The workflow for self-encode task based on variational graph auto-encoder.

Encoder: We designed an encoder based on variational encoder and graph attention neural network, which can adaptively adjust the weighting between ads nodes. Latent nodes vector z_i for ads item i are further introduced as:

$$q(z_i \mid X, A) = \mathcal{N}(z_i \mid \mu_i, diag(\sigma_i^2)) \tag{5}$$

$$GAT(A, X) = \gamma_{i,i} \mathbf{W} x_i + \sum_{j \in \mathcal{N}(i)} \gamma_{i,j} \mathbf{W} x_i \tag{6}$$

$$\mu = GAT_\mu(A, X) \tag{7}$$

$$\log \sigma = GAT_\sigma(A, X) \tag{8}$$

where $x_i \in \mathbb{R}^F$. is the node vector of ads item i, and F is the dimension of each node feature. $\mathbf{W} \in \mathbb{R}^{F' \times F}$ is the trainable weighting matrix to distill useful information, and F' is the transformation size. $\mathcal{N}(i)$ is the set of neighbors of node i, and the cross-page attention weighting $\gamma_{i,j}$ is calculated as:

$$\gamma_{i,j} = \frac{exp(LeakyReLU(\alpha^\top[\mathbf{W} x_i || \mathbf{W} x_j]) + \Theta_{i,j})}{\sum_{k \in \mathcal{N}(i)} (exp(LeakyReLU(\alpha^\top[\mathbf{W} x_i || \mathbf{W} x_k]) + \Theta_{i,k})} \tag{9}$$

$$\Theta_{i,j} = softmax_j(\mathbf{A}_{i,j}) = \frac{exp(\mathbf{A}_{i,j})}{\sum_{k \in \mathcal{N}(i)} exp(\mathbf{A}_{i,k})} \tag{10}$$

$$LeakyReLU(x) = max(0, x) + negative\_slope \cdot min(0, x) \tag{11}$$

where *LeakyReLU* and *softmax* are non-linear activation functions. A feedforward neural network is proposed here and is parameterized by a weight vector $a \in \mathbb{R}^{2F'}$. $\Theta_{i,j}$ is the enhanced and normalized edge weighting between ads item i and ads item j.

Decoder: For non-probabilistic variant of the GAE model, we reconstruct the adjacent weighting matrix $\hat{\mathbf{A}}$ as:

$$\hat{\mathbf{A}} = \sigma(ZZ^\top) \tag{12}$$

where $\sigma(\cdot)$ is the ReLU non-linear activation function.

Learning: For GACE pre-training, we designed $\mathcal{L}_r$ to optimize the reconstruction of the adjacent weighting structure, and $\mathcal{L}_n$ ensures that the generated vector follows the normal distribution.

$$\mathcal{L} = \mathcal{L}_r + \mathcal{L}_n = \sum_{i=0}^{n} KL[q(\hat{\mathbf{A}}_i)\|p(\mathbf{A}_i)] + (-KL[q(z|X,\mathbf{A})\|p(Z)]) \tag{13}$$

where $KL[q(\cdot)\|p(\cdot)]$ is the Kullback-Leibler divergence [13] between distribution $q(\cdot)$ and $p(\cdot)$. We take $p(Z) = \prod_i p(z_i)$ as a Gaussian prior, and take Kullback-Leibler divergence for $\mathcal{L}_r$ as the reconstruction loss to retain the weighting distribution information. We perform full-batch gradient descent [23] and use the reparameterization trick [28] for training.

5   Experiment

5.1   Dataset

Considering cross-page information is significant, we elaborately selected the public datasets that have page-level information. Experiments are conducted on the public dataset AliEC [25] from RecBole [29] which has two pages (regarded as different resources PageID (PID) location) and a real-world CTR prediction dataset collected from three pages of Alipay. We designed offline comparison experiments, evaluated performance on different pages respectively, and then conducted an online A/B test. The statistics of experimental datasets are shown in Table 1

Table 1. Statistics of RecBole Public Dataset AliEC and Alipay Real-World Offline Experimental Dataset

DataSet	AliEC		Alipay Real-World Offline Experimental Dataset		
Page	Page1	Page2	Page1	Page2	Page3
Old Ads Num	579323	752454	6238	4575	4306
User Num	403042	726392	1681441	2179180	5971524
Samples for Warming Embedding and Training	8760940	14235512	16082049	22217967	7776079
Sample for Old Ads Testing	1324123	2237386	4018443	5554002	1943008
New Ads Num	–	–	610	415	407
Samples for New Ads Testing	–	–	6885571	4316894	7835890

5.2   Experimental Settings

Main CTR Prediction Models. Because GACE is the pre-training model for item ID embedding generation. They can be applied to various CTR prediction models which require item embeddings. We conduct experiments on the real-world datasets on the following base CTR prediction models:

Wide &Deep: Wide&Deep Model has been widely accepted in industrial applications. It combined joint training with a wide DNN (for memorization) and a deep DNN (for generalization). We follow the practice in [11] to take the cross-product of user behaviors and candidates as wide inputs.

DCN: Deep&Cross Model. It combined cross layers with DNN in Wide&Deep.

DeepFM: DeepFM Model. It applies factorization machines as "wide" DNN module in Wide&Deep.

Bert4Rec: Bert4Rec Model. It employs deep bidirectional self-attention to model user behavior sequences and is combined with DNN to process user and item candidates' features.

Item ID Embedding Models. For each CTR prediction model, we evaluate the following embedding models, which generate initial embeddings for new and old ads.

RndEmb: It uses randomly generated embedding for new ads and looks up item knowledge base for old ads representation.

NgbEmb: It aggregates the initial embedding of neighbor items and generates the ads embedding z_ngb as follows:

$$z\_ngb_i = x_i + \sum\nolimits_{j \in \mathcal{N}(i)} x_j \tag{14}$$

NgbEmb serves as a baseline that only considers neighbor information.

N2VEmb: Node2vec is a graph embedding method that comprehensively considers DFS and BFS neighborhoods. It can be regarded as an extension of deep-walk, which combines DFS and BFS random walk.

GACEEmb: Graph-based cross-page ad embedding. It generates ad item embedding for both new and old ads considering the cross-page and semantic knowledge.

Parameter Setting. We set the dimension of the ads embedding as 15. The MLP part of Wide&Deep, DeepFM, DCN and Bert4Rec is implemented using the same architecture: two-layer fully connected layers with [256, 128, 64] hidden units. For all attention layers in the above models, the number of hidden units is set to 128. All models are implemented in TensorFlow and optimized using AdamW optimizer. The grid-search strategy is applied to find the optimal hyper-parameters such as batch size (among 256, 512, 1024) and learning rate (1e-2, 1e-3, 5e-4, 1e-4). The experiment of each model with the optimal hyper-parameters is conducted for three times and the average result is reported.

Evaluation Scheme. Model performance is evaluated on different pages and model performance between new and old ads is also compared. We used the following evaluation matrix in our recommendation tasks. AUC and Loss are adopted for offline comparison experiments to evaluate models' performance.

AUC: AUC measures the entire two-dimensional area underneath the ROC curve. It is widely used for evaluating the classification model. It reflects the probability that the model will rank a randomly chosen positive example more highly than a randomly chosen negative example. Higher AUC indicates better model performance.

Loss: We use the cross-entropy loss for the click-through rate task on the test set to evaluate the learning process.
Smaller loss indicates better model performance.

$$loss = \frac{1}{Y} \sum\nolimits_{y \in Y} [-\mathcal{Y}\log\hat{\mathcal{Y}} - (1 - \mathcal{Y})\log(1 - \hat{\mathcal{Y}})] \tag{15}$$

5.3   Result from Model Comparison on Public AliEC Dataset

In this section, we conduct model comparison experiments on the public AliEC Dataset and evaluate model performance on two different page scenarios.

Table 2 shows the results from the public AliEC Dataset. Models with deep feature interactions perform better than the original wide&deep models. GACEEmb stands out significantly among all the other item-embedding competitors, which shows in the two pages evaluated results respectively. We owe this to the neighbor information aggregation generation mechanism considering the similar content and similar distribution pages. Through improved variational graph auto-encoding, GACE obtained the enhanced representation of ad items.

5.4   Result from Model Comparison on Alipay Offline Experimental Dataset

In this section, we conducted offline comparison experiments and evaluate model performance on different Alipay pages respectively. Typically, we evaluated the impact of the GACE embedding generation framework on the recommendation effect of new and old ads in different CTR prediction models.

Effectiveness of Cold-Start Ads. Table 3 exhibits the performance of various embedding models based on different CTR prediction models for cold-start ads. It can be observed that NgbEmb performs better than RndEmb, which means that neighbors' related attributes can contribute useful information and alleviate the cold-start problem. N2VEmb improves better than NgbEmb, indicating that just considering the simple average pre-training neighbor ID embedding is insufficient, and considering more neighbors' information is proved effective. GACE leads to a more stable and effective performance improvement than NgbEmb and N2VEmb. It shows that the marginal effect of aggregating neighbor information can be effectively improved by considering the content and page information.

Table 2. Test AUC and loss for Public Dataset Aliec on **Existing old ads**. Pred Model: CTR Prediction Model. Emb Model: Embedding Generation Model.

Pred Model	Embed Model	Page1		Page2	
		AUC	Loss	AUC	Loss
Wide & Deep	RndEmb	0.520	0.511	0.505	0.640
	NgbEmb	0.537	0.489	0.537	0.562
	N2VEmb	0.578	0.453	0.590	0.428
	GACEEmb	**0.588**	**0.346**	**0.595**	**0.337**
DCN	RndEmb	0.573	0.352	0.511	0.290
	NgbEmb	0.578	0.339	0.570	0.272
	N2VEmb	0.586	0.328	0.597	0.255
	GACEEmb	**0.595**	**0.315**	**0.602**	**0.252**
DeepFM	RndEmb	0.557	0.345	0.572	0.295
	NgbEmb	0.582	0.314	0.600	0.260
	N2VEmb	0.593	0.292	0.605	0.257
	GACEEmb	**0.599**	**0.278**	**0.616**	**0.252**
Bert4Rec	RndEmb	0.579	0.318	0.559	0.282
	NgbEmb	0.591	0.309	0.590	0.261
	N2VEmb	0.602	0.299	0.608	0.254
	GACEEmb	**0.610**	**0.258**	**0.627**	**0.245**

Effectiveness of Old Ads. Various embedding models' performances for old ads embedding warming up are compared in Table 4. The item warm-up graph contains both old and new ads so that we can provide the embeddings of old ads' from GACE learning. It is observed that GACE performs best in the cold-start phase and the warm-up phase, indicating that aggregating information considering the physical and semantic neighbors can effectively improve old ads' knowledge representation. The pre-training loss design with double KL divergence helps the embedded generation in the pre-training retain the information between ads to the maximum extent and map it to a higher space of information expression.

5.5   Result from Online A/B Testing

In order to address the limitations of offline evaluation and demonstrate the practical value of GACE, we further deployed the GACE model in an actual online environment, that owns tens of millions of users access it daily.

An online A/B test is designed to further evaluate the performance of GACE. We conducted a week-long rigorous A/B testing with three objectives: to cover more users, cover more cold start items, and collect more reliable results. Bert4Reg with N2VEmb is the main model already serving in the recommendation system. We allocated 10% of Bert4Rec traffic with N2VEmb as the baseline

Table 3. Test AUC and loss for Alipay Real-World Dataset on **Cold-start ads**. Pred Model: CTR Prediction Model. Emb Model: Embedding Generation Model.

Pred Model	Embed Model	Page1		Page2		Page3	
		AUC	Loss	AUC	Loss	AUC	Loss
Wide & Deep	RndEmb	0.700	1.074	0.531	0.339	0.344	0.997
	NgbEmb	0.733	0.817	0.551	0.331	0.463	0.449
	N2VEmb	0.741	0.759	0.566	0.326	0.541	0.386
	GACEEmb	**0.790**	**0.645**	**0.608**	**0.264**	**0.892**	**0.247**
DCN	RndEmb	0.610	1.185	0.569	0.343	0.345	0.873
	NgbEmb	0.641	1.133	0.597	0.314	0.446	0.467
	N2VEmb	0.702	0.868	0.609	0.295	0.511	0.429
	GACEEmb	**0.789**	**0.594**	**0.635**	**0.279**	**0.706**	**0.380**
DeepFM	RndEmb	0.557	0.727	0.764	0.564	0.416	0.331
	NgbEmb	0.582	0.759	0.854	0.611	0.351	0.445
	N2VEmb	0.593	0.772	0.691	0.649	0.294	0.513
	GACEEmb	**0.796**	**0.536**	**0.663**	**0.273**	**0.874**	**0.251**
Bert4Rec	RndEmb	0.744	0.791	0.565	0.417	0.380	1.125
	NgbEmb	0.758	0.787	0.593	0.335	0.428	0.397
	N2VEmb	0.764	0.727	0.613	0.292	0.557	0.371
	GACEEmb	**0.783**	**0.548**	**0.676**	**0.244**	**0.911**	**0.225**

and compared 10% of Belt4Rec traffic with GACEEmb. We evaluated the new and old items that appeared this week separately.

Table 5 illustrates the cumulative relative improvement of the experimental model compared to the baseline model. The average CTR of using GACE has increased by 3.61%, 2.13%, and 3.02% respectively. For cold-start ads distribution, the CTR has increased by 9.96%, 7.51%, and 8.97% respectively. It is worth noting that both improvements have statistical significance (p-value less than 0.05). This practice-oriented experiment demonstrates the effectiveness of our model in real-world recommendation scenarios.

Table 4. Test AUC and loss for Alipay Real-World Dataset on **Existing old ads.** Pred Model: CTR Prediction Model. Emb Model: Embedding Generation Model.

Pred Model	Embed Model	Page1		Page2		Page3	
		AUC	Loss	AUC	Loss	AUC	Loss
Wide & Deep	RndEmb	0.705	0.811	0.562	0.651	0.863	0.573
	NgbEmb	0.727	0.776	0.598	0.572	0.881	0.508
	N2VEmb	0.783	0.719	0.656	0.435	0.921	0.253
	GACEEmb	**0.803**	**0.549**	**0.714**	**0.343**	**0.939**	**0.207**
DCN	RndEmb	0.796	0.558	0.621	0.295	0.903	0.215
	NgbEmb	0.802	0.537	0.693	0.277	0.912	0.156
	N2VEmb	0.813	0.520	0.726	0.259	0.938	0.137
	GACEEmb	**0.826**	**0.499**	**0.731**	**0.256**	**0.952**	**0.112**
DeepFM	RndEmb	0.793	0.548	0.664	0.300	0.902	0.165
	NgbEmb	0.829	0.498	0.697	0.265	0.911	0.156
	N2VEmb	0.844	0.463	0.702	0.261	0.945	0.133
	GACEEmb	**0.867**	**0.441**	**0.727**	**0.256**	**0.961**	**0.124**
Bert4Rec	RndEmb	0.838	0.505	0.671	0.287	0.922	0.150
	NgbEmb	0.856	0.490	0.708	0.266	0.939	0.144
	N2VEmb	0.872	0.475	0.730	0.258	0.946	0.125
	GACEEmb	**0.883**	**0.409**	**0.753**	**0.249**	**0.963**	**0.112**

Table 5. Cumulative relative improvement of the experimental model compared to the baseline model for a week. Baseline: Bert4Rec + N2VEmb.

Model	Eval scope	Page1	Page2	Page3
Bert4Rec +GACEEmb	All ads	+3.61%	+2.13%	+3.02%
	Cold-start ads	+9.96%	+7.51%	+8.97%

6 Conclusion

This paper addresses the CTR prediction problem for cross-page ads whose ID embeddings still need well-learned. Graph-based cross-page ad embedding (GACE) can effectively learn how to generate desirable ad item embedding using cross-page data based on graph neural networks. It takes into account the page-level similarity relationship and semantic-based content relationship simultaneously, establishes a graph to connect all ads across pages, and adaptively extracts the information from cross-page adjacent ads. Because of the characteristics of the generative model, it can generate efficient embedding representations of new and old ads simultaneously. The experiment results show that GACE can effectively improve the CTR task performance for both new and old ads on four major deep-learning-based models. In the future, we will consider enhancing neighbor representation, try other methods to retrieve more information from neighbors and extend it to CVR and GMV Recall application scenarios.

References

1. Barkan, O., Koenigstein, N.: Item2vec: neural item embedding for collaborative filtering. arXiv (2016)
2. Chen, J., Sun, B., Li, H., Lu, H., Hua, X.S.: Deep ctr prediction in display advertising. In: Proceedings of the 24th ACM international conference on Multimedia, pp. 811–820 (2016)
3. Chen, Q., Zhao, H., Li, W., Huang, P., Ou, W.: Behavior sequence transformer for e-commerce recommendation in alibaba (2019)
4. Cheng, H.T., et al.: Wide & deep learning for recommender systems. In: Proceedings of the 1st Workshop on Deep Learning for Recommender Systems, pp. 7–10 (2016)
5. Chowdhary, P.: Fundamentals of Artificial Intelligence. Fundam. Artifi. Intell. (2020)
6. Church, W.K.: Word2vec. Nat. Lang. Eng. **23**(01), 155–162 (2017)
7. Devika, R., Vairavasundaram, S., Mahenthar, C.S.J., Varadarajan, V., Kotecha, K.: A deep learning model based on bert and sentence transformer for semantic keyphrase extraction on big social data. IEEE Access **9**, 165252–165261 (2021)
8. Feng, Y., et al.: Deep session interest network for click-through rate prediction (2019)
9. Goh, K.L., Singh, A.K., Lim, K.H.: Multilayer perceptrons neural network based web spam detection application. In: IEEE China Summit & International Conference on Signal & Information Processing (2013)
10. Guo, H., Chen, B., Tang, R., Li, Z., He, X.: Autodis: automatic discretization for embedding numerical features in ctr prediction (2020)
11. Guo, H., Tang, R., Ye, Y., Li, Z., He, X.: Deepfm: a factorization-machine based neural network for ctr prediction. arXiv preprint arXiv:1703.04247 (2017)
12. Hamilton, W.L., Ying, R., Leskovec, J.: Inductive representation learning on large graphs (2017)
13. Joyce, J.M.: Kullback-leibler divergence. In: International Encyclopedia Of Statistical Science, pp. 720–722. Springer (2011). https://doi.org/10.1007/978-3-642-04898-2_327
14. Khawar, F., Poon, L., Zhang, N.L.: Learning the structure of auto-encoding recommenders. In: Proceedings of The Web Conference 2020, pp. 519–529 (2020)
15. Kipf, T.N., Welling, M.: Variational graph auto-encoders. arXiv preprint arXiv:1611.07308 (2016)
16. Lipmaa, H., Rogaway, P., Wagner, D.: Ctr-mode encryption. In: First NIST Workshop on Modes of Operation, vol. 39, Citeseer, MD (2000)
17. Okura, S., Tagami, Y., Ono, S., Tajima, A.: Embedding-based news recommendation for millions of users. In: Proceedings of the 23rd ACM SIGKDD International Conference on Knowledge Discovery and Data Mining, pp. 1933–1942 (2017)
18. Ouyang, W., et al.: Deep spatio-temporal neural networks for click-through rate prediction. In: Proceedings of the 25th ACM SIGKDD International Conference on Knowledge Discovery & Data Mining, pp. 2078–2086 (2019)
19. Perozzi, B., Al-Rfou, R., Skiena, S.: Deepwalk: online learning of social representations. ACM (2014)

20. Pettie, S., Ramachandran, V.: A shortest path algorithm for real-weighted undirected graphs. SIAM J. Comput. **34**(6), 1398–1431 (2005)
21. Liu, Q., Lu, J., Zhang, G., Shen, T., Zhang, Z., Huang, H.: Domain-specific meta-embedding with latent semantic structures - sciencedirect. Inform. Sci. (2020)
22. Song, W., Shi, C., Xiao, Z., Duan, Z., Tang, J.: Autoint: automatic feature interaction learning via self-attentive neural networks. In: The 28th ACM International Conference (2019)
23. Soodabeh, A., Manfred, V.: A learning rate method for full-batch gradient descent. Műszaki Tudományos Közlemények **13**(1), 174–177 (2020)
24. Sun, F., Liu, J., Wu, J., Pei, C., Lin, X., Ou, W., Jiang, P.: Bert4rec: sequential recommendation with bidirectional encoder representations from transformer. In: Proceedings of the 28th ACM International Conference on Information and Knowledge Management, CIKM 2019, pp. 1441–1450. Association for Computing Machinery, New York (2019). https://doi.org/10.1145/3357384.3357895,https://doi.org/10.1145/3357384.3357895
25. Tianchi: (2018). https://tianchi.aliyun.com/dataset/dataDetail?dataId=56
26. Veličković, P., Cucurull, G., Casanova, A., Romero, A., Lio, P., Bengio, Y.: Graph attention networks. arXiv preprint arXiv:1710.10903 (2017)
27. Wang, R., Fu, B., Fu, G., Wang, M.: Deep & cross network for ad click predictions. In: Proceedings of the ADKDD 2017, pp. 1–7 (2017)
28. Wilson, J.T., Moriconi, R., Hutter, F., Deisenroth, M.P.: The reparameterization trick for acquisition functions. arXiv preprint arXiv:1712.00424 (2017)
29. Zhao, W.X., et al.: Recbole 2.0: towards a more up-to-date recommendation library. In: Proceedings of the 31st ACM International Conference on Information & Knowledge Management, pp. 4722–4726 (2022)
30. Zhou, G., et al.: Deep interest network for click-through rate prediction. In: Proceedings of the 24th ACM SIGKDD International Conference on Knowledge Discovery & Data Mining, pp. 1059–1068 (2018)

TEZARNet: TEmporal Zero-Shot Activity Recognition Network

Pathirage N. Deelaka[1], Devin Y. De Silva[1],
Sandareka Wickramanayake[1(✉)], Dulani Meedeniya[1],
and Sanka Rasnayaka[2]

[1] Department of Computer Science and Engineering, University of Moratuwa,
Moratuwa, Sri Lanka
{nipun.18,devin.18,sandarekaw,dulanim}@cse.mrt.ac.lk
[2] School of Computing, National University of Singapore, Singapore, Singapore
sanka@nus.edu.sg

Abstract. Human Activity Recognition (HAR) using Inertial Measurement Unit (IMU) sensor data has practical applications in healthcare and assisted living environments. However, its use in real-world scenarios has been limited due to the lack of comprehensive IMU-based HAR datasets covering various activities. Zero-shot HAR (ZS-HAR) can overcome these data limitations. However, most existing ZS-HAR methods based on IMU data rely on attributes or word embeddings of class labels as auxiliary data to relate the seen and unseen classes. This approach requires expert knowledge and lacks motion-specific information. In contrast, videos depicting various human activities provide valuable information for ZS-HAR based on inertial sensor data, and they are readily available. Our proposed model, TEZARNet: TEmporal Zero-shot Activity Recognition Network, uses videos as auxiliary data and employs a Bidirectional Long-Short Term IMU encoder to exploit temporal information, distinguishing it from current work. The proposed model outperforms the state-of-the-art accuracy by 4.7%, 7.8%, 3.7%, and 9.3% for benchmark datasets PAMAP2, DaLiAc, UTD-MHAD, and MHEALTH, respectively. The code is available at https://github.com/nipdep/TEZARNet

Keywords: Human Activity Recognition · Zero-shot Learning · Inertial Measurement Unit Data

1 Introduction

Human activity recognition (HAR) has numerous applications in healthcare and ambient assisted living, such as fall detection [8], remote elder care, and fitness tracking [1]. Further, the growing usage of wearable devices has enabled Inertial Measurement Unit (IMU) sensor data-based HAR as an alternative to video-based HAR. However, owing to numerous human activities, collecting sensor-based datasets containing many activities costs time and effort. The existing IMU datasets [4,10,11,16] only cover a limited number of activities, e.g., 20

B. Luo et al. (Eds.): ICONIP 2023, CCIS 1969, pp. 444–455, 2024.
https://doi.org/10.1007/978-981-99-8184-7_34

activities. As a result, supervised models trained on such datasets have limited usage in the real world because they cannot identify activities not present during the training phase. Zero-Shot Human Activity Recognition (ZS-HAR) solves the problem of recognizing previously unseen activities by relating seen and unseen classes via a semantic space built using auxiliary data.

Most existing ZS-HAR models based on IMU data utilize handcrafted attributes, word embeddings of the class labels, or word descriptions as the auxiliary data to build the semantic space [13,20]. However, defining attributes for human activities requires expert knowledge, while word descriptions may not correctly detail the activities. Besides, these auxiliary data lack motion-specific information such as attitude differences, orientations, velocity, acceleration, and repetitions. As an alternative, videos of human activities can be used to construct the semantic space as they are widely available and contain motion-specific information. Recently, Tong et al. [17] have proposed a ZS-HAR model that builds the semantic space based on video data. However, they do not utilize the temporal information in IMU data to construct the feature vectors for IMU data.

This paper proposes TEZARNet: TEmporal Zero-shot Activity Recognition Network. TEZARNet consists of a Bi-Directional Long-Short Term Memory (Bi-LSTM) network [12] based IMU encoder to extract the temporal information encoded in the multivariate IMU signals to improve the Zero-shot classification accuracy. We build the semantic space using the videos of activity classes. Furthermore, we utilize a K-Nearest Neighbour (KNN) algorithm to infer the activity class for unseen data, further improving accuracy. Extensive evaluations using four widely used IMU datasets demonstrate that the TEZARNet improves Zero-shot classification accuracy. TEZARNet outperforms the state of the art accuracy by 4.7%, 7.8%, 3.7% and 9.3% for datasets PAMAP2 [16], DaLiAc [11], UTD-MHAD [4] and MHEALTH [2] respectively.

2   Related Work

Zero-Shot Human Activity Recognition (ZS-HAR) involves determining the class labels of activities without any labeled instances during training. Early work in sensor-based ZS-HAR has relied heavily on expert-determined attribute maps. For example, Cheng et al. [6] proposed a Support Vector Machine-based approach to predict binary attributes and determine the class. This work was then extended using a conditional random field and the 1-nearest neighbor (1NN) classifier in [5]. In contrast, recent work by Matsuki et al. in [13] and Wu et al. in [20] have used word embeddings based on the Word2Vec model [7] to learn the semantic space. However, word embeddings lack the motion-related information necessary for distinguishing IMU features. A recent study by Catherine et al. [17] introduced a novel video embedding space, achieving state-of-the-art performance.

Current approaches in IMU-based ZS-HAR have primarily focused on processing IMU signals as stationary features. However, processing IMU signals as temporal signals have been shown to improve the accuracy of supervised sensor-based HAR [21,23]. Despite the inherent single reference point bias, almost all

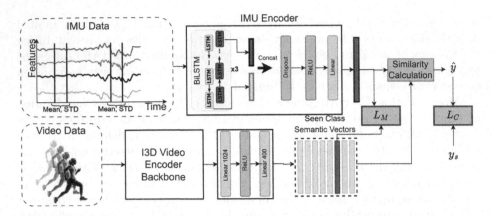

Fig. 1. The overview of the training process of the proposed TEZARNet model.

existing studies [13,17,19,20] have employed 1NN-based inference for unseen prediction. Therefore, this work explores temporal modeling for IMU-based ZSL for HAR and uses the KNN algorithm to infer class labels for unseen classes.

3 Proposed TEZARNet

Let's denote the set of seen classes as C_s and the unseen classes as C_u where $C_s \cap C_u = \emptyset$. Suppose the training dataset, which comprises the samples from seen classes, is denoted by D_s, whereas the test dataset, which consists of samples from unseen classes, is indicated by D_u.

The IMU encoder is the main component of TEZARNet. The input to the IMU encoder is a sample of IMU data, which is a multivariate time-series $x \in \mathcal{R}^{n \times d}$, where n is the sequence length and d is the feature dimension. The IMU data comes from various sensors placed on different human body parts. For example, the PAMAP2 dataset [16], a popular resource in this field, was collected by gathering data from 9 individuals, each of whom wore three IMUs on their wrist, chest, and ankle. These IMUs use multiple sensors to measure and report physical motion and orientation. An IMU includes an accelerometer, gyroscope, magnetometer, and barometer sensors, producing multivariate time series data.

The multivariate time series is first divided into a sequential set of windows along the temporal axis before being passed to the IMU encoder. In each window, the mean and standard deviation for each of the d dimensions are calculated. The resulting multivariate series is then fed into the IMU encoder.

The IMU encoder consists of a stack of Bi-LSTM layers followed by a dropout layer, a ReLu layer, and a linear layer, as shown in Fig. 1. Each Bi-LSTM layer processes the input signal over the temporal dimension in both forward and backward directions utilizing auto-correlation while maintaining LSTM-gated memory to capture temporal dependencies accurately [14]. The output of the forward LSTM layer of one Bi-LSTM layer is passed to the forward LSTM layer of the next Bi-LSTM layer, while the output of the backward LSTM layer of one Bi-LSTM layer is passed to the backward LSTM layer of the next Bi-LSTM layer. The final embedding is produced by concatenating the last hidden states from the forward and backward LSTM layers of the last Bi-LSTM layer. This embedding is then passed through a Dropout layer, ReLu layer, and a fully connected linear block to generate a high-level feature vector z.

We derive a semantic representation for each class $c \in C_s \cup C_u$ using a set of videos of the c collected from online sources such as Pexel and Youtube. It is worth noting that these videos do not align with the IMU data but contain useful information about each activity to learn the mapping between seen and unseen classes. We use a video encoder of the I3D model [3] to derive this semantic representation. The I3D model takes a 4D tensor as input, which captures RGB values at a fixed resolution frame over a fixed number of time steps. The I3D model processes input in a spatiotemporal manner, using inflated convolutional layers and a region proposal layer to engineer an optimal representative feature vector for the given video. We feed a set of videos corresponding to the activity c to the I3D model and obtain a set of video feature vectors from an internal layer of the I3D model. We extract features from an inner layer rather than the logit layer because they contain more information. The average of these vectors v_c, is the semantic vector of the class c.

TEZARNet learns to make predictions during training by mapping IMU samples to their corresponding class semantic vectors. A training sample of TEZARNet consists of an IMU sample and the corresponding activity label, $(x_s, y_s) \in D_s$ where $y_s \in C_s$. Given x_s, the IMU encoder produces z_s. Further, we use m videos per class to derive class semantic vectors using the I3D video encoder. Let the set of class semantic vectors be $V = \{v_1, v_2, ..., v_{|C_s|}\}$ and the class semantic vector of y_s be v_{y_s}. TEZARNet learns to map the IMU sample to its corresponding class by minimizing the L2 distance between z_s and v_{y_s}. This objective function is defined as L_M:

$$L_M = \|z_s - v_{y_s}\|_2 \tag{1}$$

We consider the class of the class semantic vector that is most similar to z_s as the activity class of x_s. To calculate the similarity, we first project z_s onto the unit vector of each class semantic vector. Let the similarity between z_s and $v_k \in V$ be β_k where $k \in 1, 2, .., |C_s|$.

$$\beta_k = z_s \cdot \frac{v_k}{\|v_k\|_2} \tag{2}$$

Then we apply SoftMax to derive the class with the highest similarity.

$$p(y_k|z_s) = \frac{exp(\beta_k)}{\sum_{k \in |C_s|} exp(\beta_k)} \tag{3}$$

The classification objective, L_C is defined using the multi-class cross-entropy loss.

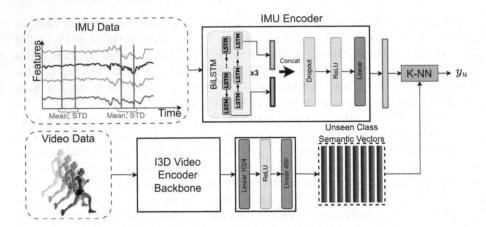

Fig. 2. Overview of unseen class prediction process of the proposed TEZARNet model.

$$L_C = - \sum_{k \in |C_s|} y_k \log(p(y_s = c_s | x_s)) \tag{4}$$

The final objective function of TEZARNet is defined as

$$L = L_M + \lambda L_C \tag{5}$$

, where λ is a hyper-parameter.

We use the KNN algorithm to derive the activity class labels for unseen instances. Given an IMU sample $x_u \in D_u$ and m videos for each class in C_u, TEZARNet produces z_u and a set of semantic vectors $\bar{V} = \{\{\bar{v}_1^1, \bar{v}_2^1, ..., \bar{v}_m^1\}, ..., \{\bar{v}_1^{|C_u|}, \bar{v}_2^{|C_u|}, ..., \bar{v}_m^{|C_u|}\}\}$. $\bar{V}$ sets the neighborhoods for each unseen class in C_u to check whether a new sample belongs to a particular class. We predict the class y_u by considering the k-nearest neighbors in the fitted neighborhood using the KNN algorithm. The inference phase of TEZARNet is shown in Fig 2.

4   Experimental Study

4.1   Datasets

We conduct extensive experiments evaluating the proposed TEZARNet with four IMU datasets commonly used for benchmarking ZS-HAR: namely, PAMAP2 [16], DaLiAc [11], UTD-MHAD [4] and MHEALTH [2]. All these datasets are publicly available. The PAMAP2 contains IMU data from 9 subjects performing 18 daily activities, where each subject wears three IMUs on their wrist, chest, and ankle. DaLiAc comprises IMU data collected from 19 subjects performing 13 daily activities, each wearing four IMUs at different locations. UTD-MHAD is a multi-modal dataset that includes RGB videos and depth information, in addition to IMU data from 8 subjects performing 27 classes of actions. The

Table 1. IMU Dataset Characteristics

Dataset	Activities	Subjects	Samples	Features	Folds
PAMAP2 [16]	18	9	5169	54	5
DaLiAc [11]	13	19	21844	24	4
UTD-MHAD [4]	27	8	861	6	5
MHEALTH [2]	12	10	2774	12	4

MHEALTH contains IMU and ECG data from 10 subjects performing 12 activities. A summary of the datasets is shown in Table 1.

TEZARNet uses video data to learn the relationship between seen and unseen classes. However, only the UTD-MHAD dataset contains a corresponding video dataset. Hence, for the other datasets, we collected videos from online sources such as Pexel and YouTube and HMDB51 [10] dataset. We manually selected high-quality videos with one human performing the activity of interest in the foreground and from various angles. We also ensured that videos of the same class had similar primary movements. To ensure consistency, we fixed the video duration at 3 s, the frame rate at 30 fps, and the resolution at 1280×1024 HD. The videos used for TEZARNet experiments are available at https://bit.ly/3pvnvXy.

4.2 Evaluation Metric

Following the current work in ZS-HAR, we use the *average accuracy per class* as the evaluation metric in our experiments. The average accuracy per class compensates for the class imbalance problem in most of the standard IMU-based HAR datasets. Suppose the number of correct predictions for a unseen class c_u is $N_{c_u}^{correct}$ and number of total instances for c_u is $N_{c_u}^{total}$. Then the average accuracy per class is defined as

$$\text{Average Accuracy per Class (AAC)} = \frac{1}{|C_u|} \sum_{c_u \in C_u} \frac{N_{c_u}^{correct}}{N_{c_u}^{total}} \qquad (6)$$

4.3 Implementation

In this study, we adopt a k-fold evaluation approach [14] to split the activity classes into seen and unseen sets over k disjoint folds. For the DaLiAc and UTD-MHAD datasets, the fold separation strategy introduced in [17] is followed. For the MHEALTH dataset, three unseen classes are randomly selected in each fold. The activity classes are divided according to their activity superclass, e.g., static, dynamic, and sports. Activities are randomly selected from each superclass to create the set of unseen classes to ensure activity type balance in the seen-unseen splits. The k-fold class separation ensures that each fold's seen and unseen class sets include at least one sample from each activity superclass. The seen dataset is split within each fold into 90% of training and 10% of validation.

We implement TEZARNet using PyTorch framework [15], and it is trained on an NVIDIA Tesla T4 GPU or an NVIDIA GeForce RTX 2040 GPU. The ADAM optimizer [9] with a learning rate of 10^{-3} is used in training. The training is carried out for 50 epochs with a batch size of 64. We use ten videos for each class, and λ in Eq. 5 is set to 0.001 based on hyper-parameter tuning [See Table 2]. The hidden size of an LSTM unit is set to 128, and each LSTM layer in Bi-LSTM outputs an embedding of size 400. The dropout rate of stacked Bi-LSTM layers is set to 0.1. During the KNN inference, five neighbors are employed, utilizing distance metrics to assign weights to the neighbors' contributions.

Table 2. Comparison of λ value in the loss function with PAMAP2 dataset

λ-value	1-fold	2-fold	3-fold	4-fold	5-fold	Average
0.0001	77.6	73.8	68.5	21.3	25.1	55.6
0.001	82.1	77.8	67.3	21.3	25.2	58.2
0.01	76.2	66.6	65.8	21.3	25.2	54.1
0.1	55.1	69.8	60.1	21.4	25.1	48.8
1	56.0	76.4	60.9	21.2	25.1	50.7

4.4 Comparative Study

We compare TEZARNet with the following state-of-the-art IMU-based ZS-HAR models.

- **MLCLM** [13]: Multi-Layer Perceptron(MLP) feature extractor on stationary IMU data for projection-based ZSL on Word Embedding semantic space.
- **VbZSL** [17]: MLP feature extractor on stationary IMU data for projection-based ZSL on Video Embedding semantic space.

We use the accuracy values reported in [13] for the MLCLM model. In contrast, we implement the VbZSL model as their code, and the video dataset used in experiments is not publicly available. For a fair comparison, for PAMAP2, UTD-MHAD, and MHEALTH datasets, we train VbZSL with the video datasets used in TEZARNet training. For the DaLiAc dataset, we use video data coming with it for the training of both VbZSL and TEZARNet models.

The results are shown in Table 3. The table presents the AAC over k-folds for different datasets and the percentage improvement (*% Improvement*) compared to existing methods. We calculate the % Improvement as follows:

$$\% \text{ Improvement} = \frac{\text{TEZARNet AAC} - \text{Current Best AAC}}{\text{Current Best AAC}} \quad (7)$$

TEZARNet consistently achieves the highest AAC for all four datasets. In contrast, VbZSL falls short as it doesn't utilize the temporal information in

Table 3. Comparison of percentage Average Accuracy per Class over k-folds for different datasets.

Model	Semantic Space	PAMAP2	DaLiAc	UTD-MHAD	MHEALTH
MLCLM	Word Emb	54.93	–	–	–
VbZSL	Video Emb	42.20	70.60	24.84	38.80
	Word Emb	47.70	60	32.40	–
TEZARNet	Video Emb.	**58.27**	**76.10**	**32.60**	**40.40**
% Improvement		+4.86	+7.79	+3.70	+9.27

Table 4. Comparison of percentage Average Accuracy per Class when using different IMU encoders for TEZARNet.

IMU Encorder	1-fold	2-fold	3-fold	4-fold	5-fold	Average
TEZARNet with LSTM	68.22	74.5	53.06	30.39	25.57	52.84
TEZARNet with Transformer Encoder	70.23	80.75	45.48	10.41	25.10	49.58
TEZARNet with Attention LSTM	64.95	69.74	49.10	30.00	25.53	50.08
TEZARNet	66.82	78.23	56.34	29.3	25.32	**58.27**

IMU data. Further, TEZARNet has outperformed MLCLM and VbZSL with word embedding, demonstrating that using video data as the auxiliary data in IMU-based ZS-HAR leads to better performance.

Further, we conduct a paired t-test on average α values achieved by TEZARNet and VbZSL for PAMAP2 and DaLiAc datasets. In this experiment, we run each model 10 times for each dataset. The null hypothesis is that TEZARNet and VbZSL have identical average α values for the given dataset. The p-value for the PAMAP2 dataset is $1.82e-13$, and for the DaLiAc dataset, the p-value is $7.602e-9$. Since the p-value is below 0.05 for both datasets, these results indicate that the improvement achieved by TEZARNet is statistically significant.

4.5 Ablation Study

Next, we conduct experiments to demonstrate the effectiveness of our design choices. In particular, in the following experiments, we show that employing Bi-LSTM and using k-NN based inference leads to better unseen-accuracy. We use PAMAP2 dataset for these experiments.

In the first experiment, we use four widely used temporal models to extract features from IMU data. Namely, LSTM, Bi-LSTM, Transformer [18], and Multi-Headed Attention LSTM Encoder [22]. We replace the IMU encoder of TEZARNet with each of these models and build a variation of TEZARNet, keeping all the configurations fixed. The results are shown in Table 4. The results indicate that Bi-LSTM achieves the best overall unseen performance. On the other hand, some temporal models (e.g., Transformer) win in some sets of unseen classes but fail miserably for some unseen classes.

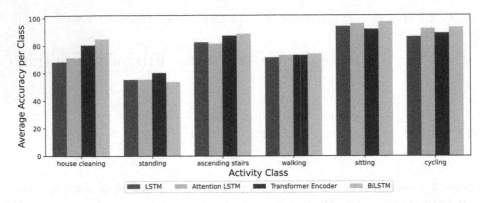

Fig. 3. Average per Class Accuracy of different IMU encoders for sample unseen classes

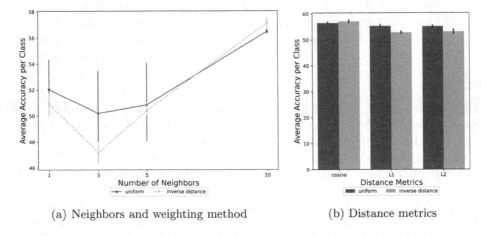

(a) Neighbors and weighting method   (b) Distance metrics

Fig. 4. Performance variation of KNN inference

Figure 3, show Average per Class Accuracy for a set of sample unseen classes achieved by TEZARNet when different IMU encoders are used. The graph indicates that Bi-LSTM consistently achieves the best accuracy.

In the next experiment, we evaluate the impact of different KNN parameters on the unseen accuracy: namely, the number of neighbors, neighbor weighting methods, and distance metrics. Using a trained TEZARNet model, we infer unseen accuracy by changing one parameter at a time. Figure 4a shows the average accuracy per class of TEZARNet on the PAMAP2 dataset for different numbers of nearest neighbors. The results indicate that the accuracy decreases when the number of nearest neighbors is 3 or 5 for both uniform and inverse distances. However, the accuracy keeps increasing when the number of neighbors is assigned a value higher than 5. In our experiment, the best accuracy is achieved when the number of neighbors is 10. Further, Additionally, using inverse distance-based

Table 5. Performance Reproducibility analysis

Model	Mean	Standard Error	P-Value
VbZSL	39.44	0.51	0.51
TEZARNet with 1-NN Inference	52.78	1.20	0.31
TEZARNet with 10-NN Inference	55.03	0.61	0.04

neighbor weight leads to higher confidence, and a higher neighbor count leads to lower performance variation, as demonstrated by the error bar.

Figure 4b shows the effect of different distance metrics on the unseen accuracy. We observe that using cosine distance, we can achieve a superior accuracy of 57.11% compared to $L1$ and $L2$ distance. In this experiment, inverse distance provides better performance. In conclusion, we can get the best performance when the number of neighbors is ten and inverse cosine distance is used.

4.6 Reproducibility of TEZARNet

Demonstrating the reproducibility of the experimental results is imperative to ensure the reliability of the experiment and transparency. To show the reproducibility of our results, we conduct a paired t-test. We ran the TEZARNet ten times and obtained results. These results are randomly separated into two groups to perform the paired t-test. All model runs are conducted on identical hardware and software setups without explicitly setting the seed to analyze the effect of random states. The null hypothesis is the two sets of data are unlikely to have come from the same distribution. We conduct this experiment with TEZARNet with 1-NN inference, TEZARNet with 10-NN inference, and the VbZSL model. Table 5 shows the results of the paired t-test along with mean and standard error for average performance for the class.

TEZARNet with 10-NN has achieved the best mean performance and a p-value less than 0.05, indicating statistically significant reproducibility. In contrast, VbZSL has a p-value of 0.51, indicating that VbZSL may not always display the same performance. Further, TEZARNet with 1-NN has a higher p-value corroborating our K-NN-based inference approach.

5 Conclusions and Future Work

This paper presents a new zero-shot model, called TEZARNet, for IMU-data-based human activity recognition using video data to learn the relationship between seen and unseen classes. In contrast to the current work, we employ an IMU encoder consisting of a stack of Bi-LSTM layers to extract the features containing temporal information from IMU data. The assumption here is that online video repositories have videos clearly representing each activity in the dataset. In the absence of such videos, the performance of TEZARNet may be affected. Nevertheless, experiments on four benchmark datasets, namely

PAMAP2, DiLiAc, UTD-MHAD, and MHEALTH, demonstrate that our proposed model outperforms the state-of-the-art IMU-based zero-shot human activity recognition models using video data or word embedding to build the semantic space. In future work, we can explore alternatives for video data that contain temporal information, such as skeleton movements, to build the semantic space. Further, TEZARNet architecture can be extended to incorporate explainability into the model so that decisions made by TEZARNet are explained to the user, enhancing user trust in the model.

References

1. Ahmed, N., Rafiq, J.I., Islam, M.R.: Enhanced human activity recognition based on smartphone sensor data using hybrid feature selection model. Sensors **20**(1), 317 (2020)
2. Banos, O., et al.: mhealthdroid: a novel framework for agile development of mobile health applications. In: Proceedings of the Ambient Assisted Living and Daily Activities: 6th International Work-Conference, pp. 91–98 (2014)
3. Carreira, J., Zisserman, A.: Quo vadis, action recognition? a new model and the kinetics dataset. In: Proceedings of the IEEE Conference on Computer Vision and Pattern Recognition, pp. 6299–6308 (2017)
4. Chen, C., Jafari, R., Kehtarnavaz, N.: Utd-mhad: a multimodal dataset for human action recognition utilizing a depth camera and a wearable inertial sensor. In: Proceedings of the IEEE International Conference on Image Processing (ICIP), pp. 168–172 (2015)
5. Cheng, H.T., Griss, M., Davis, P., Li, J., You, D.: Towards zero-shot learning for human activity recognition using semantic attribute sequence model. In: Proceedings of the 2013 ACM International Joint Conference on Pervasive and Ubiquitous Computing, pp. 355–358 (2013)
6. Cheng, H.T., Sun, F.T., Griss, M., Davis, P., Li, J., You, D.: Nuactiv: recognizing unseen new activities using semantic attribute-based learning. In: Proceeding of the 11th annual international conference on Mobile Systems, Applications, and Services, pp. 361–374 (2013)
7. Church, K.W.: Word2vec. Nat. Lang. Eng. **23**(1), 155–162 (2017)
8. Khojasteh, S.B., Villar, J.R., Chira, C., González, V.M., De la Cal, E.: Improving fall detection using an on-wrist wearable accelerometer. Sensors **18**(5), 1350 (2018)
9. Kingma, D.P., Ba, J.: Adam: A method for stochastic optimization. In: Yann LeCun, Y.B. (ed.) Proceedings of the 3rd International Conference on Learning Representations, ICLR (2015)
10. Kuehne, H., Jhuang, H., Garrote, E., Poggio, T., Serre, T.: HMDB: a large video database for human motion recognition. In: Proceedings of the International Conference on Computer Vision (ICCV) (2011)
11. Leutheuser, H., Schuldhaus, D., Eskofier, B.M.: Hierarchical, multi-sensor based classification of daily life activities: comparison with state-of-the-art algorithms using a benchmark dataset. PLoS ONE **8**(10), e75196 (2013)
12. Li, Y., Wang, L.: Human activity recognition based on residual network and bilstm. Sensors **22**(2), 635 (2022)
13. Matsuki, M., Lago, P., Inoue, S.: Characterizing word embeddings for zero-shot sensor-based human activity recognition. Sensors **19**(22), 5043 (2019)

14. Meedeniya, D.: Deep Learning: A Beginners' Guide. CRC Press LLC (2023). https://www.routledge.com/9781032473246
15. Paszke, A., et al.: Pytorch: an imperative style, high-performance deep learning library. In: Wallach, H., Larochelle, H., Beygelzimer, A., d'Alché-Buc, F., Fox, E., Garnett, R. (eds.) Proceedings of the Advances in Neural Information Processing Systems, pp. 8024–8035 (2019)
16. Reiss, A., Stricker, D.: Introducing a new benchmarked dataset for activity monitoring. In: Proceedings of the 16th International Symposium on Wearable Computers, pp. 108–109 (2012)
17. Tong, C., Ge, J., Lane, N.D.: Zero-shot learning for imu-based activity recognition using video embeddings. In: Proceedings of the ACM on Interactive, Mobile, Wearable and Ubiquitous Technologies, vol. 5(4), pp. 1–23 (2021)
18. Vaswani, A., et al.: Attention is all you need. In: Advances in Neural Information Processing Systems 30 (2017)
19. Wang, Q., Chen, K.: Alternative semantic representations for zero-shot human action recognition. In: Proceedings of the Joint European Conference on Machine Learning and Knowledge Discovery in Databases, pp. 87–102 (2017)
20. Wu, T., Chen, Y., Gu, Y., Wang, J., Zhang, S., Zhechen, Z.: Multi-layer cross loss model for zero-shot human activity recognition. In: Proceedings of the Pacific-Asia Conference on Knowledge Discovery and Data Mining, pp. 210–221 (2020)
21. Zebin, T., Sperrin, M., Peek, N., Casson, A.J.: Human activity recognition from inertial sensor time-series using batch normalized deep lstm recurrent networks. In: Proceedings of the 40th annual International Conference of the IEEE Engineering in Medicine and Biology Society (EMBC), pp. 1–4 (2018)
22. Zerveas, G., Jayaraman, S., Patel, D., Bhamidipaty, A., Eickhoff, C.: A transformer-based framework for multivariate time series representation learning. In: Proceedings of the 27th ACM SIGKDD Conference on Knowledge Discovery & Data Mining, pp. 2114–2124 (2021)
23. Zhang, H., Xiao, Z., Wang, J., Li, F., Szczerbicki, E.: A novel iot-perceptive human activity recognition (har) approach using multihead convolutional attention. IEEE Internet Things J. 7(2), 1072–1080 (2019)

Tigrinya OCR: Applying CRNN for Text Recognition

Aaron Afewerki Hailu[(✉)], Abiel Tesfamichael Hayleslassie,
Danait Weldu Gebresilasie, Robel Estifanos Haile, Tesfana Tekeste Ghebremedhin,
and Yemane Keleta Tedla

Mainefhi College of Engineering and Technology, Mainefhi, Eritrea
aaron29afewerki@gmail.com, yemane@jnlp.org

Abstract. Tigrinya is a language predominantly spoken in Eritrea and the Tigray region of Ethiopia. It is classified as a low resourced language when it comes to Natural Language Processing (NLP) and its documents are not widely accessible due to the lack of printed material. Although the language has a rich cultural heritage, its literature hasn't been exposed to large-scale automated digitization when compared to other widely-spoken languages. In this paper, we design an end-to-end CRNN (Convolutional Recurrent Neural Network) to recognize machine-printed Tigrinya text from document images. This will help Tigrinya documents to be more accessible and also bridge the gap with languages that are rich in NLP resources. We have included all the 304 characters in Tigrinya and the network is trained on a total of over a million text-line images constructed from different domains. The majority of the data was synthesized to augment the limited real data to help the model generalize better. We employed two external datasets (ADOCR and GLOCR) in addition to ours to train the network. Furthermore, to improve the performance of the model, extensive parameter tuning was conducted. Without the use of post processing techniques, the model has achieved a 2.32% Character Error Rate (CER). The learning curve shows that given more data, the model can improve the CER. We finally managed to get a lightweight model that achieves comparable results to state-of-the-art results. This result implies that augmenting low resource data with synthetic data can significantly reduce the error rate in text recognition and proper hyperparameter tuning can find us lightweight models without compromising much accuracy.

Keywords: Tigrinya · OCR · Text Recognition · CRNN · Pattern Recognition

1 Introduction

1.1 OCR

Optical Character Recognition (OCR) is the electronic identification and digital encoding of characters by means of optical scanners and specialized software [1]. The aim of this technology is to digitize vast amounts of paper data records and then store them in an editable, searchable and compact format.

B. Luo et al. (Eds.): ICONIP 2023, CCIS 1969, pp. 456–467, 2024.
https://doi.org/10.1007/978-981-99-8184-7_35

OCR systems provide a tremendous opportunity in handling repetitive, boring, labor-intensive and time-consuming processes for human beings. At present, the recognition of Latin-based characters from well-conditioned documents can be considered as a relatively feasible technology due to the number of ongoing research and software in those characters. On the other hand, languages such as Tigrinya have little to none exposure to OCR. And hence, the bulk of Tigrinya literature is not electronically available and if available, it is limited to scanned pictures of text. In this study, we have come up with a novel system for reading Tigrinya text from scanned or photographed documents. A short introduction to the Tigrinya language follows in Sect. 1.2, previous works in the research space are described in Sect. 2. Section 3 deals with the architecture of the proposed system and the data used to train it, while Sect. 4 explains the experimentation phase. The results of the experiment are discussed in Sect. 5 and the final conclusions of our work are presented in Sect. 6.

1.2 The Tigrinya Language

The Tigrinya language (ትግርኛ), often written as Tigrigna, is a member of the Semitic branch of the Afro-Asiatic languages that includes Arabic, Amharic, Ge'ez, Hebrew, Aramaic and Maltese [2]. Tigrinya is spoken by over seven million people. It is a widely spoken language in Eritrea and in the northern part of Ethiopia. In Eritrea, Tigrinya serves as a working language in offices along with Arabic [3]. Tigrinya originates from the Ge'ez language. Although Tigrinya differs markedly from Ge'ez, there is a strong influence of Ge'ez on Tigrinya literature, especially with terms that relate to Christian life, Biblical names, and so on [4].

The Tigrinya writing system is a slight variant of the writing system used for Ge'ez. The Ge'ez script is an abugida written left-to-right and the symbols are organized in groups of similar symbols on the basis of both the consonant and the vowel [5]. As a result, the writing system is usually displayed as a two-dimensional matrix in which the rows contain units beginning with the same consonant and the columns contain units ending in the same vowel. The columns are traditionally known as "orders" [6]. Some of the alphabets that are used in Ge'ez have been lost in the Eritrean Tigrinya [7].

However, these less-used series have been included in our system for the sake of comprehensiveness with archaic and modern Tigrinya. With respect to numbers, the general practice (in Eritrea) is to use Arabic numerals to express numerical values [8].

2 Related Works

Optical Character Recognition has been attracting a lot of interest from researchers lately. Geez scripts have mostly been studied from the perspective of the Amharic and Geez languages. To the best of our knowledge, there has only been one published study on Geez scripts' OCR from the Tigrinya perspective [9]. Moreover, it was very recently that a synthetically generated dataset was made publicly available [10].

According to [11], standard end-to-end text recognition systems consist of the following three steps:

- Manual labelling of data to create ground truth

- Specific preprocessing operations
- Training of recognizer using text images and their labels.

To come to Geez-based scripts, Amharic was the first to be exposed to OCR technology in 1997 [12]. Worku [12] implemented stage-by-stage segmentation algorithms, thinning of characters, binary tree-structured classification to recognize characters. While this was the first research in Amharic OCR, it had a limited number of handcrafted features and used a single font type and size. Soon afterwards, Ermias [13] improved Worku's work by adding underline detection and removal. Another extension of Worku's work was also found in Teferi [14], where he employed noise detection and removal by using contour analysis. These research all suffered from having a limited number of handcrafted features.

Berhanu [15] also experimented with neural networks to classify Amharic characters based on Worku's stage-by-stage segmentation. This was the first use of neural networks in Amharic OCR, and it only experimented with a very small number of characters. Following this attempt, Meshesha [16], Nigussie [17], Mesay [18], Wondwossen [19] and Assabie [20] based their works on Worku's segmentation. These efforts all suffered from the limitation of handcrafted features. Worku and Fuchs [21] again came up with a new approach that used Hidden Markov Random Fields to develop an Amharic bank check recognition system. Although this was the first attempt which didn't use [12], it could not be applicable owing to the wide variety of Amharic checks in use. Cowell and Hussain [22] tried using templates to recognize characters, however, the recognition was highly dependent on the quality of the image.

Meshesha and Jawahar [23] employed PCA (Principal Component Analysis) followed by LDA (Local-Density Approximation) for feature extraction and used a DAG (Directed Acyclic Graph) classifier with SVM (Support Vector Machines). Assabie and Bigun used Directional Field Tensors for feature extraction and a special tree along with a knowledge base to recognize Ethiopic characters [24]. Assabie and Bigun further improved this work by feeding the features to an ANN classifier [25] and using template matching for confusing characters [26]. Moreover, these two researchers applied this architecture to Ethiopic handwriting recognition [27]. Assabie and Bigun [28] then experimented with Hidden Markov Models to recognize Amharic handwriting. This research was the first Geez-based OCR that tried to recognize words instead of characters. Other notable contributions utilized ANN [29], and SVM [30–32]. Similarly, Fitsum [9] analyzed Histogram of Oriented Gradients (HOG) to extract features and fed those features to an SVM classifier. This study by Fitsum is the only one which specifically studied Tigrinya characters. There was no mention of the dataset used in the study, but an accuracy of 98.72% was reported. Furthermore, Fitsum only dealt with the recognition of isolated Tigrinya characters, which is not commonly found in real-world scenarios. In comparison, our approach has been to take text-lines instead of isolated characters to recognize the text so as the model could be suitable for real applications.

All the previous researches employed classical and statistical machine learning techniques, namely, SVM and shallow networks. It wasn't until Siranesh's work [33] that deep learning was adopted in Geez-based recognition systems. Siranesh trained a deep neural network to predict handwritten Geez characters, but didn't implement regularization. Belay [34] and Kesito [35] worked with a CNN (Convolutional Neural Network) to recognize Amharic and Amharic-Latin characters, respectively. Meanwhile, Addis [36] used LSTM (Long short-term memory) networks to recognize Ethiopic (Tigrinya, Amharic and Geez) text-lines. Those three researchers generated synthetic datasets on top of real-life data for training and research purposes. Belay got an accuracy of 92.71% and Addis got a Character Error Rate (CER) of 2.12%. Belay later improved this work by using Factored CNN [37] to decrease the number of classes from 231 to 40. This approach, hence, got a better accuracy of 94.97%. Although these deep learning approaches have increased the recognition accuracy for Geez-based scripts, they needed segmentation up to the character level for them to work in real life. As the segmentation process is erroneous and not perfect, most researchers resort to text-lines or word recognition, which is much easier to implement.

Therefore, with the exception of Addis' [36] and Assabie's [28] works, most of those works are not easily feasible to implement. In [38], Belay created the first publicly available Amharic dataset and also got an average CER of 4.23% using LSTMs [39] and Connectionist Temporal Classification (CTC) [40] for alignment of characters with ground truth and calculating loss. Belay further improved the CER to 2.46% in [41] by adding CNN layers before the BLSTM (Bidirectional LSTM) and CTC architecture.

Our work is the first of its kind to use neural networks for recognition of a the low-resourced language Tigrinya. We've chosen a CRNN architecture for our model because of the higher recognition rate when using CNNs and LSTMs together. Our model utilizes a backbone similar to Belay's work [41], but we've used automatic parameter tuning to keep the number of parameters used in this work to a minimum. This will help the adoption of this work in devices where resources are limited. Furthermore, we plan to make our newly constructed dataset publicly available so as to help future improvement of Tigrinya text-recognition.

3   Methodology

3.1   Data

The unavailability of large, standard, free and publicly available datasets has hindered text recognition research for Geez-based scripts. Belay et al. prepared the first publicly available Amharic OCR (known as ADOCR) in 2019 [42]. We used this dataset because Tigrinya and Amharic share a common writing system with some minor differences. ADOCR is organized from two types of documents. The dataset contains 40,929 printed text-lines while the second document contains 296,403 synthetic text-lines for a total of 337,332 text-lines images. We further excluded the text-lines which had characters not included in our character set.

The GLOCR, a purely Tigrinya dataset by Gaim, was recently made publicly available [10]. This dataset was later employed in our research as the use of additional data is of immense importance in enhancing the performance of OCR systems. Even though the dataset was synthetic, it was purely based on Tigrinya language text, it had multiple fonts in it and it was comprised entirely of Tigrinya characters. We used the "News text-lines", "Biblical text-lines" and "Top-150K" parts of this dataset. One limitation of this dataset is that it lacks the representation of real-world text-images (which can be noisy, blurry, distorted, etc.) even after multiple stages of pre-processing. Hence, we decided to use this dataset to augment Belay's data. Gaim uses 326 characters in his dataset, hence we had to exclude some of the characters that we didn't have in our character set (like Geez numerals) and resize everything to suit our requirements of 32 × 128-sized images.

After we collected those two datasets, we did a frequency analysis of the characters in them to determine if they had a good distribution of the characters. But the results were not satisfactory, hence we had to generate our own dataset to solve the underrepresentation of characters in the datasets (Fig. 1).

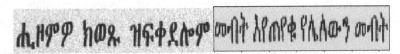

Fig. 1. Sample images from GLOCR (left) and ADOCR (right) datasets

Our Dataset. As noted by [38], it is a time-consuming task to manually annotate images of printed text. Hence, we generated synthetic Tigrinya text-lines to complement the aforementioned two datasets. We had to include some characters which were underrepresented in those datasets so that our model would work well on all our 304 characters. We chose trdg[1] and OCRopus [43] for generating the synthetic Tigrinya dataset. Those synthesizers needed a TrueType-Font (ttf) of the font and a utf-8-encoded string to generate images. At the direction of the printing presses we consulted, we chose five most commonly used Tigrinya fonts (namely, Geez Able, Geez Mahtem Unicode, Geez Wookianos, Geez Zemen and Nyala) and obtained their font files. With regard to the text to be fed to the synthesizer, we obtained a publicly available Tigrinya wordlist[2] and filtered the words which included the underrepresented characters. We then created randomly concatenated 3-word strings from our filtered wordlist. We generated 406,414 training images and 27,679 test images in this manner. Both TRDG and OCRopus allowed us to use ample noise (Gaussian noise, transformations, skewing, jitter, blurring, degradation, etc.). The output of those synthesizers did not meet our image-width requirement of 128 pixels, so we had to add empty white space after the images if the image was smaller or resize the image if it was wider than 128 pixels. We manually discarded the images that were horizontally distorted past the point of recognition. The details of our dataset are provided in Table 1.

3.2 Proposed Architecture

Recurrent Neural Networks (RNN) have been found to be most suited to sequence recognition problems [44], while Convolutional Neural Networks (CNN) perform best in pattern/image recognition problems [45–48]. As OCR is an image-based sequence recognition task, it would be better to combine CNN, RNN and CTC to create a CRNN [41, 49]. The proposed architecture uses CNN for automatic feature extraction, the

---

[1] E. Belval's TextRecognitionDataGenerator, available at GitHub.

[2] Compiled by Dr. Biniam Ghebremichael, available at www.cs.ru.nl/biniam/geez/crawl.php

Table 1. Composition of our dataset

Font type	Geez Able	Geez Mahtem Unicode	Geez Wookianos	Geez Zemen	Nyala
Samples	74,290	73,909	75,011	74,273	80,802
Test-samples	5,705	6,091	4,989	5,277	5,617
Train-samples	79,995	80,000	80,000	80,000	86,419

BLSTM for sequence learning and CTC for transcribing. Those three modules are built one after the other and are trained end-to-end with CTC loss. This network is inspired by [41] which served as the baseline for this study. The proposed architecture is shown in Fig. 3 (Fig. 2).

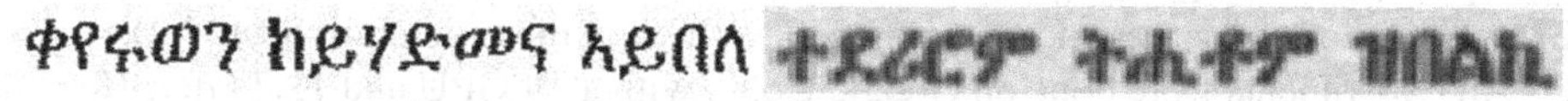

ቀየፉወን ከይሃይድ፞መና እይበለ ተደሪርም ትሒፈም ዝበለኪ

Fig. 2. Sample images generated using OCRopus (left) and TRDG (right)

CNN. We went with the baseline feature extraction architecture and further experimented on it by optimizing the hyperparameters and training with a larger dataset. Finally, the most optimal hyperparameters were chosen after tuning of the hyperparameters as described in Sect. 4.2. Our network, after hyperparameter tuning, is described below.

The CNN is generated by using seven convolutional layers with 3×3 kernels each, except for the one at the end of the network with a 2×2 kernel. There is also a max-pooling layer after each of the first, second, fourth and sixth convolutional layers. We take pool sizes of 2×2 along with 2×2 strides in the first two pooling layers but we have taken pool sizes of 2×1 and the default stride in the other two pooling layers. We have also applied batch normalization (Batch Norm) [50]. 'SeLU' (Scaled Exponential Linear Unit) [51] activation is used to activate each of the convolutional layers and the same padding is used across all the convolutional layers except in the last one. The number of feature maps starts from 16 and is gradually increased to 64.

BLSTM. The four-dimensional output of the CNN is then squeezed to conform to the three-dimension input for the BLSTM. This network consists of two bidirectional LSTM layers with 128 units each. We varied the number of BLSTM layers, increased the number of hidden units in the layers, tried different dropout values but the best results were obtained with the baseline BLSTM architecture.

CTC. CTC is an objective function that adopts dynamic programming algorithms to directly learn the alignment between the input and output sequences [40]. Hence, CTC allows us to forget about the alignment of each character to its location in the input image. "The CTC algorithm takes as an input probability distributions of all the characters which are outputted from SoftMax activation layer at the end of the RNN sequence modeling unit and finds the maximum probable label sequence for a given input word image" [52]. The output of the SoftMax layer is then used to calculate CTC-decoded by removing repeated characters and CTC-blank character. We have used best path decoding method, as proposed in [40] to find the most probable character at every step.

The output of these two layers is then fed into a fully connected layer with Soft-Max activation to predict using 307 nodes (corresponding to each character, a space character, an unknown character and the CTC blank symbol). The network predicts a character for each of the 32 time-steps.

4 Experiments

During the training phase, we fed the neural network text-line images of Tigrinya text (32 × 128 pixel dimensions) along with their annotations and we received the output as a string of 32 characters maximum length. Additionally, we experimented with various hyperparameter configurations in order to determine the most optimal ones for Tigrinya text recognition (hyperparameter tuning is described in more detail in Sect. 4.2).

All experiments were done on Google Colab with a T4 GPU and 12.7 GB of RAM.

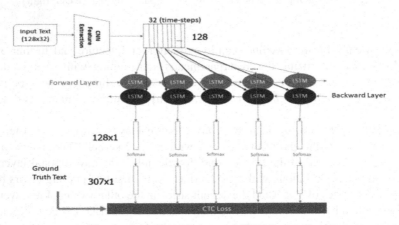

Fig. 3. Overall architecture of CRNN network

In all the experiments, the entire network is trained using Adam optimizer [53]. The model was left to train for 50 epochs, but we had an early-stopping rule to stop the training if there was no improvement in validation loss for 5 epochs. Therefore, the model was trained for 18 epochs with a batch size of 256 before it was early-stopped.

Moreover, in order to decrease our memory footprint while training we extracted images from ADOCR. After that, we loaded only a single batch in memory instead of the entire training data.

4.1 Evaluation Metrics

We used the Character Error Rate (CER) metric to evaluate how the model performs on the test data. The CER defines the minimal Levenshtein edit operations between predicted text and the ground truth transcription of the word image [52]. Moreover, we have also taken the number of parameters of a model to determine the size of the model.

4.2 Hyperparameter Tuning

We have done extensive hyperparameter tuning to decide the hyperparameters that provide the highest performance in our end-to-end network. We used the Weights & Biases API when searching for hyperparameters.

The parameters we fine-tuned include different learning rates, optimizers (Adam, NAdam, SGD), activations (ReLU, ELU, SeLU), number of units in convolutional layer (16 up to 2048), dropout in BLSTM (0.2–0.5) and batch sizes (32–2048).

To find the optimal hyperparameters, we first performed random search over our very large search space. Then, based on the results we got, we narrowed down the search space and we further performed Bayesian optimization over this narrower search space (Fig. 4).

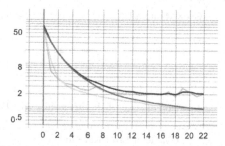

Fig. 4. Loss (vertical axis) vs epochs (horizontal axis) graph. Training and validation losses are indicated by orange and blue lines respectively

5   Results

Although our model took more epochs to train compared to the baseline model, it has learned to predict over two distinct datasets. Overall, the model tends to recognize synthetic text better than printed text. This may be due to the presence of a larger number of synthetic images in the training dataset.

5.1   Comparison of Error Rate

This problem, among others, is a reason for the recognition errors. We have tested our model on GLOCR and our own dataset, and the results are shown in Tables 2 and 3. As we could not find any other research which was benchmarked on GLOCR, we have only provided our own CER in Table 2.

As noted in [41], there are some misalignments between the ground truth and the images resulting from the methods used to generate the dataset. Our results are not directly comparable to works using the ADOCR dataset (because we have excluded a significant portion of images from the dataset both), nevertheless we have evaluated our model's performance by training solely on this 'safe' part of ADOCR. The results obtained are depicted in Table 4 shown below.

Table 2. Test results on GLOCR

Technique	Character Error Rate in %		
	Tigrinya Bible	Tigrinya-150kw	Tigrinya-News
Our method	0.21	0.45	0.19

Table 3. Test results on our dataset.

Technique	Character Error Rate in %				
	Geez Able	Geez Mahtem Unicode	Geez Wookianos	Geez Zemen	Nyala
Our method	0.28	2.38	1.44	1.57	0.78

Table 4. Comparison of state-of-the-art results on ADOCR[3].

Technique	Character Error Rate in %		
	Printed PowerGeez	Synth. PowerGeez	Synth. VisualGeez
LSTM + CTC [38]	8.54	4.24	2.28
CNN + LSTM + CTC [41]	1.56	3.73	1.05
Attention Encoder-Decoder [54]	1.54	5.21	1.17
Blended Attention-CTC [55]	1.04	3.57	0.93
Our method	5.99	-[a]	0.74

[a] We couldn't test with synthetic PowerGeez due to misalignment between labels and images.

5.2 Comparison of Model Size

For the purpose of comparison, we only took the model proposed in [41] as it is the most similar to ours. The architecture proposed in [41] uses 6,675,737 parameters while our model uses only 823,603 parameters.

Hence, it can be seen that our architecture is substantially efficient as it decreases the number of parameters used in the model by 87.66% without trading off much accuracy.

---

[3] Printed PowerGeez, Synthesized PowerGeez and Synthesized VisualGeez are parts of ADOCR.

This could potentially make the deployment of this model easier in mobile devices where storage and memory are scarce. Not only has our model decreased the number of parameters significantly but it has also surpassed state of the art results in the Synthetic VisualGeez part of ADOCR.

6 Conclusion

In this study, we have investigated a CRNN approach for Tigrinya text recognition. In general, the Tigrinya language, lacks access to modern digitization techniques. This study proposes an efficient CRNN network for Tigrinya text recognition. The proposed architecture compromises a negligible amount of accuracy in order to reduce the size of the network and make it feasible for most devices. The architecture was determined by extensive tuning for the most optimal hyperparameters both in terms of network size and accuracy. We have also improved state-of-the-art VisualGeez results in Belay's work [55] by improving the error rate by 0.19%. Furthermore, we have also demonstrated the generalization of this model over three distinct datasets.

In the future, the CER achieved in this study can be further reduced by using correctly labeled, real-life datasets. Using deep-learning based word-detection would also help to correct segmentation errors encountered prior to recognition. The error rate can also be further diminished by utilizing a language model for rectifying the errors of the optical model. Recently, transformer-based models have achieved better results than CNNs in solving computer vision problems. Hence, we plan to experiment with a transformer-based text-recognition model for both handwritten and printed Tigrinya text.

Acknowledgements. We would like to express our gratitude to Dr. Birhanu Hailu Belay and Fitsum Gaim for their valuable feedback and datasets. We would also like to thank Daniel Yacob for providing us with books, Dawit Habtegergish for his help in setting up our development infrastructure and most importantly to the Mainefhi College of Engineering and Technology (Eritrea) for permitting us to use their facilities.

References

1. OCR. www.yourdictionary.com/optical-character-recognition. Accessed 25 Aug 2021
2. Voigt, R.: Tigrinya, an African-semitic language. Folia Linguistica Historica **30**, 391–400 (2009)
3. Tigrinya Language. www.ucl.ac.uk/atlas/tigrinya/language.html. Accessed 10 Dec 2020
4. United Bible Society: The Bible in Tigrigna (1997)
5. Ge'ez script. https://en.wikipedia.org/wiki/Ge%CA%BDez_script. Accessed 21 Dec 2020
6. Tigrigna Writing System. www.ling.upenn.edu/courses/ling202/WritingSystem.html. Accessed 8 Dec 2020
7. Tigrinya language. https://en.wikipedia.org/wiki/Tigrinya_language#Writing_system. Accessed 15 Dec 2020
8. Microsoft: Tigrigna Style Guide. Accessed 20 Jan 2021
9. Fitsum, K.T., Patel, Y.: Optical character recognition for Tigrigna Printed Documents using HOG and SVM. In: 2nd International Conference on Inventive Communication and Computational Technologies (ICICCT), pp. 1489–1494. IEEE (2018)

10. Gaim, F.: GLOCR: GeezLab OCR Dataset, Harvard Dataverse (2021)
11. Fischer, A., et al.: Automatic transcription of handwritten medieval documents, pp. 137–142 (2009)
12. Alemu, W.: The Application of OCR Techniques to the Amharic Script. Addis Ababa University, Addis Ababa, Ethiopia (1997)
13. Abebe, E.: Recognition of Formatted Amharic Text Using Optical Character Recognition (OCR) Techniques. Addis Ababa University, Addis Ababa, Ethiopia (1998)
14. Teferi, D.: Optical Character Recognition of Typewritten Amharic Text. Addis Ababa University, Addis Ababa, Ethiopia (1999)
15. Aderaw, B.: Amharic Character Recognition Using Artificial Neural Networks. Addis Ababa University, Addis Ababa, Ethiopia (1999)
16. Meshesha, M.: A Generalized Approach to Optical Character Recognition (OCR) of Amharic Texts. Addis Ababa University, Addis Ababa, Ethiopia (2000)
17. Taddesse, N.: Handwritten Amharic Text Recognition Applied to the Processing of Bank Checks. Addis Ababa University, Addis Ababa, Ethiopia (2000)
18. Hailemariam, M.: Line Fitting to Amharic OCR: The Case of Postal Address (2003)
19. Mulugeta, W.: OCR for Special Type of Handwritten Amharic Text ("Yekum Tsifet"): Neural Network Approach. Addis Ababa University, Addis Ababa, Ethiopia (2004)
20. Yaregal, A.: Optical Character Recognition of Amharic Text: An Integrated Approach. Addis Ababa University, Addis Ababa, Ethiopia (2002)
21. Alemu, W., Fuchs, S.: Handwritten Amharic bank check recognition using hidden Markov random field. In: Computer Vision and Pattern Recognition Workshops (CVPRW), Madison, Wisconsin, USA (2003)
22. Cowell, J., Hussain, F.: Amharic Character Recognition using a fast signature-based algorithm. In: Seventh International Conference on Information Visualization, London (2003)
23. Meshesha, M., Jawahar, C.: Recognition of printed Amharic documents. In: International Conference on Document Analysis and Recognition (ICDAR), Seoul, Korea (2005)
24. Yaregal, A., Bigun, J.: Ethiopic character recognition using direction field tensor. In: International Conference on Pattern Recognition (ICPR), Hong Kong (2006)
25. Yaregal, A., Bigun, J.: A neural network approach for multifont and size-independent recognition of Ethiopic characters. In: IWAPR (2007)
26. Yaregal, A., Bigun, J.: A hybrid system for robust recognition of Ethiopic script. In: International Conference on Document Analysis and Recognition, Curitiba, Brazil (2007)
27. Yaregal, A., Bigun, J.: Writer-independent offline recognition of handwritten Ethiopic characters. In: International Conference on Frontiers in Handwriting Recognition (2008)
28. Yaregal, A., Bigun, J.: HMM-based handwritten Amharic word recognition with feature concatenation. In: ICDAR, Barcelona (2009)
29. Teshager, A.: Amharic character recognition system for printed real-life documents. Addis Ababa University, Addis Ababa, Ethiopia (2010)
30. Fitsum, D., Zewidie, F.D.: Developing optical character recognition for Ethiopic scripts. Dalarna University, Dalarna, Sweden (2011)
31. Shiferaw, T.: Optical character recognition for Ge'ez scripts. University of Gonder, Ethiopia (2017)
32. Reta, B.Y., Rana, D., Bhalerao, G.V.: Amharic handwritten character recognition using combined features and support vector machine. In: ICOEI (2018)
33. Siranesh, G.: Ancient Ethiopic manuscript recognition using deep learning artificial neural networks. Addis Ababa University, Addis Ababa, Ethiopia (2016)
34. Belay, B.H., Habtegebrial, T., Stricker, D.: Amharic character image recognition. In: IEEE International Conference on Communication Technology, pp. 1179–1182 (2018)
35. Kesito, A.A.: Character recognition of bilingual Amharic-Latin printed documents. Addis Ababa University, Addis Ababa, Ethiopia (2018)

36. Addis, D., Liu, C., Ta, V.: Printed Ethiopic script recognition by using LSTM networks. In: ICSSE, Taipei, Taiwan (2018)
37. Belay, B.H., Habtegebrial, T., Liwicki, M., Belay, G., Stricker, D.: Factored convolutional neural network for Amharic character image recognition. In: ICIP, pp. 2906–2910 (2019)
38. Belay, B.H., Habtegebrial, T., Liwicki, M., Belay, G., Stricker, D.: Amharic text image recognition: database, algorithm, and analysis. In: International Conference on Document Analysis and Recognition (ICDAR). IEEE CPS 2019
39. Greff, K., Srivastava, R.K., Koutnik, J., Steunebrink, B.R., Schmidhuber, J.: LSTM: a search space odyssey. IEEE Trans. Neural Netw. Learn. Syst. **28**, 2222–2232 (2017)
40. Graves, A., Fernandez, S., Gomez, F., Schmidhuber, J.: Connectionist temporal classification: labelling unsegmented sequence data with recurrent neural networks. In International Conference on Machine Learning (ICML), Pittsburgh, Pennsylvania, USA (2006)
41. Belay, B.H., Habtegebrial, T., Meshesha, M., Liwicki, M., Belay, G., Stricker, D.: Amharic OCR: an end-to-end learning. Appl. Sci. **10**(1117), 1–13 (2020)
42. Belay, B.H., Habtegebrial, T., Liwicki, M., Belay, G., Stricker, D.: Amharic text image. In: International Conference on Document Analysis and Recognition, Sydney (2019)
43. Breuel, T. M.: The OCRopus open source OCR system. In: Proceedings of SPIE-IS&T Electronic Imaging, SPIE, Kaiserslautern, Germany, vol. 6815, pp. 68150F (2008)
44. Ghosh, R., Vamshi, C., Kumar, P.: RNN based online handwritten word recognition in Devanagari and Bengali scripts using horizontal zoning. Pattern Recogn. **92**, 203–218 (2019)
45. Ciresan, D., Meier, U., Schmidhuber, J.: Multi-column deep neural networks for image classification. In Computer Vision and Pattern Recognition (CVPR), Providence, Rhode Island, USA (2012)
46. Ciresan, D., Meier, U., Masci, J., Gambardella, L. M., Schmidhuber, J.: Flexible, high performance convolutional neural networks for image classification. In: Twenty-Second International Joint Conference on Artificial Intelligence (2011)
47. Russakovsky, O., et al.: ImageNet large scale visual recognition challenge. Int. J. Comput. Vis. **115**, 211–252 (2014)
48. Lawrence, S., Giles, C.L., Tsoi, A.C., Back, A.D.: Face recognition: a convolutional neural network approach. Trans. Neural Netw. **8**(1), 98–113 (1997)
49. Shi, B., Bai, X., Yao, C.: An end-to-end trainable neural network for image-based sequence recognition and its application to scene text recognition. Trans. Pattern Anal. Mach. Intell. **39**(11), 2298–2304 (2017)
50. Ioffe, S., Szegedy, C..: Batch normalization: accelerating deep network training by reducing internal covariate shift. CoRR (2015)
51. Klambauer, G., Unterthiner, T., Mayr, A., Hochreiter, S.: Self-normalizing neural networks. In: Advances in Neural Information Processing Systems, vol. 30 (2017)
52. Abdurahman, F., Sisay, E., Fante, K.A.: AHWR Net: offline handwritten Amharic word recognition using convolutional recurrent neural network. SN Appl. Sci. **3**, 760 (2021). https://doi.org/10.1007/s42452-021-04742-x
53. Kingma, D.P., Ba, J.: Adam: a method for stochastic optimization (2017)
54. Belay, B.H., Habtegebrial, T., Meshesha, M., Liwicki, M., Belay, G., Stricker, D.: Learning by injection: attention embedded recurrent neural network for Amharic text-image recognition. Preprints (2020)
55. Belay, B.H., Habtegebrial, T., Liwicki, M., Belay, G., Stricker, D.: Blended attention-CTC for Amharic text-image recognition: sequence-to-sequence-learning. In: International Conference on Pattern Recognition Applications and Methods (ICPRAM) (2021)

Generating Pseudo-labels for Car Damage Segmentation Using Deep Spectral Method

Nonthapaht Taspan[1], Bukorree Madthing[1], Panumate Chetprayoon[2],
Thanatwit Angsarawanee[2], Kitsuchart Pasupa[1(✉)] [iD],
and Theerat Sakdejayont[2]

[1] School of Information Technology, King Mongkut's Institute of Technology
Ladkrabang, Bangkok 10520, Thailand
{63070228,63070234,kitsuchart}@it.kmitl.ac.th
[2] Kasikorn Labs, Nonthaburi 11120, Thailand
{panumate.c,thanatwit.a,theerat.s}@kbtg.tech

Abstract. Car damage segmentation, an integral part of vehicle damage assessment, involves identifying and classifying various types of damages from images of vehicles, thereby enhancing the efficiency and accuracy of assessment processes. This paper introduces an efficient approach for car damage assessment by combining pseudo-labeling and deep learning techniques. The method addresses the challenge of limited labeled data in car damage segmentation by leveraging unlabeled data. Pseudo-labels are generated using a deep spectral approach and refined through merge and flip-bit operations. Two models, i.e., Mask R-CNN and SegFormer, are trained using a combination of ground truth labels and pseudo-labels. Experimental evaluation of the CarDD dataset demonstrates the superior accuracy of our method, achieving improvements of 12.9% in instance segmentation and 18.8% in semantic segmentation when utilizing a 1/2 ground truth ratio. In addition to enhanced accuracy, our approach offers several benefits, including time savings, cost reductions, and the elimination of biases associated with human judgment. By enabling more precise and reliable identification of car damages, our method enhances the overall effectiveness of the assessment process. The integration of pseudo-labeling and deep learning techniques in car damage assessment holds significant potential for improving efficiency and accuracy in real-world scenarios.

Keywords: Pseudo-Label · Car Damage Segmentation · Deep Spectral Method

1 Introduction

Car damage assessment is vital in automotive, insurance, and accident investigations. It traditionally relied on manual inspection, which is slow, costly, and subjective. AI, deep learning, and computer vision advancements now offer automated and improved car damage assessment, saving time, reducing costs, and

© The Author(s), under exclusive license to Springer Nature Singapore Pte Ltd. 2024
B. Luo et al. (Eds.): ICONIP 2023, CCIS 1969, pp. 468–482, 2024.
https://doi.org/10.1007/978-981-99-8184-7_36

minimizing subjective judgments. The integration of AI into car damage assessment offers a range of advantages. Firstly, it saves time by automating image or video analysis, leading to prompt repairs and improved customer satisfaction. Secondly, AI reduces costs by minimizing manual labor and optimizing resource utilization. Additionally, AI provides consistent and objective evaluations, eliminating biases in human judgment.

Car damage assessment has widespread applications across various domains, such as the car insurance industry, where it plays a pivotal role in reducing the cost of insurance claims [11], determining insurance package costs [1], and streamlining the insurance inspection process [18]. Additionally, it is utilized in the car leasing industry [10] and car crash forensic analysis [3].

However, a major challenge in developing models for car damage assessment is the scarcity of labeled data. Obtaining accurate segmentation labels for damaged regions in car images is labor-intensive and costly. Furthermore, Zhang et al. [18] highlight that the cost of annotating image segmentation is five times higher compared to object detection. Similarly, Lee et al., [8] explain that the annotation time for segmenting one image is approximately four minutes, whereas bounding box annotation takes only 38 s. These findings underscore the challenging nature of segmentation annotation tasks and the additional time and effort required compared to other annotation methods.

To address this problem, pseudo-labeling [7] has emerged as a promising technique. It utilizes AI to generate pseudo-labels for unannotated data, expanding the labeled dataset and allowing the utilization of larger quantities of unlabeled data. However, traditional pseudo-labeling models often generate pseudo-labels that are not of sufficient quality for accurate car damage assessment. This can lead to suboptimal performance and reduced reliability of AI models in segmenting and assessing car damages.

Therefore, this paper aims to propose a method for generating pseudo-labels specifically designed for car damage segmentation. This paper makes the following key contributions:

- **Pseudo-Label Generation.** We propose a novel method to generate pseudo-labels specifically tailored for car damage segmentation.
- **Application to CarDD Dataset.** We demonstrate the effectiveness of our approach by applying it to the CarDD dataset [14], which comprises a comprehensive collection of car images with varying degrees of damage.
- **Comparative Evaluation.** We conduct evaluations to compare the similarity between the generated pseudo-labels and ground truth annotations. Additionally, we compare the performance of models trained with both types of labels, providing valuable insights into our approach's effectiveness.

2   Related Work

In this section, we discuss relevant research on car AI damage assessment, which serves as our main task, and pseudo-labeling, the technique we used to address real-world car damage data scarcity. Additionally, we introduce two semantic segmentation models that we utilized to evaluate the quality of our generated pseudo-labels.

2.1 Car AI Damage Assessment

Patil et al. [11] introduce an intuitive damage classification model that uses convolutional neural networks (CNNs) to classify various types of damage based on the affected parts, such as bumper dent, door dent, and glass shatter. Kyu and Woraratpanya [6] present a straightforward yet effective car damage classification model. The model performs an initial assessment to determine if a car is damaged or not, followed by a comprehensive analysis that includes damage severity classification into three levels (minor, moderate, severe) and damage location classification into three regions (front, rear, side). Shaikh et al. [12] propose a damage binary classification model utilized in a mobile application for car price prediction. Unlike other approaches, Zhang et al. [18] utilize videos instead of photos to interact with users to make the whole procedure as simple as possible by using a two-stage detection model.

2.2 Pseudo-Labeling

Lee et al. [7] introduces a straightforward approach for semi-supervised learning using pseudo-labeling. The method involves iteratively training deep neural networks with both labeled and unlabeled data, assigning pseudo-labels to the unlabeled samples based on model predictions. It achieves competitive results with a reduced number of labeled samples, making it a simple and efficient approach for leveraging unlabeled data in deep learning tasks. MixMatch [2] employs data augmentation, consistency regularization, and ensemble models with cross-entropy loss and mean squared error, while FixMatch [13] enhances this approach by simplifying the process, using pseudo-labeling for easy augmentations and incorporating hard augmentations for prediction. These methods are evaluated on the task of image classification. For the segmentation task, MaskContrast [4] is the leading unsupervised semantic segmentation method that utilizes saliency detection to identify object segments and learns pixel-wise embeddings through a contrastive objective. However, it heavily depends on a saliency network initialized with a pre-trained network and assumes that all foreground pixels in an image belong to the same object category, which may not hold true for images with multiple objects like car damage segmentation task.

Therefore, in our paper, we apply the Deep Spectral Method [9], which is suitable for car damage segmentation tasks, and introduce original contributions to the generation of pseudo-labels. We assess the generated pseudo-labels by training and evaluating the semantic segmentation models, which will be discussed in the following section.

2.3 Segmentation Tasks

In this section, we consider two types of segmentation tasks: instance segmentation and semantic segmentation. For the instance segmentation task, we select

Mask-RCNN as the base model because of its widespread use in the field and its reference in [14]. To further validate the effectiveness of our generated pseudo-labels, we also evaluate them in a semantic segmentation task, using SegFormer as the base model.

Mask-RCNN. Mask R-CNN [5] is a framework for object instance segmentation that combines object detection and image segmentation. It extends the Faster R-CNN architecture by adding a parallel branch for predicting pixel-level segmentation masks alongside bounding box recognition. This approach efficiently detects objects in an image while generating high-quality segmentation masks for each instance. Mask R-CNN is simple to train, adds minimal overhead to Faster R-CNN, and achieves real-time performance. It has achieved top results in various challenges, surpassing existing methods in tasks like instance segmentation, object detection, and person keypoint detection. The availability of the code has made it a solid baseline and has contributed to easing future research in instance-level recognition.

SegFormer. SegFormer [16] is a baseline semantic segmentation framework that effectively combines Transformers with lightweight multilayer perceptron (MLP) decoders. It tackles two significant challenges in semantic segmentation with innovative solutions. Firstly, SegFormer introduces a hierarchically structured Transformer encoder that overcomes the need for positional encoding while generating multiscale features. This eliminates the performance degradation that occurs when the testing resolution differs from the training resolution. Secondly, the framework utilizes a straightforward MLP decoder that leverages both local and global attention mechanisms to create powerful representations. This unique design enables efficient segmentation of Transformers, delivering impressive results.

3 Methodology

This research consists of two modules as follows: (i) Generating pseudo-labels from bounding box annotations using the deep spectral method: The extracted bounding boxes are fed into the deep spectral method (described in Sect. 3.1) to extract relevant features. These features are then subjected to the proposed post-processing techniques (described in Sects. 3.2 and 3.3) in order to generate categorical masks. (ii) Utilizing pseudo-labels to train a car damage segmentation model: Once the pseudo-labels are obtained, they are used to train the car damage segmentation models and are evaluated in various scenarios. The overview of this work is illustrated in Fig. 1.

3.1 Deep Spectral Methods

The Deep Spectral [9] is an approach that tackles the challenges of unsupervised localization and segmentation in computer vision. Unlike existing deep

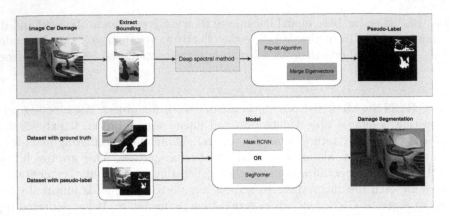

Fig. 1. System Overview: Generating Pseudo-Labels (Top) and Utilizing Pseudo-Labels for Training Car Damage Segmentation Model (Bottom)

learning-based methods, this approach draws inspiration from traditional spectral segmentation techniques. It begins by extracting dense features from an image using a neural network and constructing a semantic affinity matrix to capture relationships between image regions. By fusing this matrix with color information, the method preserves fine-grained details. The image is then decomposed into soft segments using eigenvectors derived from the Laplacian matrix of the affinity matrix. Notably, these eigenvectors already provide meaningful segments and can be utilized for various downstream tasks. For object localization, the eigenvector with the smallest nonzero eigenvalue accurately identifies the region of interest. Experimental results on complex datasets demonstrate that it outperforms existing techniques by a significant margin, delivering outstanding performance in unsupervised localization and segmentation, and offers versatility for complex image editing tasks like background removal and compositing.

3.2   Flip-Bit Algorithm

Flip-bit algorithm to refine the binary masks obtained from thresholding the selected eigenvectors. The algorithm aims to adjust the pixel values in the corners of the binary image, considering that these corners might not accurately represent the foreground (car damage) due to the nature of the pseudo-labeling generated from damage bounding boxes. This post-processing step aims to improve the accuracy and consistency of the segmentation results. The proposed algorithm is shown in Algorithm 1.

3.3   Merge Eigenvectors

In the merge eigenvectors process, we expand the information used for thresholding by considering the top three eigenvectors from the Deep Spectral algorithm. We believe that additional eigenvectors capture complementary features that

Algorithm 1. Flip-bit Algorithm

Require: I: Binary image
Require: *checksum*: Threshold value for flipping the pixel values
1: Initialize an empty list x
2: **for** each corner pixel in I **do**
3:     Add the pixel value to the list x
4: **end for**
5: $sum \leftarrow$ sum of the values in x
6: **if** $sum \geq checksum$ **then**
7:     Flip all pixel values in the binary image (from 1 to 0 and from 0 to 1)
8: **end if**

---

further improve the segmentation results, as shown in Fig. 2. We individually threshold each eigenvector and merge the resulting binary images using the logical *OR* operation. This allows us to create a fused representation of the car damage regions based on the collective information from multiple eigenvectors.

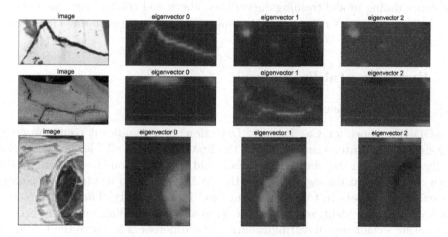

Fig. 2. The top three eigenvectors extracted using the Deep Spectral method capturing complementary features

Once we obtain the pseudo-labels that achieve the desired level of mean Intersection over Union (mIoU). These pseudo-labels—binary masks—will be converted to category masks. This can be done by replacing the bounding box annotation with its corresponding pseudo-label.

3.4   Utilizing Pseudo-Labels for Training

Once the pseudo-labels for car damage segmentation are generated using the proposed method, they can effectively train models. The integration of pseudo-labels aims to overcome the limitations posed by the scarcity of annotated data in car damage segmentation and improves the performance and reliability of the trained models.

A weight penalty loss is introduced to optimize the training process and account for potential noise or inconsistencies in the pseudo-labels,

$$\mathcal{L}_{weight} = \sum_{i=1}^{N} w_i \cdot \mathcal{L}_{seg}(q_j, z_j) + \sum_{j=1}^{M} \mathcal{L}_{seg}(p_i, y_i), \qquad (1)$$

where N represents the number of samples that possess pseudo-labels; M denotes the number of samples that possess ground truth labels; w_i refers to the weight assigned to each sample that possesses a pseudo-label; p_i and y_i represents the predicted and ground truth segmentation mask for the i-th ground truth sample, respectively; q_i and z_i represents the predicted and ground truth segmentation mask for the j-th pseudo-labeled sample, respectively; $\mathcal{L}_{seg}(q_j, z_j)$ and $\mathcal{L}_{seg}(p_i, y_i)$ represents the segmentation loss calculated for samples with pseudo-labels and for samples with ground truth labels, respectively. It can be seen that this loss assigns weights to the pseudo-labeled samples, thereby controlling their influence during model training.queryPlease check and confirm the edit made in the sentence "where N represents the number of samples ...pseudo-labeled labeled sample...". Amend if necessary.

4   Experimental Framework

4.1   CarDD Dataset

CarDD [14], short for Car Damage Detection, is a significant contribution to the field of computer vision, offering the first publicly available dataset specifically designed for car damage detection and segmentation. It comprises 4,000 high-resolution car damage images with over 9,000 labeled instances, surpassing all existing datasets in terms of volume and image quality. The dataset covers six damage types: dent, scratch, crack, glass shatter, tire flat, and lamp broken, presenting a challenge in distinguishing subtle differences between them. CarDD supports multiple tasks, including classification, object detection, instance segmentation, and salient object detection, and provides annotations for damage type, location, and magnitude, making it a practical resource for advancing the development of car damage assessment algorithms.

4.2   Experiment Setting

Generated Pseudo-Labels from Deep Spectral. In this section, the focus is on generating pseudo-labels from Deep Spectral for car damage segmentation. The process begins with the extraction of car damage bounding boxes from the images. These bounding boxes are then used as input to the deep spectral to generate binary masks corresponding to the car damage regions.

To refine and improve the binary masks generated by Deep Spectral, a proposed post-processing technique is employed. This technique involves two steps:

flip-bit and merge eigenvectors. In the flip-bit step, a checksum operation is performed on the binary mask, with the value changing from 1 to 4. This operation helps enhance the clarity and accuracy of the binary mask. Next, the merge eigenvectors step is executed, where the first three eigenvectors are merged using the *OR* operation. This merging process aims to capture and incorporate the underlying structure and patterns in the data, resulting in an improved binary mask representation. After applying the proposed post-processing technique, the refined binary mask is assigned as the pseudo-label. To evaluate the quality and performance of these pseudo-labels, they are compared against the ground truth masks using the mIoU:

$$mIoU = \frac{1}{N} \sum_{i=1}^{N} \frac{TP_i}{TP_i + FP_i + FN_i}, \tag{2}$$

where N is the total number of classes, TP_i is the true positive count for class i, FP_i is the false positive count for class i, and FN_i is the false negative count for class i.

Training Segmentation with Pseudo-Labels. Here, we utilize a set of pseudo-labels that achieve the best performance for each class in our previous experiment. We focus on investigating the impact of varying the ratio between ground truth labels and pseudo-labels in the training set for car damage segmentation. Specifically, we set different ground truth ratios, including $1/2$, $1/4$, $1/8$, $1/16$, and 1 (no pseudo-labels) [15,17] to understand how the availability of ground truth annotations influences the model's performance and evaluate the model with the test set.

As described in the previous section, the weight penalty loss is employed in the training process. When a training sample is labeled with a pseudo-label, it is assigned a weight penalty loss (w_i) of 0.5. This weight penalty loss serves as a regularization mechanism to control the influence of the pseudo-labels on the optimization process. By assigning a lower weight penalty to the pseudo-labeled samples, the model is encouraged to prioritize the ground truth annotations in its learning.

Mask-RCNN. In our experimental setup, we utilize the Mask-RCNN model with a ResNet-50 backbone for training the car damage segmentation task. The Mask-RCNN architecture is a widely used framework for instance segmentation, capable of simultaneously detecting and segmenting objects within an image.

The overall loss function of the Mask-RCNN model includes three components: the bounding box regression loss ($\mathcal{L}_{bbox}$), the classification loss ($\mathcal{L}_{cls}$), and the mask segmentation loss for ground truth ($\mathcal{L}_{mask}^{(y)}$), and the mask segmentation loss for pseudo-label ($\mathcal{L}_{mask}^{(z)}$). The total loss is calculated as the sum of these individual losses:

$$\mathcal{L}_{Mask-RCNN} = \mathcal{L}_{bbox} + \mathcal{L}_{cls} + \mathcal{L}_{mask}^{(y)} + w \cdot \mathcal{L}_{mask}^{(z)}. \tag{3}$$

To evaluate the performance of the Mask-RCNN model, we employ Box Average Precision (Box AP) and Mask Average Precision (Mask AP) metrics. This metric measures the accuracy of both bounding box and mask predictions by considering the IoU between the predicted masks and the ground truth masks at various IoU thresholds. A higher Mask AP score indicates better segmentation performance.

During training, we set the hyperparameters as follows: the initial learning rate is set to 1e−4, the batch size is set to 1, and we train the model for a total of 50 epochs. We utilize the AdamW optimizer with default parameters for optimization. The optimal epoch is determined based on the criterion of achieving the lowest validation loss.

SegFormer. In our experimental setup, we utilize the SegFormer-b2 model for training the car damage segmentation task in the semantic segmentation framework. It is based on the transformer architecture for semantic segmentation, which focuses on capturing contextual information and global dependencies within an image.

Here, we employed the weighted penalty loss that incorporates pseudo-labels into the training process:

$$\mathcal{L}_{SegFormer} = \mathcal{L}^{(y)}_{mask} + w \cdot \mathcal{L}^{(z)}_{mask} \tag{4}$$

To assess the performance of the SegFormer model, we utilize the mIoU metric. The hyperparameters for training the SegFormer model with the weight penalty loss are set as follows: the learning rate is initialized to 5e−5, the batch size is set to 1, and the number of training epochs is 60. We use the AdamW optimizer with default parameters for optimization. Early stopping is applied with a patience of 10 epochs, monitoring the maximum mIoU on the validation set.

5   Results and Discussion

5.1   Evaluate Generated Pseudo-Labels from Deep Spectral

In this section, we evaluate the effectiveness of the proposed method for generating pseudo-labels in car damage segmentation. We analyze the quality of the generated pseudo-labels by comparing them with the ground truth annotations. The evaluation is primarily based on the IoU metric, which measures the overlap between the pseudo-labels and the corresponding ground truth masks.

The results presented in Table 1 compare the performance of the proposed method with the baseline conventional deep spectral method. The evaluation is based on the mIoU between the generated pseudo-labels and the ground truth segmentation masks for different car damage types.

According to Table 1, the baseline method, utilizing the first eigenvector, achieves mIoU values ranging from 0.171 for the dent to 0.462 for the tire flat.

Table 1. Comparison of mIoU performance between bounding box pseudo-labels and ground truth annotations from the CarDD Dataset using the deep spectral baseline method and our proposed method. bold—best

Merge Eigenvector	flip-bit checksum value	dent	scratch	crack	lamp broken	glass shatter	tire flat
Baseline	No	0.171	0.220	0.444	0.448	0.377	0.462
1st	1	0.611	0.546	0.078	0.627	0.592	0.482
	2	0.527	0.518	0.266	0.630	0.592	0.478
	3	0.311	0.422	0.471	0.613	0.586	0.477
	4	0.202	0.289	**0.472**	0.580	0.538	0.472
1st OR 2nd	No	0.369	0.442	0.376	0.567	0.560	0.510
	1	0.655	0.584	0.214	0.718	0.757	0.840
	2	0.640	0.585	0.274	0.760	0.784	0.843
	3	0.537	0.550	0.367	0.759	0.775	0.845
	4	0.422	0.493	0.386	0.735	0.731	0.837
1st OR 2nd OR 3rd	No	0.486	0.501	0.351	0.641	0.674	0.601
	1	**0.669**	0.589	0.256	0.737	0.801	0.828
	2	0.667	**0.597**	0.284	0.779	0.820	0.834
	3	0.618	0.580	0.321	**0.784**	**0.821**	**0.846**
	4	0.536	0.539	0.351	0.759	0.790	0.842

When we apply the flip-bit algorithm to the first eigenvector, we observe improvements in mIoU values for all categories except for cracks. In the case of cracks, there is a significant drop in mIoU, decreasing from 0.444 to 0.078 for a flip-bit checksum of 1 and to 0.266 for a flip-bit checksum of 2. These results suggest challenges in accurately segmenting cracks using the generated pseudo-labels.

By merging two eigenvectors, we noticeably improve the performance of the model in 5 out of 6 classes, except for the class with cracks. When we apply the flip-bit algorithm, we further observe improvements in mIoU values for all classes except for cracks, specifically on flip-bit checksums of 1 to 3. These results indicate that incorporating additional eigenvectors and refining the binary masks through the flip-bit algorithm contribute to enhancing the overall segmentation performance.

The highest mIoU values are obtained when merging three eigenvectors and applying the flip-bit algorithm, resulting in improved performance across most categories. However, in the case of cracks, the best performance is achieved with only one eigenvector and a flip-bit checksum of 4. This finding suggests that utilizing more eigenvectors captures complementary features and generally improves the accuracy of segmentation.

5.2 Evaluate by Training the Model

We conduct a performance evaluation of two models for car damage segmentation using a semi-supervised approach. The first model, Mask R-CNN, is trained on the CarDD dataset specifically for instance segmentation. The second model, SegFormer, is also trained on the same dataset but focuses on semantic segmentation. To harness the advantages of semi-supervised learning, we utilize the pseudo-labels generated from the previous task. We select the pseudo-label generated for each class, based on the optimal mIoU of each class obtained from the settings mentioned in Sect. 5.1, as the ground truth to train the models. Examples of pseudo-labels are shown in Fig. 3.

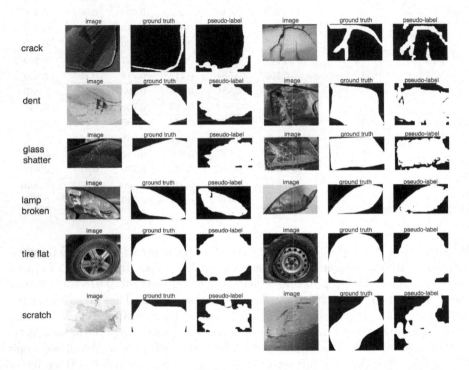

Fig. 3. Examples of the pseudo-labels generated by the proposed method

Mask-RCNN. According to Table 2, when we use only samples with the ground truth labels (y), the Mask AP gradually improves as the ground truth ratio increases. This indicates the importance of having a larger number of ground truth labels for training, as it provides more accurate and reliable supervision.

When we incorporate samples with pseudo-labels generated from the deep spectral method (z_{DS}), the performance is lower compared to using only the ground truth labels. This observation suggests that the pseudo-labels may introduce some noise or inconsistencies that adversely affect the model's ability to segment car damage instances accurately.

However, when we utilize the pseudo-labels generated from the deep spectral method with merge and flip-bit operations (z_{DSMF}), the performance improves,

Table 2. Comparison of the usage of ground truth and pseudo-labels on Mask Average Precision (Mask AP) at IoU $= 0.5$ by varying different sizes of ground truth ratios in the training set with Mask R-CNN (number in parenthesis indicates the number of ground truth labels). The labels considered include ground truth labels (y), pseudo-labels from deep spectral (z_{DS}), and pseudo-labels from deep spectral with merge and flip-bit operation (z_{DSMF}). bold—best

Label	Ground Truth Ratio				
	1/16 (176)	1/8 (352)	1/4 (704)	1/2 (1408)	1 (2816)
y	0.335	0.422	0.464	0.487	0.584
$y+z_{DS}$	0.170	0.319	0.446	0.550	-
$y+z_{DSMF}$	0.510	0.589	0.550	**0.621**	-

especially at a ground truth ratio of 1/2 (1408 labels). This indicates that the additional processing steps applied to the pseudo-labels help refine and enhance their quality, leading to better segmentation results.

Overall, the results demonstrate that incorporating pseudo-labels, especially with additional post-processing steps, can complement the limited availability of ground truth labels and improve the performance of the Mask-RCNN model in car damage instance segmentation tasks.

In order to make a direct comparison with the CarDD paper, we evaluate the results obtained from training the Mask R-CNN model. The table presents the evaluation metrics, including Box AP and Mask AP, at different ground truth ratios and corresponding pseudo-label ratios, as illustrated in Table 3. When the ground truth ratio is 1/2, meaning that only half of the training data consists of ground truth annotations, the model still achieves a high Mask AP of 0.621 and a respectable Box AP of 0.651. Even in scenarios where the pseudo-label ratio is high, such as 15/16, the model maintains a Mask AP of 0.510 and a Box AP of 0.584. The results demonstrate that the incorporation of pseudo-labels does not lead to a significant degradation in the model's performance. Regardless of the ground truth ratios, which vary from 1/16 to 1, the model consistently achieves competitive results in terms of both Box AP and Mask AP. This indicates that the pseudo-labels provide valuable information that complements the ground truth annotations, resulting in accurate and precise segmentation predictions.

SegFormer. Table 4 illustrates the outcomes of the semantic segmentation task performed using SegFormer, where a combination of ground truth and pseudo-labels at different ratios is utilized. The performance of SegFormer demonstrates a consistent trend that aligns with the findings from Mask-RCNN in the instance segmentation task. This similarity indicates the effectiveness of the pseudo-labels generated by our proposed algorithm. It is worth noting that employing 100% ground truth data still yields the best performance. Nonetheless, the results underscore the valuable contribution of the pseudo-labels in complementing the ground truth annotations and achieving effective semantic segmentation outcomes.

Table 3. Performance of the model trained on a combination of ground truth and pseudo-labels generated from our proposed technique, with varying proportions, evaluated with the test set. The number in parentheses indicates the number of samples used in each combination.

Ground Truth	Psuedo-label	Box AP$_{@50}$	Mask AP$_{@50}$	Box AP$_{@75}$	Mask AP$_{@75}$
1/16 (176)	15/16 (2640)	0.584	0.510	0.387	0.341
1/8 (352)	7/8 (2464)	0.645	0.589	0.447	0.403
1/4 (704)	3/4 (2112)	0.594	0.550	0.381	0.354
1/2 (1408)	1/2 (1408)	0.651	0.621	0.435	0.416
1 (2816)	0 (0)	0.613	0.584	0.419	0.431

Table 4. Comparison of the usage of ground truth and pseudo-labels on Mean Intersect Over Union (mIoU) by varying different sizes of ground truth ratios in the training set with SegFormer (number in parenthesis indicates the number of ground truth labels). The labels considered include ground truth labels (y), pseudo-labels from deep spectral (z_{DS}), and pseudo-labels from deep spectral with merge and flip-bit operation (z_{DSMF}). bold—best

Label	Ground Truth Ratio				
	1/16 (176)	1/8 (352)	1/4 (704)	1/2 (1408)	1 (2816)
y	0.545	0.536	0.600	0.625	**0.693**
$y+z_{DS}$	0.346	0.389	0.402	0.558	-
$y+z_{DSMF}$	0.612	0.598	0.621	0.663	-

We further investigate the mIoU achieved by SegFormer when using a combination of ground truth and pseudo-labels at different ratios, specifically for different damage categories in the semantic segmentation task, as shown in Table 5. This analysis allows us to understand the performance of SegFormer in accurately segmenting each damage category and assess the impact of the pseudo-labels on the segmentation quality.

The results show an improvement in performance, with an increase in mIoU as the ratio of ground truth data increases. Comparing the results between the ground truth ratios of 1 and 1/4, using 100% ground truth data results in mIoU of 0.693, while using only 25% ground truth data leads to a performance drop to 0.621, representing a 10.4% worsening in performance. It is important to note that this drop in performance occurs despite the reduction in the number of ground truth samples by 75.0%. In addition, the observed 10.4% drop in performance is primarily driven by a significant reduction in the crack class, which decreases from 0.578 in the case of using 100% ground truth data to 0.152 when only 25% ground truth data is used.

Table 5. Performance comparison of SegFormer with varying ground truth (y) ratios and the inclusion of pseudo-labels (z_{DSMF}) for different damage types in the semantic segmentation task.

Label		mIoU						
y	z_{DSMF}	Total	dent	scratch	crack	lamp broken	glass shatter	tire flat
1/16 (176)	15/16 (2640)	0.612	0.784	0.689	0.150	0.950	0.792	0.919
1/8 (352)	7/8 (2464)	0.598	0.670	0.733	0.049	0.976	0.808	0.953
1/4 (704)	3/4 (2112)	0.621	0.712	0.721	0.152	0.953	0.889	0.920
1/2 (1408)	1/2 (1408)	0.663	0.771	0.735	0.446	0.924	0.822	0.946
1 (2816)	0 (0)	0.693	0.733	0.701	0.578	0.984	0.894	0.961

6   Conclusion

In conclusion, this paper addresses the need for efficient car damage assessment using AI technologies and highlights the significant benefits it offers, including time savings, cost reductions, and the elimination of biases associated with human judgment. Furthermore, by leveraging the power of pseudo-labeling, we successfully overcome the challenge of limited labeled data, enabling us to harness the vast potential of unlabeled data for training AI models. Our proposed method for generating pseudo-labels in car damage segmentation outperforms the baseline approach on the CarDD dataset, demonstrating its superior accuracy. With an improvement of 12.9% in the instance segmentation task and 18.8% in the semantic segmentation task when using a 1/2 ground truth ratio, our method enables more precise and reliable identification of car damages, enhancing the overall effectiveness of the assessment process.

In future work, we aim to extend the application of pseudo-labeling techniques in car damage segmentation by exploring real-world unlabeled datasets, thereby harnessing the potential of unlabeled data for training AI models in practical scenarios. Additionally, we will validate and generalize our approach by exploring diverse car damage segmentation datasets beyond CarDD, considering variations in car models, damage types, image qualities, and environmental conditions.

References

1. Balasubramanian, R., Libarikian, A., McElhaney, D.: Insurance 2030–the impact of AI on the future of insurance. McKinsey & Company (2018)
2. Berthelot, D., Carlini, N., Goodfellow, I.J., Papernot, N., Oliver, A., Raffel, C.: MixMatch: a holistic approach to semi-supervised learning. In: Annual Conference on Neural Information Processing Systems 2019, NeurIPS 2019, pp. 5050–5060 (2019)
3. Chen, Q., et al.: A deep neural network inverse solution to recover pre-crash impact data of car collisions. Transp. Res. Part C Emerg. Technol. **126**, 103009 (2021)

4. Gansbeke, W.V., Vandenhende, S., Georgoulis, S., Gool, L.V.: Unsupervised semantic segmentation by contrasting object mask proposals. In: 2021 IEEE/CVF International Conference on Computer Vision, ICCV 2021, pp. 10032–10042 (2021)
5. He, K., Gkioxari, G., Dollár, P., Girshick, R.B.: Mask R-CNN. In: IEEE International Conference on Computer Vision, ICCV 2017, pp. 2980–2988 (2017)
6. Kyu, P.M., Woraratpanya, K.: Car damage detection and classification. In: The 11th International Conference on Advances in Information Technology, IAIT 2020 (2020)
7. Lee, D.H.: Pseudo-label: the simple and efficient semi-supervised learning method for deep neural networks. In: Workshop on Challenges in Representation Learning, International Conference on Machine Learning Workshop 2013, pp. 1–6 (2013)
8. Lee, J., Yi, J., Shin, C., Yoon, S.: BBAM: bounding box attribution map for weakly supervised semantic and instance segmentation. In: IEEE Conference on Computer Vision and Pattern Recognition, CVPR 2021, pp. 2643–2652 (2021)
9. Melas-Kyriazi, L., Rupprecht, C., Laina, I., Vedaldi, A.: Deep spectral methods: a surprisingly strong baseline for unsupervised semantic segmentation and localization. In: IEEE/CVF Conference on Computer Vision and Pattern Recognition, CVPR 2022, pp. 8354–8365 (2022)
10. Pal, N., Arora, P., Kohli, P., Sundararaman, D., Palakurthy, S.S.: How much is my car worth? A methodology for predicting used cars' prices using random forest. In: The 2018 Future of Information and Communication Conference, FICC 2018, pp. 413–422 (2019)
11. Patil, K., Kulkarni, M., Sriraman, A., Karande, S.: Deep learning based car damage classification. In: 16th IEEE International Conference on Machine Learning and Applications, ICMLA 2017, pp. 50–54 (2017)
12. Shaikh, M.K., Zaki, H., Tahir, M., Khan, M.A., Siddiqui, O.A., Rahim, I.U.: The framework of car price prediction and damage detection technique. Pak. J. Eng. Technol. **5**(4), 52–59 (2022)
13. Sohn, K., et al.: FixMatch: simplifying semi-supervised learning with consistency and confidence. In: Annual Conference on Neural Information Processing Systems 2020, NeurIPS 2020 (2020)
14. Wang, X., Li, W., Wu, Z.: CarDD: a new dataset for vision-based car damage detection. IEEE Trans. Intell. Transp. Syst. **24**(7), 7202–7214 (2023)
15. Wang, Y., et al.: Semi-supervised semantic segmentation using unreliable pseudo-labels. In: IEEE/CVF Conference on Computer Vision and Pattern Recognition, CVPR 2022, pp. 4238–4247 (2022)
16. Xie, E., Wang, W., Yu, Z., Anandkumar, A., Álvarez, J.M., Luo, P.: SegFormer: simple and efficient design for semantic segmentation with transformers. In: Annual Conference on Neural Information Processing Systems 2021, NeurIPS 2021, pp. 12077–12090 (2021)
17. Yang, L., Qi, L., Feng, L., Zhang, W., Shi, Y.: Revisiting weak-to-strong consistency in semi-supervised semantic segmentation. In: IEEE/CVF Conference on Computer Vision and Pattern Recognition, CVPR 2023, pp. 7236–7246 (2023)
18. Zhang, W., et al.: Automatic car damage assessment system: reading and understanding videos as professional insurance inspectors. In: The Thirty-Fourth AAAI Conference on Artificial Intelligence, AAAI 2020, pp. 13646–13647 (2020)

Two-Stage Graph Convolutional Networks for Relation Extraction

Zhiqiang Wang[1,2(✉)], Yiping Yang[1], and Junjie Ma[3]

[1] Institute of Automation, Chinese Academy of Sciences, Beijing, China
{wangzhiqiang2019,yiping.yang}@ia.ac.cn
[2] University of Chinese Academy of Sciences, Beijing, China
[3] Department of Computer Science and Technology, Tsinghua University, Beijing, China
junjiema@tsinghua.edu.cn

Abstract. The purpose of relation extraction is to extract semantic relationships between entities in sentences, which can be seen as a classification task. In recent years, the use of graph neural networks to handle relation extraction tasks has become increasingly popular. However, most existing graph-based methods have the following problems: 1) they cannot fully utilize dependency relation information; 2) there is no consistent criterion for pruning dependency trees. To address these issues, we propose a two-stage graph convolutional networks for relation extraction. In the first stage of the model, the node representation, dependency relation type representation and dependency type weight jointly generate new node representations, fully utilizing the dependency relation information. In the second stage, with the help of the adjacency matrix derived from the dependency tree, the graph convolution operation is performed. In this way, the model can automatically complete the pruning operation. We evaluated our proposed method on two public datasets, and the results show that our model outperforms previous studies in terms of F1 score and achieves the best performance. Further ablation experiments also confirm the effectiveness of each component in our proposed model.

Keywords: Relation extraction · Knowledge graph · Graph neural networks

1 Introduction

The goal of relation extraction (RE) is to extract entities from given text and identify the semantic relationships between them, forming relation triples in the form of (subject, relation, object). It plays a significant role in many downstream tasks in the field of natural language processing (NLP), including information extraction [6], knowledge graph construction [14], user profiling [16], sentiment analysis [2], automatic question answering [10] and so on. Therefore, it has gained tremendous interest in recent years.

In recent years, several studies have been dedicated to addressing the problem of relation triple extraction. Looking at the timeline, people have mainly gone through three stages in solving the relation extraction task. The first stage is feature-based methods [6]. In these methods, a large number of words, grammatical, and semantic features are combined with machine learning algorithms. However, feature selection becomes crucial in these methods, and different feature combinations can have a significant impact on the model's performance. The second stage is kernel-based methods [18]. These methods specify a similarity measure between two given samples. Unfortunately, the design of the kernel function has a significant influence on the performance of these methods. More recently, neural network-based methods have achieved great success in relation extraction tasks [9], which has attracted increasing attention from researchers. Many scholars have shifted from traditional research approaches to deep learning approaches. In these methods, models such as graph convolutional networks have played an important role and have been widely developed in recent years. For example, Xu et al. [15] proposed a method called the shortest dependency path, which integrates two segmented LSTM models to handle relation extraction tasks. Guo et al. [4] directly adopt graph convolutional neural networks for relation extraction. Tian et al. [13] have further considered the information on the edges between nodes. In the process of constructing the adjacency matrix $\mathbf{A}$, they have employed a multi-head attention mechanism. The purpose of this operation is to convert the 0 and 1 in the adjacency matrix into real numbers between 0 and 1.

However, all these methods share some common shortcomings: 1) When constructing the adjacency matrix based on dependency relations, they only consider whether there is a dependency relationship between two words but do not consider the specific type of dependency relationship. 2) They do not consider the varying impact of different types of dependency relationships. 3) Different pruning operations are performed at different levels during the use of dependency relationships, but there is no unified standard for these pruning operations.

To address the aforementioned issues, in this paper, we propose a two-stage graph convolutional neural network (TS-GCN) for relation extraction. In our approach, given an input sentence, we first perform dependency analysis to obtain the dependency tree. While constructing the adjacency matrix $\mathbf{A}$, we also construct a dependency relation type matrix, denoted as $\mathbf{D}$, which labels the dependency relation types between words. To track the impact of different dependency relations on the semantics of the sentence, we construct a corresponding dependency type weight matrix, denoted as $\mathbf{W}$. In the first stage of the model, the node representation, dependency relation type representation and dependency type weight jointly generate new node representations, fully utilizing the dependency relation type information. In the second stage of the model, with the help of the adjacency matrix $\mathbf{A}$, the graph convolution operation is performed. In this way, the model automatically completes the pruning operation by the joint action of the dependency relation type matrix $\mathbf{D}$ and the dependency type weight matrix $\mathbf{W}$. We conduct experiments on two public datasets,

SemEval 2010 Task-8 and PubMed, and the results show that our proposed model achieves improvements of 0.71% and 10.8% on F1-scores, respectively, compared to the current state-of-the-art models. Further experiments demonstrate the effectiveness of our proposed method.

The contributions of this paper are as follows:

- We incorporate dependency relation types and dependency type weights, making full use of the information provided by dependency syntactic analysis.
- We divide the operations of the graph neural network into two stages, enabling more comprehensive information interaction between nodes.
- We conduct extensive experiments on public datasets, achieving state-of-the-art results and demonstrating the effectiveness of our proposed method.

This paper is structured as follows. In Sect. 2, we briefly review the background of this work. In the Sect. 3, we provide a detailed explanation of the proposed method. The experimental settings and results are described in Sect. 4. The last Sect. 5 concludes the paper.

2   Related Work

AGGCN is an earlier work that proposes graph convolutional networks for relation extraction [4]. However, it does not distinguish between the types of dependencies, that is, all dependencies are treated equally, which makes the performance of the model not high. Tian et al. [13] pay attention to the type of dependency. When calculating the adjacency matrix $\mathbf{A}$, the traditional $\mathbf{A}$ is replaced by the result deduced by the type and two dependency words, at the same time, a generalized attention mechanism is used. In the generated dependency matrix, four different pruning strategies are verified, they are 'Local', 'Global', 'Full' and 'Local+Global'. But when calculating 'Global', The shortest path strategy is adopted, which is to find the shortest dependent path between two entities. It should be noted that when solving the shortest dependency path, whether using BFS algorithm or DFS algorithm, both are recursive strategies, which can cause significant computational cost, especially when sentence length increases, the computational efficiency will significantly decrease. Xu et al. [15] also adopt a combination of shortest path and LSTM model, they believe that when the shortest path between two entities is found, the skeleton of a sentence is obtained. As long as the backbone of the sentence is extracted, it is crucial for sentence analysis and entity relation extraction. The skeleton of a sentence is divided into two parts by the root word, and the authors use an LSTM model for each part. This is the first time using the shortest path. But similarly, for long sentence, to find the shortest path between two entities is a costly task. Especially, sometimes entities are not just individual words. Zhang et al. [19] argue that pruning the dependency tree too harshly can lead to the deletion of negative meanings in sentence and incomplete expression of information, which damages the robustness of the model. Therefore, they propose three pruning strategies. Yu et al. [17] also suggest that manual pruning strategies may lead to the omission of useful information. For each word in an entity, other method [1] first extract two types of dependency information related to it, considering two types of dependency information: one is "intra entity" dependency, and the other is "cross entity" dependency, which is the dependency path between two entities. The shortest dependency path is also used here.

Our work is based on previous research. Unlike previous work, our method considers both the words in the sentence and the types of dependency relationships between words. In order to maximize the preservation of the semantic information contained in the sentence, we did not adopt pruning operations.

3 Our Proposed Approach

In this section, we will first introduce the task definition, followed by a detailed description of the proposed TS-GCN model. The overall architecture of our model is illustrated in Fig. 1.

3.1 Task Definition

Given a pre-defined relation set $\mathcal{R} = \{r_1, r_2, \cdots, r_m\}$ contains m types, and a sentence $s = \{x_1, x_2, \cdots, x_n\}$ composed of n words. Note that the entities in the sentence have been labeled. The goal of the task is to identify the semantic relationship between the two entities and form a triple (subject, relation, object). An entity consists of one or more consecutive words in the sentence.

3.2 BERT Encoder

To obtain the context-aware representation of each word in the input sentence $s = \{x_1, x_2, \cdots, x_n\}$. We choose BERT [3] as our sentence encoder. This process can be expressed using the following formula:

$$\{\mathbf{h}_1, \mathbf{h}_2, \cdots, \mathbf{h}_n\} = \text{BERT}(\{x_1, x_2, \cdots, x_n\} \tag{1}$$

where $\mathbf{h}_i \in \mathbb{R}^d$ is the contextual representation of the ith word in sentenc s and d is the embedding size.

3.3 Two-Stage GCN

Standard Graph Convolutional Networks. The standard graph convolutional networks can be expressed by the following formula:

$$\mathbf{h}_i^l = \sigma \left(\sum_{j=1}^{n} a_{i,j} \left(\mathbf{W}^l \mathbf{h}_j^{l-1} + \mathbf{b}^l \right) \right) \tag{2}$$

where $\mathbf{h}_i^l$ and $\mathbf{h}_i^{l-1}$ denote the contextual representations of the ith word in the lth and $(l-1)^{th}$ layers, respectively, $\mathbf{W}^l$ and $\mathbf{b}^l$ are trainable matrix and bias, $a_{i,j}$ denote the entry at the ith row and j^{th} column of the adjacency matrix $\mathbf{A}$ and σ denotes activation function, such as ReLU. Graph convolutional networks can be seen as nodes in the lth layer gather information from their neighboring nodes in the $(l-1)^{th}$ layer to represent themselves.

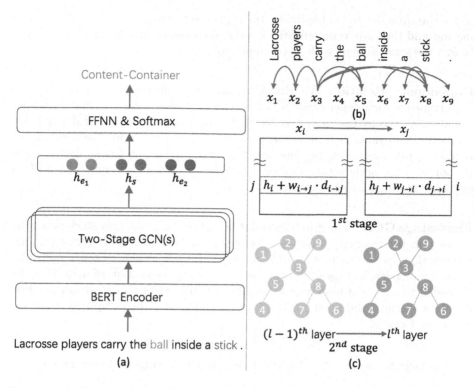

Fig. 1. Overview of our model architecture illustrated with an example sentence (the blue and red fonts indicate two entities). (a) represents the overall framework of the model, (b) denotes the dependency analysis results of the sentence, we omitted the dependency relations here, and (c) refers to the details of the two-stage graph convolutional networks. (Color figure online)

Adjacency Matrix. Given a sentence $s = \{x_1, x_2, \cdots, x_n\}$ with n words, we first use the existing natural language processing toolkit CoreNLP[1] to perform dependency parsing, which identifies the dependency relationships between words, as shown in Fig. 1(b). Based on the generated dependency tree, we can form an adjacency matrix $\mathbf{A} = (a_{i,j})_{n \times n}$, where the entry $a_{i,j} = a_{j,i} = 1$ if there is a relationship between words x_i and x_j; otherwise, $a_{i,j} = a_{j,i} = 0$. Note that the matrix $\mathbf{A}$ is symmetric.

Dependency Relation Type Matrix. Because the adjacency matrix $\mathbf{A}$ can only reflect whether there is an dependency relationship between two words but cannot capture other information such as dependency relation type, so we construct a corresponding matrix $\mathbf{D}$ for representing dependency relation types. In this matrix, each entry is an dependency relation type, such as 'nsubj', 'amod'

---

[1] https://stanfordnlp.github.io/CoreNLP/.

and so on. In order to enable interaction between the dependency relation information and the node representations, we embed each dependency relation type
to $\mathbb{R}^d$, the same dimens with word embedding.

Dependency Relation Weight Matrix. To explicitly capture the varying
importance of each dependency relation type in representing information within
a sentence, we construct a matrix $\mathbf{W}$ that corresponds to $\mathbf{D}$, called dependency
relation weight matrix. It serves as a weighting mechanism for the dependency
relation types, controlling the interaction between the dependency relation information and the node representations. Similar to $\mathbf{D}$, we also embed each of its
elements to d-dimensional space.

First-Stage GCN. The graph convolution operation in the first stage primarily
focuses on the interaction between node information and dependency relation
information, as shown in Fig. 1(c). Since each word in the sentence can have
dependency relationships with other words, the representation of node i can be
extended to an $n \times d$ matrix. Specifically, if there exists a dependency relation
d between node i and node j, the jth row of node i can be computed using the
following formula:

$$\tilde{\mathbf{h}}_{i,j}^{l-1} = \mathbf{h}_i^{l-1} + \mathbf{w}_{i \rightarrow j} \odot \mathbf{d}_{i \rightarrow j} \tag{3}$$

Similarly, for node j, its ith row can be computed using the following formula:

$$\tilde{\mathbf{h}}_{j,i}^{l-1} = \mathbf{h}_j^{l-1} + \mathbf{w}_{j \rightarrow i} \odot \mathbf{d}_{j \rightarrow i} \tag{4}$$

where $\mathbf{h}_i^{l-1}$, $\mathbf{h}_j^{l-1}$ denote the ith and j^{th} node representation in the $(l-1)^{th}$ layer,
respectively, $\mathbf{w}_{i \rightarrow j}$, $\mathbf{w}_{j \rightarrow i}$, $\mathbf{d}_{i \rightarrow j}$ and $\mathbf{d}_{j \rightarrow i}$ denote the embedding of relation d and
weight w, respectively, $\odot$ denotes the element-wise multiplication.

Through this operation, the model fully leverages both node information and
dependency relation information.

Second-Stage GCN. The second stage is primarily dedicated to performing
convolutional operation, as illustrated in Fig. 1(c). This process can be expressed
using the following formula:

$$\mathbf{h}_i^{(l)} = \sigma \left(\sum_{j=1}^{n} a_{i,j} \left(\mathbf{W}^{(l)} \tilde{\mathbf{h}}_j^{(l-1)} + \mathbf{b}^{(l)} \right) \right) \tag{5}$$

the relevant parameters are similar to Eq. 2.

3.4   Relation Classification

Entities and Sentence Representation. To obtain the entity representation,
we first obtain the span of two entities in the sentence, and then we use the

maximum pooling operation to obtain the vector representation of the entity, as shown below:

$$\mathbf{h}_{e_k} = \text{MaxPooling}\left(\{\mathbf{h}_i \mid x_i \in e_k\}\right) \tag{6}$$

where e_k is the kth entity and $k \in \{1, 2\}$.

In the same way, we can obtain the representation of the sentence, as shown below:

$$\mathbf{h}_{sent} = \textbf{MaxPooling}\left(\{\mathbf{h}_i \mid x_i \notin e_k\}\right) \tag{7}$$

Relation Extraction. Following [19], we concatenate the sentence representation and entity representations to get the final representation for classification.

$$\mathbf{h}_o = \mathbf{h}_{e_1} \oplus \mathbf{h}_{sent} \oplus \mathbf{h}_{e_2} \tag{8}$$

$$\mathbf{y} = \mathbf{W}_o \mathbf{h}_o + \mathbf{b}_o \tag{9}$$

where $\mathbf{W}_o \in \mathbb{R}^{(3 \times d) \times |\mathcal{R}|}$ and $\mathbf{b}_o \in \mathbb{R}^{|\mathcal{R}|}$ are trainable weight matrix and bias vector, respectively.

In the relation extraction stage, we use feed-forward neural network and then use softmax operation to select the label corresponding to the value with the largest output probability as the predicted relation, as shown below:

$$\hat{r} = \arg\max_r \frac{\exp\left(\mathbf{y}^r\right)}{\sum_{i=1}^{|\mathcal{R}|} \exp\left(\mathbf{y}^r\right)} \tag{10}$$

where $|\mathcal{R}|$ denotes the size of relation set and $\mathbf{y}^r$ denotes the rth element of $\mathbf{y}$, the result $\hat{r}$ indicates that these two entities have the rth relation in the relation set $\mathcal{R}$.

3.5 Loss Function

In the training phase, we optimize the following objective

$$J(\theta) = -\frac{1}{N} \sum_{i=1}^{N} \sum_{r=1}^{|\mathcal{R}|} \mathbf{p}_i^r \log \mathbf{y}_i^r \tag{11}$$

where N denotes the number of training examples, $|\mathcal{R}|$ denotes the size of relation set, $\mathbf{p}$ denotes the golden distribution, $\mathbf{y}$ indicates the predicted distribution, $\mathbf{p}^r$ and $\mathbf{y}^r$ represent the rth element in $\mathbf{p}$ and $\mathbf{y}$, respectively.

4 Experiments

In this section, we will design a series of experiments to verify the effectiveness of our proposed model.

4.1   Datasets

We evaluate the performance of our method on two public datasets: SemEval 2010 Task 8 [5]and PubMed dataset [4]. The SemEval dataset contains 10717 samples on 19 relations in total, including 8000 training samples, 2717 test samples. The PubMed contains 6087 samples and there are five relationship types. But in our experiment, like previous research, we consider two classifications, i.e. the category of 'None' is regarded as a "neg" class, while other categories are regarded as "pos" class. Note that this dataset does not give the partition of training set and test set. Therefore, in our experiment, we use 10-fold cross validation to carry out the experiment, and take the average of the experimental results as the final result.

4.2   Evaluation Metrics

For the SemEval dataset, we use the officially given evaluation metric[2], i.e. macro-Precision, macro-Recall and macro-F1 score. For the PubMed dataset, in order to compare with previous work easily, we use standard micro-Precision, micro-Recall and micro-F1 score as our evaluation indicators.

4.3   Implementation Details

In our experiment, we use Adam [7] as our optimizer to train the model and bert-base-uncased as our encoder. For the parameters in BERT, we take values from 1e-5 to 1e-4. For other parameters, we take values from 1e-3 to 1e-4. Dropout is set to 0.1, and the total number of training epochs is set to 40. For the SemEval dataset, the batchsize is set to 16. For the PubMed dataset, the batchsize is set to 8. All experiments are conducted with an NVIDIA GeForce RTX 3090 GPU.

4.4   Main Results

For each dataset, we conduct experiments using one, two and three convolutional layers, respectively. The experimental results on both datasets are shown in Table 1.

Table 1. Precision (%), Recall (%) and F1-score (%) on both the SemEval and PubMed dataset

Model	SemEval			PubMed		
	Precision	Recall	F1-score	Precision	Recall	F1-score
1 convolutional layer	90.17	88.36	89.21	95.81	95.71	95.75
2 convolutional layers	**90.65**	89.17	**89.87**	**96.42**	**96.38**	**96.4**
3 convolutional layers	89.38	**89.22**	89.39	96.32	96.2	96.25

---

[2] http://semeval2.fbk.eu/scorers/task08/SemEval2010_task8_scorer-v1.2.zip.

It can be observed that having more convolutional layers does not necessarily lead to better performance. For example, in our experiments, when the number of convolutional layers is two, we achieve the highest precision and F1-score on the SemEval dataset, while the recall is different and reaches its maximum with three convolutional layers. For the PubMed dataset, the highest precision, recall and F1-score are all obtained with two convolutional layers. Furthermore, for the SemEval dataset, from the F1-score perspective, using a model with two convolutional layers outperforms models with one layer and three layers by 0.66% and 0.48%, respectively. The performance difference between one and two convolutional layes is particularly evident in the PubMed dataset. Additionally, we observe that there is variation in model performance between the two different datasets, which is expected since binary classification is generally considered easier compared to multi-class classification.

4.5 Comparison with Other SOTA

We also compared our results with the previous state-of-the-art (SOTA) models, and the results are shown in Table 2.

As we can see from the table, our model exceeds all previous methods on both the two datasets. Specifically, for the SemEval dataset, our method is 0.71% higher than the previous SOTA model A-GCN [13], and our model has greater advantages than AGGCN [4], which is 4% higher. For PubMed dataset, compared with the previous SOTA model AGGCN, our model has reached 96.4% in the same task, exceeding AGGCN by 10%. This fully demonstrates the effectiveness of our proposed model.

Table 2. Comparison (%) with other baselines. Note that the experimental results of all baselines are derived from the original literatures.

	Models	SemEval	PubMed
1	Peng et al. [8]	–	76.5
2	Peng et al. [8]	–	76.7
3	Song et al. [12]	–	83.6
4	Zhang et al. [19]	84.8	83.7
5	Guo et al. [4]	85.7	85.6
6	Xu et al. [15]	83.7	–
7	Soares et al. [11]	89.5	–
8	Yu et al. [17]	86.4	–
9	Tian et al. [13]	89.16	–
10	TS-GCN (Ours)	**89.87**	**96.4**

4.6 Ablation Study

In order to further explore the importance of each component module of our proposed model, we conduct ablation study on two datasets and the experimental results are shown in Table 3.

Table 3. Results (%) of ablation study on F1-score.

	Model	SemEval	PubMed
1	baseline	**89.87**	**96.40**
2	w/o **W**	88.56	94.57
3	w/o **D** & **D**	86.78	91.89

It can be seen that when we consider the dependency relation type matrix **D** but do not consider the dependency relation weight matrix **W**, the F1 score decreases by 1.31% and 1.83% on the two datasets, respectively. When we neither consider **W** nor **D**, the F1 score decreases even more, on the two datasets, it reaches 3.09% and 4.51%, respectively. This further demonstrates the effectiveness of each module in the model.

4.7 Extraction Speed

In order to explore the effect of sentence length on extraction performance, we compare our method with the method of adding the shortest dependency path [15] for extraction speed, and the results are shown in the Fig. 2.

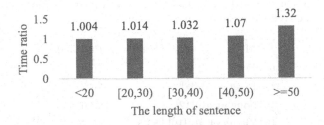

Fig. 2. Comparison of extraction speed. Under the condition of different sentence lengths, the ratio of the average extraction time between our model and the variant with the shortest dependent path.

It can be observed from the figure that as the length of the sentence increases, the use of the shortest dependency path method becomes slower. This is because the shortest dependency path is obtained through recursive operations, which require high computational overhead on the computer. Our method does not involve recursive operations. Therefore, for long sentences, our method does not exhibit significant changes in speed.

5   Conclusion

In this paper, we propose a two-stage graph convolutional network that utilizes dependency relationship type information and dependency relationship type weights for relation triple extraction. Our approach integrates dependency information, node information, and weight information. We evaluate our proposed method on two public datasets, and the experimental results on two benchmark datasets demonstrate the effectiveness of our method and all its components.

References

1. Chen, G., Tian, Y., Song, Y., Wan, X.: Relation extraction with type-aware map memories of word dependencies. In: Zong, C., Xia, F., Li, W., Navigli, R. (eds.) Findings of the Association for Computational Linguistics: ACL/IJCNLP 2021, Online Event, 1–6 August 2021. Findings of ACL, vol. ACL/IJCNLP 2021, pp. 2501–2512. Association for Computational Linguistics (2021). https://doi.org/10.18653/v1/2021.findings-acl.221

2. Denecke, K.: Sentiment Analysis in the Medical Domain. Springer (2023). https://doi.org/10.1007/978-3-031-30187-2

3. Devlin, J., Chang, M., Lee, K., Toutanova, K.: BERT: pre-training of deep bidirectional transformers for language understanding. In: Burstein, J., Doran, C., Solorio, T. (eds.) Proceedings of the 2019 Conference of the North American Chapter of the Association for Computational Linguistics: Human Language Technologies, NAACL-HLT 2019, Minneapolis, MN, USA, 2–7 June 2019, Volume 1 (Long and Short Papers), pp. 4171–4186. Association for Computational Linguistics (2019). https://doi.org/10.18653/v1/n19-1423

4. Guo, Z., Zhang, Y., Lu, W.: Attention guided graph convolutional networks for relation extraction. In: Proceedings of the 57th Annual Meeting of the Association for Computational Linguistics, pp. 241–251 (2019)

5. Hendrickx, I., et al.: Semeval-2010 task 8: Multi-way classification of semantic relations between pairs of nominals. In: Erk, K., Strapparava, C. (eds.) Proceedings of the 5th International Workshop on Semantic Evaluation, SemEval@ACL 2010, Uppsala University, Uppsala, Sweden, 15–16 July 2010, pp. 33–38. The Association for Computer Linguistics (2010). https://aclanthology.org/S10-1006/

6. Kambhatla, N.: Combining lexical, syntactic, and semantic features with maximum entropy models for information extraction. In: Proceedings of the 42nd Annual Meeting of the Association for Computational Linguistics, Barcelona, Spain, 21–26 July 2004 - Poster and Demonstration. ACL (2004). https://aclanthology.org/P04-3022/

7. Kingma, D.P., Ba, J.: Adam: a method for stochastic optimization. In: Bengio, Y., LeCun, Y. (eds.) 3rd International Conference on Learning Representations, ICLR 2015, San Diego, CA, USA, 7–9 May 2015, Conference Track Proceedings (2015)

8. Peng, N., Poon, H., Quirk, C., Toutanova, K., Yih, W.T.: Cross-sentence n-ary relation extraction with graph lstms. Trans. Assoc. Comput. Ling. **5**, 101–115 (2017)

9. Pérez-Pérez, M., Ferreira, T., Igrejas, G., Fdez-Riverola, F.: A deep learning relation extraction approach to support a biomedical semi-automatic curation task: the case of the gluten bibliome. Expert Syst. Appl. **195**, 116616 (2022). https://doi.org/10.1016/j.eswa.2022.116616

10. Sheng, S., et al.: Human-adversarial visual question answering. Adv. Neural. Inf. Process. Syst. **34**, 20346–20359 (2021)
11. Soares, L.B., Fitzgerald, N., Ling, J., Kwiatkowski, T.: Matching the blanks: Distributional similarity for relation learning. In: Proceedings of the 57th Annual Meeting of the Association for Computational Linguistics, pp. 2895–2905 (2019)
12. Song, L., Zhang, Y., Wang, Z., Gildea, D.: N-ary relation extraction using graph-state lstm. In: Proceedings of the 2018 Conference on Empirical Methods in Natural Language Processing, pp. 2226–2235 (2018)
13. Tian, Y., Chen, G., Song, Y., Wan, X.: Dependency-driven relation extraction with attentive graph convolutional networks. In: Proceedings of the 59th Annual Meeting of the Association for Computational Linguistics and the 11th International Joint Conference on Natural Language Processing (Volume 1: Long Papers), pp. 4458–4471 (2021)
14. Wang, X., et al.: Learning intents behind interactions with knowledge graph for recommendation. In: Proceedings of the Web Conference 2021, pp. 878–887 (2021)
15. Xu, Y., Mou, L., Li, G., Chen, Y., Peng, H., Jin, Z.: Classifying relations via long short term memory networks along shortest dependency paths. In: Proceedings of the 2015 Conference on Empirical Methods in Natural Language Processing, pp. 1785–1794 (2015)
16. Yan, Q., Zhang, Y., Liu, Q., Wu, S., Wang, L.: Relation-aware heterogeneous graph for user profiling. In: Proceedings of the 30th ACM International Conference on Information & Knowledge Management, pp. 3573–3577 (2021)
17. Yu, B., Mengge, X., Zhang, Z., Liu, T., Yubin, W., Wang, B.: Learning to prune dependency trees with rethinking for neural relation extraction. In: Proceedings of the 28th International Conference on Computational Linguistics, pp. 3842–3852 (2020)
18. Zelenko, D., Aone, C., Richardella, A.: Kernel methods for relation extraction. J. Mach. Learn. Res. **3**, 1083–1106 (2003). http://jmlr.org/papers/v3/zelenko03a.html
19. Zhang, Y., Qi, P., Manning, C.D.: Graph convolution over pruned dependency trees improves relation extraction. In: Proceedings of the 2018 Conference on Empirical Methods in Natural Language Processing, pp. 2205–2215 (2018)

Multi-vehicle Platoon Overtaking Using NoisyNet Multi-agent Deep Q-Learning Network

Lv He[1], Dongbo Zhang[1(✉)], Tianmeng Hu[2], and Biao Luo[2(✉)]

[1] Xiangtan University, Xiangtan 411105, China
zhadonbo@sina.com
[2] Central South University, Changsha 410083, China
biao.luo@hotmail.com

Abstract. With the recent advancements in Vehicle-to-Vehicle communication technology, autonomous vehicles are able to connect and collaborate in platoon, minimizing accident risks, costs, and energy consumption. The significant benefits of vehicle platooning have gained increasing attention from the automation and artificial intelligence areas. However, few studies have focused on platoon with overtaking. To address this problem, the NoisyNet multi-agent deep Q-learning algorithm is developed in this paper, which the NoisyNet is employed to improve the exploration of the environment. By considering the factors of overtake, speed, collision, time headway and following vehicles, a domain-tailored reward function is proposed to accomplish safe platoon overtaking with high speed. Finally, simulation results show that the proposed method achieves successfully overtake in various traffic density situations.

Keywords: Multi-vehicle platoon · overtake · multi-agent reinforcement learning · mixed traffic

1 Introduction

In recent years, autonomous vehicles (AVs) and their technologies have received extensive attention worldwide. Autonomous driving has stronger perception and shorter reaction time compared to human driving. There is no human driver behavior such as fatigue driving, and it is safer for long-distance driving. By using advanced Vehicle-to-Vehicle communication technologies, AVs are able to share information with each other and cooperate in dynamic driving tasks. Through sharing information about the environment [1,2], locations, and actions, it will improve the driving safety [3], reduce the traffic congestion [4], and decrease the energy consumption [5]. Multi-vehicle collaboration and overtaking are two important topics for AVs.

1.1 Multi-vehicle Collaboration

The cooperation of multiple vehicles is a promising way to improve traffic efficiency and reduce congestion, and many research works have been reported. By

B. Luo et al. (Eds.): ICONIP 2023, CCIS 1969, pp. 495–509, 2024.
https://doi.org/10.1007/978-981-99-8184-7_38

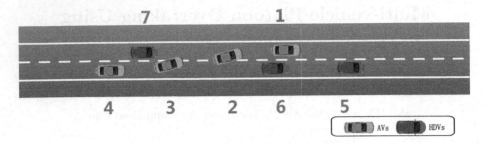

Fig. 1. Illustration of the considered platoon overtaking traffic scenario. AVs (blue) and HDVs (green) coexist in the straight lane. (Color figure online)

integrating CARLA [6] and SUMO [7], OpenCDA [8,9] was proposed, which supports both cooperative driving automation prototyping and regular autonomous driving components. Based on OpenCDA, a series of multi-vehicle collaboration works focused on various fields were studied, e.g., collaborative perception [1,2,10–12], planning [13], localization, and safety system [14]. Coordinated strategies between autonomous vehicles could improve transportation efficiency and reduce unnecessary waiting time for passengers [15]. So that people have more time to do more meaningful things. Reasonable coordination of multiple vehicles in different road environments can improve traffic safety [16], improve traffic efficiency [17], and reduce fuel consumption [18]. These optimized indexes are of great significance to social traffic operations [19]. Based on the development of autonomous multi-vehicle coordination, in order to improve the transportation capacity of multiple vehicles, and reduce fuel consumption. A lot of useful techniques have been developed, and one of the very useful techniques is platoon [20], which has been studied in great detail. Platoon driving refers to the situation where multiple vehicles coordinate and the rear vehicles follow the front vehicles at a short distance.

Reinforcement learning is a powerful method for decision-making, which has been applied to address autonomous driving problems in recent years. Multi-agent reinforcement learning algorithms can coordinate agents effectively and explore a large number of potential different environments quickly. It not only enables multi-agent to adapt to the dense and complex dynamic driving environment but also enables multi-agents to make effective collaborative decisions [21].

1.2   Reinforcement Learning for Overtaking

Overtaking is an important way for AVs to improve driving efficiency, especially in mixed-traffic environments that contain AVs and human-driven vehicles (HDVs). There are some works [22–24] using reinforcement learning to handle single-vehicle overtaking problems. When making overtaking decisions, the agent needs to consider that other vehicles are in the vicinity of the agent and that different vehicles among them are traveling at different speeds. This requires agents to have multiple abilities to deal with overtaking problems [24]. Experienced human drivers can handle overtaking problems better, [25] used curriculum reinforcement learning to make the agent perform overtaking operations comparable to experienced human drivers. When the platoon cannot travel at a relatively high or expected speed in the traffic flow, it will lead to a reduction in the efficiency of vehicle transportation. The platoon needs to speed up to overtake the slow vehicles in front and reach the destination faster. As shown in Fig. 1, platoon overtaking requires not only close coordination among members of the platoon but also the prevention of collision with other human vehicles around the platoon during the process of overtaking. The reinforcement learning algorithm with the more effective exploration of the environment can realize and improve the platoon's performance during overtaking.

However, to the best of our knowledge, using RL, especially multi-agent RL, for the AV platoon overtaking problem has rarely been studied. It is still an open and challenging problem, which motivates our studies in this paper. Inspired by NoisyNet [26], we added the parameterized factorised Gaussian noise to the linear layer networks weights of the multi-agent deep Q-learning network, which induced stochasticity in the agent's policy that can be used to aid efficient exploration. The factorised Gaussian noise parameters are learned by gradient descent along with the weights of the remaining networks. The longitudinal vehicle following distance in a platoon cannot be designed to be a fixed value. AVs at the end of a fixed-distance platoon may collide with nearby HDVs in overtaking passes, which will make the platoon much less safe when overtaking. Self-driving platoon faces the possibility of encountering other HDVs in the process of driving and overtaking, and need a suitable safety distance to adjust the self-driving vehicles policy to reduce the risk of collision. With the consideration of the above factors, a domain-tailored customized reward function is designed to achieve high-speed safe platoon overtaking. In order to reduce fluctuations arosen from the vehicle following reward, coefficients are added to the same lane following reward and the following distance interval reward, respectively. The total reward curve is more likely to converge after adding the coefficients, which is extremely helpful for multi-agents to learn stable policy. The contributions of the paper are summarized as follows:

- The NoisyNet based multi-agent deep Q network (NoisyNet-MADQN) algorithm is developed for multi-vehicle platoon overtaking. By adding the parameterized factorised Gaussian noise to the linear layer networks weights of the multi-agent deep Q-learning network, which induced stochasticity in the agent's policy that improves the exploration efficiency. The parameters of the factorised Gaussian noise are learned with gradient descent along with the remaining network weights. To reduce this computational overload, we select factorised Gaussian noise, which reduces the computational time for generating random numbers in the NoisyNet-MADQN algorithm.
- By considering the factors of overtaking, speed, collision, time headway and vehicle following, the domain-tailored customized platoon overtaking is designed. The safety distance is designed in the vehicle-following reward to

reduce the risk of collisions between the platoon and nearby HDVs while straight driving and overtaking. It is able to reduce collision rate of the AVs in the safety distance by the reward adaptively adjusting the distance they maintain from the AV in front.

The rest of the paper is organized as follows. In Sect. 2, the preliminary works of RL and the NoisyNet are presented. Section 3 presents the reward function's design and the platoon overtaking algorithm. Section 4 presents the experiments and results. Finally, Sect. 5 concludes our work.

2   Problem Formulation

2.1   Preliminary of Reinforcement Learning

In reinforcement learning, the agent's goal is to learn the optimal policy π^* that maximizes the cumulative future rewards $R_t = \sum_{k=0}^{T} \gamma^k r_{t+k}$, where t is the time step, r_{t+k} is the reward at time step $t + k$, and $\gamma \in (0, 1]$ is the discount factor that quantifies the relative importance of future rewards. At time step t, the agent observes the state $s_t \in S \subseteq \mathbb{R}^n$, selects an action $a_t \in A \subseteq \mathbb{R}^m$, and receives a reward signal $r_t \subseteq \mathbb{R}$. n represents the total number of agents, $i \in n$.

Action Space. An agent's action space A_i is defined as a set of high-level control decisions. Decision-making behaviors include turning left, turning right, idling, speeding up, and slowing down.

State Space. The state of agent i, S_i, is defined as a matrix of dimension $N_{\mathcal{N}_i} \times m$, where $N_{\mathcal{N}_i}$ is the number of observed vehicles and m is the number of features. *Ispresent* is a binary variable that indicates whether there are other observable vehicles in the vicinity of 150 m from the ego vehicle. x is the observed longitudinal position of the vehicle relative to the ego vehicle. y represents the lateral position of the observed vehicle relative to the ego vehicle. v_x and v_y represent the longitudinal and lateral speeds of the observed vehicle relative to the ego vehicle, respectively.

In the highway simulator, we assume that the ego vehicle can only obtain information about neighboring vehicles within 150 m of the longitudinal distance of the ego vehicle. In the considered two-lane scenario (see Fig. 1), the neighboring vehicle is located in the lane and its neighboring lanes closest to the ego vehicle, with the 2th AV as the ego vehicle and its neighboring vehicles as the 1th AV, 3th AV, 6th HDV, and 7th HDV.

Reward Distribution. In this paper, the NoisyNet multi-agent deep Q-learning is developed. As a multi-agent algorithm, since the vehicles in the platoon are the same type of vehicles, we assume that all the agents share the same network structure and parameters. Our algorithm aims to maximize the overall reward. To solve the communication overhead and credit assignment problems [27], we use the following local reward design [28]. So, the reward for the ith agent at time t is defined as:

$$r_{i,t} = \frac{1}{|V_i|} \sum_{j \in V_i} r_{j,t}, \tag{1}$$

where $|V_i|$ denotes the cardinality of a set containing the ego vehicle and its close neighbors. This reward design includes only the rewards of the agents most relevant to the success or failure of the task [29].

2.2   Multi-agent Reinforcement Learning (MARL) and NoisyNets

This subsection will focus on the NoisyNets and MARL. In the NoisyNet, its neural network weights and biases are perturbed by a function of noise parameters. These parameters are adjusted according to gradient descent [26]. They assume that $y = f_\theta(x)$ is a neural network parameterized by a vector of noise parameters θ that accepts input x and output y. In our experiments, we assume that there are n AVs in the experimental environment. Each AV represents an agent, the ith AV represents the ith agent, $i \in n$. x_i represents the observed state of the ith agent, y_i represents the action of the ith agent. The noise parameter θ_i is denoted as $\theta_i \overset{def}{=} \mu_i + \Sigma_i \odot \varepsilon_i$, $\zeta_i \overset{def}{=} (\mu_i, \Sigma_i)$ is a set of learnable parameter vectors, ε_i is a zero-mean noise vector with fixed statistics, and $\odot$ denotes element multiplication. The loss of the neural network is wrapped by the expectation of the noise $\varepsilon_i : \bar{L}(\zeta_i) \overset{def}{=} \mathbb{E}[L(\theta_i)]$. Then, the set of parameters ζ_i is optimized. Consider the linear layers of the neural networks with p inputs and q outputs in these experiments, represented by

$$y_i = w_i x_i + b_i, \tag{2}$$

where $x_i \in \mathbb{R}^p$ are the layers inputs, $w_i \in \mathbb{R}^{q \times p}$ the weight matrix, and $b_i \in \mathbb{R}^q$ the bias. The corresponding noisy linear layers are defined as:

$$y_i \overset{def}{=} (\mu_i^{w_i} + \sigma_i^{w_i} \odot \varepsilon_i^{w_i}) x_i + \mu_i^{b_i} + \sigma_i^{b_i} \odot \varepsilon_i^{b_i}, \tag{3}$$

where $\mu_i^{w_i} + \sigma_i^{w_i} \odot \varepsilon_i^{w_i}$ and $\mu_i^{b_i} + \sigma_i^{b_i} \odot \varepsilon_i^{b_i}$ replace correspondingly w_i and b_i in Eq.(2). The parameters $\mu_i^{w_i} \in \mathbb{R}^{q \times p}$, $\mu_i^{b_i} \in \mathbb{R}^q$, $\sigma_i^{w_i} \in \mathbb{R}^{q \times p}$, $\sigma_i^{b_i} \in \mathbb{R}^q$, are learnable whereas $\varepsilon_i^{w_i} \in \mathbb{R}^{q \times p}$ and $\varepsilon_i^{b_i} \in \mathbb{R}^q$ are noise random variables. Deep-Mind introduced two types of Gaussian noise: independent Gaussian noise and factorised Gaussian noise. The computation overhead for generating random numbers in the algorithm is particularly prohibitive in the case of single-thread agents. To reduce the computation overhead for generating random numbers in the multi-agent deep Q-learning network, we selected factorised Gaussian noise.

We factorize $\varepsilon_{j,k}^w$, use p unit Gaussian variables ε_j for the noise of the inputs and q unit Gaussian variables ε_k for the noise of the outputs. Each $\varepsilon_{j,k}^w$ and ε_k^b can then be written as:

$$\varepsilon_{j,k}^w = f(\varepsilon_j) f(\varepsilon_k), \tag{4}$$

$$\varepsilon_k^b = f(\varepsilon_k), \tag{5}$$

where f is a real-valued function. In this experiment, we used $f(x) = \text{sgn}(x)\sqrt{|x|}$. We can obtain the loss of multiple noise networks. $\bar{L}_i(\zeta_i) = \mathbb{E}[L_i(\theta_i)]$, present the expectation of multiple gradients can be obtained directly from:

$$\nabla \bar{L}_i(\zeta_i) = \nabla \mathbb{E}[L_i(\theta_i)] = \mathbb{E}[\nabla_{\mu_i, \Sigma_i} L(\mu_i + \Sigma_i \odot \varepsilon_i)]. \tag{6}$$

Using a Monte Carlo approximation to the above gradients, taking samples ξ_i at each step of optimization:

$$\nabla \bar{L}_i(\zeta_i) \approx \nabla_{\mu_i, \Sigma_i} L(\mu_i + \Sigma_i \odot \xi_i). \tag{7}$$

In this work, we will no longer use ε-greed, The policy greedily optimizes the (randomised) action-value function. Then the fully connected layers of the value network are parameterized to the noisy network, where the parameters are extracted from the noisy network parameter distribution after each replay step. Before each action, the noisy network parameters will be resampled, so that each action step of the algorithm can be optimized. In the target networks, the parameterized action-value function $Q(s_i, a_i, \varepsilon_i; \zeta_i)$ and $Q(s_i, a_i, \varepsilon_i'; \zeta_i^-)$ can be regarded as a random variable when the linear layers in the network are replaced by the noisy layers. The outer expectation is with respect to the distribution of the noise variables ε for the noisy value function $Q(s_i, a_i, \varepsilon_i; \zeta_i)$ and the noise variable ε' for the noisy target value function $Q(s_i, a_i, \varepsilon_i'; \zeta_i^-)$. So the NoisyNet-MADQN loss:

$$\bar{L}_i(\zeta_i) = \mathbb{E}\left[\mathbb{E}_{(s,a,r,s_{t+1})\sim D}[r + \gamma \mathbf{max} Q(s_i, a_i, \varepsilon_i'; \zeta_i^-) - Q(s_i, a_i, \varepsilon_i; \zeta_i)]^2\right]. \tag{8}$$

3   NoisyNet-MADQN for Platoon Overtaking

3.1   Reward Function Design

This subsection proposes a novel reward function for reinforcement learning algorithms to implement platoon overtaking. Reward functions are crucial for reinforcement learning models. By designing the reward function, we can guide the learning of RL agents to achieve our purpose.

The Overtake and Speed Evaluation. The vehicles will choose to drive at high speed driven by the reward, which will improve efficiency and allow more vehicles to reach their destination faster. When the speed of the front HDVs is less than the AVs, the platoon leader will increase the speed to overtake the front slower vehicles to get more rewards for completing the overtaking behavior. The other AVs in the platoon will also overtake the low-speed HDVs in front of them because of the following reward and speed reward. They follow the leader closely to form a platoon overtake. So the speed reward can also be seen as an overtaking reward. Therefore we still define the overtaking and speed reward for this vehicle as follows:

$$r_{os} = \frac{v_t - v_{min}}{v_{max} - v_{min}}, \tag{9}$$

where v_t, $v_{min} = 20\,\text{m/s}$, and $v_{max} = 30\,\text{m/s}$ are the current, minimum, and maximum speeds of the ego vehicle, respectively.

The Collision Penalty Design. Safety is the most critical factor in autonomous driving: if a collision occurs, the collision evaluation r_c is set to -1. If there is no collision, r_c is set to 0. The collision evaluation is defined as

$$r_c = \begin{cases} -1, & \text{collision}; \\ 0, & \text{safety}. \end{cases} \tag{10}$$

The Time Headway Evaluation. The time headway evaluation is defined as

$$r_h = log\frac{d_{headway}}{t_h v_t}, \tag{11}$$

where $d_{headway}$ is the distance headway and t_h is a predefined time headway threshold. As such, the ego vehicle will get penalized when the time headway is less than t_h and rewarded only when the time headway is greater than t_h. In this paper, we choose t_h as 1.2 s as suggested in [30].

The Vehicles Following Evaluation. To keep the AVs in the platoon. So, the vehicles following evaluation is defined as

$$r_f = \begin{cases} 0.3k_1|C[i+1].p[0] - C[i].p[0]|/(t_h v_{max}), & |C[i+1].p[0] - C[i].p[0]| \le 2v_{max}; \\ 0.7k_2, & C[i].p[1] = C[i+1].p[1], \end{cases} \tag{12}$$

where $c[i]$ represents the ith AV, and $p[0]$,$p[1]$ represents the longitudinal and lateral coordinates of the AV, respectively. When the longitudinal distance between the AVs and the vehicle ahead of it in the platoon at time t is kept within 60 m, the reward obtained at this time is $0.3k_1|C[i+1].p[0] - C[i].p[0]|/(t_h v_{max})$. According to the time headway, we defined the danger distance as the distance range when the vehicle travels in a straight line at maximum speed for 1.2 s. The safety distance is defined as the distance range of 1.2–2 s when the vehicle travels in a straight line at the maximum speed. As shown in Fig. 2. When the AVs are at the safety distance, the AVs get more rewards. When the distance between AVs is at the danger distance, the closer the two AVs are to each other, the less reward they will receive. They also get the penalty because of the time headway evaluation. The situations encountered by the self-driving platoon are classified into two types: platoon straight driving and platoon overtaking. HDVs can have an impact on the safety of the platoon while straight driving and overtaking. In Fig. 2, when the 7th HDV enters between the 2th AV and the 3th AV, the fixed following distance of the rear AVs in order to follow the front AVs when the self-driving platoon leader accelerates will increase the risk of collision between the 3th AV and the 7th HDV. When the platoon leader in the self-driving platoon overtakes the 6th HDV to obtain a greater overtake and speed reward. The AVs behind it follow closely. The 2th AV and the 6th HDV are very close to each other, and if the following distance is designed as a constant value, then they have a very high chance of collision. Based on the above, so we designed the safety distance to reduce the risk of following when overtaking. AVs can automatically adjust their following distance through training. After the training, the self-driving platoon vehicles will choose the safer and more effective action at time t when faced with such situations.

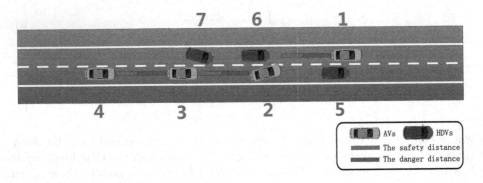

Fig. 2. Illustration of the considered platoon overtaking traffic scenario. AVs (blue) and HDVs (green) coexist in the straight lane. The long green bar and the long red bar represent the safety distance and the danger distance respectively (Color figure online)

When the rear AVs in the platoon follow the front AVs in the same lane, at time t the reward obtained is 0.7. Set the reward for keeping the AVs in the same lane to be larger than the reward for keeping the safety distance between the platoon. This is to ensure that when the platoon leader changes lanes to overtake, the following AVs can also change lanes in time to overtake.

In our training, an episode has 100 steps, and the total reward accumulated by the acquisition fluctuates too much, which is not conducive to the agent obtaining a stable policy. So we add the weight k_1 to the reward obtained by keeping the distance with the vehicle ahead and the weight k_2 to the reward function of keeping in the same lane with the vehicle ahead in the platoon, respectively. k_1 and k_2 have values of 0.25 and 0.3, respectively. The fluctuation of the total reward becomes smaller after we add the weights, and it is easier for the agent to obtain a stable policy. Designing the weight k_2 to be slightly larger than k_1 can guide the vehicles behind the platoon leader to keep up with the platoon leader when the platoon leader overtakes. The smaller k_1 enables AVs to take advantage of the safety distance to avoid the risk of collision when AVs in a platoon encounter nearby HDVs while overtaking.

Total Reward. The reward function r_i is necessary for training multiple agents to behave as we desire. Since our goal is to keep the platoon while overtaking other vehicles. Therefore, the reward for the ith agent at time step t is defined as follows:

$$r_{i,t} = w_c r_c + w_{os} r_{os} + w_h r_h + w_f r_f. \tag{13}$$

Among them, w_c, w_{os}, w_h, and w_f are the positive weight scalars corresponding to the collision assessment w_c, the overtake and speed evaluation w_{os}, the time headway evaluation w_h, and AVs following evaluation w_f, respectively. Since safety is the most important criterion, we made the w_c heavier than others. w_f is second only to w_c and higher than the other two weights. The coefficients w_c, w_{os}, w_h, and w_f for the reward function are set as 200, 1, 4, and 5.

3.2 Noisy Network Multi-Agent Deep Q-Learning Network

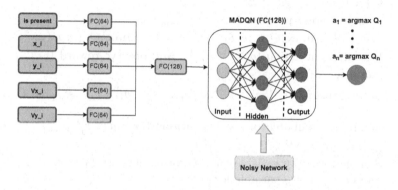

Fig. 3. The structure of the NoisyNet-MADQN network design is proposed, $i = 1, 2, ..., N$. And the numbers in parentheses indicate the size of the layers.

The leader of the platoon needs to choose the optimal policy π^* when maintaining the platoon with the followers during overtaking, which improves the safety and efficiency of driving during overtaking. In the platoon overtaking scenario, we added the parameterized factorised Gaussian noise to linear layer network weights of the multi-agent deep Q-learning network. Factorised Gaussian noise parameters are learned by gradient descent together with the remaining network weights. The computational overhead associated with single-thread agents is especially prohibitive. To overcome this computational overhead we select factorised Gaussian noise, which reduces the computational time for generating random numbers in the multi-agent deep Q-learning network. Figure 3 shows the network structure of our algorithm, in which states separated by physical units are first processed by separate 64-neuron fully connected (FC) layers. Then all hidden units are combined and fed into the 128-neuron FC layer. Based on this method, each agent in the multi-agent improves the exploration of the environment. Our algorithm can obtain a larger optimal Q-value function for the same level of traffic compared to the original algorithm to achieve a better policy. More rewards for the platoon also show that our algorithm can do better than the baseline in platoon overtaking. Algorithm 1 is the detailed procedure of NoisyNet-MADQN.

Algorithm 1. NoisyNet-MADQN

1: Initialize replay buffers $D_1, ..., D_n$ to the capacities $N_1, ...N_n$, action-value functions $Q_1, ...Q_n$ and target action-value functions $\hat{Q}_1, ..., \hat{Q}_n$, ε set of random variables of the network, ζ initial network parameters, ζ^- initial target network parameters, N_T training batch size, N^- target network replacement frequency

2: **for** $episode = \{1, ...M\}$ **do**

3:     Initialize state $s_0^{(1),...,(n)} \sim Env$

4:     **for** $t = \{1, ...T\}$ **do**

5:         Set $s^{(1),...,(n)} \leftarrow s_0^{(1),...,(n)}$

6:         Sample a noisy network $\xi_1 \sim \varepsilon_1, ...,\xi_n \sim \varepsilon_n$

7:         Select $a_t^1 = \mathbf{max}_a Q_1(s, a, \xi_1; \zeta_1), ..., a_t^{(n)} = \mathbf{max}_a Q_n(s, a, \xi_n; \zeta_n)$

8:         Execute joint action $(a_t^{(1)}, ..., a_t^{(n)})$ and observe reward $(r_t^{(1)}, ..., r_t^{(n)})$, and new state $(s_{t+1}^{(1)}, ..., s_{t+1}^{(n)})$

9:         Store transition$(s_t^{(1)}, a_t^{(1)}, r_t^{(1)}, s_{t+1}^{(1)}, ..., s_t^{(n)}, a_t^{(n)}, r_t^{(n)}, s_{t+1}^{(n)})$ in $D_1, ..., D_n$

10:         Sample a minibatch of N_T transitions $((s_j^{(1)}, a_j^{(1)}, r_j^{(1)}, s_{j+1}^{(1)}) \sim D_1)_{j=1}^{N_T}$, $...,((s_j^{(n)}, a_j^{(n)}, r_j^{(n)}, s_{j+1}^{(n)}) \sim D_n)_{j=1}^{N_T}$

11:         Sample the noisy variable for the online network $\xi_1 \sim \varepsilon_1, ...,\xi_n \sim \varepsilon_n$

12:         Sample the noisy variable for the target network $\xi_1^{'} \sim \varepsilon_1, ...,\zeta_n^{'} \sim \varepsilon_n$

13:         **for** $j \in \{1, ...,N_T\}$ **do**

14:             **if** s_j is a terminal state **then**

15:                 $\hat{Q} \leftarrow r_j$

16:             **else**

17:                 $\hat{Q} \leftarrow r_j + \gamma \mathbf{max}\, Q(s_j, a_j, \xi^{'}\, ; \zeta^-)$

18:             **end if**

19:             Perform a gradient descent step on $(\hat{Q}_1 - Q_1(s_j^{(1)}, a_j^{(1)}, \xi_1; \zeta_1))^2, ..., (\hat{Q}_n - Q_n(s_j^{(n)}, a_j^{(n)}, \xi_n; \zeta_n))^2$

20:         **end for**

21:         **if** $t \equiv 0 \ (modN^-)$ **then**

22:             Update the target network: $\zeta_1^- = \zeta_1 , ..., \zeta_n^- = \zeta_n$

23:         **end if**

24:     **end for**

25: **end for**

---

4   Experiments and Discussion

In this section, the effectiveness of the proposed method is verified by simulation. Some implementation details of the experiments are given, and we evaluate the performance of the proposed MARL algorithm in terms of training effectiveness for overtaking in the considered road scenario shown in Fig. 1. The experimental results are also discussed.

For cost and feasibility considerations, we conducted experiments in the simulator. We use an open-source simulator developed on highway-env [31] and modify it as needed. The simulator is capable of simulating the driving environment and vehicle sensors. These vehicles randomly appear on the highway with different initial speeds of 20–30 m/s. And take random actions.

4.1   Experimental Settings

In order to fully demonstrate the effectiveness of our proposed method. Three traffic density levels were used to evaluate the effectiveness of the proposed method, corresponding to low, middle, and high levels of traffic congestion. We train the POMARL algorithm for 200 episodes by applying two different random seeds. The same random seed is shared among agents. These experiments were performed on an ubuntu server with a 2.7 GHz Intel Core i5 processor and 16GB of RAM. The number of vehicles in different traffic modes is shown in Table 1.

Table 1. Traffic density modes.

Density	AVs	HDVs	Explanation
1	4	1–2	low level
2	4	2–3	middle level
3	4	3–4	high level

Figure 4 shows the comparison between our algorithm and the baseline algorithm. It shows that NoisyNet-MADQN performs better than MADQN. We put the rewards obtained by the two algorithms respectively under low-level, mid-level, and high-level to make a comprehensive curve comparison. We tabulate these reward values and divide the rewards within 200 episodes into five intervals. The value of each interval is their average. In Table 2, the NoisyNet-MADQN obtains higher rewards than MADQN most of the time. Figure 5 shows snapshots of the platoon overtaking at low, middle, and high levels, respectively. It shows that our method can be applied to solve platoon overtaking. From the comparison of NoisyNet-MADQN and baseline algorithm rewards, it can be found that the proposed new algorithm obtains better results in platoon overtaking. It shows that the platoon has achieved better results in both driving efficiency, safety, and convoy coordination.

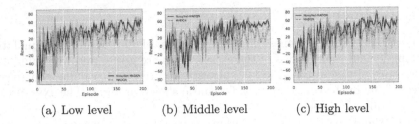

(a) Low level   (b) Middle level   (c) High level

Fig. 4. Evaluation curves during training with different algorithms for different traffic levels.

Table 2. Reward compare

Density	Method	1–40	41–80	81–120	121–160	161–200
low level	NoisyNet-MADQN	−14.31	**28.11**	**36.03**	**44.32**	**52.51**
	MADQN	**−5.02**	23.47	28.02	27.95	33.48
middle level	NoisyNet-MADQN	**−6.13**	12.33	**34.38**	**45.61**	**50.53**
	MADQN	−15.75	**28.51**	19.77	17.48	29.01
high level	NoisyNet-MADQN	**−4.48**	19.28	**33.52**	**51.58**	**52.34**
	MADQN	−19.13	**28.62**	26.60	31.72	31.29

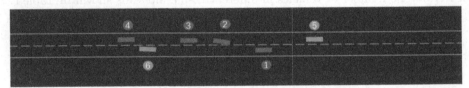

(a) Platoon overtaking at low level

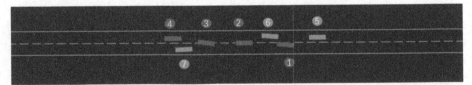

(b) Platoon overtaking at middle level

(c) Platoon overtaking at high level

Fig. 5. Platoon overtaking in the simulation environment

5   Conclusions

This paper proposes the NoisyNet multi-agent deep Q-learning network. The computation overhead for generating random numbers in the algorithm is particularly prohibitive in the case of single-thread agents. To reduce the computational overhead for generating random numbers in the multi-agent deep Q-learning network we selected factorised Gaussian noise. By adding the parameterized factorised Gaussian noise to the linear layer networks weights of the multi-agent deep Q-learning network, the induced randomness of the agent's policy can be used to help effective exploration. The parameters of the factorised Gaussian noise are learned with gradient descent along with the remaining network weights. To prove the effectiveness of our proposed algorithm, the proposed algorithm is compared with the baseline in the platoon overtaking tasks. By considering overtake, speed, collision, time headway and following vehicles factors, a domain-tailored reward function is proposed to accomplish safe platoon overtaking with high speed. The safety distance in the vehicle following evaluation allows the vehicles in the platoon to adjust the following distance to avoid collision when the platoon faces the nearby HDVs inserting into the platoon. It also allows the rear vehicle in the platoon to avoid collision with the nearby HDVs when following the overtaking vehicle. We compared it with the existing baseline algorithm at three different traffic densities and showed that it performs better than the baseline, achieving reasonable results.

References

1. Xu, R., Tu, Z., Xiang, H., Shao, W., Zhou, B., Ma, J.: Cobevt: cooperative bird's eye view semantic segmentation with sparse transformers. arXiv preprint arXiv:2207.02202 (2022)
2. Chen, W., Xu, R., Xiang, H., Liu, L., Ma, J.: Model-agnostic multi-agent perception framework. arXiv preprint arXiv:2203.13168 (2022)
3. Javed, M.A., Hamida, E.B.: On the interrelation of security, qos, and safety in cooperative its. IEEE Trans. Intell. Transp. Syst. **18**(7), 1943–1957 (2017)
4. Wang, Y., Sarkar, E., Li, W., Maniatakos, M., Jabari, S.E.: Stop-and-go: exploring backdoor attacks on deep reinforcement learning-based traffic congestion control systems. IEEE Trans. Inf. Forens. Secur. **16**, 4772–4787 (2021)
5. Zhang, Y., Ai, Z., Chen, J., You, T., Du, C., Deng, L.: Energy-saving optimization and control of autonomous electric vehicles with considering multiconstraints. IEEE Trans. Cybernet. **52**(10), 10869–10881 (2022)
6. Dosovitskiy, A., Ros, G., Codevilla, F., Lopez, A., Koltun, V.: Carla: an open urban driving simulator. In: Conference on Robot Learning, pp. 1–16. PMLR (2017)
7. Behrisch, M., Bieker, L., Erdmann, J., Krajzewicz, D.: Sumo-simulation of urban mobility: an overview. In: Proceedings of SIMUL 2011, the Third International Conference on Advances in System Simulation, ThinkMind (2011)
8. Xu, R., Guo, Y., Han, X., Xia, X., Xiang, H., Ma, J.: OpenCDA: an open cooperative driving automation framework integrated with co-simulation. In: 2021 IEEE International Intelligent Transportation Systems Conference (ITSC), pp. 1155–1162. IEEE (2021)

9. Xu, R., et al.: The opencda open-source ecosystem for cooperative driving automation research. arXiv preprint arXiv:2301.07325 (2023)

10. Xu, R., Xiang, H., Xia, X., Han, X., Li, J., Ma, J.: Opv2v: an open benchmark dataset and fusion pipeline for perception with vehicle-to-vehicle communication. In: 2022 International Conference on Robotics and Automation (ICRA), pp. 2583–2589. IEEE (2022)

11. Xu, R., Xiang, H., Tu, Z., Xia, X., Yang, M.-H., Ma, J.: V2X-ViT: vehicle-to-everything cooperative perception with vision transformer. In: Avidan, S., Brostow, G., Cissé, M., Farinella, G.M., Hassner, T. (eds.) Computer Vision – ECCV 2022: 17th European Conference, Tel Aviv, 23–27 October 2022, Proceedings, Part XXXIX, pp. 107–124. Springer, Cham (2022). https://doi.org/10.1007/978-3-031-19842-7_7

12. Xu, R., Li, J., Dong, X., Yu, H., Ma, J.: Bridging the domain gap for multi-agent perception. arXiv preprint arXiv:2210.08451 (2022)

13. Han, X., et al.: Strategic and tactical decision-making for cooperative vehicle platooning with organized behavior on multi-lane highways. Transp. Res. Part C Emerg. Technol. **145**, 103952 (2022)

14. Xiang, H., Xu, R., Xia, X., Zheng, Z., Zhou, B., Ma, J.: V2xp-asg: generating adversarial scenes for vehicle-to-everything perception. arXiv preprint arXiv:2209.13679 (2022)

15. Chu, K.-F., Lam, A.Y.S., Li, V.O.K.: Joint rebalancing and vehicle-to-grid coordination for autonomous vehicle public transportation system. IEEE Trans. Intell. Transp. Syst. **23**(7), 7156–7169 (2022)

16. Wang, F., Chen, Y.: A novel hierarchical flocking control framework for connected and automated vehicles. IEEE Trans. Intell. Transp. Syst. **22**(8), 4801–4812 (2021)

17. Goulet, N., Ayalew, B.: Distributed maneuver planning with connected and automated vehicles for boosting traffic efficiency. IEEE Trans. Intell. Transp. Syst. **23**(8), 10887–10901 (2022)

18. Validi, A., Olaverri-Monreal, C.: Simulation-based impact of connected vehicles in platooning mode on travel time, emissions and fuel consumption. IEEE Intell. Veh. Symp. (IV) **2021**, 1150–1155 (2021)

19. Zhang, Y., Hao, R., Zhang, T., Chang, X., Xie, Z., Zhang, Q.: A trajectory optimization-based intersection coordination framework for cooperative autonomous vehicles. IEEE Trans. Intell. Transp. Syst. **23**(9), 14674–14688 (2022)

20. Sturm, T., Krupitzer, C., Segata, M., Becker, C.: A taxonomy of optimization factors for platooning. IEEE Trans. Intell. Transp. Syst. **22**(10), 6097–6114 (2021)

21. Bai, Z., Hao, P., ShangGuan, W., Cai, B., Barth, M.J.: Hybrid reinforcement learning-based eco-driving strategy for connected and automated vehicles at signalized intersections. IEEE Trans. Intell. Transp. Syst. **23**(9), 15850–15863 (2022)

22. Yu, Y., Lu, C., Yang, L., Li, Z., Hu, F., Gong, J.: Hierarchical reinforcement learning combined with motion primitives for automated overtaking. IEEE Intell. Veh. Symp. (IV) **2020**, 1–6 (2020)

23. Kaushik, M., Prasad, V., Krishna, K.M., Ravindran, B.: Overtaking maneuvers in simulated highway driving using deep reinforcement learning. IEEE Intell. Veh. Symp. (IV) **2018**, 1885–1890 (2018)

24. Ngai, D.C.K., Yung, N.H.C.: A multiple-goal reinforcement learning method for complex vehicle overtaking maneuvers. IEEE Trans. Intell. Transp. Syst. **12**(2), 509–522 (2011)

25. Song, Y., Lin, H., Kaufmann, E., Dürr, P., Scaramuzza, D.: Autonomous overtaking in gran turismo sport using curriculum reinforcement learning. IEEE Int. Conf. Robot. Automat. **2021**, 9403–9409 (2021)

26. Fortunato, M., et al.: Noisy networks for exploration. In: International Conference on Learning Representations (2018)
27. Sutton, R.S., Barto, A.G.: Reinforcement Learning: An Introduction. MIT Press (2018)
28. Chen, D., Li, Z., Wang, Y., Jiang, L., Wang, Y.: Deep multi-agent reinforcement learning for highway on-ramp merging in mixed traffic. arXiv preprint arXiv:2105.05701 (2021)
29. ElSayed-Aly, I., Bharadwaj, S., Amato, C., Ehlers, R., Topcu, U., Feng, L.: Safe multi-agent reinforcement learning via shielding. arXiv preprint arXiv:2101.11196 (2021)
30. Ayres, T., Li, L., Schleuning, D., Young, D.: Preferred time-headway of highway drivers. In: 2001 IEEE Intelligent Transportation Systems. Proceedings (Cat. No. 01TH8585) (ITSC 2001), pp. 826–829. IEEE (2001)
31. Leurent, E.: An environment for autonomous driving decision-making (2018). https://github.com/eleurent/highway-env

Correction to: Multiclass Classification and Defect Detection of Steel Tube Using Modified YOLO

Deepti Raj Gurrammagari⃝, Prabadevi Boopathy⃝,
Thippa Reddy Gadekallu⃝, Surbhi Bhatia Khan,
and Mohammed Saraee

Correction to:
Chapter 32 in: B. Luo et al. (Eds.): *Neural Information Processing*, **CCIS 1969,**
https://doi.org/10.1007/978-981-99-8184-7_32

In an older version of this paper, the names of the first two authors were displayed incorrectly. The initials "G" and "B" respectively were in the wrong position. This has been corrected.

The updated version of this chapter can be found at
https://doi.org/10.1007/978-981-99-8184-7_32

Author Index

B. Luo et al. (Eds.): ICONIP 2023, CCIS 1969, pp. 511–513, 2024.
https://doi.org/10.1007/978-981-99-8184-7